CORONAVIRUSES
Molecular Biology and Virus–Host Interactions

ADVANCES IN EXPERIMENTAL MEDICINE AND BIOLOGY

Recent Volumes in this Series

A Continuation Order Plan is available for this series. A continuation order will bring delivery of each new volume immediately upon publication. Volumes are billed only upon actual shipment. For further information please contact the publisher.

CORONAVIRUSES
Molecular Biology and Virus−Host Interactions

Edited by

Hubert Laude and Jean-François Vautherot
Institut National de la Recherche Agronomique (INRA)
Jouy-en-Josas, France

SPRINGER SCIENCE+BUSINESS MEDIA, LLC

Library of Congress Cataloging-in-Publication Data

Coronaviruses : molecular biology and virus-host interactions / edited
  by Hubert Laude and Jean-François Vautherot.
       p.   cm. -- (Advances in experimental medicine and biology ; v.
  342)
    "Proceedings of the Fifth International Symposium on
  Coronaviruses, held September 13-18, 1992, in Chantilly, France"-
  -T.p. verso.
    Includes bibliographical references and index.
    ISBN 978-1-4613-6305-7   ISBN 978-1-4615-2996-5 (eBook)
    DOI 10.1007/978-1-4615-2996-5
    1. Coronaviruses--Congresses.   I. Laude, Hubert. II. Vautherot,
  Jean-François. III. International Symposium on Coronaviruses (5th :
  1992 : Chantilly, France) IV. Series.
    [DNLM: 1. Coronaviridae--congresses. 2. Coronavirus Infections-
  -congresses. 3. Receptors, Virus--congresses.   W1 AD559 v.342 1993
  / QW 168.5.C8 C822 1992]
  QR399.C67 1993
  576'.6484--dc20
  DNLM/DLC
  for Library of Congress   93-46630
                                                                  CIP

Proceedings of the Fifth International Symposium on Coronaviruses, held September 13–18, 1992, in
Chantilly, France

ISBN 978-1-4613-6305-7

© 1993 Springer Science+Business Media New York
Originally published by Plenum Press, New York in 1993
Softcover reprint of the hardcover 1st edition 1993

PREFACE

Coronaviruses represent a major group of viruses of both molecular biological interest and clinical significance in animals and humans. During the past two decades, coronavirus research has been an expanding field and, since 1980, an international symposium was held every 3 years. We organized the Vth symposium for providing an opportunity to assess important progresses made since the last symposium in Cambridge (U.K.) and to suggest areas for future investigations. The symposium, held in September 1992, in Chantilly, France, was attended by 120 participants representing the majority of the laboratories engaged in the field. The present volume collects 75 papers which were presented during the Vth symposium, thus providing a comprehensive view of the state of the art of Coronavirology.

The book is divided into 7 chapters. The first chapters gather reports dealing with genome organization, gene expression and structure-function relationships of the viral polypeptides. New sequence data about as yet poorly studied coronaviruses - canine coronavirus CCV and porcine epidemic diarrhoea virus PEDV - are presented. Increasing efforts appear to be devoted to the characterization of products of unknown function, encoded by various open reading frames present in the coronavirus genomes or derived from the processing of the large polymerase polyprotein. Due to the extreme size of their genome, the genetic engineering of coronaviruses through the production of full length cDNA clones is presently viewed as an unachievable task. A high point in this meeting was the presentation of two alternative approaches, both taking advantage of the high RNA recombination rate of coronaviruses, which allowed the targeting of site-specific modifications into the genome of murine hepatitis virus MHV. This symposium also provided a timely opportunity to sum up the common and distinctive features existing between the coronaviruses and two other virus groups : the toroviruses, which now form a second genus within the *Coronaviridae*, and the arteriviruses, a so far unclassified genus. Another high point was the presentation of new data concerning two kinds of cell surface molecules, aminopeptidase N and carcinoembryonic antigens, which are used as a major receptor by coronaviruses. Increased investigations about these molecules might have interesting implications for a better understanding of pathogenesis and the development of new antiviral approaches. The long-lasting interest in rodent coronaviruses as a model for virus-induced neurological disorders was renewed by several presentations, which collectively point to a possible involvement of coronaviruses in chronic infections of the human CNS. Finally, substantial progresses have been reported in the identification of the components of the immune system which modulate the disease course or contribute to protection, and this largely relies upon the use of recombinant polypeptides as antigens or immunogens.

We gratefully acknowledge the substantial support from Institut National de la Recherche Agronomique, Rhône-Mérieux France and Smithkline Beecham Companies as well as corporate contribution from Intervet, Sanofi Santé Animale, Ambico and Virbac Companies, which were all instrumental in the success of the meeting. We also take the opportunity to thank the FEMS, who sponsored the attendance of a number of young scientists. Finally, we wish to express our personal thanks to Brigitte Schaeffer, member of the Molecular Virology and Immunology Unit (Jouy-en-Josas), for her dedicated assistance in the management of the meeting.

It was agreed by the participants that a VIth International Symposium on *Corona- and Related Viruses* will be held in Quebec (Canada) in 1994, organized by Drs. P. Talbot and G. Levy.

Hubert Laude

Jean-François Vautherot

Institut National de la Recherche Agronomique

France

CONTENTS

1-CORONAVIRUS GENES: COMPARATIVE ASPECTS

2-TRANSCRIPTION, REPLICATION AND GENOME ENGINEERING

3-CHARACTERIZATION AND FUNCTIONS OF VIRAL PROTEINS

4-CORONAVIRUSES, TOROVIRUSES AND ARTERIVIRUSES: COMMON AND DISTINCTIVE FEATURES

5-CELLULAR RECEPTORS FOR CORONAVIRUSES

6-PATHOGENESIS: MECHANISMS AND DIVERSITY

7-IMMUNE RESPONSE AND PROTECTION

Chapter 1

Coronavirus Genes: Comparative Aspects

SEQUENCE ANALYSIS OF CCV AND ITS RELATIONSHIP TO FIPV, TGEV AND PRCV

Brian C. Horsburgh and T. David K. Brown

Division of Virology
Department of Pathology
University of Cambridge
Tennis Court Road
Cambridge CB2 1QP
UK

INTRODUCTION

Canine coronavirus (CCV), a causative agent of enteritis in newborn dogs, was first identified in 1971[1]. The disease is characterized by infection of the absorptive epithelium of the villi and the onset of diarrhoea followed by villus atrophy[2]. The CCV virion is known to contain at least three protein species viz. the 204 kDa spike glycoprotein, S, the 32 kDa membrane glycoprotein, M, and the 50 kDa nucleocapsid protein, N[3].

CCV belongs to one of the major antigenic groups of coronaviruses[4,5] and is serologically related to feline infectious peritonitis virus (FIPV), feline enteric coronavirus (FECV), transmissible gastroenteritis virus (TGEV), and porcine respiratory coronavirus (PRCV)[6]. These viruses have been distinguished mainly by their host species of origin. It has been reported however, that some strains of CCV can also infect cats[7,8] and swine[9]. Likewise TGEV can also infect the other member species[10,11], and FIPV can infect swine[12]. This close relationship indicates that the viruses may have a common ancestor[13,6].

Molecular analysis has helped to elucidate some of the aspects of this phylogenetic relationship and some of the mechanisms involved in pathogenesis. TGEV, PRCV, and FIPV have been characterized in some detail and the genes encoding the structural proteins have been cloned and sequenced[14,15,16,17,18,19]. A comparison of the FIPV amino acid sequences with the corresponding sequences of TGEV and PRCV has revealed that the structural genes are very closely related. For S the identities were 81.6% (TGEV) and 76% (PRCV), for M 84.4% and 85.9%, and for N, 77% and 75.6% respectively. This contrasts greatly with the relationship to murine hepatitis virus (MHV), a prototypic coronavirus from another antigenic group, where the identities for these polypeptides are 24%, 30% and 27% respectively[20,21,22]. Despite this high degree of homology amongst the structural proteins of these three viruses there are, nevertheless, differences at the 3'-end of their viral genomes and in their subgenomic message organization.

The close relationship between TGEV, PRCV, FIPV and CCV has considerable epidemiological implications. Clearly, more information is needed in order to better understand the taxonomic relationship of this antigenic group of viruses. CCV is the least characterized virus from this antigenic group; it was hoped that cloning, sequencing and subsequent analyses would help to illuminate the taxonomic relationship of this family of viruses.

Coronaviruses, Edited by H. Laude and J.F. Vautherot
Plenum Press, New York, 1994

METHODS

The materials and methods are described in detail by Horsburgh et al.[23]. Briefly, virus was pelleted from the supernatant of CCV-Insavc-1 infected A72 cells, and the pellet homogenized in guanidinium isothiocyanate solution. The mixture was layered onto a CsCl pad, and the viral RNA pelleted by centrifugation. The RNA was dissolved in TE, oligo d(T) selected, and the concentration determined (A260). A cDNA library was produced from approximately 25μg poly (A)+ tailed RNA using oligo d(T) and random pentanucleotides as primers[24]. The cDNA was blunted ended using T4 DNA polymerase and ligated into the SmaI site of pUC119. Portions of the ligation mixture were transformed into E.coli strain TG-1 and clones identified by colour selection. Positive colonies were probed with radio-labelled, randomly primed CCV cDNA. 'Minipreps' of plasmid DNA were prepared from colonies which gave the strongest signals. A number of the colonies contained inserts of 1.8 kb or over and were selected for further analysis.

PCR-amplified fragments were obtained using cDNA:RNA heteroduplexes as templates and oligonucleotides 7 (5' GTT GCA ATT GCG GCC GCA CAG TTA TTA TTG TTG) and 8 (5' CCC ATT GGC AAC GCG GCC GCT GTC ACC AAA ATT GGC), each of which was modified to contain a NotI site, as primers. Amplification of the cDNA was performed as described by Sambrook et al.[25] using Taq polymerase. The generated fragment was cleaved with NotI, gel purifed and ligated into the NotI site of pKL1. (pKL1 is a pUC based vector with a modified polylinker and was a gift from Dr. K Law, University of Cambridge).

The nucleotide sequence of the overlapping cDNA clones at the 3'-end of the genome (BH5 - 10) was determined using the M13/dideoxynucleotide method[26]. Briefly, insert DNA was excised from vector sequences, self ligated and sonicated. The sonicated DNA fragments were end-repaired with the Klenow fragment of E.coli DNA polymerase and T4 DNA polymerase prior to size selection on a 1.2% agarose gel. Fragments in the size range 300-500 bps were purified and subcloned into SmaI digested, phosphatased M13mp8, from which single stranded DNA was prepared and used as template for sequencing reactions.

The nucleotide sequence data obtained was analyzed using the programs of Staden[27] on a VAX 8350 computer.

RESULTS

To clone the 3'-end of the CCV genome we prepared a cDNA library from CCV genomic RNA. Inserts from recombinant clones of approximately 1.8kb or greater in size were selected for further analysis. In order to map the clones, we took advantage of the suspected nucleotide sequence homology between the genomes of CCV and TGEV[28]. Double stranded sequencing of recombinant plasmid DNA, using the pUC forward and reverse primers, revealed approximately, 150 nts of sequence at each end of the viral insert. Comparison of these sequences with published TGEV sequences disclosed identity in excess of 95%. This permitted initial alignment of the CCV clones with respect to the TGEV genome. This approach proved fruitful in that five clones (BH5, 7, 8, 9 and 10) were identified which spanned in total, some 8.5 kb at the 3'-end (see Fig. 1). A region at the 3'-end for which large clones were not present in the library was prepared by PCR amplification (BH6; Fig. 1). Briefly, partial sequence information obtained from the 3'-end of BH7 and the 5'-end of BH5, allowed the design of primers 7 and 8, each of which was modified to contain a NotI site. PCR-amplified fragments were obtained using cDNA:RNA heteroduplexes as template and oligonucleotides 7 and 8 as primers. The generated DNA fragment was cloned into the NotI site of pKL1 and transformed into E.coli strain TG-1. A positive clone, pBH6 was selected for analysis. The relationships between putative overlapping clones were confirmed by Southern hybridization.

The consensus sequence determined from the six overlapping cDNA clones, 9580 bps long (exclusive of the poly (A) tail), is presented elsewhere[23]; EMBL/Genbank accession number D13096). This consensus sequence was analyzed using the SAP programs of Staden[27]. Analysis revealed the presence of 10 open reading frames (ORFs) over 50 amino acids in length (see Fig. 2). Pairwise alignment of these ORFs with their likely counterparts from other members of this coronavirus group disclosed very high levels of identity and indicated that the CCV structural proteins, S, M and N are encoded by ORFs 2, 5 and 6 respectively.

With respect to subgenomic mRNA synthesis, it is known that the conserved signal for transcription in this coronavirus group, CTAAAC, is identical in TGEV, PRCV and FIPV and is therefore likely to be conserved in CCV (as reviewed by Spaan et al.[5]). Indeed, analysis of the CCV sequence revealed that this sequence was present upstream of all the

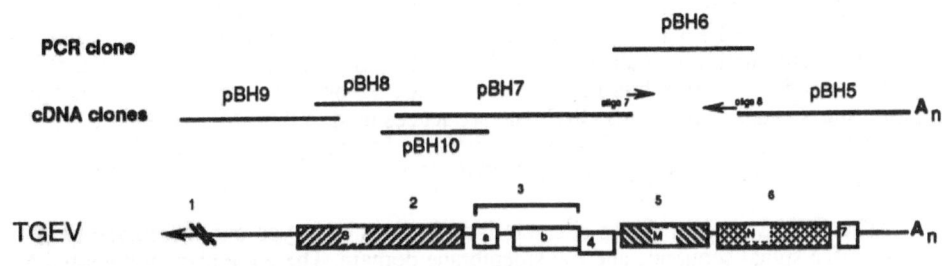

Fig. 1

Alignment of CCV cDNA clones with respect to the TGEV genome using partial sequence information.

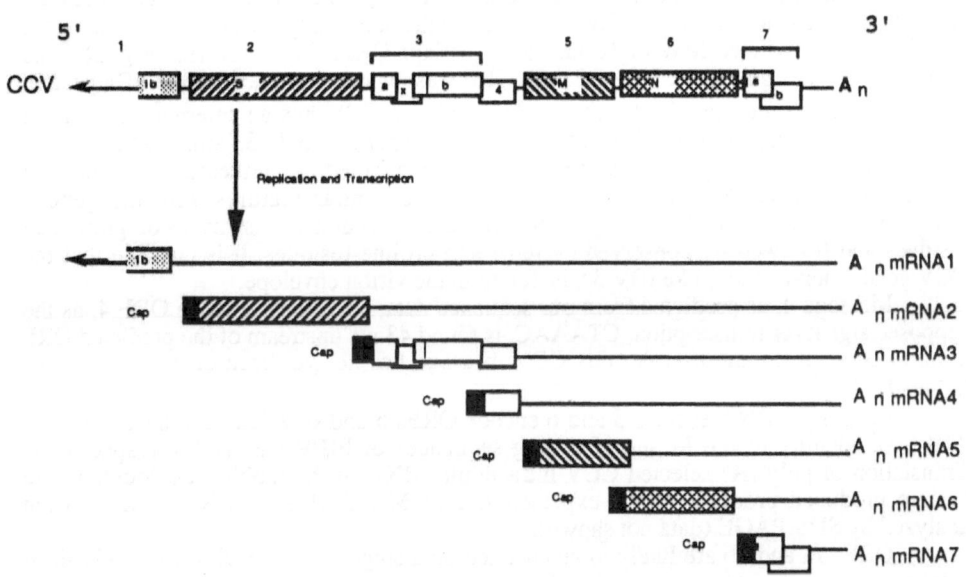

Fig. 2

Gene and subgenomic message organization predicted from the sequence data. The vertical line in ORF 3b represents a stop codon and black boxes represent leader sequences. Numbers represent ORFs encoded by that message.

ORFs with the exception of 1b, 3x and 7b. As ORF 1 is incomplete (see below), an additional CTAAAC sequence is presumably located at the 5'-end of the genomic RNA. When we analyzed intracellular RNAs produced during CCV infection of canine A72 cells: seven species of mRNA were observed[23]. Taking into account the predicted size of each mRNA and the known location of the CTAAAC sequences, we predict a subgenomic message organization as depicted in Fig. 2. This organization is very similar to that observed by Groot et al.[29] for FIPV except that CCV has an extra message (mRNA4) which has not yet been detected in FIPV-infected cells. The ORFs encoded by each mRNA are described below.

ORF 1 is incomplete, it has no ATG start codon, consists of 168 amino acids and terminates in a UGA stop codon at position 510. A comparison of this ORF with TGEV strain FS772/70 shows 99.2% similarity to 1b and 47% and 52.7% identity to gene 1b of IBV and MHV respectively[30,31,32]. Thus this ORF represents the 3'-end of the putative polymerase encoding region of genome mRNA1.

ORF 2, located immediately downstream of the polymerase gene would be translated from the 9.1 kb subgenomic message 2. This ORF, 4356 nts long representing 1452 amino acids with a calculated MW of 160 kDa, is preceded by the potential RNA polymerase-leader complex binding site, CTAAAC, 32 bps upstream of the translation initiation site. Comparison of this ORF with sequences held in the EMBL database reveals remarkably high identity to the FIPV spike glycoprotein coding sequence (91.2%) and to a lesser degree, the porcine S genes (79% TGEV; 75% PRCV), identifying this ORF as the CCV S gene.

The CCV S protein shares characteristic features of a type one membrane protein viz. a putative signal sequence and transmembrane domain. There are also 30 potential N-glycosylation sites; glycosylation probably accounts for the higher apparent MW of the S protein found in the virion[3].

There are four ORFs distal of the S gene coding sequence which are likely to be encoded by mRNAs 3 and 4 (Fig. 2). Three of these have high identity to their porcine counterparts, TGEV and PRCV, and have been named 3a (8.6 kDa; 83.5%/48.8%), 3b (28.4 kDa; 92.7%/92.6%) and 4 (9.3 kDa; 88.4%/88.4%). The fourth ORF; which to date has not been detected in this family of viruses could potentially encode a 71 amino acid protein with a predicted MW of 10 kDa and overlaps ORFs 3a and 3b (see Fig. 2). This ORF has been designated 3x. The CCV 3b ORF was expected to encode a 28 kDa protein like its TGEV counterpart[33], however, this strain of CCV has an internal termination codon, TAA, which would result in a truncated polypeptide of only 33 amino acids. Direct sequencing of the viral genomic and mRNAs has confirmed the authenticity of this stop codon (data not shown). The IBV 3c ORF shares similar features with the gene 4 polypeptide of CCV, such as a hydrophobic core preceded by aspartate or glutamate residues and followed by conserved cysteine and proline residues. It is possible that the CCV gene 4 polypeptide, like IBV 3c, is found in the virion envelope.

Message 4, as predicted from our sequence data, could only encode ORF 4, as the supposed signal for transcription, CTAAAC, is found 43 nts upstream of the predicted ORF 4 start codon. A counterpart for this CCV message has not yet been detected in FIPV-infected cells[29].

Messenger RNA species 5 and 6 encode ORFs 5 and 6 which have 84%/88% and 77%/90% identity to the M and N coding sequences of FIPV and TGEV respectively. Translation of poly (A) selected CCV intracellular RNA in the rabbit reticulocyte lysate system produced products of the expected size of M (25 kDa) and N (45 kDa) when analyzed by SDS-PAGE (data not shown).

ORFs 7a and 7b are likely to be encoded on a single RNA species (mRNA7) since smaller messages were not seen on Northern blots (data not shown), nor is another message predicted from the sequence data. Furthermore, an equivalent RNA in FIPV is thought to be bicistronic[34] and the levels of identity between the 7a and 7b ORFs of CCV and the 6a and 6b ORFs of FIPV are 78.4% and 57% respectively. Alignment of this region of CCV with the related regions of the TGEV and PRCV revealed that the 7a ORF of the porcine coronaviruses has undergone a deletion of 69 nts and furthermore, they have no counterpart to ORF 7b. Nevertheless, the CCV structural ORFs, with the exception of S, have higher identities to TGEV than to FIPV.

The CCV sequence contains the octameric sequence, GGAAGAGC, at the 3'-end of the genome, upstream of the poly (A) site, which is conserved in all coronavirus sequences to date.

DISCUSSION

In this study approximately 9.6 kb from the 3' genomic end of CCV strain Insavc-1 was cloned and sequenced. This region is likely to include all of the viral genes excluding the polymerase gene for which only the 3' terminal 168 amino acids have been determined. Therefore a substantial part of the virus' genetic information was available for comparison with the other antigenically related coronaviruses, namely TGEV, PRCV and FIPV. The respective genetic organizations are shown in Fig. 3.

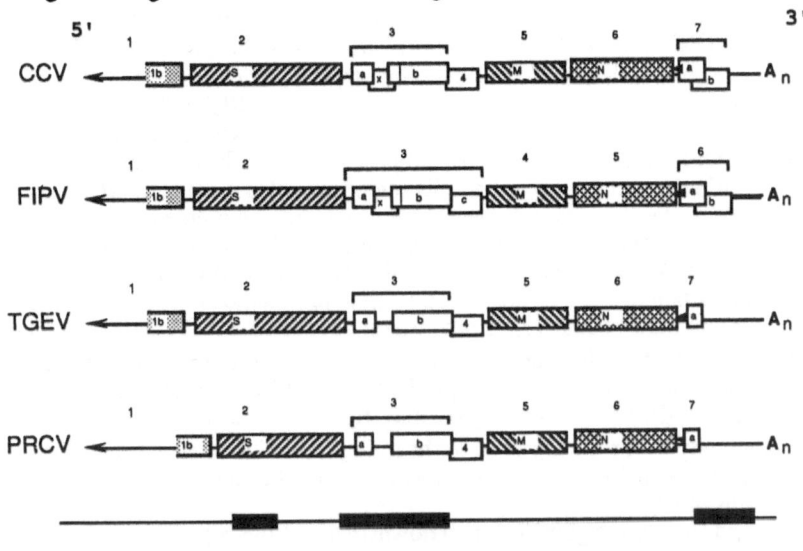

Fig. 3
Genomic organization of CCV, FIPV, TGEV and PRCV. The black boxes represent putative 'hot spot' regions.

Fig. 3
Genomic organization of CCV, FIPV, TGEV and PRCV. The black boxes represent putative 'hot spot' regions.

From antigenic data and cross infectivity studies, the viruses within this group have been termed 'host range mutants'[13]. This close relationship is emphasized by our analyses of the CCV sequence data. The CCV spike is closely related to the other spikes and has the features typical of coronavirus peplomer glycoproteins. Any variation in the sequence of this protein within the group presumably reflect changes in cell tropism, genetic drift and selection from the host's immune system. Similarly, interspecies comparison of the other structural proteins M and N, again revealed very high levels of identity (see results).

The genome organizations of both CCV and FIPV are virtually identical (Fig. 3), implying that CCV is more related to the feline coronaviruses than to the porcine coronaviruses. Nevertheless, the four viruses have strikingly similar genetic organizations. However, it would appear that there are 3 'hot spot' regions where recombination, insertions or usually deletions occur at a higher frequency relative to the surrounding sequences. These are within S, between S and M, and downstream of N (Fig. 3). The dynamism of the coronavirus genome is well documented[35,36] and may be related to the propensity of the replicase complex to fall off its template and then to reinitiate RNA replication on the same or different template.

The PRCV S gene has acquired a deletion which may be responsible for the change in virus tropism. The polymorphism of S found in MHV strains with differing passage histories is mainly due to deletions in that gene which can be up to 159 amino acids. This can have an effect on pathogenicity, as deletions in the MHV-4 S coding sequence apparently result in a loss of ability to induce fatal encephalitis and the acquisition of the capacity to produce a non-fatal demyelinating disease in mice[37].

Polymorphism is also observed in the second 'hot spot' region, between the S and M genes (Fig. 3). The CCV ORFs which lie between the spike and membrane genes like the other ORFs so far analyzed have high identities to their porcine counterparts and presumably perform similar functions. The degree of variability in the lengths of the non coding sequences that lie upstream and downstream of ORF 3a in members of this antigenic group is striking. The lengths of these sequences range from 40bp to over 200bp.

Alignment of the ORF 3a amino acid sequences reveals that in addition, variation is found at the ends of these coding sequences. Moreover, another deletion in the PRCV genome results in the loss of a functional 3a protein.

A previously undetected ORF, 3x was identified which could potentially encode a 10 kDa polypeptide. However codon usage and base preference programs of Staden[27] suggest that this ORF is not a coding sequence. Furthermore, the proximal ATG is in a poor context for translation initiation[38], and the only other ATG is found at the very 3'-end of the coding sequence. Therefore it is very unlikely that this ORF will be expressed in this strain of CCV; it may represent an evolutionarily redundant sequence which is no longer required by the virus or may contain signals important for translation of downstream ORFs. Analysis of TGEV genomic sequence in this region revealed a counterpart for this canine pseudogene; 92 nucleotides have however been deleted. This deletion also results in a frame-shift in the sequence which explains why this ORF has not hitherto been noticed. In addition to the likely non-functionality of ORF 3x, it is also unlikely that ORF 3b is expressed in this strain of CCV as there is an internal termination codon (TAA; represented by a vertical bar in Fig.3) some 93 nts downstream of the first ATG and subsequent ATG codons are in poor contexts for ribosome binding[38]. *In vitro* transcription and translation of this ORF did not yield any discernible products on SDS-polyacrylamide gels (data not shown). Further evidence that this region may function as a recombinational 'hot spot' comes from a study by Cavanagh *et al.*[39]. In it they reported an IBV strain (Port/322/85) which appears to have arisen as a result of recombination between the M and S genes from two other strains of IBV.

The third "hot spot region" is found downstream of the N gene (Fig. 3). The porcine coronaviruses have a 69 nt deletion in ORF 7a whilst ORF 7b is not present[34]. This phenomenon is not unique to the coronaviruses from this antigenic group. Deletions of up to 170 nts are found downstream of the N gene in some strains of IBV[40]. It is interesting to note that the CCV ORF 7b has 57% identity to FIPV ORF 6b. This ORF is the least conserved between the two viruses.

In conclusion, sequencing and subsequent analyses stress the very close relationship CCV has to the other viruses within its antigenic group. We must however, be careful when generalizing about the CCV sequence data from this limited information. Coronavirus genomes are dynamic, subject to recombination, insertion and deletion, and as a consequence strains may differ genetically. Clearly, there is a need to clone and sequence other strains in order to build a consensus picture of the CCV genome.

REFERENCES

1. Binn, L.N., Lazar, E.C., Keenan, K.P., Huxsoll, D.L., Marchwicki, R.M. & Scott, F.W. Proc.U.S. Animal Health Assoc. 78:359-366 (1974).
2. Keenan, K.P., Jervis, H.R., Marchwicki, R.H. & Binn, L.N. Amer. J. Vet. Res. 37: 247-256 (1976).
3. Garwes, D.J. & Reynolds, D.J. J. gen. Virol. 52: 153-157 (1981).
4. Siddell, S., Wege, H. & ter Meulen, V.. J. gen. Virol. 64: 761-776 (1983).
5. Spaan, W., Cavanagh, D. & Horzinek, M.C. J. gen. Virol. 69: 2939-2952 (1988).
6. Sanchez, C.M., Jimenez, G., Laviada, M.D., Correa, I., Sune, C., Bullido, M.J., Gebaues, F., Smerdou, C., Callebaut, P., Escribano, J.M. & Enjuanes, L. Virol. 174: 410-417 (1990).
7. Barlough, J.E., Stoddart, C.A., Sorresso, G.P., Jacobson, R.H. & Scott, F.W. Lab. Animal Sci. 34: 592-597 (1984).
8. Stoddart, C.A., Barlough, J.E., Baldwin, C.A. & Scott, F.W. Res. Vet. Sci. 45: 383-388 (1988).
9. Woods, R.D. & Wesley, R.D. Amer. J. Vet. Res. 47: 1239-1242 (1986).
10. Norman, J.O., McClurkin, A.W. & Stark, S.L. Can. J. Comp. Med. 34: 115-117 (1970).
11. Woods, R.D. & Pedersen, N.C. Vet. Microbiol. 4: 11-16 (1979).
12. Woods, R.D., Cheville, N.F. & Gallagher, J.E. Amer. J. Vet. Res. 42: 1163-1169 (1981).
13. Horzinek, M.C., Lutz, H. & Pedersen, N. Infect. and Immun. 37: 1148-1155 (1982).
14. Groot, R.J. de, Maduro, J., Lenstra, J.A., Horzinek, M.C., Zeijst, B.A.M. van der & Spaan, W.J.M. J. gen. Virol. 68: 2639-2646 (1987).
15. Rasschaert, D. & Laude, H. J. gen. Virol. 68: 1883-1890 (1987).

16. Britton, P., Carmenes, R.S., Page, K.W., Garwes, D.J. & Parra, F. Mol. Microbiol. 2: 89-99 (1988).
17. Britton, P., Carmenes, R.S., Page, K.W. & Garwes, D.J. Mol. Microbiol. 2: 497-505 (1988).
18. Rasschaert, D., Duarte, M. & Laude, H. J. gen. Virol. 71: 2599-2607 (1990).
19. Vennema, H., Groot, R.J. de., Harbour, D.A., Horzinek, M.C. & Spaan, W.J.M. Virol. 181: 327-335 (1991).
20. Skinner, M.A. & Siddell, S.G. Nucl. Acids Res. 11: 5045-5054 (1983).
21. Armstrong, J., Neimann, H., Smeekens, S., Rottier, P. & Warren, G. Nature 308: 751-752 (1984).
22. Schmidt, I., Skinner, M. & Siddell, S.G. J. gen. Virol. 68: 47-56 (1987).
23. Horsburgh, B.C., Brierley, I. & Brown, T.D.K. J. gen. Virol. 73: 2849-2862 (1992).
24. Gubler, U. & Hoffman, B.J. Gene 25: 263-269 (1983).
25. Sambrook, J., Fritsch, E.F. & Maniatis, T. Molecular cloning: a laboratory manual 2nd edn., Cold Spring Harbour Laboratory, New York 1989).
26. Bankier A.T., Weston, K.W. & Barrell, B.G. Methods in Enzymology 155: 51-93 (1987).
27. Staden, R. Nucl. Acids Res. 14: 217-231 (1986).
28. Shockley, L.J., Kapke, P.A., Lapps, W., Brian, D.A., Potgeiter, L.N.D. & Woods, R. J. Clin. Microbiol. 25: 1591-1596 (1987).
29. Groot, R.J. de, Haar, R.J. ter, Horzinek, M.C. & Zeijst, B.A.M van der . J. gen. Virol. 68: 995-1002 (1987).
30. Boursnell, M.E.G., Brown, T.D.K., Foulds, I.J., Green, P.F., Tomley, F.M. & Binns, M.M. J. gen. Virol. 68: 57-77 (1987).
31. Bredenbeek, P.J., Pachuk, C.J., Noten, A.F.H., Charite, J., Lutyes, W., Weiss, S.R. & Spaan, W.J.M. Nucl. Acids Res. 18: 1825-1832 (1990).
32. Britton, P, & Page, K.W. Virus Res. 18: 71-80 (1990).
33 Jacobs, L., Zeijst, B.A.M. van der & Horzinek, M.C. J. Virol. 57: 1010-1015 (1986).
34. Groot, R.J. de, Andeweg, A.C., Horzinek, M.C. & Spaan, W.J.M. Virol. 167: 370-376 (1988).
35. Keck, J.G., Matsushima, G.K., Makino, S., Fleming, J.O., Vannier, D.M., Stohlman, S.A. & Lai, M.M.C. J. Virol. 62: 1810-1813 (1988).
36. Kusters, J.G., Jager, E.J. & Zeijst, B.A.M. van der. Nucl. Acids Res 17: 6726-6729 (1989).
37. Parker, S.E., Gallagher, T.M. & Buchmeier, M.J. Virol. 173: 664-673 (1989).
38. Kozak, M. Adv.Virus Res. 31: 229-292 (1986).
39. Cavanagh, D., Davis, P, Cook, J. & Li, D. Adv. Exptl. Med. Biol. 276: 369-372 (1990).
40. Collisson, E.W., Williams, A.K., Haar, R.V., Li, W. & Sneed, L.W. Adv. Exptl. Med. Biol. 276: 373-377 (1990).

GENOMIC ORGANIZATION AND EXPRESSION OF THE 3' END OF THE CANINE AND FELINE ENTERIC CORONAVIRUSES

H. Vennema, J.W.A. Rossen, J. Wesseling, M.C. Horzinek and P.J.M. Rottier

Department of Virology, Faculty of Veterinary Medicine, University of Utrecht, Yalelaan 1, P.O. Box 80.165, 3508 TD Utrecht, The Netherlands

ABSTRACT

The genomic organization at the 3' end of canine coronavirus (CCV) and feline enteric coronavirus (FECV) was determined by sequence analysis and compared to that of feline infectious peritonitis virus (FIPV) and transmissible gastroenteritis virus (TGEV) of swine. Comparison of the latter two has previously revealed an extra open reading frame (ORF) at the 3' end of the FIPV genome, lacking in TGEV, now designated ORF 6b. Both CCV and FECV possess 6b-related ORFs. The CCV ORF 6b is colinear with that of FIPV, but the predicted amino acid sequences are only 58% identical. The FECV ORF 6b contains a large deletion compared to that of FIPV, reducing the colinear part to 60%. The sequence homologies were highest between CCV and TGEV on the one hand and between FECV and FIPV on the other. The expression product of the CCV and the FECV ORF 6b can be detected in infected cells by immunoprecipitation.

INTRODUCTION

Canine coronavirus (CCV), feline enteric coronavirus (FECV), feline infectious peritonitis virus (FIPV), and transmissible gastroenteritis virus (TGEV) of swine belong to one antigenic cluster (1). Sequence analysis revealed a close genetic relatedness between FIPV and TGEV (2, 3, 4). FIPV contains an extra open reading frame (ORF) in the 3'-terminal region of its genome (3). It is the second ORF of mRNA 6, currently designated ORF 6b. The 6b gene product was detected in FIPV-infected cells (5). It is a secreted nonstructural glycoprotein. In cats it induces antibodies during FIPV infections. The 6b protein provides an antigenic distinction between FIPV and TGEV. The aim of the present study was to examine whether this distinction could be extended to CCV and FECV.

MATERIALS AND METHODS

Cells and viruses

FIPV strain 79-1146, FECV strain 79-1683 (6), and CCV strain K378 (Dutch field isolate) were grown in *Felis catus* whole fetus cells (fcwf-D). Recombinant vaccinia virus vTF7-3 (7) infections were carried out in HeLa cells. Cells were maintained in Dulbecco's modified Eagle's medium (GIBCO Laboratories) containing 5% fetal bovine serum.

Cloning and sequence analysis of the 3' end of CCV

cDNA libraries were prepared of intracellular poly(A)$^+$ RNA from CCV infected fcwf-D cells as described elsewhere (J. Wesseling manuscript in preparation). Clones containing sequences derived from the 3' end of the genome were selected by colony hybridization with restriction fragments of FIPV cDNA clone B12 (3) as probes. Nucleotide sequencing was performed on double stranded DNA using a bacteriophage T7 DNA polymerase based kit (Pharmacia LKB). Sequence data were analyzed using the computer programs of Devereux et al. (8).

cDNA synthesis and PCR amplification of the 3' end of FECV

Synthesis of cDNA on total RNA isolated from FECV infected fcwf-D cells was performed as described (9) by priming specifically with synthetic oligonucleotide 5'-CCAGTTTTAGACATCGGG-3' which binds to a sequence in the 3' non-coding region of FIPV, downstream of ORF 6b. Oligonucleotide 5'-GATCCAGACGTTAGCTC-3', was used to prime cDNA synthesis from a position closer to the 3' end. Amplification of cDNA was performed by the polymerase chain reaction (PCR) as described (9), after the addition of synthetic oligonucleotide 5'-GATGACACACAGGTTGAG-3', which is located at the 3' end of the nucleocapsid (N) protein gene of FIPV. PCR amplified cDNA fragments of FECV were cloned after homopolymer tailing in dG-tailed pUC9 (Pharmacia LKB) and sequenced as described above.

Radio immunoprecipitation assays (RIPA)

Lysates from coronavirus-infected fcwf-D cells or recombinant vaccinia virus-infected HeLa cells were prepared after metabolic labeling with L-[^{35}S]cysteine (ICN Biomedicals, Inc.). Lysis, RIPA with ascites fluid from a field case of FIP and endo-ß-N-acetylglucosaminidase H (endo H; Boehringer Mannheim Biochemicals) treatment were carried out as described (10). Analysis by SDS-polyacrylamide gel electrophoresis (SDS-PAGE) was performed as described (11).

RESULTS

Sequence analysis of the 3' end of the CCV genome

CCV cDNA clones were prepared, selected and sequenced as described under materials and methods. We obtained a contiguous sequence of 2.5 kb from the 3' end. The sequence data are available under the accession number X66717. Translation of the nucleotide sequence revealed three ORFs corresponding to the N protein and ORFs 6a and 6b of FIPV. The ORF 6b predicts a short hydrophobic amino-terminus, which may function as a signal sequence. In contrast to the FIPV ORF 6b, the CCV ORF 6b contains no potential N-glycosylation site.

PCR amplification of cDNA derived from the 3' end of the FECV genome

We performed cDNA synthesis followed by PCR amplification on total RNA from FECV- and CCV-infected cells using primers which flanked the coding region of mRNA 6. The CCV PCR product had the expected size of 1 kbp. The FECV product was considerably smaller, being approximately 750 bp. The controls with RNA from mock-infected cells and without RNA were both negative. Sequence analysis revealed that the FECV ORF 6b extended into the sequences used to design the PCR primers. Therefore, cDNA-PCR was repeated with a primer hybridizing 36 nucleotides upstream of the poly(A)-tail. This resulted in a fragment of approximately 950 bp (data not shown).

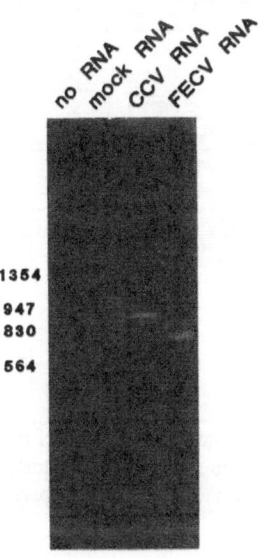

Figure 1. Agarose gel analysis of PCR amplified CCV and FECV cDNA. The RNA sources used for cDNA preparation are indicated above the lanes. Marker indicated on the left is bacteriophage lambda DNA digested with EcoRI and HindIII.

Analysis of genomic sequences of FECV near the 3' end

The PCR products of FECV were cloned and sequenced, leading to a contiguous sequence of 957 nucleotides. The sequence data are available under the accession number X66718. Comparison with the corresponding sequence of FIPV showed a single deletion of 238 nucleotides and an overall sequence identity of 93.6%. Translation of the nucleotide sequence revealed the presence of two ORFs similar to the FIPV ORFs 6a and 6b. The deletion is located in ORF 6b. The FECV and FIPV 6b sequences are colinear for the amino-terminal 123 amino acid residues. The deletion results in a shift to the -1 reading frame which extends 53 codons. The ORF specifies a polypeptide with a total length of 176 amino acid residues and a predicted mol wt of 20,300. A short hydrophobic amino-terminus, probably acting as a signal sequence, and one N-glycosylation site are predicted.

Sequence comparison and genomic organization

The genomic organizations of the CCV and FECV 3' ends are similar to FIPV, containing an ORF 6b and the extra 69 nucleotides in ORF 6a as compared to TGEV ORF 7 (Fig. 2). Paired alignments of the colinear parts of the amino acid sequences revealed that CCV is closely related to TGEV and FECV to FIPV (Table 1). The same was found when the nucleotide sequences were compared. The 6b amino acid sequences of CCV and FIPV were only 58% identical and several small insertions in the CCV sequence were found. Nevertheless, the hydrophobicity plots of the putative 6b proteins were remarkably similar (not shown). The same plot for FECV 6b showed divergence in the C-terminal 40%.

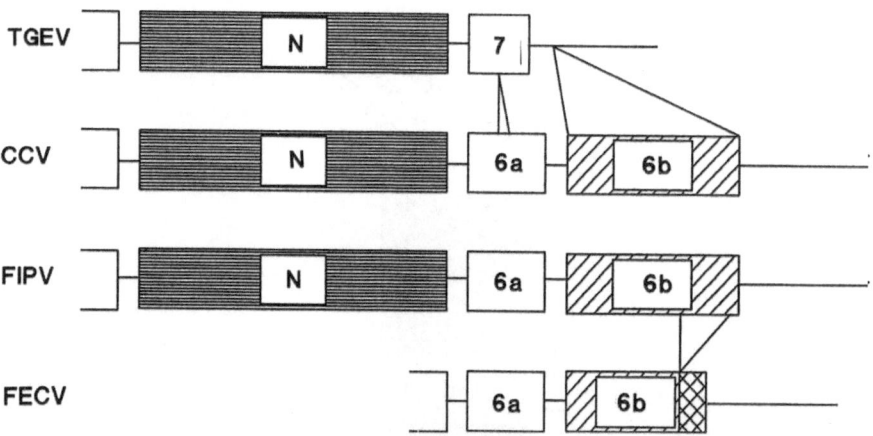

Figure 2. Schematic comparison of the genomic organization of the 3' terminal regions of TGEV, CCV, FIPV and FECV. Boxes represent the relevant open reading frames. The cross hatched box at the C-terminal end of FECV ORF 6b indicates the divergent part.

Table 1. Paired comparisons of the colinear parts of the amino acid sequences, in percentages identical residues.

		FIPV		CCV			TGEV	
		6a	6b	N	6a	6b	N	7
FECV	6a	99	-	-	80	-	-	77
	6b		89	-	-	51	-	-
FIPV	N			78	-	-	76	-
	6a				79	-	-	77
	6b					58	-	-
CCV	N						93	-
	6a							96

Identification of the CCV and FECV 6b proteins

The FIPV 6b protein was readily detected in lysates of FIPV-infected cells (5). To identify the CCV and FECV 6b proteins their 6b ORFs were recloned in a T7 expression vector. The resulting constructs pTC6b and pTE6b, respectively, and pTF6b containing the FIPV 6b gene (5) were used to transfect HeLa cells infected with recombinant vaccinia virus vTF7-3, which produces T7 RNA polymerase (7). The expression products were analyzed by metabolic labeling with [^{35}S]cysteine, RIPA, and endo H treatment followed by SDS-PAGE (Fig. 3). The CCV and FECV 6b proteins appeared to be slightly smaller than the FIPV 6b protein, the FECV 6b protein being the smallest. Digestion with endo H which cleaves high mannose N-linked oligosaccharides, resulted in an approximately 2,000 mol wt reduction of the FECV and FIPV 6b protein. The CCV 6b protein, was not affected and was also insensitive to digestion by endoglycosidase F, which cleaves complex N-linked sugars (data not shown). This indicates that the 6b proteins of FECV and FIPV are glycoproteins while CCV 6b is not. The shift in molecular weight of the FECV and FIPV 6b proteins is consistent with the removal of 1 sugar side chain (12). This is in agreement with the predicted numbers of glycosylation sites in the amino acid sequences. The observed molecular weights of the CCV 6b protein and of the FECV 6b protein after deglycosylation are also in agreement with those predicted from the amino acid sequences. Similar proteins were detected in CCV-, FECV-, and FIPV-infected cells, analyzed in the same way (Fig. 3). The lanes of FECV were overexposed to reveal the 6b protein band. This indicates that the expression level of the FECV 6b protein was lower than that of CCV and FIPV.

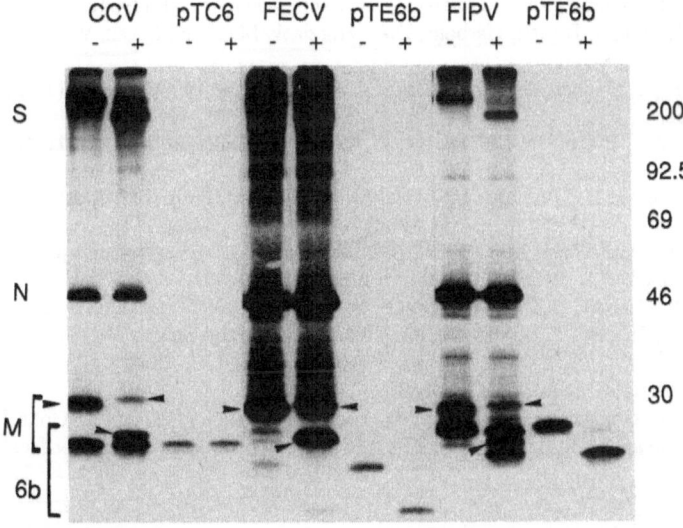

Figure 3. Radio immunoprecipitation and SDS-PAGE analysis of lysates from CCV-, FECV-, FIPV-, and vTF7-3-infected cells. Recombinant vaccinia virus vTF7-3-infected cells were transfected with the plasmid DNAs pTC6b, pTE6b and pTF6b, as indicated above the lanes. One half of each sample was treated with endo H, the other half was mock treated (indicated with + and -, respectively). Structural proteins (S, N and M) are indicated. In addition, the M protein bands are indicated with arrowheads. The region of the gel in which the 6b protein bands appear overlaps with that of M protein bands.

DISCUSSION

The genomic organization of the 3' end of FIPV differs from that of TGEV in that it contains an additional ORF (3). Recently, we identified the expression product of this extra gene, designated 6b (5). These observations prompted us to study CCV and FECV of the same antigenic cluster. Sequence analysis showed that their genomic organization in the 3' terminal region is similar to that of FIPV. Therefore, the presence rather than the absence of ORF 6b appears to be the common theme, suggesting that TGEV has lost the corresponding ORF by deletion. The same inference probably holds true for the 69 nucleotides that are present in all 6a ORFs but not in the corresponding ORF 7 of TGEV. Alignment of nucleotide and amino acid sequences of TGEV, CCV, FIPV and FECV allowed division into two pairs on the basis of their homologies; TGEV and CCV on the one hand and FIPV and FECV on the other.

Both in CCV- and in FECV-infected cells 6b proteins are produced. This observation appears to compromise their antigenic distinction from FIPV. However, the differences between the 6b proteins may allow discrimination using e.g. monoclonal antibodies. The deletion of 238 nucleotides in FECV 79-1683 was revealed by cDNA-PCR, allowing discrimination from CCV (Fig. 1) and FIPV 79-1146 (data not shown). It remains to be determined whether this is a universal distinguishing property of FIPV and FECV.

REFERENCES

1. Siddel, S., Wege, H., and Meulen, V. ter. (1983). *J. Gen. Virol.* 64, 761-776.
2. Jacobs, L., Groot, R.J. de, Horzinek, M.C., Zeijst, B.A.M. van der, and Spaan W.J.M. (1987). *Virus Res.* 8, 363-371.
3. Groot, R.J. de, Andeweg, A.C., Horzinek, M.C., and Spaan, W.J.M. (1988). *Virology* 167, 370-376.
4. Vennema, H., Groot, R.J. de, Harbour, D.A., Horzinek, M.C., and Spaan, W.J.M. (1991). *Virology* 181, 327-335.
5. Vennema, H., L. Heijnen, P.J.M. Rottier, M.C. Horzinek, and W.J.M. Spaan. (1992) *J. Virol.* 66:4951-4956.
6. McKeirnan, A.J., Evermann, J.F., Hargis, A., Miller, L.M., and Ott, R.L. (1981). *Feline Pract.* 11, 16-20.
7. Fuerst, T.R., Niles, E.G., Studier, F.W., and Moss, B. (1986). Proc. Natl. Acad. Sci. USA 83, 8122-8126.
8. Kawasaki, E.S., and Wang, A.M. (1989). *In* PCR technology: Principles and applications for DNA amplification, ed. H.A. Erlich, Stockton Press, New York, 89-97.
9. Devereux, J., Haeberli, P., and Smithies, O. (1984). *Nucleic Acids Res.* 12, 387-395.
10. Vennema, H., Heijnen, L., Zijderveld, A., Horzinek, M.C., and Spaan, W.J.M. (1990). J. Virol. 64, 339-346.
11. Laemmli, U.K. (1970). *Nature* 227, 680-685.
12. Neuberger, A., Gottschalk, A., Marshall, R.O., and Spiro, R.G. (1972). *In* A. Gottschalk (ed.), The glycoproteins: Their composition, structure and function., Elsevier, Amsterdam.

CLONING AND SEQUENCE ANALYSIS OF THE SPIKE GENE FROM SEVERAL FELINE CORONAVIRUSES

A. Paul Reed, Sharon Klepfer, Timothy Miller and Elaine Jones

Department of Molecular Biology
SmithKline Beecham Animal Health
P.O. Box 1539, L34
King of Prussia, Pennsylvania 19406-0939

ABSTRACT

The DNA sequence encoding the spike gene from the DF2 strain of Type II feline infectious peritonitis virus (FIPV), a temperature sensitive FIPV virus (TS-DF2) and an isolate of feline enteric coronavirus (FECV 1683) were determined. Comparison of the published WSU 1146 and DF2 FIPV S genes showed that the viruses shared a high degree of homology (99.6%). Likewise, the S gene of the virulent DF2 FIPV virus was closely conserved to that isolated from the vaccine virus strain, TS-DF2 FIPV. In contrast, the FECV S gene had numerous DNA and amino acid differences when compared to the virulent FIPV sequences. Sequence differences among the feline coronavirus isolates were localized to the amino-terminus region of the S gene.

INTRODUCTION

Feline Infectious Peritonitis Virus (FIPV) is a member of the coronavirus family. The virus contains three major structural proteins: 25-30 kD matrix protein (M); 50 kD nucleocapsid protein (N); and the 200 kD surface glycoprotein spike (S). The spike protrudes from the surface of the virus and is the principal target of serum neutralizing antibodies.

FIPV causes a complex and highly fatal disease in cats. Although the mechanism of pathogenesis is not well understood, FIPV is generally considered to cause an immune complex disease associated with high circulating antibody titers and deposition of antigen-antibody complexes in many major organs (1, 2, 3). Vaccination attempts have largely been unsuccessful since previous exposure to the virus is associated with accelerated death after challenge (4, 5, 6, 7). The spike protein appears sufficient to induce sensitization to the virus in that kittens immunized with vaccinia recombinants expressing the S gene developed FIP disease and died more rapidly than non vaccinated animals following virulent FIPV challenge (8, 9).

Recently, a temperature sensitive FIPV (TS-DF2) vaccine has been developed which, when administered intranasally, is efficacious and safe upon FIPV challenge (10). This virus was derived from a virulent DF2 FIPV strain but differs in its temperature specificity, plaque size and structural protein expression (11).

Cats are often infected with another related feline coronavirus, feline enteric coronavirus (FECV), which causes only mild enteritis following isolation. Although the pathogenicity of the two viruses is distinct, FECV shares serological identity with FIPV.

To determine if alterations in the spike genes of these viruses had occurred, gene amplification techniques were used to amplify, clone and DNA sequence the S gene from the wild type DF2 FIPV strain, the temperature sensitive variant (TS-DF2) and FECV. Comparison of these S genes with the published sequence of the virulent WSU 1146 FIPV spike suggests regions which could be used to develop a diagnostic assay to distinguish the viruses serologically.

METHODS

Virus strains.

Several feline coronavirus strains were used in this study: FECV 1683 and FIPV WSU 1146 (Washington State University) (12), wildtype DF2 FIPV virus (11) and TS-DF2 FIPV (11).

RNA purification.

Roller bottles of confluent NLFK cells were infected with a feline coronavirus at MOI = 0.1 in 50 ml of BME supplemented with 2% FBS. DF2 infections were performed in serum-free medium. The virus was absorbed for 2 hours and then 250 ml of growth medium (BME + 2% FBS) added. The cultures were monitored for cytopathic effect and typically harvested at 24 - 36 hours post-infection. Total cytoplasmic RNA was prepared from the infected monolayers by guanidine isothiocyanate extraction (13).

Oligonucleotide Design and Synthesis.

Oligonucleotides were designed from the published WSU 1146 S gene of 4500 nucleotides and contained restriction sites to facilitate cloning into the pBluescript vector (Stratagene). The primers were synthesized on an Applied Biosystem Model 380B DNA Synthesizer using the phosphoramidite method and gel purified prior to use.

PCR Amplification.

PCR amplified S gene fragments were generated using the following procedure. cDNA from total cellular RNA infected with a specific coronavirus (1 μg) was first synthesized (14). Amplification of the cDNA was performed essentially according to the method of Saiki et al. (1985) (15) using the Taq polymerase. PCR products were analyzed by electrophoresis of 5.0 μl of the reaction on a 1.2 % agarose gel run 16-17 hours. Bands were visualized by ethidium bromide staining the gel and fluorescence by UV irradiation at 256 nm.

DNA Sequencing and Analysis.

DNA sequence was determined from overlapping cloned regions of the S gene from each virus strain. Nested set deletions were prepared and the sequence determined from both strands using the chain termination method (16) (Lark Sequencing Technologies, Houston, TX). DNA sequence analysis was performed using Beckman Microgenie programs on an IBM Model PS2 Model 70 or the University of Wisconsin GCG package of programs (17).

RESULTS AND DISCUSSION

The S genes from DF2, TS-DF2 and FECV were 4362 bp in length and encoded an open reading frame of 1454 aa (160 kD molecular weight). As predicted for a surface glycoprotein, all three S genes contained a signal sequence at the amino terminus, a C-terminal transmembrane domain and multiple N-glycosylation sites. Some of these potential N-glycosylation sites are common to both viruses.

The S gene of DF2 FIPV shares 99.8% nucleotide homology and 99.6% amino acid homology with another virulent strain of Type II FIPV virus, WSU 1146. Interestingly, the DF2 S gene sequence contains 6 extra nucleotides (positions 351 - 356) when compared to the published WSU 1146 S gene sequence. These two additional amino acids at positions 119 and 120 increase the size of the DF2 spike gene to 1454 aa in contrast to 1452 aa of WSU 1146. Six other amino acids differed between these virulent Type II FIPV strains, four of which were located in the amino terminal half of the gene.

The TS-DF2 FIPV vaccine virus was derived after 99 passages in vitro and mutagenesis of the virulent DF2 strain. Comparison of the S genes from the parental DF2 and the temperature sensitive TS-DF2 vaccine strain did not reveal substantial differences. Eighteen nucleotide changes were observed (99.6% homology) which resulted in 13 amino acid differences (99.1% homology). Nine of these 13 changes were located in the amino terminus but were not clustered, making it difficult to identify a region which could be used to differentiate TS-DF2 from its progenitor, DF2. One change at nucleotide #1346 alters the sequence from Asn-Ala-Thr to Asn-Thr-Thr but does not affect glycosylation. The overall conservation of S sequence between these isolates suggests that these viruses are relatively stable and have not undergone recombination at this gene locus.

Although DF2 FIPV and FECV are serologically related, comparison of the S genes showed 67 amino acid differences between them (94.3% nucleotide homology, 95.4% amino acid homology). FECV appears similar to DF2 in that its S gene contains the two additional amino acids at positions 119 and 120 missing from WSU 1146. The majority of sequence changes in FECV were localized to the amino terminus of the gene (50 amino acid differences in the region 1-748). In addition, significant differences exist between FECV and the FIPV strains in specific regions of the sequence, i.e. 400 - 450 bp, suggesting that may be feasible to generate diagnostic probes to distinguish these serologically related viruses.

Table 1. Feline coronavirus nucleotide and amino acid[1] homology matrix comparison.

STRAIN	DF2	TS-DF2	FECV
Nucleotide			
WSU 1146	99.8	99.8	94.3
DF2		99.6	94.3
TS-DF2			94.1
Amino Acid			
WSU 1146	99.6	99.5	95.5
DF2		99.1	95.4
TS-DF2			95.3

[1] Value represents percent amino acid identity (actual match) between strains.

In general, spike genes of feline coronaviruses are highly conserved (>95.0% average amino acid homology). No clear differentiation between the S genes of virulent FIPV strains (DF2 or WSU 1146) and an attenuated vaccine strain (TS-DF2) or avirulent enteric virus (FECV) could be made by this analysis. When compared to the serologically related porcine

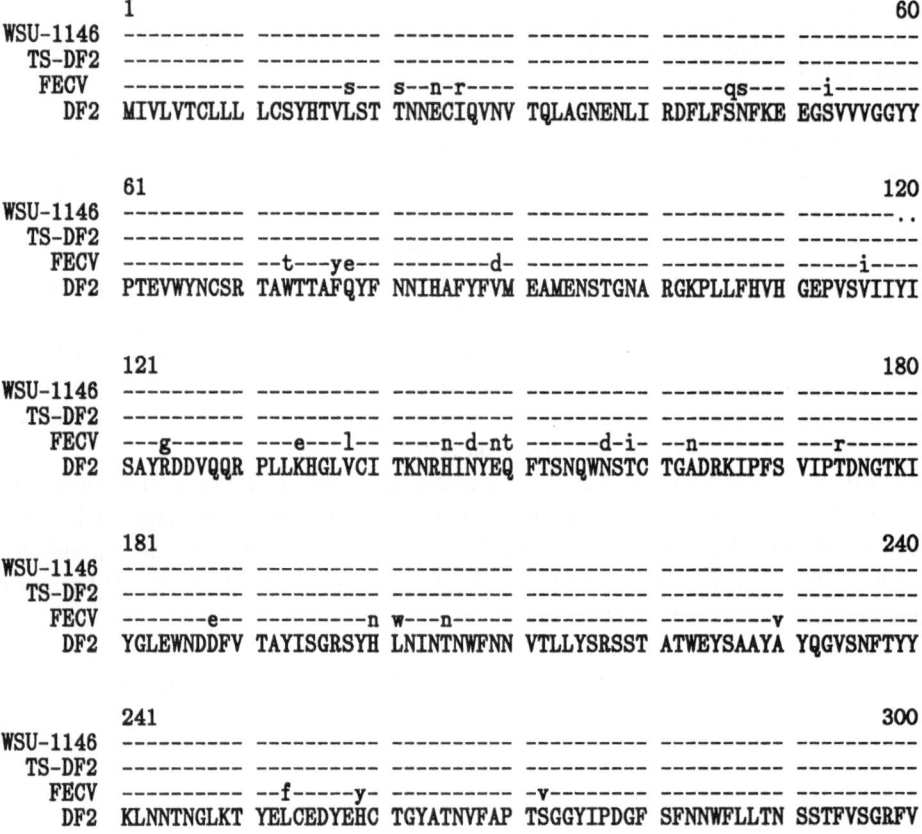

```
          1                                                               60
WSU-1146  ---------- ---------- ---------- ---------- ---------- ----------
TS-DF2    ---------- ---------- ---------- ---------- ---------- ----------
FECV      ---------- -------s-- s--n-r---- ---------- -----qs--- --i-------
DF2       MIVLVTCLLL LCSYHTVLST TNNECIQVNV TQLAGNENLI RDFLFSNFKE EGSVVVGGYY

          61                                                              120
WSU-1146  ---------- ---------- ---------- ---------- ---------- --------..
TS-DF2    ---------- ---------- ---------- ---------- ---------- ----------
FECV      ---------- --t---ye-- ---------d- ---------- ---------- -----i----
DF2       PTEVWYNCSR TAWTTAFQYF NNIHAFYFVM EAMENSTGNA RGKPLLFHVH GEPVSVIIYI

          121                                                             180
WSU-1146  ---------- ---------- ---------- ---------- ---------- ----------
TS-DF2    ---------- ---------- ---------- ---------- ---------- ----------
FECV      ---g------ ---e---l-- ----n-d-nt -------d-i- --n------- ---r------
DF2       SAYRDDVQQR PLLKHGLVCI TKNRHINYEQ FTSNQWNSTC TGADRKIPFS VIPTDNGTKI

          181                                                             240
WSU-1146  ---------- ---------- ---------- ---------- ---------- ----------
TS-DF2    ---------- ---------- ---------- ---------- ---------- ----------
FECV      -------e-- ----------n w---n----- ---------- ----------v ----------
DF2       YGLEWNDDFV TAYISGRSYH LNINTNWFNN VTLLYSRSST ATWEYSAAYA YQGVSNFTYY

          241                                                             300
WSU-1146  ---------- ---------- ---------- ---------- ---------- ----------
TS-DF2    ---------- ---------- ---------- ---------- ---------- ----------
FECV      ---------- --f-----y- ---------- -v-------- ---------- ----------
DF2       KLNNTNGLKT YELCEDYEHC TGYATNVFAP TSGGYIPDGF SFNNWFLLTN SSTFVSGRFV
```

Figure 1. Comparison of the amino acid sequence of the first 300 aa of the S gene from WSU 1146 (12), DF2 FIPV, TS-DF2 FIPV and FECV.

coronavirus, Transmissible Gastroenteritis virus (TGEV), the S genes of the DF2 FIPV and FECV feline coronaviruses were 82.0% homologous at the amino acid level (18). Not surprisingly, homology continued to decline as the feline S genes were compared to those published from Mouse Hepatitis Virus (MHV-A59) and Avian Infectious Bronchitis Virus (IBV) - 29.2% and 30.7% amino acid homology, respectively (19, 20).

REFERENCES

1. N.C. Pedersen and J.F. Boyle, <u>Am. J. Vet. Res.</u> 41:868-876 (1980).
2. H. Lutz, B. Hauser, and M.C. Horzinek, <u>J. Small Animal Pract.</u> 27:108 (1986).
3. N.C. Pedersen, <u>Adv. Exp. Med. Biol.</u> 218:529-550 (1987).
4. R.D. Woods and N.C. Pedersen, <u>Vet. Microbiol.</u> 4:11-16 (1979).
5. N.C. Pedersen and J.W. Black, <u>Am. J. Vet. Res.</u> 44:229-234 (1983).
6. C.A. Stoddart, J.E. Barlough, C.A. Baldwin and F.W. Scott, <u>Res. Vet. Sci.</u> 45:383-388 (1988).
7. J.E. Barlough. C.A. Stoddart, G.P. Jacobson and F.W. Scott, <u>Lab. Anim. Sci.</u> 34:592-597 (1984).
8. H. Vennema, R.J. De Groot, D.A. Harbour, M. Dalderup, T. Gruffyd-Jones, M.C. Horzinek and W.J.M. Spaan, <u>J. Virol.</u> 64:1407-1409 (1990).
9. H. Vennema, R.J. De Groot, D.A. Harbour, M.C. Horzinek and W.J.M. Spaan, <u>Virology</u> 181:327-335 (1991).
10. J.D. Gerber, J.D. Ingersoll, A.M. Gast, K.K. Christianson, N.L. Selzer, R.M. Landon, N.E. Pfeiffer, R.L. Sharpee and W.H. Beckenhauer, <u>Vaccine</u> 8:536-542 (1990).
11. K.K. Christianson, J.D. Ingersoll, R.M. Landon, N.E. Pfeiffer and J.D. Gerber, <u>Arch. Virol.</u> 109:185-196 (1989).
12. R.J. De Groot, J. Maduro, J.A. Lenstra, M.C. Horzinek, B.A.M. Van Der Zeijst and W.J.M. Spaan, <u>J. Gen. Virol.</u> 68:2639-2646 (1987).
13. J.M. Chirgwin, A.E. Przybyla, R.J. MacDonald and W.J. Rutter, <u>Biochemistry</u> 18:5294-5301 (1979).
14. E.S. Kawasaki. Amplification of RNA, in "PCR Protocols. A Guide to Methods and Applications," M.A. Innis, D.H. Gelfand, J.J. Sninsky and T.J. White, eds. Academic Press, San Diego, CA (1990).
15. R.K. Saiki, S. Scharf, F.A. Faloona, K.B. Mullis, G.T. Horn, H.A. Erlich and N. Arnheim, <u>Science</u> 230:1350-1354 (1985).
16. F. Sanger, S. Nicklen and A.R. Coulson, <u>Proc. Natl. Acad. Sci. USA</u> 74:5463-5467 (1977).
17. J. Devereux, P. Haberlie and O. Smithies, <u>Nucleic Acids Research</u> 12:387-395 (1984).
18. L. Jacobs, R.J. De Groot, M.C. Horzinek, B.A.M. Van der Zeijst and W.J.M. Spaan, (FIPV), <u>Virus Res.</u> 8:363-3712 (1987).
19. W. Luytjes, L.S. Sturman, P.J. Bredenbeek, J. Charite, B.A.M. Van Der Zeijst, M.C. Horzinek and W.J.M. Spaan, <u>Virology</u> 161:479-487 (1987).
21. J.G. Kusters, H.G.M. Niesters, J.A. Lenstra, M.C. Horzinek, and B.A.M. Van Der Zeijst, <u>Virology</u> 169:217-221 (1989).

GENOMIC ORGANISATION OF A VIRULENT TAIWANESE STRAIN OF TRANSMISSIBLE GASTROENTERITIS VIRUS

C.-M. Chen[1], D.H. Pocock[2] and P. Britton[2]

[1]Pig Research Institute, Taiwan, PO Box 23 Chunan, Miaoli, Taiwan, Republic of China. [2]Division of Molecular Biology, A.F.R.C., Institute for Animal Health, Compton, Newbury, Berkshire RG16 0NN, United Kingdom

ABSTRACT

Transmissible gastroenteritis (TGE) infection causes 65% of infectious piglet diarrhoea in Taiwan. A virulent Taiwanese strain, TFI, of transmissible gastroenteritis virus (TGEV) from a field outbreak was isolated in cell culture and plaque purified. Phenotypic differences were observed in the ability of TFI to infect certain cell lines. TGEV strains TLM-83 (PRCV Belgium), TO-163 (TGEV Japan) and Purdue-115 (TGEV USA) infected both ST (swine testis) and RPTG (pig kidney) cell lines whereas TFI infected ST but not RPTG cells. To investigate this phenotypic variation cDNA was generated from TFI genomic and amplified by PCR with oligonucleotides derived from published TGEV sequence data. An 8.4kb cDNA derived from the 3'-end of the TFI genome was sequenced. Eight ORFs, corresponding to the three structural protein genes, four potential genes and the 3'-end of an incomplete ORF whose amino acid sequence corresponded to the carboxyl end of the 1b subunit of the polymerase gene, were identified on the TFI sequence. The overall sequence similarity of TFI with the other TGEV strains was over 97%. However, several deletions, insertions and point mutations were found on the TFI sequence when compared with other TGEV strains. The TFI S protein was found to contain 1449 amino acids, as also identified for the FS772/70 and Miller TGEV strains, but two amino acids longer than the Purdue S protein. The TFI ORF-3a gene encodes 72 amino acids, however, a 37 nucleotide deletion was found 16 nucleotides downstream of the TFI ORF-3a stop codon. The TFI M, N, ORF-4 and ORF-7 genes are highly conserved in comparison to other TGEV strains.

INTRODUCTION

A virulent Taiwanese strain of TGEV, TFI, was isolated and purified from a field outbreak of TGE in Taiwan (1). The TFI strain was found to be phenotypically different from two other TGEV strains, Purdue-115 (avirulent strain), and TO-163 (a Japanese vaccine strain), and a variant virus of TGEV porcine respiratory coronavirus (PRCV), TLM-83. TFI was found to be highly cytopathic to ST cells and produce large plagues, whereas the other three strains were moderately cytopathic to ST cells and produce medium to small plaques. When TFI was used to infect RPTG cells, a cell line derived from swine kidney cells, neither a cytopathic effect (CPE) nor plaques were observed. In contrast, the two other TGEV strains and PRCV produced either medium or small plaques and moderate to mild CPE in RPTG cells. The ST^+RPTG^- phenotype was proposed for TFI and the ST^+RPTG^+ phenotype for the other three viruses. It was postulated that the non-permissiveness of TFI in RPTG cells could result from either TFI not attaching to RPTG cells or the inability of TFI to replicate in these cells (1).

The genomic organisation of TGEV has been shown to be Pol-S-3a-3b-4-M-N-7. Three structural proteins of TGEV, S (spike protein), N (nucleoprotein) and M (membrane protein) have been identified and recently the ORF-4 encoded protein was identified and shown to be associated with the virion (2). The S protein is responsible for binding the virus to the cell receptor, aminopeptidase N, a membrane protein of M_r 150K, found in ST cells and porcine small intestinal enterocytes (3). TGEV S protein encodes a polypeptide of 1449 amino acids in the FS772/70 and Miller strains and 1447 amino acids in the Purdue strain. Glycosylation of the S protein is essential for maintaining the conformational structure (4); also the major antigenic sites are localized within the N-terminal half of the protein (5). Although PRCV was found to have 224 amino acids deleted from the N-terminal end of the S protein, the major neutralisation sites are maintained, but this region may be associated with the loss of PRCV tropism for enterocytes (6, 7). The sequences of the TGEV ORF-3a and ORF-3b genes vary between strains, a difference that may be important in pathogenesis. A segment of TFI genome from the 3'-end of the polymerase to the 3'-end was sequenced in order to identify any sequence variation that may contribute to the change in the TFI phenotype.

METHODS

Preparation of partially purified viral genomic RNA

TFI was grown in ST cells as described previously (1). TFI genomic RNA was prepared from lyophilized virus, 10^9 PFU, using 1ml of guanidine lysis buffer (4 M guanidine isothiocyanate; 0.5% N-lauryl sarcosine; 25mM sodium citrate, pH 7.0; 0.7% β-mercaptoethanol; 0.1% antifoam-A) and centrifugation through 5.7M CsCl containing 0.1M EDTA, at 38,000 rpm for 17h at 15°C in a TL-100 Beckman ultracentrifuge. The RNA pellets were resuspended in 100μl 10mM Tris-HCl; 1mM EDTA (pH 8.0) and ethanol precipitated.

Amplification the TFI genomic cDNA by polymerase chain reaction (PCR)

For cDNA synthesis 1μg of viral RNA, heated at 65°C for 3 min, was added to 40 units RNasin (Promega), 50mM Tris-HCl (pH 8.3), 75mM KCl, 1mM DTT, 15mM $MgCl_2$, 1mM dNTPs (Boehringer Mannheim), 400 units Moloney murine leukaemia virus reverse transcriptase (BRL), and 200ng of primer. The primers used were derived from published TGEV sequences. The reaction mixture was incubated at 42°C for 2h and 5µl aliquots were subsequently used as template cDNA for PCR amplifications. The PCR amplifications were carried out according to the manufacture's instructions (Perkin Elmer Cetus) and subjected to 35 cycles in a Techne programmable heating block (1 minute at 94°C, 2 minutes at 50°C, 3 minutes at 72°C with a final extension time of 9 minutes at 72°C).

Cloning of PCR products

Purified PCR cDNA fragments were end repaired by the Klenow enzyme, phosphorylated using T4 polynucleotide kinase, ligated into SmaI digested and calf intestinal phosphatase treated pGemini-3Z vector (Promega) and transformed into DH5α competent *E.coli* cells. The *E.coli* cells were plated on LB agar plates containing 100µg/ml ampicillin, 50ng/ml X-Gal, 300ng/ml IPTG.

Screening of TFI cDNA fragments and analysis of DNA sequences

Potential recombinants were screened as described previously (8) using ^{32}P-labelled oligonucleotides derived from FS772/70 TGEV sequence data. Plasmid DNA was isolated from recombinants, purified by CsCl/ethidium bromide gradient centrifugation and sequenced by the dideoxy chain termination method using Sequenase™ (United States Biochemical Corporation). The TFI sequence and its comparison with other TGEV strains was analysed using the UWGCG (9) package.

RESULTS AND DISCUSSION

Eight ORFs were identified in the 8394bp cDNA sequence derived from the TFI genome (Table 1). The TFI S gene encoded a polypeptide of 1449 amino acids (Fig. 1) the same as for the FS772/70 and Miller TGEV strains but two amino acids larger than the Purdue strain. There were 34 potential N-linked glycosylation sites within the TFI S protein sequence, the same number as found for the Miller strain but two more than the Purdue strain and one more than the FS772/70 strain. However, amino acid substitutions within the TFI sequence had changed the potential N-linked glycosylation profile (Table 2). TFI was found to contain two unique N-glycosylation sites at residues 123 and 480, with another potential site at residue 562 also found in the Miller S protein. A potential N-glycosylation site at residue 94, common to the other three TGEV strains, was absent in TFI. All the alterations in the TGEV N-glycosylation profiles were located within the N-terminal half of the S protein, coinciding with the region containing the four major TGEV antigenic sites (10). Changes in the folding and assembly of simian influenza virus haemagglutinin-neuraminidase was shown to be due to changes in N-glycosylation (11), therefore the changes identified in the TFI S N-glycosylation profile may potentially alter the protein conformation.

Several amino acid substitutions were found within the N-terminal half of the TFI S sequence. The Try → Cys substitution at residue 226 could result in an extra intramolecular disulphide bond. The Cys-131 residue in TGEV strains TFI, FS772/70 and Miller was identified as a Met residue in the Purdue strain. All the virulent TGEV strains, TFI, FS772/70 and Miller contain the same two amino acid insert, residues 375 and 376, when compared with the Purdue strain. TFI and Purdue both contain a Pro at residue 48 whereas FS772/70 and Miller contain a Ser residue. This substitution occurred within the

Table 1. Number of amino acids within the coding sequences for four TGEV strains

	TFI	PURDUE	FS772/70	MILLER
S	1449	1447	1449	1449
ORF-3a	72[1,2]	71[1]	62[1]	72[1,2]
ORF-3b	244[3]	165[3]	244[3]	244
ORF-4	81	81	81	81
M	262	262	262	-
N	382	382	382	-
ORF-7	78	78	78	-

[1]Same start site
[2]Same stop site
[3]Leader RNA binding site was mutated (Purdue and FS772/70) or deleted (TFI)

Table 2. Potential N-glycosylation sites within TFI S protein showing differences with three other TGEV strains

TFI	Purdue	FS772/70	Miller
94 DYATENIIWNHRQR	94 DYATENITWNHRQR	94 DYATENSTWNHKQR	94 DYATENITLNHKQR
123 TTRNFNSSEGAIIC	123 TTRNFNCAEGAIYM	123 TTRNFNSAEGAIIC	123 TTRNFNSAEGAIIC
375 EISCYNDIVSDSSF	EISCY**TVSDSSFFS	375 EISCYNDTVSDSSF	375 EISCYNDTVSDSSF
480 NTAITNVTYCNSYV	478 NTAITKVTYCNSYV	480 NTAITKVTYCNSHV	480 NTAITKVTYCNSYV
562 LPVQDNNTDVYCIR	560 LPMQDHNTDVYCIR	562 LPMQDNNIDVYCIR	562 LPMQDNNTDVYCIR

Potential N-glycosylation motifs are underlined
Numbers show either position of N-glycosylation site, modified residue or the position of the Asn within the altered motif
* Padding character

monoclonal antibody (MAb) binding site C of the Purdue strain (12). In contrast the carboxyl half of the TFI S protein was found to be highly conserved between TGEV strains and showed similarities with the S proteins from other coronaviruses. This region of the S protein has been postulated to form the stalk structure and contain the transmembrane domain (13). Alterations in the structure of the TFI S protein could result from amino acid substitutions or insertions at critical points within the sequence. For example a Thr → Ile change at residue 284 of the influenza C haemagglutinin glycoprotein altered its ability to bind with its cell receptor, Neu5,9Ac$_2$, allowing the virus to infect a non-permissive cell line (14).

It has been proposed that coronavirus mRNA species are synthesised by leader primed transcription. Two mRNA species of 3.9kb (mRNA 3) and 3.0kb (mRNA 4), produced from the RNA-leader binding sites ACTAAAC and CTAAAC respectively, have been observed in FS772/70 or Purdue infected cells. The RNA-leader binding sites upstream

```
                                .         .         .         .         .    80
MKKLFVVLVIIPLIYGDNFPCSKLTNRTIGNHWNLIETFLLNYSSRLPPNSDVVLGDYFPTVQPWFNCIRNNSNDLYVTL
                                .         .         .         .         .   160
ENLKALYWDYATENIIWNHRQRLNVVVNGYPYSITVTTTRNFNSSEGAIICICKGSPPTTTTESSLTCNWGRECRLNHKF
                                .         .         .         .         .   240
PICPSNSEANCGNMLYGLQWFADAVVAYLHGASYRISFENQWSGTVTLGDMRSTTLETAGTLVDLCWFNPVYDVSYYRVN
                                .         .         .         .         .   320
NKNGTIVVSNCTDQCASYVANVFTTQPGGFIPSDFSFNNWFLLTNSSTLVSGKLVTKQPLLVNCLWPVPSFEEAASTFCF
                                .         .         .         .         .   400
EGADFDQCNGAVLNNTVDVIRFNLNFTTNVQSGKGATVFSLNATGGVTLEISCYNDIVSDSSFSSYGEIPFGVTDGPRYC
                                .         .         .         .         .   480
YVLYNGTALKYLGTLPPSVKEIAISKWGHFYINGYNFFSTFPIDCISFNLTTGDSDVFWTIAYTSYTEALVQVENTAITN
                                .         .         .         .         .   560
VTYCNSYVNNIKCSQLTANLNNGFYPVSSSEVGLVNKSVVLLPSFYTHTIVNITIDLGMKRSGYGQPIASTLSNITLPVQ
                                .         .         .         .         .   640
DNNTDVYCIRSDQFSVYVHSTCKSALWDNVFKRNCTDVLDATAVIKTGTCPFSFDKLNNYLTFNKFCLSLSPVGANCKFD
                                .         .         .         .         .   720
VAARTRPNDQVVRSLYVIYEEGDNIVGVPSDNSGLHDLSVLHLDSCTDYNIYGRTGVGIIRQTNRTLLSGLYYTSLSGDL
                                .         .         .         .         .   800
LGFKNVSDGVIYSVTPCDVSVQAAVIDGTIVGAVTSINSELLGLTHWTTTPNFYYYSIYNYTNDRTRGTAIDSNDVDCEP
                                .         .         .         .         .   880
VITYSNIGVCKNGALVFINVTHSDGDVQPISTGNVTIPTNFTISVQVEYIQVYTTPVSIDCSRYVCNGNPRCNKLLTQYV
                                .         .         .         .         .   960
SACQTIEQALAMGARLENMEVDSMLFVSENALKLGSVEAFNSSETLDPIYKEWPNIGGSWLEGLKYILPSDNSKRKYRSA
                                .         .         .         .         .  1040
IEDLLFSKVVTSGLGTVDEDYKRCTGGYDIADLVCAQYYNGIMVLPGVANADKMTMYTASLAGGITLGALGGGAVAIPFA
                                .         .         .         .         .  1120
VAVQARLNYVALQTDVLNKNQQILASAFNQAIGNITQSFGKVNDAIHQTSRGLATVAKALAKVQDVVNTQGQALSHLTVQ
                                .         .         .         .         .  1200
LQNNFQAISSSISDIYNRLDELSADAQVDRLITGRLTALNAFVSQTLTRQAEVRASRQLAKDKVNECVRSQSQRFGFCGN
                                .         .         .         .         .  1280
GTHLFSLANAAPNGMIFFHTVLLPTAYETVTAWAGICALDGDRTFGLVVKDVQLTLFRNLDDKFYLTPRTMYQPRVATSS
                                .         .         .         .         .  1360
DFVQIEGCDVLFVNATVSDLPSIIPDYIDINQTVQDILENFRPNWTVPELTFDIFNATYLNLTGEIDDLEFRSEKLHNTT
                                .         .         .         .         .  1440
VGLAILIDNINNTLVNLEWLNRIETYVKWPWYVWLLIGLVVIFCIPLLLFCCCSTGCCGCISCLGSCCHSICSRRQFENY

EPIEKVHVH*
```

Figure 1. Amino acid sequence of TFI S protein, potential N-glycosylation sites are double underlined

of all the TGEV genes has been identified as ACTAAAC except for mRNA 4 which has the sequence, CTAAAC. The shorter length of the RNA-leader binding site of mRNA 4 may be responsible for the lower amounts of this mRNA compared to the other mRNA species. TGEV mRNA 3 has been shown to contain two potential ORFs, ORF-3a and ORF-3b, and mRNA 4 one potential ORF, ORF-4, at the 5'-ends (15, 16). However, the Miller strain was observed to produce an additional 3.7kb mRNA species (mRNA 3-1) in infected cells (17), due to an extra RNA-leader binding site upstream of the ORF-3b gene. This new RNA-leader binding site was produced by a point mutation T → C resulting in the sequence CTAAAC rather than CTAAAT upstream of ORF-3b. This mRNA species was also produced in lower amounts than the other mRNA species. A 37 nucleotide deletion in the non-coding region between the ORF-3a/3b genes was observed in the TFI sequence 16 nucleotides downstream of TFI ORF-3a stop codon. This deletion resulted in the loss of the region containing the RNA-leader binding site identified in the Miller strain.

The potential TGEV gene, ORF-3a, in the TFI and Miller strains encode 72 amino acid proteins, M_r 7,700, 10 amino acids longer than the FS772/70 ORF-3a gene product. The FS772/70 ORF-3a gene product is shorter due to the presence of a stop codon within the gene. This stop codon in FS772/70 was TAA whereas in TFI and Miller the sequence was GAA. Due to an insertion within the 3'-end of the ORF-3a gene of the Purdue strain, the last six C-terminal residues of the Purdue ORF-3a protein were different from the TFI and Miller proteins.

Comparison of the TGEV ORF-3b genes showed that Miller, FS772/70 and TFI have potential gene products of M_r 27,600, with only minor amino acids substitutions.

However, the Purdue strain has a smaller ORF-3b gene encoding a product of M_r 18,800, due to a point mutation in the ORF-3b start codon and the gene potentially initiating at an ATG codon 237 nucleotides downstream, resulting in the loss of 79 amino acids (Table 1; 16).

The ORF-4, M, N and ORF-7 gene products of TFI varied only slightly when compared to the other TGEV strains.

REFERENCES

1. C.M. Chen, S.C. Tien, J.L. Wu and R.M. Chu, *J. Chin. Soc. Vet. Sci.* **15**:11-18 (1989).
2. M. Godet, R. L'Haridon, J.-F. Vautherot and H. Laude, *Virology* **188**:666-675 (1992).
3. B. Delmas, J. Gelfi, R. L'Haridon, L.K. Vogel, H. Sjostrom, O. Noren and H. Laude, *Nature* **357**:417-420 (1992).
4. B. Delmas and H. Laude, *Virus Res.* **20**:107-120 (1991).
5. B. Delmas, D. Rasschaert, M. Godet, J. Gelfi and H. Laude, *J. Gen. Virol.* **71**:1313-1323 (1990).
6. P. Britton, K.L. Mawditt and K.W. Page, *Virus Res.* **21**:181-198 (1991).
7. D. Rasschaert, M. Duarte and H. Laude, *J. Gen. Virol.* **71**:2599-2607 (1990).
8. C.M. Chen, L.H. Hwang, S.T. Liu and R.M. Chu, *J. Chin. Soc. Vet. Sci.* **15**:1-9 (1989).
9. J. Devereux, P. Haeberli and O. Smithies, *Nucl. Acids Res.* **12**:387-395 (1984).
10. B. Delmas, J. Gelfi and H. Laude, *J. Gen. Virol.* **67**:1405-1418 (1986).
11. D.T.W. Ng, S.W. Hiebert and A.R. Lamb, *Mol. Cell Biol.* **10**:1989-2001 (1990).
12. L. Enjuanes, I. Correa, G. Jimenez, M.P. Melogosa and M.J. Bullido, *Adv. Exp. Med. Biol.* **218**:351-363 (1987).
13. H. Laude, D. Rasschaert, B. Delmas, M. Godet, J. Gelfi and B. Charley, *Vet. Microbiol.* **23**:147-154 (1990).
14. S. Szepanski, H.J. Gross, R. Brossmer, H.-D. Klenk and G. Herrler, *Virology* **188**:85-92 (1992).
15. P. Britton, C.L. Otin, J.M.M. Alonso and F. Parra, *Arch. Virol.* **105**:165-178 (1989).
16. D. Rasschaert, J. Gelfi and H. Laude, *Biochimie* **69**:591-600 (1987).
17. R.D. Wesley, A.K. Cheung, D.D. Michael and R.D. Woods, *Virus Res.* **13**:87-100 (1989).

THE USE OF PCR GENOME MAPPING FOR THE CHARACTERISATION OF TGEV STRAINS

P. Britton[1], S. Kottier[1], C.-M. Chen[1], D.H. Pocock[1], H. Salmon[2] and J.M. Aynaud[2]

[1]Division of Molecular Biology, A.F.R.C., Institute for Animal Health, Compton, Newbury, Berkshire, RG16 0NN, United Kingdom. [2]I.N.R.A., Laboratoire de Pathologie Infectieuse et Immunologie, Nouzilly, 37380, France

ABSTRACT

Previous studies on different transmissible gastroenteritis virus (TGEV) strains, including porcine respiratory coronavirus (PRCV), have identified regions within the genome that are polymorphic as regards insertions and deletions. For example the 672 base deletion within the S gene and multiple deletions 5', within and 3' of the ORF-3a gene were detected in strains of PRCV. The presence of deletions may be associated with a change in the virulence, attenuation or tissue tropism of the isolate. The Nouzilly (188-SG) TGEV vaccine strain was attenuated by passage of a cell culture adapted virulent isolate D-52 188 times through swine testis cells after treatment with gastric juice. PCR amplification with oligonucleotides, corresponding to known TGEV sequences, were used to analyse D-52 and 188-SG for genetic variation. Results with several pairs of oligonucleotides within the first 1565 nucleotides of the S gene did not identify a deletion within this region of the genome from either strain. However, oligonucleotides directed against the ORF-3a/3b region detected a deletion of about 250 nucleotides within the 188-SG genome but not in the D-52 genome. Since all the attenuated TGEV strains so far sequenced, PRCV, Miller SP and 188-SG, contained deletions within the ORF-3a/3b, it would suggest that this region of the TGEV genome is involved in regulating viral virulence.

INTRODUCTION

TGEV was first isolated in 1946 by Doyle and Hutchings[1] and has since been shown to belong to the family *Coronaviridae*. TGEV normally infects via the oral route, although nasal or airborne infections can occur[2], causing diarrhoea in pigs of all ages. Mortality in neonates is high mortality, often 100%. The virus is capable of resisting the low pH of the gastric juices, proteolytic enzymes and bile to infect villous enterocytes of the lower small intestine causing their destruction resulting in diarrhoea. Piglets are passively protected from disease by secretory IgA antibodies in the dam's milk and colostrum[3]. Following a TGEV infection of the dam sensitised lymphocytes in the Peyer's Patches migrate via the mesenteric lymph nodes to the mammary lymph nodes and result in the production of secretory IgA antibodies in colostrum and milk.

Several attempts have been made to develop vaccine strains of TGEV by the attenuation of virulent viruses. However, the levels of protection afforded following either oral infection or intramuscular injections of attenuated TGEV strains[3,4,5] were low and dependant on the levels of IgG antibodies in milk. It appeared that attenuated strains of TGEV were unable to induce IgA antibody production in the mammary gland. One reason why attenuated strains fail to induce an IgA response may be their lability in gastric juices resulting in limited replication in the gut. To circumvent this a virulent TGEV strain (D-52) was treated with gastric juice and passaged 188 times through a swine testis cell line[6,7]. The resulting attenuated strain 188-SG, Nouzilly vaccine strain, was more resistant to low pH and proteolytic enzymes and was able to provide some protection to piglets but again only through elevated levels of IgG present in milk[7]. The characteristics of the 188-SG strain include; (i) acid resistance, (ii) small plaque variant and (iii) decreased growth in cell culture with delayed RNA synthesis. The aim of the project was to identify any changes in the virus genome that may be responsible for the attenuation of the virus.

METHODS

RNA Isolation and Purification

TGEV strains FS772/70, D-52 or 188-SG were grown in the pig kidney cell line LLC-PK1 in the presence of actinomycin D as described previously[8,9]. Total cellular RNA, including viral RNA, was isolated using a Stratagene RNA isolation kit, from LLC-PK1 cells infected with TGEV .

Production and Analysis of TGEV cDNA

Synthesis of First Strand cDNA. Oligonucleotides used for cDNA synthesis and PCR amplifications, derived from published TGEV sequence data[9,10,11] (Table 1), were synthesised by the phosphoramidite method on an Applied Biosystem 381A DNA synthesizer. Total RNA (5 μg), isolated from virus infected cells, was dissolved in H_2O heated at 60°C for 5 min, added to 160 ng of oligonucleotide 51 or oligonucleotide 92, incubated at 70°C for 10 min and cooled on ice. First strand cDNA synthesis was carried out on the annealed RNA/primer using a Superscript^{TM} RNase H$^-$ reverse transcriptase cDNA synthesis kit (BRL) at 45°C for 90 min.

PCR Amplification. The ssDNA was PCR amplified, with the oligonucleotide primers shown in Tables 1, 2 and on Fig. 1, following a protocol supplied with the Taq Polymerase (Boehringer), in a Techne PHC-1 programmable thermal cycler using 35 cycles of 94°C for 1min, 55°C for 2min and 72°C for 3min with a final elongation step of 72°C for 9min. Samples were electrophoresed into 1% agarose gels.

Cloning of cDNA. PCR cDNA fragments from TGEV 188-SG, synthesised using oligonucleotides 52 and 92, and D-52, synthesised with oligonucleotides 51 and 49, were purified by agarose gel electrophoresis and isolated from the gel with Geneclean^{TM}. The PCR fragment from 188-SG was 5'-phosphorylated using T_4 polynucleotide kinase and any incomplete ends repaired with DNA polymerase Klenow fragment, prior to ligation into SmaI-cut pGEM®-3Z. The PCR fragment from D-52 was digested with NsiI and ligated into NsiI-cut pGEM®-7Zf(-).

Sequencing of cDNA. Plasmids containing the TGEV cDNA fragments were CsCl purified and sequenced using the Sequenase™ (USB) alkaline denaturation protocol with oligonucleotide primers derived from TGEV sequence data. Sequence data was analysed as described previously[12] using the programs of[13,14].

Table 1. Oligonucleotide primers used for cDNA synthesis and PCR amplifications.

Oligonucleotide	Sequence (5' → 3')	Sense
8	TATAGCACTAACAACCTGAT	-
10	AGCAAAGTATGCACAATC	-
15	CTAAGTAGGCGAATCTTAAA	-
49	TCTTGAATGGCTCAATAG	+
51	CTGTCCTTCCTAAATTGCAACACACCATGCATAGC	-
52	GGCCTTGGTATGTGTGGCTACTAATAGGC	+
59	ACAGTTAGATCAGCATG	-
60	GTGTCGGCATCTTAATG	+
75	CCTTTTAAAGTAAAGTGAGT	+
84	ACACGTTCTTGTTGTATATGATGCACATTTGCTGTG	-
92	ACACCAGTTGGCACACCT	-

RESULTS and DISCUSSION

PCR Analysis of ORF-3a/3b Genomic Region of TGEV 188-SG

The size of the PCR fragment, 1193bp (Fig 1; Table 2), generated from 188-SG with oligos 51 and 52 was smaller than the fragment, 1451bp (Table 2) from FS772/70 and this was confirmed using other oligonucleotide pairs (Fig 1; Table 2). These results indicated a deletion of about 250 bases occurred within the potential ORF-3a gene region of the 188-SG genome (Fig 1). PCR mapping studies with oligo 75, complementary to the 5'-end of the TGEV leader RNA[9], indicated that a leader RNA binding site was present upstream of ORF-3a as found for TGEV strains FS772/70[10], Purdue-115[15], TFI[16] and Miller-PP3[17]. Both Miller-PP3[17], PRCV-137 and PRCV-135[18] also have a leader binding site upstream of the ORF-3b gene, no evidence was found for a leader RNA binding site in this position on the 188-SG genome. The PCR analyses could not determine the exact position of the deletion, except that the sequence corresponding to oligonucleotide 10, within ORF-3a, was present.

PCR Analysis of ORF-3a/3b Genomic Region of TGEV D-52

Similar PCR amplifications were carried out on cDNA generated from RNA isolated from D-52 infected cells. Comparison of the sizes of the generated fragments to those from 188-SG and FS772/70 indicated that there was no deletion within the ORF-3a/3b region of the D-52 genome (Table 2). Some of the D-52 generated fragments were slightly larger than FS772/70 generated fragments indicating that there were some insertions present in the D-52 genome when compared to FS772/70 (Table 2).

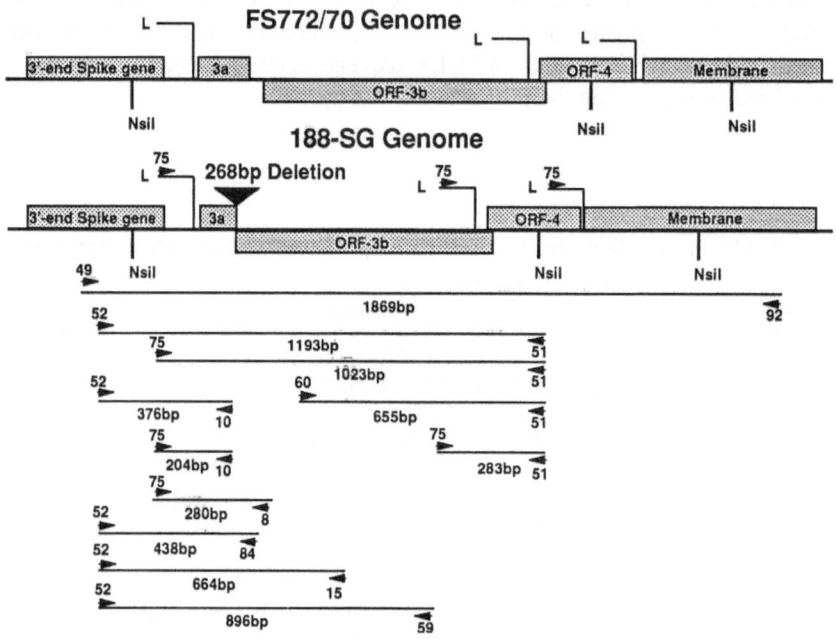

Fig 1. Position of the oligonucleotides and the corresponding PCR fragments are shown for TGEV 188-SG. The genome corresponding to TGEV FS772/70 is shown above.

Table 2. Sizes of PCR fragments from TGEV strains FS772/70, D-52 and 188-SG.

PCR Primers	Length of PCR Amplified Fragments (bp)		
	FS772/70	D-52	188-SG
52 + 51	1451	1490	1193
75 + 51	1281	1310	1023
	283	283	283
75 + 10	204	204	204
75 + 8	538	567	280
52 + 10	366	376	376
52 + 84	696	735	438
52 + 15	922	961	664
52 + 59	1154	1193	896
60 + 51	655	655	655

Note: All PCR fragments synthesised from cDNA produced with oligo 51.

Sequence Comparison of 188-SG and D-52

PCR fragments generated from 188-SG and D-52 cDNA were cloned and sequenced. Neither sequence contained the nine base insertion identified within the FS772/70 sequence[10] immediately downstream of the S gene (Fig. 2). A three base insertion was identified immediately downstream of the S gene in both 188-SG and D-52. A similar insert was found in the Miller-PP3[19] and Purdue-115[15] but not in the FS772/70[10] or TFI[16] strains of TGEV or in the two British isolates of PRCV 137 and 135[18] and the American Ind89 isolate[20], however, the French RM4 PRCV isolate[21] contained two of the three bases. A 16 base insertion, also found in the Miller small plaque mutant[22] and Purdue-115[15] but not in Miller-PP3[19] or in the PRCV strains[18,20,21], was present in both 188-SG and D-52. Both the TGEV RNA leader binding site, ACTAAAC, and the ORF-3a initiation codon were present in 188-SG. However, a 268 base deletion was found in the 188-SG corresponding to nucleotide 97 in ORF-3a to nucleotide 56 in ORF-3b of D-52 (Fig 2). This deletion accounted for the difference in the size of the PCR fragments generated from 188-SG when compared with FS772/70 and D-52 and resulted in the loss of both ORFs-3a and 3b. Translation of the remaining ORF-3a and 3b sequences showed that the two frames were out of phase and that a fusion protein comprising the 5'-end of ORF-3a and the rest of ORF-3b was unlikely. A 29 base insertion at the end of ORF-3a was found in the D-52 sequence. A similar insertion was found in the Purdue-115[15] sequence resulting in a longer ORF-3a product when compared to that of FS772/70. This region of the TGEV genome has been found to vary between the different strains sequenced. The 38 base difference between the D-52 and FS772/70 accounts for the size differences of the PCR fragments (Table 2).

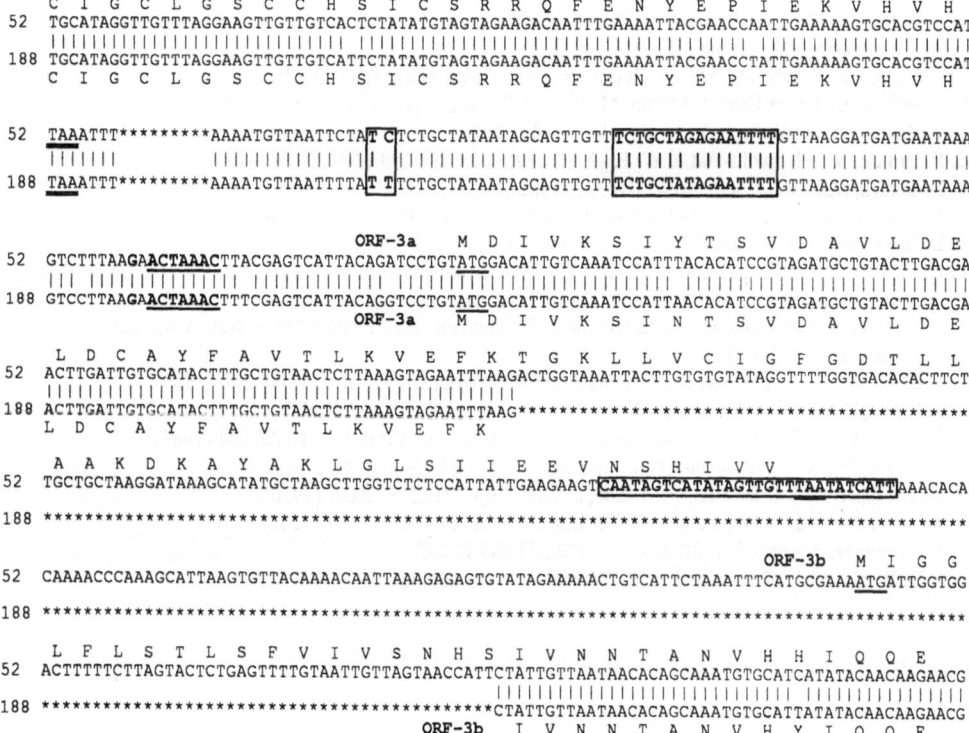

Fig 2. Alignment of the ORF-3a/3b regions of D-52 and 188-SG. Insertions are indicated by boxes and deletions by an *. The positions of the ORF-3a and 3b genes are indicated. Initiation codons are underlined and termination codons double underlined.

The small plaque phenotype of the 188-SG virus was similar to that identified for the attenuated small plaque variant of Miller-PP3[24]. Sequence analysis of the Miller-PP3 variant[20] indicated that there was a 491 base deletion. The deletion caused the loss of the RNA leader binding site upstream of ORF-3a and also resulted in the loss of the first 151 bases of ORF-3b[20]. It was interesting to note that the 16 base insert described above, present in TGEV strains Purdue-115, D-52 and 188-SG but absent from Miller-PP3 was present in the Miller small plaque variant. PCR mapping of the 5'-end of the 188-SG S gene showed no differences in fragment sizes between 188-SG and FS772/70. Thus no observable deletion was found in the 188-SG S gene in contrast to the 672 base deletion observed in the PRCV S gene[12,21,24]. No data is available whether the small plaque variant of Miller-PP3 contains a deletion in the S gene. The presence of the deletion within the ORF-3a and ORF-3b genes of 188-SG may result in the small plaque phenotype observed and may be responsible for the attenuation of the virus.

REFERENCES

1. L.P. Doyle and L.M. Hutchings, *J. Amer. Vet. Med. Assoc.* 108:257-259 (1946).
2. N.R. Underdahl, C.A. Mebus and A. Torres-Medina, *Amer. J. Vet. Res.* 36:1473-1476 (1975).
3. E.H. Bohl, R.K.P. Gupta, M.V. Olquin and L.J. Saif, *Infect. Immun.* 6:289-301 (1972).
4. E.H. Bohl and L.J. Saif, *Infect. Immun.* 11:23-32 (1975).
5. E.H. Bohl, G.T. Frederick and L.J. Saif, *Amer. J. Vet. Res.* 36:267-271 (1975).
6. J.-M. Aynaud, T.D. Nguyen, E. Bottreau, A. Brun and P. Vannier, *J. Gen. Virol.* 66:1911-1917 (1985).
7. J.M. Aynaud, S. Bernard, E. Bottreau, I. Lantier, H. Salmon and P. Vannier, *Vet. Microbiol.* 26:227-239 (1991).
8. P. Britton, D.J. Garwes, K. Page and J. Walmsley, Expression of porcine transmissible gastroenteritis virus genes in *E.coli* as β-galactosidase chimaeric proteins, *in:* "Coronaviruses, Advances in Experimental Medicine and Biology," M.M.C. Lai and S.A. Stohlman, eds., Plenum Press, New York and London (1987).
9. K.W. Page, P. Britton and M.E.G. Boursnell, *Virus Genes* 4:289-301. (1990).
10. P. Britton, C. Lopez Otin, J. Martin Alonso and F. Parra, *Arch. Virol.* 105:165-178 (1989).
11. P. Britton and K.W. Page, *Virus Res.* 18:71-80 (1990).
12. P. Britton, K.L. Mawditt and K.W. Page, *Virus Res.* 21:181-198 (1991).
13. R. Staden, *Nucl. Acids Res.* 10:4731-4751 (1982).
14. J. Devereux, P. Haeberli and O. Smithies, *Nucl. Acids Res.* 12:387-395 (1984).
15. D. Rasschaert, J. Gelfi and H. Laude, *Biochimie* 69:591-600 (1987).
16. C.-M. Chen, D.H. Pocock and P. Britton, Genomic organisation of a virulent Taiwanese strain of TGEV, *in:* "Coronaviruses:Molecular Biology and Pathogenesis, Advances in Experimental Medicine and Biology," H. Laude and J.F. Vautherot, eds., Plenum Press, New York and London (This book).
17. R.D. Wesley, A.K. Cheung, D.D. Michael and R.D. Woods, *Virus Res.* 13:87-100 (1989).
18. K.W. Page, K.L. Mawditt and P. Britton, *J. Gen. Virol.* 72:579-587 (1991).
19. R.D. Wesley, A.K. Cheung, D.D. Michael and R.D. Woods, *Virus Res.* 13:87-100 (1989).
20. R.D. Wesley, R.D. Woods and A.K. Cheung, *J. Virol.* 64:4761-4766 (1990).
21. D. Rasschaert, M. Duarte and H. Laude, *J. Gen. Virol.* 71:2599-2607. (1990).
22. R.D. Wesley, R.D. Woods and A.K. Cheung, *J. Virol.* 64:4761-4766 (1990).
23. R.D. Woods, *J. Amer. Vet. Med. Assoc.* 173:643-647 (1978).
24. R.D. Wesley, R.D. Woods and A.K. Cheung, *J. Virol.* 65:3369-3373 (1991).

EVOLUTION AND TROPISM OF TRANSMISSIBLE GASTROENTE-RITIS CORONAVIRUS

Luis Enjuanes,[1] Carlos Sánchez,[1] Fátima Gebauer,[1] Ana Méndez, [1]
Joaquín Dopazo,[2] and María L. Ballesteros[1]

[1]Centro Nacional de Biotecnología, CSIC. Campus Universidad
Autónoma, Canto Blanco, 28049 Madrid, Spain
[2]Departamento de Sanidad Animal, INIA, Embajadores, 68
28012 Madrid, Spain

ABSTRACT

Transmissible gastroenteritis coronavirus (TGEV) is an enteropathogenic coronavirus isolated for the first time in 1946. Nonenteropathogenic porcine respiratory coronaviruses (PRCVs) have been derived from TGEV. The genetic relationship among six European PRCVs and five coronaviruses of the TGEV antigenic cluster has been determined based on their RNA sequences. The S proteins of six European PRCVs have an identical deletion of 224 amino acids starting at position 21. The deleted area includes the antigenic sites C and B of TGEV S glycoprotein. Interestingly, two viruses (NEB72 and TOY56) with respiratory tropism have the S protein with a similar size to the enteric viruses. NEB72 and TOY56 viruses have 2 and 15 specific amino acid differences with the enteric viruses, respectively. Four of the residues changed are located within the deletion present in the PRCVs and may influence the enteric tropism of TGEV *in vivo*. A receptor binding site (RBS) used by the virus to infect ST and other cell types might be located between sites A and D of the S glycoprotein, since monoclonal antibodies (MAbs) specific for these sites inhibit the binding of the virus to ST cells. An evolutionary tree relating 13 enteric and respiratory isolates has been proposed. According to this tree, a main virus lineage evolved from a recent progenitor which was circulating around 1941. From this, secondary lineages originated PUR46, NEB72, TOY56, MIL65, BRI70, and the PRCVs, in this order. Least squares estimation of the origin of TGEV-related coronaviruses showed a significant constancy in the mutation fixation rate. This rate was $7\pm2 \times 10^{-4}$ nucleotide substitutions per site and per year and falls in the range reported for other RNA viruses. Point mutations and probably recombination events have occurred during TGEV evolution.

INTRODUCTION

TGEV replicates in both villus epithelia cells of the small intestine and in lung cells. A nonenteropathogenic coronavirus related to TGEV appeared in 1984 in Europe[1]. This virus replicates in the respiratory tract and undergoes only limited replication in unidentified submucosal cell types of the small intestine. Both enteric and respiratory viruses cross-react[2]. The analysis of the genetic relationship among these respiratory isolates and others with respiratory tropism will allow to determine the molecular bases of their tropism and evolution. In this paper, we describe the genetic homology among eight respiratory and five enteric isolates of the TGEV antigenic cluster, which identified amino acids potentially involved in receptor binding sites, and conserved areas of the S gene. Based on these viral sequences, an evolutionary tree, and mechanism for TGEV evolution, have been proposed.

MATERIALS AND METHODS

Cells, Viruses, and proteins

All viruses were grown on swine testis (ST) cells and their characteristics have been previously described[2]. Viral proteins were obtained from purified virions, incubated in the presence of proteinase inhibitors and deglycosylated with N-glycosidase F as described[3,4].

RNA Sequencing

RNA extracted from purified virions[3] was sequenced by oligodeoxynucleotide primer extension and dideoxynucleotide chain termination procedure[5].

Evolutionary Tree

Sequence information has been analyzed following standard phylogenetic methods[6]. The phylogenetic tree was obtained by the neighbor-joining[7] and the least squares methods. The reliability of the tree, i. e., the confidence levels in the branching order, was determined by the bootstrap method[8,9]. The origin of the phylogenetic tree was estimated by a linear least squares procedure[10].

RESULTS AND DISCUSSION

Structural Proteins and Sequence analysis of Enteric and Respiratory Porcine Coronaviruses

Enteric and respiratory TGEVs have been studied. According to the apparent size of the S protein of these viruses they could be grouped into two clusters, one including the enteric (PUR46) and the respiratory viruses (NEB72 and TOY56) with an S protein of standard size, and a second group, including the PRCVs with an S protein of smaller size (Figure 1). The first group of viruses have antigenic sites B and C, while in the second these antigenic sites were not detected by radioimmunoassay[2]. These results indicate that isolates with an almost exclusive respiratory tropism (NEB72 and TOY56) do not have a reduction in the molecular weight of the spike protein as the one detected in the PRCVs isolates.

To determine the relationship among the enteric and respiratory isolates, the nucleotide sequences of the complete S gene or of the half 5'-end were determined in 8 viral isolates and compared with sequences published for other isolates. The deduced amino acid sequences of 13 isolates were aligned, and the first 720 aminoacids are presented (Figure 2) and the results were diagrammatically summarized (Figure 3A). The amino acid positions reported in this manuscript refer to the location of equivalent residues in the sequence of MIL65 virus. All the PRCVs showed a deletion of 224 amino acids starting in the same position (aa 21), suggesting that all them have a common ancestor. In contrast, two North American PRCVs recently isolated (ISU-1, also designated IND89, and AR310)[11,12] have deletions of different size (227 and 207aa) starting in residues 23 and 17, respectively. These results indicate that the American and the European PRCVs have being independently originated. Although the NEB72 and TOY56 isolates have respiratory tropism, they do not have the large deletion on the S protein.

These viruses have only 2 and 15 specific amino acid differences with the enteric virus[4]. Interestingly, the NEB72 (Table 1) and TOY56 isolates have each one aminoacid dif-

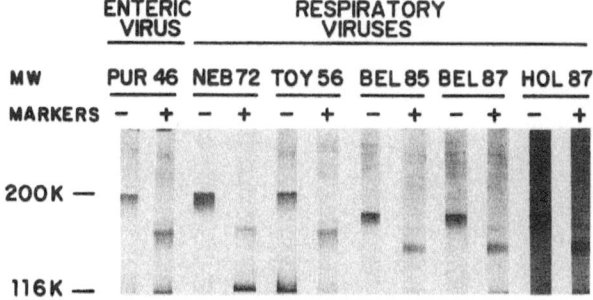

FIGURE 1. PAGE analysis of the spike protein of TGEV related coronaviruses before and after deglycosylation. Purified viruses were dissociated (1 µg/20µl) in 0.1 M sodium acetate, pH 7, with 0.5% SDS and protease inhibitors, and incubated overnight at 37° in the presence (+) or absence (-) of protein N-glycosidase F (0.04 U/µl). The proteins were separated by 7.5% PAGE in the presence of 0.1% SDS and 2-mercaptoethanol and detected using silver staining[13]. Only the gel area corresponding to the S glycoprotein is shown.

TABLE 1. Sequence differences between the S genes of enteric (PUR46) and respiratory (NEB72-RT) strain of TGEV

NUCLEOTIDE CHANGED		AMINO ACID CHANGED		CHANGED IN OTHER ENTERIC VIRUSES
214	G → A	70	Asp → Asn	+
655	G → T	219	Arg → Leu	-
2104	C → A	701	Gln → Lys	-
3263	A → R		-	-
3267	A → R		?	-
3270	N → X		?	-

ference in the S protein in positions 219 and 218, respectively. This changes are located within the deletion presented by the PRCVs, and could be responsible of the lack of enteric tropism shown by these viruses.

A RBS in the S glycoprotein of TGEV that interacts with ST cells probably maps between sites A and D, since TGEV binding to ST cells is best inhibited by MAbs specific for these sites[14]. Candidate domains for the localization of this RBS could be the highly conserved area identified between amino acids 405 and 465 (Figure 3B), although other domains around this area can not be ruled out. This RBS may mediate the infection of ST, and other cells growing in culture, by viruses with enteric and respiratory tropism, since it is present in all of them, and could interact with the aminopeptidase N described as the main TGEV receptor present on ST cells[15]. *In vivo*, a second RBS might be required to infect enteric cells. This RBS could be located around either aa 92, 94, and 218 or aa 219, changed in the TOY56 or in the NEB72 isolates, respectively. The two RBSs could mediate cell binding and fusion between the virus and cell membranes. A possibility, is that there is a unique RBS in the S protein of enteric and respiratory TGEV isolates, located between sites A and D. In this case, the deletion present in the PRCVs, or the residue changes within this area, described above, could modify this RBS resulting in viruses unable to infect epithelial intestinal cells *in vivo*. An alternative possibility is that the modifications around residues 218 and 219 could influence other viral or cellular regulatory mechanism affecting essential steps of virus replication, other than the virus-to-cell binding.

Protein sequence alignment (figure, rotated 90°).

Block 1 (positions 1–110)

Position markers: 10, 30, 50, 70, 90, 110

PUR46-MAD MKKLFVVLVVMPLIYGDNPFPCSKLTNRTIGNQWNLIETPLIANYSSRLPNSDFVLGDYPPTVQPWFNCIRNDSNDLYVTLENLKALYWDYATENITHRQRLNVVVNGYPYSITVTTR

Site C   Site B

Strain	Substitutions
NEB72	
TOY56	H ... K
BRI70-FS	H S N K T X
MIL65-AME	H S N H N S L K X
BEL85-83	K
FRA86-RM	K
ENG86-I	K
ENG86-II	K
HOL87	K
BEL87-31	K

Block 2 (positions 130–230)

Position markers: 130, 150, 170, 190, 210, 230

PUR46-MAD NFNSAEGAIICICKGSPPTTTESLTCNWGSECRLNHKFPICPSNSEANCGNNLYGLQWFADEVVAYLHGASYRISFENQWSGTVTFGDMRATTLEVAGTLVDLAWFNPVDVSYRVN

Site B   Site B

Strain	Substitutions
NEB72	
TOY56	A L S
BRI70-FS	A L I
MIL65-AME	A L T
BEL85-83	
FRA86-RM	
ENG86-I	
ENG86-II	
HOL87	
BEL87-31	

Block 3 (positions 250–350)

Position markers: 250, 270, 290, 310, 330, 350

PUR46-MAD NRKNGTTVVSNCTDQCASYVANVFTTQPGGFIPSDFSFNNWFLLTNSSTLVSGKLVTKQPLLVNCLWPVFSFEEAASTFCFEGAGFDQCNGAVLNNTVDVIRPNLNFTTNVQSGKGATVFS

Strain	Substitutions
NEB72	I
TOY56	K
BRI70-FS	
MIL65-AME	
BEL85-83	S
FRA86-RM	S IL N V L X
ENG86-I	S D
ENG86-II	S D
HOL87	S D P
BEL87-31	S D

Block 4 (positions 370–470)

Position markers: 370, 390, 410, 430, 450, 470

PUR46-MAD LNTTGGVTLEISCY...TVSDSSFFSYGEIPFGVTDGPRYCVHYNGTALKYLGTLPPSVKEIAISKWGHFYINGYNFPSTFPIDCISFNLTTGDSDVFWTIAYTSYTEALVQVENTAITK

Site D

Strain	Substitutions
NEB72	A NDI F L
TOY56	ND S L N
BRI70-FS	ND S L N
MIL65-AME	ND S M L NE N
BEL85-83	ND S L N N
FRA86-RM	ND A L N D N
ENG86-I	ND S L N N
ENG86-II	ND S L N N
HOL87	ND S S L N N
BEL87-31	ND S N⁻ L N

```
              490            510            530            550            570            590
PUR46-MAD  VTTYCNSHVNNIKCSQITANLNRGPYPVSSSEVGLVNKSVVLLPSFYHTVNITIGLG M SG K PIASTLSNITLPMODHNTDVVCIRSDQFSVVVHSTCKSALW IFH N CTDVLD
NEB72                                                              Site A                                                    Site A
TOY56         Y                                                                                  D                  Q
BRI70-FS      Y         L                    P                                                          N I                        V
MIL65-AME     Y         L                    T         S                                               N                  S        V
BEL85-83      Y         L                              S                                               N        V                  V
FRA86-RM      Y         L                              S                                               N N      V                  V
ENG86-I       Y         L                              S                                               N        V        X         V
ENG86-II      Y         L                              S                                               N        V                  V
HOL87         Y         L                              S                                               N        V            H     V
BEL87-31      Y         L                              S                                               N        V                  V

              610            630            650            670            690            710
PUR46-MAD  ATAVIKTGTCPFSFDKLNNYLTFNKFCLSLSPVGANCKFDVAARTRTNEQVVRSLYVIYEEGDNIVGVPSDNSGVHDLSVLHLDSCTDYNIYGRTGVGIIRQTNKTLSGLYYTSLSGDL
NEB72
TOY56                                        D
BRI70-FS                                     A D                        L                                    K
MIL65-AME                                      D                        L                     S
BEL85-83                                       D                    S
FRA86-RM                                       D                    S L                                                    I
ENG86-I                    H                   D  *
ENG86-II                   X                   D  *                 S L
HOL87                                          D  *
BEL87-31                   S                                        *
```

FIGURE 2. Sequence alignment of spike (S) protein of TGEVs and PRCVs. The sequence of the first 720 amino acids of the S protein of PUR46-MAD virus is shown in the first line. In the other lines, the aa changes in the sequence of the S proteins of other viruses have been indicated. In the alignment deleted residues have been filled out with points. Sequence numbers indicate the positions that the aa would have in the MIL65 virus. For simplicity, the sequences of two clones of the PUR46 isolate (PUR46-PAR and PUR46-UTR) have been omitted in this series of sequences, since they show minor changes and their sequences were previously published. The sequences of the strains PUR46-MAD, NEB72, TOY56, HOL87, FRA86-RM, MIL65-AME, BRI70-FS, PUR46-PAR, and PUR46-UTR have been previously reported[4,16,17,18,19,20]. Sequence indeterminations have been coded as X, for any amino acid. Underlined amino acids correspond to the signal peptide. Residues in boxes are involved in the indicated antigenic sites. Asterisks indicate the carboxi-terminus of the segments sequenced. Dashes indicate non-sequenced segments. The nucleotide sequence data reported in this paper have been submitted to the GenBank nucleotide sequence database and have been assigned the accesion numbers: PUR46-MAD, M94101; NEB72-RT, M94099; TOY56, M94103; HOL87, M94096; BEL85-83, M94097; BEL87-31, M94098; ENG86-I, M94100; ENG86-II, M94102.

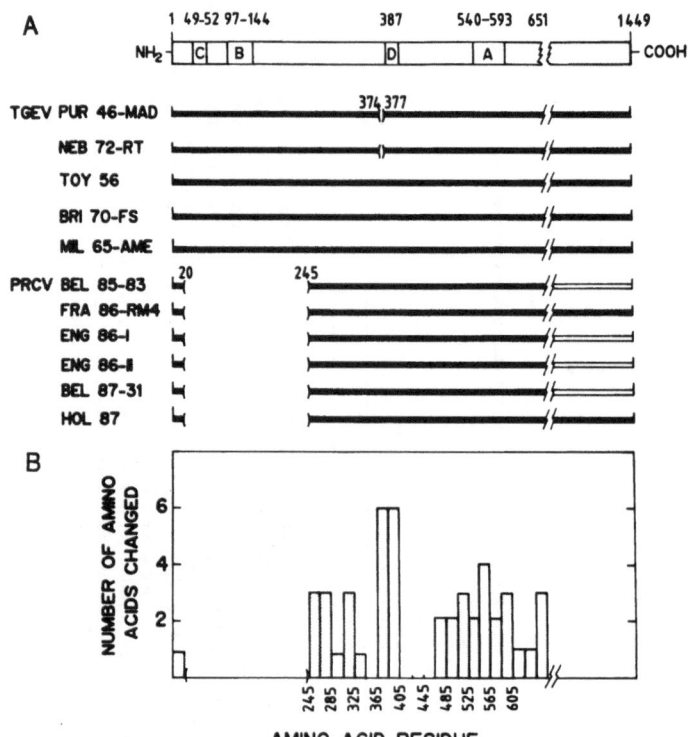

FIGURE 3. Summary of the deletions and amino acid changes present in the S glycoprotein of TGEVs and PRCVs. A. Location of the deletion. Full and empty bars indicate the sequences known and undetermined, respectively. Letters indicate the approximate location of the antigenic sites. The numbers above these letters indicate amino acid residues involved in the formation of these sites. The position of the deletions is indicated by brackets, and the numbers next to the brackets show the amino acids flanking the residues deleted. B. Number of amino acid changes in sequential fragments of 20 aa each, in relation with the PUR46-MAD virus sequence. Only the segments for which the sequences of the 13 virus strains were available have been included in the comparison. Amino acid residues have been numbered according to their position in the MIL65 virus after the alignment. The origin of the sequences of the different strains has been indicated in the legend of Fig. 2.

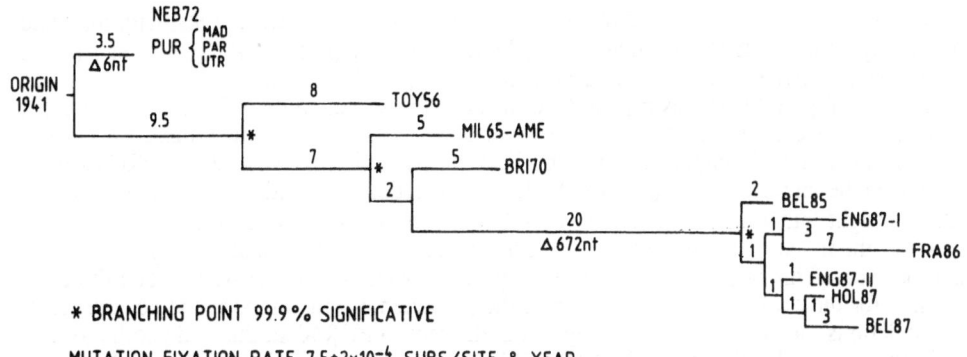

* BRANCHING POINT 99.9% SIGNIFICATIVE

MUTATION FIXATION RATE 7.5±2x10^{-4} SUBS/SITE & YEAR

FIGURE 4. Evolutionary tree of TGEV related coronaviruses. Neighbour-joining and least squares methods of tree reconstruction procedures were applied to the first 1956 nucleotides of 13 virus isolates (the 11 isolates indicated in Fig. 2, and the clones PUR46-PAR and PUR46-UTR previously reported)[17,18]. Numbers in the diagram indicate residue substitutions between branching points. Δ, indicates the introduction of a deletion between branching points. *, indicates that all the descendents of this fork have, with a probability of 99.9%, a recent common ancestor.

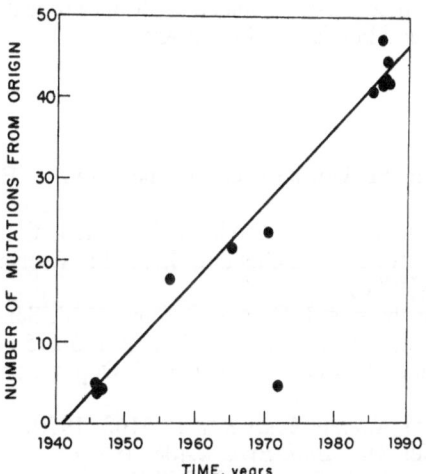

FIGURE 5. Relationship between mutation fixation rate and year of isolation. The line relating the number of mutations from origin with the year of isolation was plotted. Line and origin were estimated at the same time by linear least squares fit. The expression for the line was: d= 0.95 t - 1893, r^2= 0.97, where d is the distance to the origin, t is the time in years, and r^2 the Pearson's correlation coefficient[10]. The data correspond to the viral isolates used in the construction of the evolutionary tree (Fig. 4). The line with a minimum square error was determined and represented. The point showing minimum fitness with the line corresponds to the NEB72-RT isolate.

Evolutionary tree for the S gene of TGEVs and PRCVs

The nucleotide sequence of the S glycoprotein of eight respiratory and five enteric TGEVs (three of which were different clones of the same PUR46 virus strain) were aligned taking into account the two deletions of 6 and 672 nts present in the sequence of the PUR 46 and PRCVs, respectively, for maximum fitness. Phylogenetic analysis of the sequences (first 1956 nt) of the viruses described in Figure 2, by either the neighbor-joining or the least squares methods of tree-reconstruction procedures, gave two identical trees, with the same branching order, confidence levels, and branch lengths (Figure 4). The least squares relationship between the number of mutations from origin and the year of isolation was determined (Figure 5). The extrapolation of this line to zero mutations allowed to predict that these TGEVs were originated from a recent common ancestor circulating around 1941. Since then, from a main lineage, the PUR46, TOY56, MIL65, BRI70, and the PRCVs were derived in the indicated order (Figure 4). The accumulation of mutations with time (Figure 5) fits a straight line with high Pearsons correlation coefficient ($r^2=0.97$). From the slope of this line, the mutation fixation rate can be estimated at $7\pm2 \times 10^{-4}$ substitutions per nucleotide and per year. This rate falls in the range reported for other RNA viruses[21]. The direction defined for the evolutionary process, from the predicted origin, supports the occurrence of two deletions: one of 6 nt in the lineage from the root to PUR46 strains and another of 672 nt in the lineage leading from TGEV to PRCVs. It may be concluded that the European PRCVs have been derived by a 672-nt deletion from an enteric TGEV, since we have examined isolates preceding the PRCVs. In contrast, it cannot be guaranteed that the PUR46 emerged by a 6-nt deletion from an unknown ancestor. An alternative explanation could be that the other enteric isolates shown in Figure 3 could have been derived from PUR46 by the addition of 6 nt. If recombination has been the cause of the deletion present in the PUR46 and PRCVs, then two mechanisms of evolution would be involved in the antigenic variation of TGEV, point mutations and recombination.

Acknowledgements

This investigation has been founded by grants from the Comisión Interministerial de Ciencia y Tecnología, Comunidad Autónoma de Madrid, European Communities (BAP and Science Projects), NATO, and Fundación Ramón Areces.

REFERENCES

1. P. Callebaut, I. Correa, M. Pensaert, G. Jiménez, and L. Enjuanes, *J. Gen. Virol.* 69:1725 (1988).
2. C.M. Sánchez, G. Jiménez, M.D. Laviada, I. Correa, C. Suñé, M.J. Bullido, F. Gebauer, C. Smerdou, P. Callebaut, J. A. M. Escribano, and L. Enjuanes, *Virology* 174:410(1990).
3. F. Gebauer, W.A.P. Posthumus, I. Correa, C. Suñé, C.M. Sánchez, C. Smerdou, J. A. Lenstra, R. Meloen, and L. Enjuanes, *Virology* 183:225 (1991).
4. C.M. Sánchez, F. Gebauer, C. Suñé, A. Mendez, J. Dopazo, and L. Enjuanes, *Virology* 190:92 (1992).
5. O. Fichot, and M. Girard, *Nucleic Acids Res.* 18:6162 (1990).
6. N.M. Saitou, and M. Nei, *Mol. Biol. Evol.* 4:406 (1987).
7. W.M. Fitch, and E. Margoliash, *Science* 155:279 (1967).
8. B. Efron. "The Jackknife, the bootstrap and other resampling plans," Society for industrial and applied mathematics, Phyladelphia (1982).
9. J. Felsenstein. "PHYLIP manual version version 3.3", University of California (1990).
10. R.R. Sokal, and F. J. Rohlf. "Biometry," Freeman, New York (1981).
11. R.D. Wesley, R.D. Woods, and A.K. Cheung, *J. Virol.* 65:3369 (1991).
12. P.S. Paul, E.M. Vaughn, and P.G. Halbur, Int. Pig Vet.Soc. 15th Meeting (1992).
13. W. Ansorge, *J. Biochem. Bioph. Meth.* 11:13 (1985).
14. C. Suñé, G. Jiménez, I. Correa, M.J. Bullido, F. Gebauer, C. Smerdou, and L. Enjuanes, *Virology* 177:559 (1990).
15. B. Delmas, J. Gelfi, R.L'Haridon, L.K. Vogel, H. Sjöström, O. Norén, and H. Laude, *Nature* 357:417 (1992).
16. P. Britton, K. and W. Page, *Vir. Res.* 18:71 (1990).
17. L. Jacobs, R. de Groot, B.A.M. Van der Zeijst, M.C. Horzinek, and W. Spaan, *Vir. Res.* 8:363 (1987).
18. D. Rasschaert, and H. Laude, *J. Gen. Virol.* 68:1883 (1987).
19. D. Rasschaert, M. Duarte, and H. Laude, *J. Gen. Virol.* 71:2599 (1990).
20. R.D. Wesley, *Adv. Exp. Med. Biol.* 276:301 (1990).
21. E. Domingo, and J.J. Holland, in "RNA genetics," CRC Press, Florida (1988).

TRANSMISSIBLE GASTROENTERITIS VIRUS AND PORCINE RESPIRATORY CORONAVIRUS: MOLECULAR CHARACTERIZATION OF THE S GENE USING cDNA PROBES AND NUCLEOTIDE SEQUENCE ANALYSIS

Daral J. Jackwood, Insoo Bae, Renee J. Jackwood, and Linda J. Saif

Food Animal Health Research Program
The Ohio State University
Ohio Agricultural Research and Development Center
Wooster, Ohio 44691

ABSTRACT

Two transmissible gastroenteritis virus (TGEV, Miller strain) cDNA clones were identified and their nucleotide sequences determined. The clones were non-overlapping and were located in the 5' region of the S glycoprotein gene. The TGEV clone pE21 contained 381 bp of the S glycoprotein gene and had >98% nucleotide and amino acid sequence homology with the Purdue (P115) strain of TGEV and over 87% sequence homology with feline infectious peritonitis virus (FIPV). The TGEV clone, pD24, contained 267 bp of the S glycoprotein gene. It had >98% nucleotide and amino acid sequence homology with P115 but only a 49% nucleotide sequence homology and a 24% amino acid sequence homology with FIPV. Using dot blot hybridization, a probe prepared from pD24 could differentiate TGEV from the antigenically related coronaviruses, FIPV, feline enteric coronavirus and canine coronavirus. This probe could also differentiate TGEV from porcine respiratory coronavirus (PRCV). Using polymerase chain reaction amplified regions of PRCV isolates and nucleotide sequencing, a 681 bp deletion in the 5' region of the S gene from PRCV isolate ISU-1 was identified. This deletion was located in the area of the S glycoprotein gene identified by the pD24 probe.

INTRODUCTION

Transmissible gastroenteritis virus (TGEV) causes severe diarrhea in swine (1). Although molecular studies on this virus have advanced our knowledge greatly, an efficacious vaccine and reliable diagnostic assay have not been developed. Furthermore, the recent discovery of porcine respiratory coronavirus (PRCV) has complicated the diagnosis of TGEV infected herds (2, 3, 4).

Coronaviruses have been placed into antigenic groups based on serology (5, 6).

Coronaviruses, Edited by H. Laude and J.F. Vautherot
Plenum Press, New York, 1994

Using virus neutralization assays, TGEV and PRCV appear to be antigenically identical (7). Other viruses which are grouped antigenically with TGEV and PRCV include feline infectious peritonitis virus (FIPV), feline enteric coronavirus (FECV), canine coronavirus (CCV), and human respiratory coronavirus (HCV-299E). Monoclonal antibodies have been developed which can distinguish TGEV and PRCV strains (7, 8). A competitive inhibition ELISA using non-neutralizing monoclonal antibodies was developed to distinguish serum antibodies from TGEV and PRCV infected pigs (8).

Diagnosis of TGEV infections using cDNA probes and hybridization was first reported by Shockley and coworkers (9). These investigators demonstrated the utility of dot blot hybridization to detect TGEV in clinical samples. Later, Benfield *et al.* (10) reported on the application of this assay to detect TGEV in fecal samples. The dot blot assay was comparable in sensitivity to conventional assays such as immunofluorescence and electron microscopy.

The appearance of PRCV in the United States prompted our laboratory to examine the use of dot blot hybridization for the differentiation of TGEV and PRCV infections. A nucleotide sequence comparison between TGEV and FIPV was reported in 1987 by Jacobs and coworkers (11). This study demonstrated a variable sequence region at the 5' end of the S glycoprotein gene between these two antigenically related viruses. Therefore, we postulated that the 5' end of the TGEV S glycoprotein may also contain sequence variability compared to other antigenically related coronaviruses including PRCV.

MATERIALS AND METHODS

Viruses and Cells

The virulent Miller strain of TGEV was used for the preparation of cDNA clones. The attenuated Purdue strain (P115) was used in dot blot hybridization studies. Other viruses used in dot blot studies were FIPV and FECV. Feline viruses FIPV and FECV were propagated in Crandell feline kidney cells (CrFK) cells. These viruses and cells were obtained from J. F. Everman, Washington State University, Pullman Washington (12, 13). The ISU-1 and ISU-2 isolates of PRCV were obtained from H. Hill, Veterinary Diagnostic Laboratory, Iowa State University, Ames, Iowa (4). Swine testicular (ST) cells were used to propagate the TGEV and PRCV strains.

Production of cDNA Clones and Probes

The cDNA clones to TGEV were prepared from genomic RNA as previously described (10). The first strand cDNA reaction was primed using random calf-thymus DNA oligonucleotides and the second strand reaction was conducted using DNA polymerase I and RNase H. The cDNA was tailed using dCTP and terminal transferase and annealed into the *Pst I* site of pUC9 which was *Pst I* cut and tailed with dGTP. Clones specific for the S glycoprotein gene were identified using colony blot hybridization (14). A probe prepared from Hpa-1600 (a gift from R. Wesley, National Animal Disease Center, United States Department of Agriculture, Ames, Iowa) was used in the colony blot assay. The Hpa-1600 cDNA contained 1,600 base pares (bp) of nucleotide sequence in the S glycoprotein gene of TGEV. All probes used in this study were prepared using (^{32}P)dCTP (specific activity, >600 Ci/mmol; ICN Pharmaceuticals Inc., Irvine, California) and nick translation.

Dot Blot Hybridization Assays

Viruses were harvested from cell cultures and treated with proteinase K (100ug/ml) at 37C for 45 min. Each sample was extracted with phenol and chloroform before precipitation in ethanol at -20C. Precipitated RNA was collected by centrifugation and resuspended in 300 ul of sterile water and an equal volume of SSC-formaldehyde solution (two parts 37% formaldehyde and three parts 20X SSC; 1X SSC is 0.15 M NaCl plus 0.05 M sodium citrate). The samples were incubated at 65C for 15 min and then applied to nylon hybridization membranes (Biotrans; ICN Biochemicals, Irvine, California) using a dot blot hybridization manifold (Bethesda Research Laboratories, Inc., Gaithersberg, Maryland).

Prehybridization and hybridization were conducted in buffer containing 5X SSC, 5X Denhardt solution (0.1% [wt/vol] Ficoll, 0.1% [wt/vol] polyvinylpyrrolidone, 0.1% [wt/vol] bovine serum albumin), 50 mM sodium phosphate (pH 6.5), 0.1% SDS, 250 ug/ml salmon sperm DNA, and 50% (vol/vol) formamide. Prehybridization was conducted at 42C for 4 hr and hybridization was conducted for 18 hr at 42C. Approximately 3×10^6 cpm per membrane of each probe were added to the hybridization buffer. Following hybridization, the membranes were washed at room temperature in 2X SSC containing 0.1% (wt/vol) SDS and then at 42C in 0.1 X SSC containing 0.1% (wt/vol) SDS before hybridization.

Nucleotide Sequencing

Nucleotide sequence analysis of ISU-1 was conducted on cDNA amplified using the polymerase chain reaction (PCR). Genomic RNA from ISU-1 was purified and a first strand cDNA was prepared as described above using random calf-thymus DNA oligonucleotides. Amplification of this cDNA was accomplished using a PCR kit (Perkin-Elmer Cetus, Norwalk, Connecticut) and two primers. Primer 1 (GGTCCATCAGTTACGCCGAA) was located in the 3' region of the adjacent polymerase gene and primer 2 (AAGGAAGGGTAAGTTGCTCA) was located in the 5' region of the S glycoprotein gene (bases 1165 to 1185). The expected amplified product was 1250 bases in length. The PCR products were analyzed by agarose gel electrophoresis and Southern blot hybridization (15).

The gene 6 exonuclease (United States Biochemical Co., Cleveland, Ohio) was used to enzymatically alter the cDNA into single-stranded templates for sequencing. Dideoxy-chain termination sequence reactions were conducted using Sequenase (United States Biochemical Co.) and primers 1 and 2 (16).

RESULTS

cDNA Probes

Using the Hpa-1600 probe and colony blot hybridization, two cDNA clones were identified in the 5' region of the S glycoprotein gene. One was designated pE21 and the other pD24. The TGEV clone pE21 was 381 bp in length and encompased bases 974 to 1355 from the 5' end of the S glycoprotein gene. The TGEV cDNA clone pD24 was located closer to the 5'end of the S glycoprotein gene (bases 198 to 465) and was 267 bp in length. On the basis of sequence homologies reported for TGEV and FIPV (11), pD21 was in a region of relatively high sequence homology (87%) and pD24 was in a region of relatively low sequence heterology (49%) (Fig. 1).

48.8% identity in 258 bp overlap

```
           230       240       250       260       270       280
FIPV   TTTCAGTATTTTAATAATATACATGCCTTTTATTTTGTTATGG-AAGCCATGGAAAATAG
            :        :   ::::::::  :::  ::::::::    :::  ::  :   : ::
pD24   AATTGCATTCGCAATAATAGTAATGACCTTTATGTTACATTGGAAAATCTTAAAGCATTG
           10        20        30        40        50        60

           290       300       310       320       330       340
FIPV   CACTGGTA--ATGC-ACGTGGTAAACCATTATTATTTCATGTGCATGGTGAGCCTGTTAG
            : :::   ::   :  :   :       : :          ::::  :   :  ::
pD24   TATTGGGATTATGCTACAGAAAATATCACTTCGAATCACAAACAACGGTTAAAC-GTAGT
           70        80        90       100       110

           350       360       370       380       390       400
FIPV   TGTTATTATATCGGCTTA-TAGGGATGATGTGCAACAAAGGCCCCTTTTAAAACATGGGT
            :::: ::  : ::::   : :::::  :  :    ::    : :::::  : ::
pD24   CGTTAATGGATACCCATACTCCATCACAGTTACAACAACCCGGCAATTTTAATTC--TGCT
       120       130       140       150       160       170

           410       420       430       440       450       460
FIPV   TAGTGTGCATAACTAAAAATCGCCATATTAACTATGAACAATTCACCTCCAACCAGTGGA
            :  :::::      ::       :: :    : :   :    ::   :  :
pD24   GAAGGTGCTATTATATGCATTTGCA-AGGGCTCACCACCTACT-ACCACCACAGAATCTA
       180       190       200       210       220       230

           470       480       490
FIPV   ATTCCACATG---TACGGGTGCTGACAGAAAA
            ::  :: ::   :   :   ::: : :
pD24   GTTTGACTTGCAATTGGGGTAGTGAGTGCAGG
       240       250       260
```

87.4% identity in 382 bp overlap

```
           990      1000      1010      1020      1030      1040
FIPV   TTTAGCCAATGTAATGGTGTGTCTTTAAATAACACAGTGGATGTTATTAGATTCAACCTT
            :::   ::::::::::::  ::::::::::: :: :: ::  :: ::  :::::::::
pE21   TTTGATCAATGTAATGGTGCTGTTTTAAATAACACTGTAGACGTCATCAGGTTTAACCTT
           10        20        30        40        50        60

          1050      1060      1070      1080      1090      1100
FIPV   AATTTCACTGCAGATGTACAATCTGGTATGGGTGCTACAGTATTTTCACTGAATACAACA
            ::::: :::  :::::::: :::: ::  ::: :::::::: ::::  :::  ::::::
pE21   AATTTTACTACAAATGTACAATCAGGTAAGGGTGCCACAGTGTTTTCATTGAACACAACG
           70        80        90       100       110       120

          1110      1120      1130      1140      1150      1160
FIPV   GGTGGTGTCATTCTTGAAATTTCATGTTATAGTGACACAGTGAGTGAGTCTAGTTCTTAC
            ::::::::::: ::::::::::  :::::::: :: ::  :::::::: :: ::  :: :
pE21   GGTGGTGTCACTCTTGAAATCTCATGTTATAATGATACAGTGAGTGACTCGAGCTTTTCC
           130       140       150       160       170       180

          1170      1180      1190      1200      1210      1220
FIPV   AGTTATGGTGAAATCCCGGTTCGGCATAACTGACGGACCACGATACTGTTATGTACTTTAC
            :::: :::::::: ::::::: ::::: ::::::: :::: ::  :::: :::  ::::
pE21   AGTTACGGTGAAATGCCGGTTCGGCGTAACTGATGGACCACGGTACTGTTACGTACACTAT
           190       200       210       220       230       240

          1230      1240      1250      1260      1270      1280
FIPV   AATGGCACAGCTCTTAAATATTTAGGAACATTACCACCCAGTGTAAAGGAAATTGCTATT
            ::::::::::::::: :::::::::::::::::::::::: :::: :  :: :::::::
pE21   AATGGCACAGCTCTTAAGTATTTAGGAACATTACCACCTAGTGTCAAGGAGATTGCTATT
           250       260       270       280       290       300

          1290      1300      1310      1320      1330      1340
FIPV   AGTAAGTGGGGCCATTTTTATATTAATGGTTACAATTTCTTTAGCACATTTCCTATTGGT
            ::::::::::::::::::::::::::::::::::::::::::::::::::::::::::  :
pE21   AGTAAGTGGGGCCATTTTTATATTAATGGTTACAATTTCTTTAGCACATTTCCTATTGAT
           310       320       330       340       350       360

          1350      1360
FIPV   TGTATATCTTTTAATTTAACC
            :::::::::::::::: :::
pE21   TGTATATCTTTTAATTTGACC
           370       380
```

Figure 1. Nucleotide sequence comparison of TGEV cDNA clones pD24 and pE21 with the FIPV sequence published by Jacobs *et al.* (11).

Dot Blot Hybridization

Probes prepared using pE21 hybridized to TGEV (Miller and Purdue stains), PRCV (ISU-1 and ISU-3 strains), FECV, and FIPV. Probes prepared to pD24 however, only hybridized to the two TGEV strains. They did not hybridize to PRCV, FECV, and FIPV (Fig. 2).

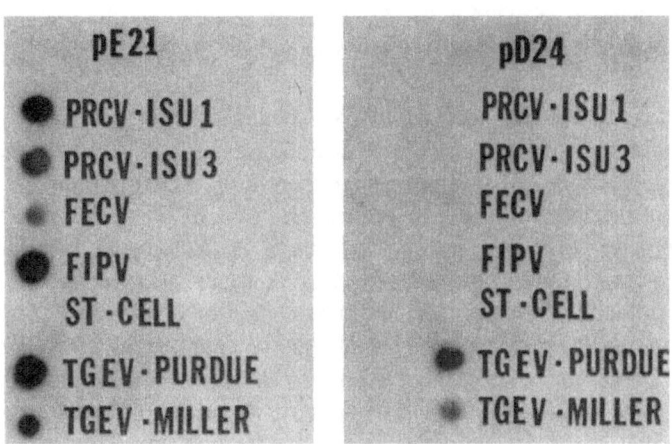

Figure 2. Dot blot hybridization using the TGEV cDNA pE21 and pD24 probes. Hybridization was conducted using 50% formamide at 42 C.

PCR Amplification and Nucleotide Sequence Comparison

The PCR products were analyzed using agarose gel electrophoresis and Southern blot hybridization. PCR amplified cDNA from the Miller TGEV strain was approximately 1250 bp in length. Using the same primers, the PCR amplified cDNA from ISU-1 was approximately 580 bp in length. These results indicated a deletion existed in this region of the PRCV isolate ISU-1. Probe pE21 hybridized to the PCR amplified products from TGEV and ISU-1 in a Southern blot hybridization assay (data not shown). Probe pD24 however, only hybridized to the TGEV PCR product.

Nucleotide sequencing of the ISU-1 PCR product confirmed the 681 base deletion (Fig. 3). This deletion begins 62 bases from the 5' end of the S glycoprotein gene.

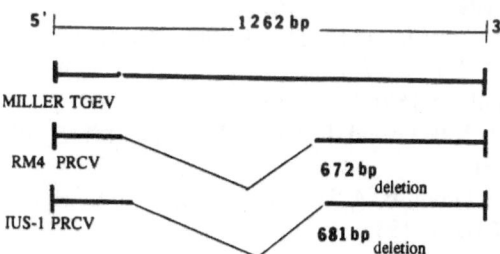

Figure 3. Schematic of the sequence comparisons among Miller strain TGEV, ISU-1 strain PRCV and RM4 strain PRCV. The deletion observed in the US PRCV isolate ISU-1, was very similar to that observed in the European RM4 PRCV isolate.

DISCUSSION

Diagnosis of TGEV infections is possible using dot blot hybridization (9). This assay has compared favorably with immunofluorescence and electron microscopy (10). We have demonstrated in this study that hybridization can also be used to differentiate TGEV from antigenically related coronaviruses including the PRCV isolates found in the US.

Our original hypothesis based on data reported by Jacobs *et al.* (11), was that the lack of hybridization of pD24 with PRCV and other antigenically related coronaviruses was due to heterologous sequences at the 5' end of the S glycoprotein gene. Nucleotide sequence data on ISU-1 demonstrated the lack of hybridization was due to a deletion in the sequence. Although European PRCV isolate RM-4 (17) was not tested in the hybridization assay, nucleotide sequence data indicates that this virus contains a similar deletion and probably could be distinguished from TGEV using the pD24 probe. Because sequence data for the S glycoprotein gene from FECV are not available, the basis for the observed lack of hybridization of probe pD24 to this virus is not known.

Since our original report in 1991 (18), others have also reported on the use of dot blot hybridization for the detection and differentiation of TGEV and PRCV isolates (19).

The dot blot hybridization assay has the potential to be an excellent diagnostic assay for TGEV infections. Currently, most assays rely on detecting virus or viral antigens in fecal samples. Shedding of the virus in fecal samples is a requirement for detecting TGEV in the dot blot hybridization assay. The length of time pigs shed virus following TGEV infection has not been studied. If the duration of shedding is short, this would limit the utility of dot blot hybridization as a practical diagnostic assay. However, the pD24 and pE21 probes may be more useful in other hybridization assays such as in situ hybridization or they could be used in combination with PCR which would increase the sensitivity of the dot blot hybridization assay.

REFERENCES

1. Saif, L. J., and E. H. Bohl. In. Diseases of swine, A. D. Leman *et al.* (ed), Iowa State University Press, Ames, Iowa. pp. 255-274, 1986.
2. Pensaert, M., P. Callebaut, and J. Vengote. Vet. Quart. 8(3):257-261, 1986.
3. Brown, I., S. Cartwright. Vet. Rec. 119:282-283, 1986.
4. Hill, H., J. B. Biwer, R. Wood, and R. Wesley. Proc. Am. Assn. of Swine Practitioners. 21:333-335, 1990.
5. Peterson, N. C., I. Ward, and W. L. Mengeling. Arch. Virol. 58:45-53, 1987.
6. Sturman, L. S., and K. V. Holmes. Adv. Virus Res. 28:35-112, 1983.
7. Callebaut, P., I. Correa, M. Pensaert, G. Jimenez, and L. Enjuanes. J. Gen. Virol. 69:1725-1730, 1988.
8. Callebaut, P., M. B. Pensaert, and J. Hooyberghs. Vet. Microbiol. 20:9-19, 1989.
9. Shockley, L. J., P. A. Kapke, W. Lapps, D. A. Brian, L. N. Potgieter, and R. Woods. J. Clin. Microbiol. 25:1591-1596, 1987.
10. Benfield, D. A., D. J. Jackwood, I. Bae, L. J. Saif, and R. D. Wesley. Arch. Vriol. 116:91-106, 1991.
11. Jacobs, L., R. J. De Groot, B. A. M. van der Zeijst, M. C. Horzinek, and W. Spaan. Virol. Res. 8:363-371, 1987.
12. McKeirnan, A. J., J. F. Evermann, A. Hargis, L. M. Miller, and R. L. Ott. Feline Pract. 11:16-20, 1981.
13. Pederson, N. C., J. F. Evermann, A. J. McKeirnan, and R. L. Ott. Am J. Vet. Res. 45:2580-2585, 1984.
14. Grunstein, M., and J. Wallis. Methods Enzymol. 68:379-388. 1979.
15. Southern, E. M. J. Mol. Biol. 98:503-517, 1975.
16. Sanger, F., S. Nicklen, and A. R. Coulson. Proc. Natl. Acad. Sci. USA 74:5463-5467, 1977.
17. Rasschaert, D. J. Duarte, and H. Laude. J. Gen. Virol. 71:2599-2607, 1990.
18. Bae, I., D. J. Jackwood, D. A. Benfield, L. J. Saif, R. D. Wesley, and H. Hill. J. Clin. Microbiol. 29:215-218, 1991.
19. Wesley, R. D., I. V. Wesley, and R. D. Woods. J. Vet. Diagn. Invest. 3:29-32, 1991.

SEQUENCE ANALYSIS OF THE NUCLEOCAPSID PROTEIN GENE OF PORCINE EPIDEMIC DIARRHOEA VIRUS

Kurt Tobler, Anne Bridgen and Mathias Ackermann

Institute for Virology
Veterinary Medical Faculty
University of Zürich
Winterthurerstrasse 266a
CH-8057 Zürich

ABSTRACT

The nucleotide (nt) sequence of 1.7 kbp cDNA representing the 3' end of the PEDV genome has been determined. Viral RNA was reverse transcribed and the cDNA was amplified by polymerase chain reaction using degenerate primers. The sequences of the primers were based on conserved regions of coronaviral genomes. A 1323 nt open reading frame (ORF) showed good homology to the nucleocapsid (N) gene of other coronaviruses. The greatest homologies at the amino acid and the nt levels were observed with Human Coronavirus 229E. A second 336 nt ORF, which might encode a leucine-rich protein, was found within the N gene. Between the 3' end of the N gene and the poly(A) tail was a sequence of eleven nt, which is conserved among the other sequenced coronaviruses. Finally, a seven base sequence similar to the conserved intergenic sequences was present 5' to the N gene. These results confirm the classification of PEDV as a coronavirus.

INTRODUCTION

Porcine epidemic diarrhoea virus (PEDV) causes diarrhoea in pigs, particularly in neonates. The biological behaviour and electron microscopic appearance of PEDV resulted in its provisional classification as a coronavirus. However, inconsistent cross-reactivity with antibodies directed against other coronaviruses[1] did not permit the classification of PEDV into a coronavirus antigenic group. This provisional classification of PEDV did, however, allow a genomic amplification strategy based on anticipated similarity to other, sequenced members of the coronavirus family. The amplified products were cloned and sequenced and the results are described in this article.

MATERIALS AND METHODS

The virus strain used in this study was the CV 777 strain of PEDV isolated by Pensaert and Debouck[2] and adapted to culture in Vero cells by Hofmann and Wyler.[3] Vero cells (ATCC 1587) were grown and after three days infected with PEDV at an moi of 0.03 as described previously.[4] RNA was extracted from both semi-purified virions and from infected cells. RNA from infected cells was harvested when the cells showed 85-90%

Coronaviruses, Edited by H. Laude and J.F. Vautherot
Plenum Press, New York, 1994

cytopathic effect (cpe) and processed by the caesium chloride method as described elsewhere.[5] For virion RNA, cells showing 95-100% cpe were disrupted by three cycles of freeze-thaw, cellular debris were removed by low speed centrifugation, and virus was pelleted by centrifugation in a Beckman SW28 rotor at 100,000 g at 4°C for 2 h. The viral pellets were washed, and the RNA was extracted.[5]

First strand cDNA was synthesized in two steps, comprising primer binding and extension, according to Wirth et al.[6] Firstly, ng amounts of purified viral RNA or μg amounts of total infected-cell RNA were annealed to 500 ng primer (see below) for 1 h at 42°C. This was followed by 2 h extension reactions, also at 42°C. The cDNA was extracted with phenol:chloroform (1:1), then with chloroform, before being precipitated with ethanol and resuspended in 20 μl of sterile distilled water.

The cDNA was amplified with the following degenerate primers, P23: 5' AAGCTTTT-ACTA(C/T)TT(A/G/T)GG(A/C/T)ACAGGACC 3' (27mer 18 fold degeneracy), P24: 5' CTCGAGCGACCCAGA(A/C)GAC(A/T)CC(G/T)TC 3' (25mer, 8 fold degeneracy) and P25: 5' GACTAGTTGGTGGAG(A/T)TTTAA(C/T)CC(A/T)GA 3' (27mer, 8 fold degeneracy), the sequences of which were based on conserved regions of coronaviral genomes.[7] P25 was also tailed with T residues (designated P25(dT)) with terminal deoxynucleotidyl exotransferase (Boehringer) for 30 min at 37°C. The buffer supplied by the manufacturer was supplemented with 1.5 mM cobalt chloride and a 50:1 molar ratio of dTTP to DNA template. First strand cDNA was primed with oligo(dT) primer (Pharmacia), P24 or P25(dT).

Polymerase chain reaction (pcr) amplifications were performed overnight using a Hybaid Intelligent Heating Block (Model IHB 2024). The pcr buffer consisted of 50 mM potassium chloride, 10 mM Tris pH 8.3, 1 mM magnesium chloride, 10 μg/ml gelatin, 0.045% NP40 and 0.045% Tween 20. Primers were used at 10 μM each and dNTPs at 208 μM each, while 0.1u taq polymerase (Cetus) and 2 μl of the cDNA were used for each 30 μl reaction. P24 and P25 were used together to amplify cDNA primed with oligo dT or with P24, while P23 and P25 were used to amplify P25(dT)-primed cDNA in a modification of the 3' RACE technique of Frohman et al.[8] With primer pair P24/P25 the samples were processed with 38 cycles of 50 sec at 94°C, 60 sec at 48°C, and 60 sec at 72°C. With primers P23/P25 40 cycles of 50 sec at 94°C, 60 sec at 47°C and 150 sec at 72°C were run. A final 5 min extension was made at 72°C for both reactions.

For cloning, the two pcr products were blunt ended with 7.5 u T_4-DNA polymerase (NEB) in the presence of 100 μM dNTPs for 15 min at 12°C and then treated with polynucleotide kinase (NEB) for 30 min at 32°C. Using T_4-DNA ligase (NEB), the pcr products were inserted into pBluescript®II KS+ (Stratagene), which had previously been digested with EcoRV and dephosphorylated. Subclones were made by standard subcloning procedures. Competent *E. coli* DH5α cells, prepared by the calcium chloride method of Maniatis et al.[9] and stored frozen, were transformed by the constructs before plating out on LB plates containing 100 μg/ml ampicillin.

The sequencing reactions were performed by the dideoxynucleotide method using the Sequenase 2 kit (United States Biochemicals). Three sequence ambiguities were resolved by use of the Vent polymerase sequencing kit (NEB) or by replacing dGTP by dITP in the Sequenase reaction. The fragments were separated by electrophoresis through 6% polyacrylamide gels containing 7 M urea. Sequences were analysed with the IntelliGenetics PC/GENE programme or with the University of Wisconsin Genetics Computer Group programmes.[10]

RESULTS AND DISCUSSION

PEDV single stranded cDNA could be amplified with degenerate primers P24/25 and P23/P25(dT) to give products of 0.7 kbp and 1.6 kbp, respectively. The two cloned pcr products, together with their subclones, were sequenced on both strands. Two additional clones derived from independent pcr reactions were also sequenced on one strand in order to take account of possible errors introduced by the taq DNA polymerase; in fact only one base difference was observed between the two 1.6 kbp clones and none between the two 0.7 kbp clones. Three regions were resequenced with dITP using the Sequenase 2 kit and with Vent polymerase to resolve observed sequence ambiguities. In one position (1189-1261 bases from the poly (A) tail) the RNA had the potential to form a stem loop structure covering 72 bases.

Analysis of the sequenced cDNA, comprising the 1.7 kb nearest to the viral poly(A) tail, revealed several open reading frames (ORFs) as is illustrated in Figure 1. The most noticeable feature shown by this figure is the large ORF of 1323 nt in length, which has the capacity to encode a protein with a predicted molecular weight of 48,967 daltons. This figure is quite similar to the 55 to 58 kilodaltons observed for the PEDV N protein.[4,11]

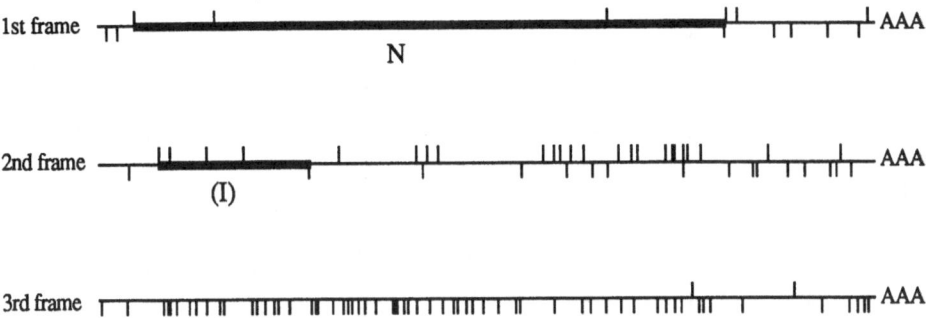

Figure 1. Representation of the 1.7 kb cDNA nearest to the 3' end of the PEDV genome. The codons for methionine (vertical lines above the baselines) and termination codons (vertical lines below the baselines) are shown in the three forward reading frames. The resulting open reading frames are depicted as bold lines and designated with N for nucleocapsid protein and (I) for internal ORF. AAA represents the poly(A) tail.

Comparison of the PEDV protein encoded by the large ORF with all the proteins in the SWISS-PROT data base showed that it was highly homologous with coronavirus N proteins. At least ten coronavirus N genes have now been sequenced, including those of Human Coronavirus (HCV) 229E,[12] Transmissible Gastroenteritis virus (TGEV) strain Purdue,[13] Feline Infectious Peritonitis virus (FIPV),[14] Murine Hepatitis virus (MHV) A59,[15] HCV OC43,[16] Bovine Coronavirus (BCV)[17] and Infectious Bronchitis virus (IBV) strain Beaudette.[18] The percentage identity of the PEDV N protein with those of other coronaviruses, as measured using the Myers and Miller method (PC/GENE), ranged from 13-17% with MHV, IBV, HCV OC43 and BCV to 32-37% with FIPV, TGEV and HCV 229E. The highest homology (37%) was observed to the HCV 229E protein, closely followed by 36% for the TGEV protein.

The predicted properties of the PEDV N protein are also consistent with those of other coronaviruses. The isoelectric points (pI) of all the coronavirus N proteins are high because of the frequency of positively charged residues, their predicted values lying between 10.1 and 10.5. The estimated pI of the N protein of PEDV is 10.5. In contrast to the complete proteins, the C-termini of coronaviral N proteins are negatively charged. This is also true for PEDV: the 50 amino acids nearest to the C-terminus of the PEDV N protein have an estimated pI of only 3.4.

The N protein sequences of all the eight coronaviruses already mentioned contain four homologous regions, as is illustrated in Figure 2. Three are close together in the first third of the sequence, the last one occurring in the last third of the protein. The distances between the third and the fourth homologous regions in TGEV, FIPV, MHV, HCV OC43, BCV and IBV are all about 140 amino acids, but they differ in their distance from the N-terminus. In contrast, the third and fourth conserved regions are separated by about 160 amino acids in HCV 229E and by more than 190 amino acids in PEDV. The intervening sequences in all eight viruses contain two hydrophilic stretches interrupted by a hydrophobic region. The more N-terminal hydrophilic regions consist of a serine-rich sequence. The secondary structure analyses made using the methods of GARNIER and GGBSM (PC/GENE) predicted an α-helix in the amino acids following this serine-rich sequence.

Figure 1 shows an additional ORF within the N gene but in the second reading frame, indicated by (I); it starts 55 bases after the ATG of the N reading frame and could encode a leucine-rich (18%) 112 amino acid protein with a predicted molecular weight of 12,211 daltons. Such a protein might be expressed since the PEDV N gene, like that of other

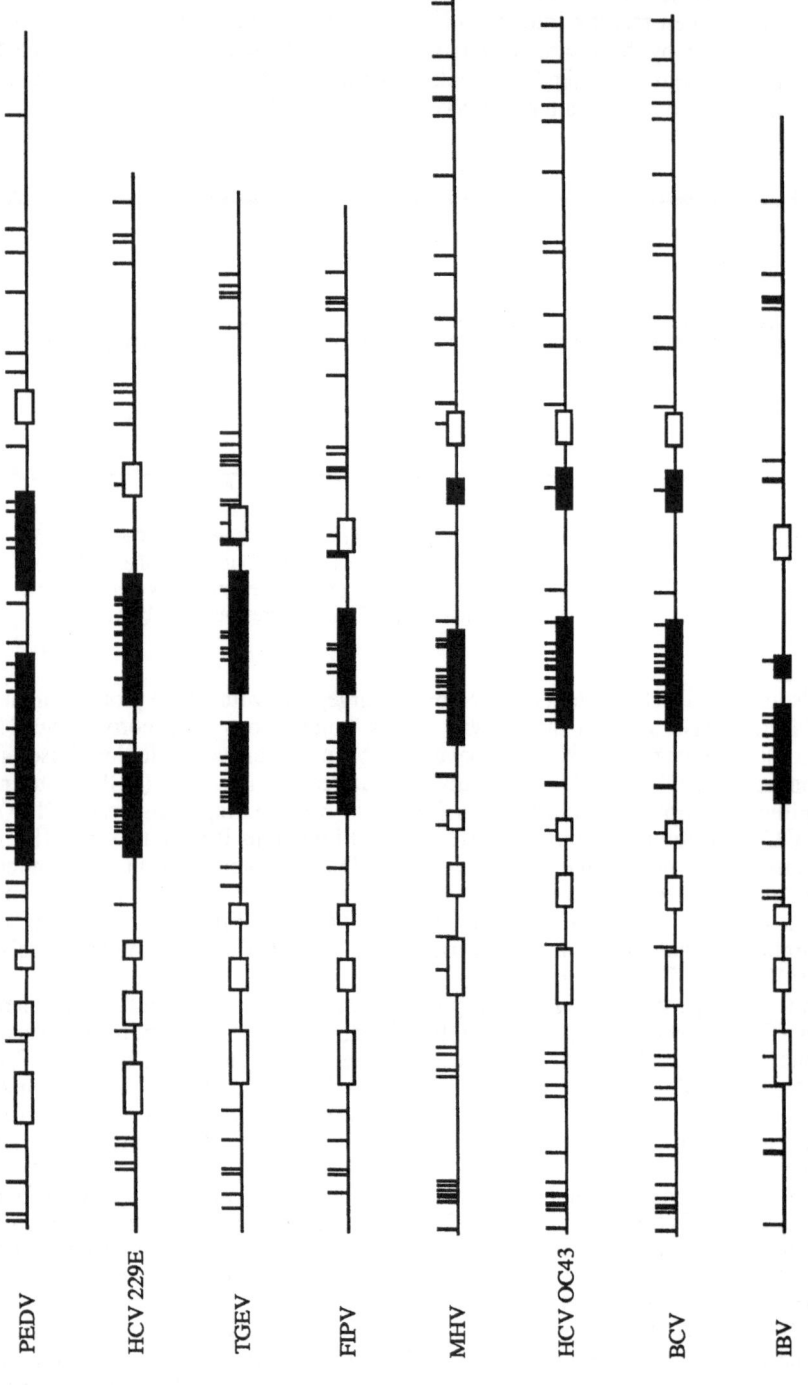

Figure 2. Comparison of different coronavirus N proteins. Serine residues are shown as vertical lines above the baselines. Homologous regions of the sequences are indicated by □ and hydrophilic regions by ■. Abbreviations of virus names are explained in the text. The consensus sequences in the homologous regions are: Gly-Tyr-Trp in the first, Gly-Thr-Gly-Pro in the second, Trp-Val-Ala in the third and Asn-Phe-Gly in the fourth conserved region.

PEDV

HCV 229E

TGEV

FIPV

MHV

HCV OC43

BCV

IBV

coronaviruses, has only a poor site for ribosomal binding.[19] The site of the starting point relative to the ORF of the N protein and the frequency of leucine in the putative protein are characteristics also seen in the I protein of BCV[20] and the ORFs of MHV A59 and HCV 229E. The open reading frames in HCV 229E and PEDV are of a similar length, which is considerably shorter than that of MHV A59 or BCV.

The PEDV RNA also shows features typical of coronaviruses. The sequence 5' UCU-AAAC 3', similar to the intergenic sequences of other coronaviruses, was found 16 bases upstream of the AUG of the N protein ORF. The intergenic sequence is thought to be the site of leader sequence addition during mRNA synthesis.[21] An eleven-nucleotide sequence, homologous to that found in other coronaviruses, was found near the 3' end of the PEDV genome; this sequence has been proposed to be a polymerase recognition site for synthesis of the RNA negative strand during viral replication.[21] These sequences, which are illustrated in Table 1, fall into two groups according to the base (G or U) before the completely conserved 5' GGAAGAGC 3' core sequence and according to the distance of this sequence from the viral poly(A) tail. Again, the PEDV sequence shows the highest similarity to the TGEV, FIPV and HCV 229E viruses.

Table 1. The 3' conserved sequences of eight coronaviruses

Virus	Sequence	Inclusive position from poly(A) tail (bases)
PEDV	UGGAAGAGCGU	74
HCV 229E	UGGAAGAGCCA	75
TGEV	UGGAAGAGCUA	76
FIPV	UGGAAGAGCUA	76
BCV	GGGAAGAGCCA	79
HCV OC43	GGGAAGAGCCA	79
IBV	GGGAAGAGCUA	81
MHV	GGGAAGAGCUC	82

modified from Schreiber et al.[12]

Thus, many features of the PEDV N protein are similar to those of other coronaviruses. One difference is the length of the protein compared with the related TGEV and HCV 229E proteins. This difference stems largely from additional amino acid residues, particularly asparagines, in the central region of the protein. The asparagine content of the protein is very high (10.9%), with a stretch of six adjacent asparagine residues occurring at one point. Although unusual, such amino acid runs do occur in nature; for example, the *Drosophila* mastermind gene contains numerous runs of asparagine, glutamine and glycine.[22] In this region of the PEDV N protein there is also an interesting motif comprising an arginine at every sixth residue, repeated five times. TGEV, but not HCV 229E, possesses a shorter such motif. Its significance is not known, but a computer search of the SWISS-PROT data base using the FASTA programme revealed that the motif is also present in numerous nucleic acid binding proteins including the Epstein Barr virus EBNA 1 and 2 nucleoproteins, yeast SNF2 nuclear protein and the Herpes Simplex Virus U_S11 protein. The last is a small, abundant, basically charged regulatory protein which binds host 60S ribosomal subunits.[23] These possible homologies are interesting in that the N protein of PEDV has been observed in the nuclei of virus-infected cells (Rosskopf, unpublished data), where it could perform a regulatory function.

In conclusion, the presented sequence data for PEDV confirms that this virus is a coronavirus since it is a polyadenylated RNA virus possessing RNA motifs distinctive to coronaviruses and a gene with high homology to the coronavirus N gene. The greatest homology was observed between the genomes of PEDV and the HCV 229E and TGEV viruses. Of these latter two viruses, the gene order of HCV 229E is more reminiscent of that of PEDV than is that of TGEV, since PEDV possesses no gene 3' to the N gene, but does possess a second ORF within the N gene with the capacity to code for a leucine rich protein. The PEDV sequence is also more homologous to that of HCV 229E at the nucleotide level. Thus, PEDV is a coronavirus with the greatest similarity to coronavirus HCV 229E.

ACKNOWLEDGEMENTS

This work was supported by Grant No 012.91.7 of the Swiss Federal Veterinary Services. We also wish to thank Dr M. Schwyzer for advice during the course of this work.

REFERENCES

1. Z. Yaling, J. Ederveen, H. Egberink, M. Pensaert, and M.C. Horzinek, *Arch. Virol.* 102:63 (1988).
2. M.B. Pensaert and P. Debouck, *Arch. Virol.* 58:243 (1978).
3. M. Hofmann and R. Wyler, *J. Clin. Microbiol.* 26:2235 (1988).
4. M. Knuchel, M. Ackermann, H.K. Müller, and U. Kihm, *Vet. Microbiol.* 32:117 (1992).
5. R.E. Kingston, "Current Protocols in Molecular Biology," section 4.1.2, Greene Publishing Associates, Brooklyn (1991).
6. U.V. Wirth, B. Vogt, and M. Schwyzer, *J. Virology* 65:195 (1991).
7. A. Bridgen, K. Tobler, and M. Ackermann, *Adv. Exp. Med. Biol.*, (this volume).
8. M.A. Frohman, M.K. Dush, and G.R. Martin, *Proc. Natl. Acad. Sci.* 85:8998 (1988).
9. T. Maniatis, E.F. Fritsch, and J. Sambrook. "Molecular Cloning: A Laboratory Manual (2nd edition)," Cold Spring Harbour Laboratory, Cold Spring Harbor NY (1989).
10. J. Devereux, P. Haeberli, and O. Smithies, *Nucl. Acids Res.* 12:387 (1984).
11. H.F.E. Egberink, J. Ederveen, P. Callebaut, and M.C. Horzinek, *Am. J. Vet. Res.* 49:1320 (1988)
12. S.S. Schreiber, T. Kamahora, and M.C. Lai, *Virology* 169:142 (1989).
13. P.A. Kapke and D.A. Brian, *Virology* 151:41 (1986).
14. H. Vennema, R.J. De Groot, D.A. Harbour, M.C. Horzinek, and W.J.M. Spaan, *Virology* 181:327 (1991).
15. M.M. Parker and P.S. Master, *Virology* 179:463 (1990).
16. T. Kamahora, L.H. Soe, and M.C. Lai, *Virus Research* 12:1 (1989).
17. W. Lapps, B.G. Hogue, and D.A. Brian, *Virology* 157:47 (1987).
18. M.E.G. Boursnell, M.M. Binns, I.J. Fould, and T.D.K. Brown, *J. Gen. Virol.* 66:573 (1985).
19. M. Kozak, *Cell* 44:283 (1986).
20. S.D. Senanayake, M.A. Hofmann, J.L. Maki, and D.A. Brian, *J. Virol.* 66:5277 (1992).
21. M.C. Lai, *Ann. Rev. Microbiol.* 44:303 (1990).
22. D. Smoller, C. Friedel, A. Schmid, D. Bettler, L. Lam, and D. Yedvobnick, *Genes Dev.* 4:1688 (1990).
23. R.J. Roller and B. Roizman, *J. Virol.* 66:3624 (1992).

GENOME ORGANIZATION OF PORCINE EPIDEMIC DIARRHOEA VIRUS

M. Duarte, J. Gelfi, P. Lambert, D. Rasschaert and H. Laude

INRA, Unité de Virologie et Immunologie Moléculaires
Jouy en Josas, France

ABSTRACT

In order to study the organization of the genome of porcine epidemic diarrhoea virus (PEDV), we constructed a cDNA library in a phage expression vector by using poly(A) RNA from PEDV-infected Vero cells. An anti-PEDV hyperimmune serum was used to probe the library. The first isolated clone mapped within the N gene and was subsequently used for rescreening the library. The selected clones allowed us to establish the sequence of the 3'-most 7.4 kb of the PEDV genome. Analysis of the cDNA sequences revealed a 3'-coterminal nested structure, which is typical of *Coronaviridae* and the presence of a hexameric sequence XUA(A/G)AC upstream of each coding region. The amino acid sequences deduced from four of the five ORFs identified showed the characteristic features of the structural proteins S, M, sM and N. Only one ORF located between the S and M genes was found to potentially encode a non-structural polypeptide. Our data lead us to conclude that PEDV is a member of *Coronaviridae* and belongs to the same genetic subset as TGEV, FIPV and HCV 229E .

INTRODUCTION

PEDV is a coronavirus responsible for a gastroenteritis similar to that caused by TGEV in pigs[1,2,3]. The molecular characterization of PEDV has been impaired due to the lack of a convenient cell system for virus propagation *in vitro*, a problem which has been solved only in 1988 ([4] and references therein). PEDV was as yet classified as a probable member of *Coronaviridae* on the basis of morphological and structural features[1,5]. Moreover, a weak serological crossreactivity between PEDV CV777 strain and FIPV nucleocapsid proteins has been reported[6]. Despite their similar tissue and species specificity, PEDV and TGEV are distinct viral entities. Thus, relevant information is anticipated from the comparison of their genetic make up. The aim of this paper is to report the present state of our knowledge about the genome of PEDV.

METHODS

Virus and cells : The British isolate 1/87 (S. Cartwright, Central Veterinary Laboratory, Weybridge), adapted to cell culture by P. Have (State Veterinary Institute for Virus

Coronaviruses, Edited by H. Laude and J.F. Vautherot
Plenum Press, New York, 1994

Research, Lindholm) was used as a virus source. The virus stock used for cDNA cloning was prepared by propagation in Vero cells, as described[4], without prior plaque purification.

cDNA synthesis and cloning : Total RNA from PEDV-infected cells was extracted by the guanidium thyocianate method and purified by ultracentrifugation through a 5.7M CsCl gradient.The lack of appropriate PEDV-specific probes, as well as the low amount of PEDV genomic RNA, available prompted us to construct a cDNA expression library into the lambda Zap II expression vector (Stratagene) according to the instruction of the manufacturer. The library was screened with hyperimmune anti-PEDV serum obtained from Station de Pathologie Porcine (CNEVA), Ploufragan. The serum sample used was incubated with an *Escherichia coli* lysate in order to reduce non-specific binding. Immunoscreening of the expression library allowed us to isolate the clone named pBE1. Nucleotide (nt) analysis of this clone established that it contained N gene-related sequences, thus indicating that it mapped in the 3' end region of the PEDV genome. The next clones were selected by successive screening of the library with a labeled oligonucleotide or a restriction fragment complementary to the 5' end of the last clone isolated. A few clones were derived by cloning RT/PCR products, using amplimers derived from the established sequence. Standard recombinant DNA procedures were used[7].

DNA sequencing and sequence analysis A series of overlapping clones from pBE1 to pBE13 were sequenced on both strands. Sequencing was done on ss or ds DNA templates using the dideoxy chain termination method with the m13 universal primer or gene-specific oligonucleotides primers. The sequence of clones pBE6 to pBE13 was performed on an automated sequencer (Applied Biosystems). Sequence data were analyzed using the WUGCG software.

RESULTS and DISCUSSION

A map of the PEDV-specific clones selected from the lambda phage library constructed using poly(A) RNA from virus-infected cells is shown in Fig.1. Sequencing and analysis of these clones and of a few additional clones derived from RT/PCR fragments, allowed us to assemble a continuous, 7.4 kb-long nt stretch, the sequence of which was determined on at least two independent clones. The sequence was deduced as corresponding to the 3' end region of the PEDV genome. A poly(A) tail was present at the end of each cDNA clone, consistent with the oligo dT-priming of the library. The clones pBE3, 8 and 13 are likely to be derived from subgenomic mRNA species, since they showed short, identical sequences at their 5' end, assumed to correspond to part of the leader sequence. The clones pBE7 and pBE11 to 13 contained a large deletion (1500 to 3500 nt). They could be derived from viral defective RNA molecules generated through repeated passage of the virus in cell cultures. However, a cloning artifact cannot be formally excluded.

Translation of the 7.4 kb revealed 5 main ORFs (Table 1). The ORFs encoding the S, sM, M and N structural proteins were identified according to their order, size, and to the characteristics of the predicted polypeptides. The calculated mw values were consistent with the apparent mw of PEDV-specific polypeptides resolved by gel electrophoresis in our laboratory (data not shown). Each of these ORFs was preceeded by a hexameric motif XUA(A/G)AC, which potentially corresponds to the conserved intergenic sequence typical of the coronavirus genome. As it can be noted (Table 1), these motifs exhibit a lesser degree of conservation than previously shown in other coronaviruses, a finding in agreement with the relative flexibility of the consensus sequence reported recently[8].

No large ORF was found in the 336 nt region located downstream from the N gene, which is in contrast with several other coronaviruses including TGEV, FIPV and CCV[9,10,11]. In this respect, PEDV appears to be more closely related to HCV 229E[12]. However, two additional conserved motifs are present downstream of the N gene : UUAAAC and CUAAAC at the respective positions 338 nt and 120 nt from the 3' end. The former is immediately followed by a small ORF having the coding capacity for a 35 amino acid-long polypeptide, which is markedly hydrophobic, similar to the amino end of the TGEV ORF7 product. Whether or not the above sequences may act for initiation of transcription remains to be seen. A 11 nt sequence UGGAAGAGCGU is present 74 nt

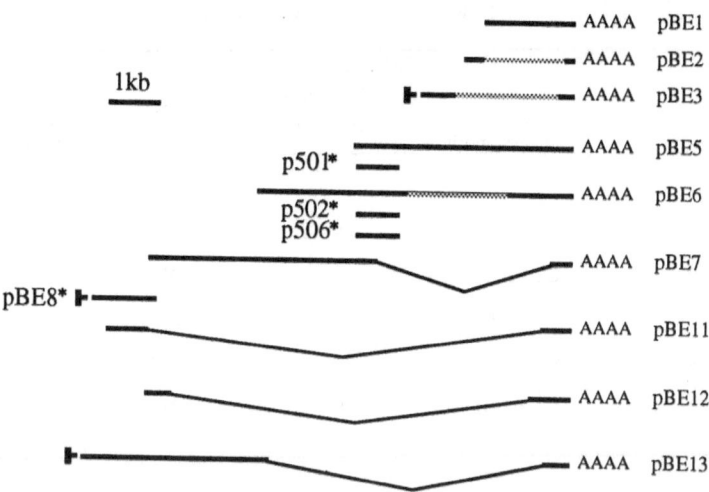

Figure 1. Map of the cDNA clones used for determination of the PEDV genome sequence. Sequenced regions : — ; observed deletions : ∨ ; clones obtained by RT/PCR : * ; partial PEDV leader sequence: ⊢

upstream from the 3' end. It is homologous to the highly conserved sequence (U/G)GGAAGAGC(C/U)(A/U) already identified in 6 other coronaviruses[12], and assumed to act as a signal for the synthesis of the minus strand RNA. However, the PEDV sequence differs from the consensus sequence by the last 2 nt . The first nt is an U, as for TGEV and HCV 229E.

The whole data lead us to propose a model for the organization of the 3' region of PEDV genome such as presented in Fig.2. The most salient feature is that a single ORF, tentatively designated ORF3, was predicted between the S and sM gene, whereas 2 or 3 ORFs have been identified in the corresponding genome region of all coronavirus sequenced so far[13]. In addition, sequence data revealed a divergence between the cDNA clones encompassing ORF3 which was confirmed by sequencing of cloned RT/PCR fragments covering the implicated region (p501, p502, p506 ; * in Fig.1). A 4 nt deletion was present in pBE5 and p501 clones, whereas a 7 nt deletion was present in pBE6, p502 and p506 clones. These deletions were mutually exclusive and distant by 87 nt. The frameshifting 4 nt deletion generated an interrupted 143 codon-long ORF. When aligning the putative polypeptide with that of ORF4a from HCV 229E[14], also located just downstream of S gene, a substantial identity (34%) was found. In addition, translation of pBE5 clone revealed that an additional ORF devoid of starting ATG was present downstream. Its potential product showed 37% identity with that of ORF4b of HCV 229E. This finding lead us to hypothesize that the authentic ORF3 gene encoded by PEDV might in fact correspond to a 672 nt long sequence such as generated by removing both of the deletions observed in either pBE5 or pBE6. As illustrated in Fig.2, the putative ORF3 would encode a 224 aa-long polypeptide homologous to the product of the fused 4a and 4b genes of HCV 229E (35% overall identity) and to the product of the whole gene 3b of TGEV[15] (30% identity). Further investigations will be needed to ascertain whether the "full-length" ORF3 is dispensable for virus replication. In any case, the apparent polymorphism associated to PEDV ORF3 is reminiscent of what has been observed with several coronavirus ORFs also believed to encode a non-structural polypeptide. In particular, alterations in the sequence of TGEV ORF3b and HCV 229E ORF4a and 4b have been reported [9,16].

As expected, the four predicted structural proteins exhibited characteristics very similar to those of other coronaviruses. The only striking feature was the presence of a putative insertion of a 45 amino acids stretch, unusually rich in Asn residues and located towards the middle of the N gene sequence (not shown). As a consequence, the size of the

Table 1. Characterization of PEDV open reading frames (ORFs).

Gene	Upstream hexameric motif	Distance between the hexameric motif and AUG codon (nt)	ORF size (nt)	Predicted mw of the product (kD)
S	GTAAAC	2	4149	151
sM	CTAGAC	4	228	8,8
M	ATAAAC	5	678	25
N	CTAAAC	10	1323	48,8
ORF3	CTAGAC	40	672	25

deduced N protein is notably larger than those of other coronavirus N proteins sequenced to date. Pairwise alignment of the structural protein amino acid sequences revealed that PEDV is more closely related to TGEV, FIPV and HCV 229E than to MHV and IBV, whichever the protein considered (Table 2). Furthermore, the PEDV S protein sequence does not contain any basic residues related to the motif found towards the middle of the cleaved coronavirus S proteins. This was consistent with our experimental data since no species with Mr 100K in addition to the 200K species could be detected by denaturing PAGE analysis of PEDV polypeptides (not shown). This observation, together with the prediction of one Asn-linked glycosylation site in the N-terminal domain of the M protein, is in support of a higher genetic relatedness of PEDV to the TGEV cluster. It also appeared from our data that, within this cluster, PEDV is closer to the respiratory human virus HCV 229E than to the porcine enteric TGEV virus. This is supported not only by the alignment data (particularly that of sM proteins ; Table 2) but also by the following features : i) no predicted signal sequence upstream of the M protein; ii) no identified ORF downstream of the N gene, such as the ORF7 of TGEV. It should be noted, however, that the degree of divergence between PEDV S and HCV or TGEV S, respectively, are not significantly different. Furthermore, a multialignment of these three sequences (not shown) revealed that both TGEV and PEDV S proteins possess an extra N-terminal domain compared to HCV 229E S protein. This can be paralleled with the large deletion found within the N-terminal domain of the S protein of PRCV, the respiratory variant of TGEV[17], and raises the intriguing possibility that an extended N-terminal domain might be one of the determinants for the expression of the enteric tropism.

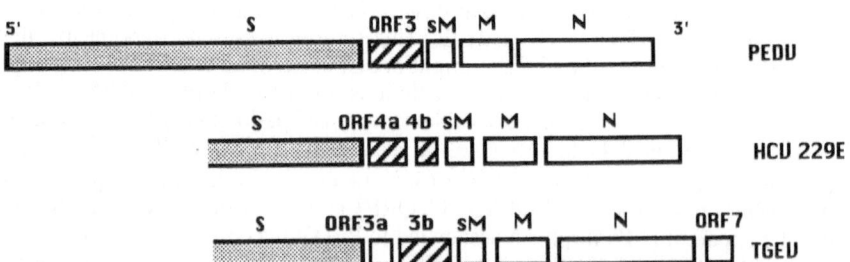

Figure 2. Compared organization of the genome 3' end of PEDV, HCV 229E and TGEV. Structural proteins : S (spike), sM (small membrane), M (membrane) and N (nucleocapsid).

Table 2. Pairwise sequence identity (%) between PEDV and HCV 299E, TGEV, FIPV, MHV and IBV.

PEDV proteins	Pairwise identity (%)				
	HCV 229E	TGEV	FIPV	MHV	IBV
S : S1 region	37	33	35	19	25
S2 region	61	60	60	36	38
sM	54	29	-	26	17
M	57	53	50	41	26
N	43	39	36	29	28

In conclusion, our whole data allowed to establish the taxonomic status of PEDV as a member of *Coronaviridae*, thus confirming earlier proposals based on structural and antigenic analyses. From the information presently available about their genome organization, it appears that the members of the coronavirus genus, except the avian virus IBV, fall into two main subgroups, tentatively designated genetic subsets, the distinctive features of which are listed in Fig.3. This view is supported by evolutionary trees constructed from the

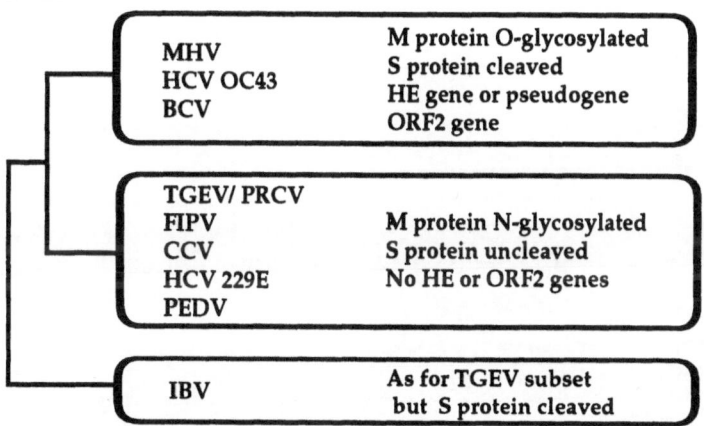

Fig 3. Evidence for two main genetic subsets within the coronavirus genus.

sequences of N and M proteins by standard phylogenetic methods (data not shown). The advantage of the notion of genetic subset over that of antigenic group introduced earlier is to acknowledge the existence of common genetic traits between viruses such as TGEV, HCV 229E and PEDV which, however, exhibit little or no antigenic relationship.

ACKNOWLEDGEMENTS

We express our thanks to Per Have and André Jestin for kindly providing the PEDV isolate and the anti-PEDV serum, respectively. The expert assistance of Nathalie Woloszyn in automated sequencing is also acknowledged.

REFERENCES

1. M.B. Pensaert, P. De Bouck. *Arch. Virol.* 58: 243-247 (1978).
2. P. De Bouck and M. Pensaert. *Am. J. Vet. Res.* 41 (n°2): 219-223 (1980).
3. M.B. Pensaert, P. De Bouck and D.J. Reynolds. *Arch. Virol.* 68: 45-52 (1981).
4. M. Hofmann and R. Wyler. *J. Clin. Microbiol.* 26: 2235-2239 (1988).
5. H.F. Egberink, J. Ederveen, P. Callebaut and M.C. Horzinek. *Am. J. Vet. Res.* 49: 1320-1324 (1988).
6. Z.Yaling, J. Edenween, H.Egberink, M. Pensaert and M. Horzinek. *Arch. Virol.* 102: 63-71 (1988).
7. J .Sambrook, E.F. Fritsch and T.Maniatis. 2nd edn. Cold Spring Harbor Laboratory, New York, (1989).
8. M. Joo and S. Makino. *J. Virol.* 66: 6330-6337 (1992).
9. D. Rasschaert, J. Gelfi and H. Laude. *Biochimie* 69: 591-600 (1987).
10. R.J. De Groot, A.C. Andeweg, M.C. Horzinek and W.J.M. Spaan. *Virology* 167: 370-376 (1988).
11. B.C. Horsburgh, I. Brierley and T.D.K. *J. gen. Virol.* 73: 2849-2862 (1992).
12. S.S. Schreiber, T. Kamahora and M.M.C. Lai. *Virology* 169: 142-151 (1989).
13. M.M.C.Lai. *Annu. Rev. Microbiol.* 44: 303-33 (1990).
14. T.Raabe and S. Siddell. *Nucl. Acids Res.* 17: 6387.(1989)
15. P.A. Kapke, F.Y.T. Tung and D.A. Brian. *Virus genes* 2: 3, 293-294 (1988).
16. P. Jouvenne, S. Mounir, J. Steward, D. Richardson, and P. Talbot. *Virus Res.* 22: 125-141 (1992).
17. D. Rasschaert, M. Duarte and H. Laude. *J. gen. Virol.* 71: 2599-2607 (1990).

CHARACTERIZATION OF THE NONSTRUCTURAL AND SPIKE PROTEINS OF THE HUMAN RESPIRATORY CORONAVIRUS OC43: COMPARISON WITH BOVINE ENTERIC CORONAVIRUS

Samir Mounir, Patrick Labonté and Pierre J. Talbot

Centre de Recherche en Virologie
Institut Armand-Frappier, Université du Québec
Laval, Québec, Canada H7N 4Z3

ABSTRACT

The nucleotide sequence of the region between the spike (S) and the membrane (M) protein genes, and sequences of the S and ns2 genes of the OC43 strain of human coronavirus (HCV-OC43) were determined. The ns2 gene comprises an open reading frame (ORF) encoding a putative nonstructural (ns) protein of 279 amino acids with a predicted molecular mass of 32-kDa. The S gene comprises an ORF encoding a protein of 1353 amino acid residues, with a predicted molecular weight of 149,918. Sequence comparison between HCV-OC43 and the antigenically related bovine coronavirus (BCV) revealed more sequence divergence in the putative bulbous part of the S protein (S1) than in the stem region (S2). The cysteine residues near the transmembrane domain and the internal predicted protease cleavage site are conserved in the HCV-OC43 S protein. Nucleotide sequence analysis of the region between the S and M gene loci revealed the presence of an unexpected intragenomic partial leader sequence and two ORFs encoding potential proteins of 12.9 and 9.5-kDa. These two proteins were identified as nonstructural by comparison with the homologous BCV genes. *In vitro* translation analyses demonstrated that the HCV-OC43 9.5-kDa protein, like its BCV counterpart, is poorly translated when situated downstream of the 12.9-kDa ORF, but is expressed in infected cells, as shown by immunofluorescence. Interestingly, two ORFs, potentially encoding 4.9 and 4.8-kDa ns proteins in BCV are absent in HCV-OC43, indicating that they are not essential for viral replication in HRT-18 cells.

INTRODUCTION

Coronaviruses are enveloped RNA viruses, isolated from a range of animal hosts (for review, see 1). They cause respiratory, gastrointestinal and neurological disorders in human and domestic animals[2,3]. Coronavirus genomic RNA directs the synthesis of a nested set of 6 to 8 subgenomic mRNAs by a leader-primed mechanism of transcription[4]. Besides the genes for known structural proteins, coronavirus genomes contain several open reading frames (ORFs) that could encode nonstructural (ns) proteins or novel structural proteins. The number and position of the genes encoding these proteins differ between coronavirus species.

Human coronavirus OC43 (HCV-OC43) comprises four major structural proteins: a 190-kDa peplomer (S) glycoprotein (cleaved into two subunits of 120 and 100-kDa), a

130-kDa hemagglutinin-esterase (HE) glycoprotein, a 55-kDa nucleocapsid (N) phospho-protein, and a 26-kDa membrane (M) glycoprotein[5,6,7]. The amino acid sequences of structural proteins HE, M and N, as well as the leader sequence of HCV-OC43 have recently been determined from cloned cDNAs[7,8,9]. Within the genome, the structural proteins are arranged in the order 5'-HE-S-M-N-3'[7,8]. Dispersed throughout the genome are genes that encode the RNA-dependent RNA polymerase and nonstructural proteins.

Here we report the nucleotide sequence of the region between the spike (S) and the membrane (M) protein genes and sequences of the S and ns2 genes of HCV-OC43, as well as Northern blot and *in vitro* translation analyses.

MATERIALS AND METHODS

Virus Growth, Isolation and Analysis of Nucleic Acids

The origin and cultivation of the HRT-18 human rectal tumor cell line and the OC43 strain of HCV, as well as the preparation, reverse transcription, polymerase chain reaction (PCR) amplification, cloning and sequencing of viral RNA (both mRNA and genomic RNA) were performed as described elsewhere[8]. For Northern blot analysis, poly (A)-containing RNA was selected with the PolyATract® mRNA isolation system (Promega, Fisher Scientific, Montréal, Québec, Canada) according to the manufacturer's instructions, was size fractionated by electrophoresis in 1% (w/v) agarose gels containing 5.3% formaldehyde, and was transferred onto Hybond-C™ extra (Amersham Canada Ltd., Oakville, Ontario, Canada) nitrocellulose filters[10]. Blots were hybridized with the random-primed ^{32}P-labeled (ICN Biomedicals Canada Ltd., Mississauga, Ontario, Canada) DNA probes at 42°C as described previously[11].

Immunofluorescence

Immunofluorescence was performed 24 h after infection of HRT-18 cells grown on coverslips with HCV-OC43 at an MOI of 0.2. Cells were examined for fluorescence with a specific polyclonal rabbit antibody prepared against the MHV-A59 9.6-kDa protein that had been expressed in *Escherichia coli* [12], or with a polyclonal rabbit antiserum against HCV-OC43.

RESULTS AND DISCUSSION

Nucleotide Sequence Analysis

The nucleotide sequence of the region between the S and M genes of HCV-OC43 contains potential open reading frames for two proteins of 12.9-kDa and 9.5-kDa (data not shown).

The 12.9-kDa protein would contain 109 residues and show an amino acid sequence identity of 96.3% with the putative 12.7-kDa protein of BCV[13] (Fig. 1). One potential N-linked glycosylation site is found at amino acid position 18, like in BCV[13]. The putative initiation codon for this protein is in a context not frequently used for initiation of protein synthesis as described by Kozak[14]. Surprisingly, a 47-nucleotide stretch that has resemblance with the 3'-half of the 82-nucleotide HCV-OC43 leader sequence[9] is found upstream of this ORF (Fig. 2). This situation is unexpected since part of the leader sequence (47 of 82 nucleotides) is found within the genomic RNA instead of being at the 5'-end of the transcript. This may have occured by a mechanism of recombination aimed at the conservation of the ORF encoding the 12.9-kDa protein, despite the loss of the 4.9- and 4.8-kDa putative ns proteins. This partial intragenomic leader sequence is preceded by a potential 11-residue ORF that could encode a 1.33-kDa polypeptide composed of 9 residues identical with the N-terminus of the BCV 4.9-kDa putative ns protein.

The 9.5-kDa ORF predicts an 84-amino acid protein that would have an amino acid sequence identity of 96.4% with the 9.5-kDa protein of BCV[13] (Fig.1). As in BCV, there is a conserved methionine residue at the putative third codon of the protein. The first putative initiation codon is not in a favorable context for initiation of protein synthesis, whereas the presence of a G residue at position +4 of the second putative initiation codon would presumably ameliorate the situation[14]. The 9.5-kDa protein contains one large hydrophobic domain that comprises more than 50% of the molecule, which suggests a transmembrane insertion (data not shown).

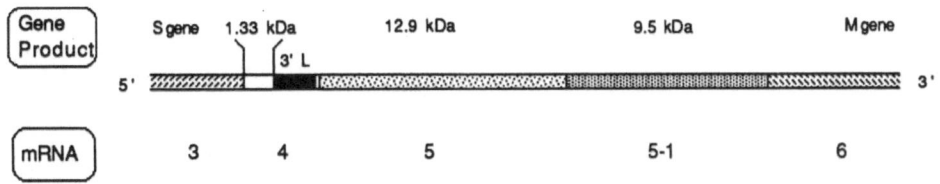

Figure 1. Organization of HCV-OC43 genome, location of ORFs and identity with BCV.
Δ: Absence of two ORFs in HCV-OC43.

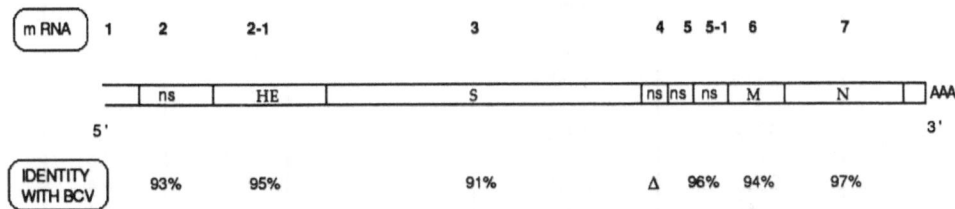

Figure 2. 3'-Leader sequence within HCV-OC43 genomic RNA.

Along with the four structural proteins and the three ORFs encoding potential non-structural proteins, we found a putative gene situated immediately upstream from the HE gene, which contains an ORF that could encode a protein of 279 residues, with a predicted molecular weight of 32,277. This protein would have an amino acid sequence identity of 93% with the BCV ns2 protein[15] (Fig. 1, Table 1). Nine consecutive amino acid residues of the ns2 HCV-OC43 C-terminus are different from those of BCV ns2 protein (data not shown).

Finally, the complete sequence of the S gene contains 4062 bases, encoding a protein of 1353 amino acid residues, with a predicted molecular weight of 149,918 (data not shown). This protein has an amino acid sequence identity of 91% with the BCV S protein (Fig. 1, Table 1), revealing more sequence divergence in the putative bulbous part of the S protein (N-terminus, S1) than in the S2 stalk region. The putative S ORF is preceded immediately upstream by the consensus intergenic sequence (UCUAAAC) thought to play a role in leader priming of coronavirus transcription. The deduced HCV-OC43 S protein reflects the properties of the other coronavirus spike proteins that have been characterized to date. There is an internal sequence of basic amino acid NRRSRG (KRRSRR for BCV) beginning with amino acid 753 which, on the basis of the pattern in BCV[16], predicts a cleavage between amino acids 758 and 759, giving an N-terminal segment of 84,640 Da (S1) and a C-terminal segment of 65,296 Da (S2).

Northern Blot Analysis, *In Vitro* Translation and Immunofluorescence

To identify the subgenomic RNAs that encode the structural and nonstructural proteins, we performed Northern blot analysis using DNA probes. Nine HCV-OC43-specific RNAs were detected (data not shown) and have been numbered 1, 2, 2-1, 3, 4, 5, 5-1, 6, 7 in order of decreasing sizes (Table 1). A 9.5-kDa probe hybridized to RNA 1 to 5-1, but not RNAs 6 and 7. A 12.9-kDa probe hybridized to RNA 1 to 5, but not RNAs 5-1, 6 and 7. A S2 probe (which extends into the putative ns4 ORF) hybridized to RNA 1 to 4 only (Fig. 3). Thus, it appears that RNAs 5 and 5-1 encode the nonstructural proteins 12.9-kDa and 9.5-kDa, respectively, a situation similar to that for the BCV homologues[13], and different from mouse hepatitis virus (MHV), where only one transcript (mRNA 5) is utilized for the synthesis of both the 13- and 9.6-kDa ns proteins[17]. Since the initiation

Table 1. HCV-OC43 mRNAs and gene product sizes

mRNA	Predicted Size of mRNAs (bases)[1]		Gene Product Designation	Gene Product Size (kDa)[2]	
	OC43	BCV		OC43	BCV
1					
2	9232	9544	ns2	32.3	32
2-1	8385	8700	HE	47.7	47.7
3	7086	7400	S	150	150.7
4	3350	3635	ns4	1.3	4.8 + 4.9
5	3032	3005	ns5	12.9	12.7
5-1	2753	2735	ns5-1	9.5	9.5
6	2345	2345	M	26.4	26.4
7	1649	1649	N	49.3	49.4

[1] Size of mRNAs predicted from the location of the intergenic nucleotide sequence.
[2] Size of polypeptides deduced from the ORF immediately downstream of the intergenic nucleotide sequence.

codon for the BCV 12.7-kDa ns protein is in a more favorable context for initiation of translation than the HCV-OC43 12.9-kDa ns protein[14] (G instead of U in position -3), we tested the translatability of both ORFs, either independently or in tandem. In BCV, *in vitro* translation experiments using a synthetic transcript containing both the 12.7- and 9.5-kDa ORFs demonstrated that the majority of the protein synthesized was the upstream 12.7-kDa protein[13]. However, when the same experiment was performed on a MHV RNA containing both ORFs (13- and 9.6-kDa), the downstream ORF (9.5-kDa) was preferentially synthesized[17]. We also observed that, like in BCV, the upstream 12.9-kDa protein was preferentially synthesized when both ORFs were present but that synthetic RNA containing only one of the two ORFs directed efficient translation of both proteins.

We also showed that the HCV-OC43 9.5-kDa protein is expressed during infection, using an antibody directed against the MHV 9.6-kDa protein[12], which specifically stained HCV-OC43-infected HRT-18 cells in an immunofluorescent assay (Fig. 4).

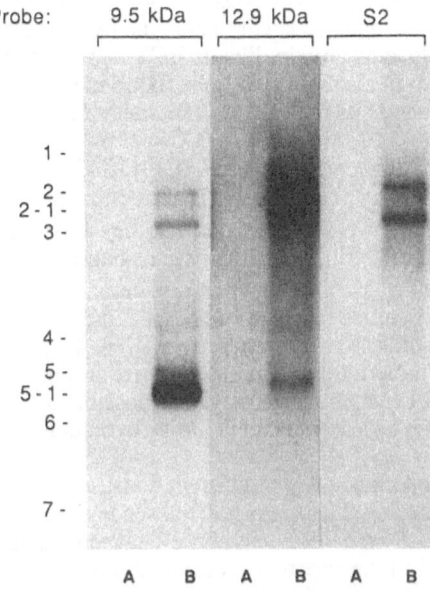

Figure 3. Northern blot analysis of RNA from uninfected cells (A) and infected cells (B). mRNAs were revealed with the probes indicated above the figure. Subgenomic mRNAs are indicated on the left.

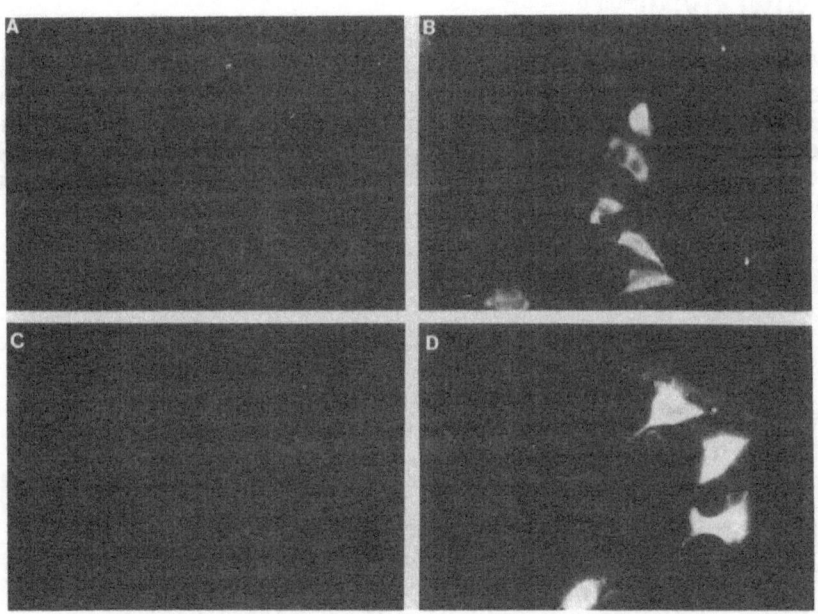

Figure 4. Immunodetection of the 9.5-kDa protein. Uninfected and HCV-OC43-infected cells revealed with an hyperimmune anti-OC43 serum (A and B, respectively), or with anti-9.5-kDa serum (C and D, respectively).

The functions of the putative 12.9-kDa and 9.5-kDa proteins in infected cells are not known. Recently, proteins analogous to the HCV-OC43 and BCV 9.5-kDa molecules were shown to be present in infectious bronchitis (IBV) and porcine transmissible gastroenteritis (TGEV) virions and were designated small membrane (SM) proteins[18,19]. We have shown that the HCV-OC43 9.5-kDa ns protein, like its BCV and MHV counterparts, is expressed in virus-infected cells. It will also most likely be found in virions. The conservation of the 12.9-kDa protein also suggests its importance in coronavirus biology.

We found a major difference between HCV-OC43 and BCV, represented by the absence of the genes potentially encoding two nonstructural proteins of 4.9- and 4.8-kDa in BCV[13]. Their absence in HCV-OC43 indicates that they are not essential for virus replication, at least in HRT-18 cells. Similarly, the ns2, ns4 and ns5a nonstructural proteins of MHV were shown not to be essential for virus replication in tissue culture[20,21]. It is interesting to note that the difference between mRNA 4 and 5 is only 320 bp, with one potential ORF of 11 amino acids. The presence of mRNA 4 in HCV-OC43-infected cells (Fig.3), in the absence, most likely, of an associated protein product suggests that this mutation may be a recent event. Alternatively, it is possible that this mRNA encodes nonessential gene products that have been deleted during evolution.

The fundamental importance of the different nonstructural proteins encoded by coronaviruses remains to be explored. However, the absence of the two putative nonstructural proteins in HCV-OC43 (4.8- and 4.9-kDa), as well as amino acid differences within the S protein could be involved in the apparent preferential respiratory tropism of HCV-OC43, which contrasts with the presumed preferential enterotropism of BCV. Indeed, the S proteins of MHV and TGEV were suggested to be important in tissue tropism[22,23]. However, replication of BCV in the respiratory tract has been reported[7] and OC43-like human enteric coronaviruses have been isolated[24]. Thus, it is premature to conclude on the molecular basis for a differential tropism of these two viruses.

ACKNOWLEDGMENTS

This work was supported by grant MT-9203 from the Medical Research Council of Canada to P.J.T., who also gratefully acknowledges salary support in the form of a University Research Scholarship from the National Sciences and Engineering Research Council of Canada (NSERC). We thank Dr. J. L. Leibowitz (University of Texas Health Science Center, Houston, Texas) for his generous gift of a polyclonal rabbit antiserum to the MHV 9.6-kDa protein.

REFERENCES

1. W. Spaan, D. Cavanagh, and M.C. Horzinek. *In*: "Immunochemistry of viruses, II. The basis for serodiagnosis and vaccines". M.H.V. van Regenmortel and A.R. Neurath, eds., p. 359, Elsevier, New York (1990).
2. K.V. Holmes. *In*: "Virology, 2nd edn.", B.N. Fields, D.M. Knipe *et al.*, eds., p. 841, Raven Press, New York (1990).
3. H. Wege, S. Siddell, and V. ter Meulen. *Curr. Top. Microbiol. Immunol.* 99: 165 (1982).
4. M.M.C. Lai. *Annu. Rev. Microbiol.* 44: 303 (1990).
5. B.G. Hogue, B. King, and D.A. Brian. *J. Virol.* 51: 384 (1984).
6. B.G. Hogue, and D.A. Brian. *Virus Res.* 5: 131 (1986).
7. X.M. Zhang, K.G. Kousoulas, and J. Storz. *Virology* 186: 318 (1992).
8. S. Mounir, and P.J. Talbot. *J. Gen. Virol.* 73: 2731 (1992).
9. T. Kamahora, L.H. Soe, and M.M.C. Lai. *Virus Res.* 12: 1 (1989).
10. E. Southern. *Methods Enzymol.* 69: 152 (1979).
11. M. Cogné, S. Mounir, J.L. Preud'homme, F. Nau, and P. Guglielmi. *Eur. J. Immunol.* 18: 1485 (1988).
12. J.L. Leibowitz, S. Perlman, G. Weinstock, J.R. DeVries, C. Budzilowicz, J.M. Weissemann, and S.R. Weiss. *Virology* 164: 156 (1988).

13. S. Abraham, T.E. Kienzle, W.E. Lapps, and D.A. Brian. *Virology* 177: 488 (1990).
14. M. Kozak. *J. Cell Biol.* 108: 229 (1989).
15. G.J. Cox, M.D. Parker, and L.A. Babiuk. *Nucl. Acids Res.* 17: 5847 (1989).
16. S. Abraham, T.E. Kienzle, W. Lapps, and D.A. Brian. *Virology* 176: 296 (1990).
17. C.J. Budzilowicz, and S.R. Weiss. *Virology* 157: 509 (1987).
18. D.X. Liu, and S.C. Inglis. *Virology* 185: 911 (1991).
19. M. Godet, R. L'Haridon, J.-F Vautherot, and H. Laude. *Virology* 188: 666 (1992).
20. B. Schwarz, E. Routledge, and S.G. Siddell. *J. Virol.* 64: 4784 (1990).
21. K. Yokomori, and M.M.C. Lai. *J. Virol.* 65: 5605 (1991).
22. D. Rasschaert, M. Duarte, and H. Laude. *J. Gen. Virol.* 71: 2599 (1990).
23. T.M. Gallagher, S.E. Parker, and M.J. Buchmeier. *J. Virol.* 64: 731 (1990).
24. G. Gerna, N. Passarani, M. Battaglia, and E.G. Rondanelli. *J. Inf. Dis.* 151: 796 (1985).

IDENTIFICATION, EXPRESSION IN *E. COLI* AND INSECT CELLS OF THE NON-STRUCTURAL PROTEIN NS2 ENCODED BY mRNA2 OF BOVINE CORONAVIRUS (BCV)

P. Boireau[1], M.F. Madelaine[2], D. Saulnier[1],
J. Laporte[2] and J.F. Vautherot[2]

[1]CNEVA-Laboratoire Central de Recherches Vétérinaires, 22 rue P. Curie
94703 Maisons-Alfort cedex, France and [2]INRA-Unité de Virologie et
Immunologie Moléculaires, C. R. J. J.-Domaine de Vilvert, 78352 Jouy en
Josas cedex, France

ABSTRACT

The coding part of mRNA 2 (ORF2) of BCV (F15 strain) was cloned and sequenced. The comparison of our sequence data with the sequence of the same ORF of BCV Quebec strain previously published revealed a major difference in the length of the C-terminal part of the NS2 protein . In vitro transcription and translation of ORF2 resulted in the synthesis of a single protein migrating with a Mr of 31 kDa. The ORF2 was fused in frame with the glutathione S transferase gene (GSH) in the pGEX vector. The fusion protein was synthezised as inclusion bodies which were concentrated and used to raise a monospecific antiserum. Alternatively the fusion protein was solubilized, purified by affinity chromatography and cleaved with Factor Xa to yield pure recombinant NS2. The ORF2 was also expressed in the baculovirus system and the recombinant proteins expressed in pro- and eukaryotic systems were compared on the basis of their size and immunoreactivity. Immunoprecipitation performed with the monospecific antiserum allowed us to identify NS2 in HRT18 infected cells, to follow its kinetic of synthesis, and to ascertain that NS2 was not incorporated in the virion as a minor structural component.

INTRODUCTION

Bovine coronavirus (BCV), an enteric coronavirus, causes acute enteritis in newborn calves and chronic infection in adult cattle. The genome of the BCV (F15 strain) is a positive stranded RNA of approximately 32 kilobases that encodes at least 4 structural proteins (1). The electrophoresis of uridine tritiated polyA RNA extracted from BCV infected HRT 18 cells followed by autoradiography revealed 8 different molecular mass species viral RNA(2, C. Crucières & J Laporte, personnal communication). Partial nucleotide sequence of BCV F15 strain has been determined and the sizes of BCV mRNA that were deduced from the location of the consensus junction sequence CC/TAAAC corresponded to the experimental data (3). The mRNA 3, 4, 7 and 8 have been shown to encode the four structural proteins: Hemagglutinin-esterase (HE), Spike (S), Membrane (M) and Nucleocapsid (N)(3, 4, 5, 6). The messengers RNA5 and 6 code for minor protein of 12 and 9.8kDa respectively (7). The messenger RNA2 of BCV Quebec strain has been shown to code for a phosphoprotein of 32 kDa (NS2)(8, 9), but the function of this protein remains to be determined. We reported here the cloning, sequencing and expression in *E. coli* and insect cells of the ORF2 of BCV F15. The use of recombinant protein expressed in bacteria allowed us to elicit specific polyclonal antibodies and to purify NS2 protein.

Coronaviruses, Edited by H. Laude and J.F. Vautherot
Plenum Press, New York, 1994

RESULTS

CLONING AND SEQUENCING OF ORF2 OF BCV F15

We obtained a cDNA library of BCV F15 using primer 532 as described (3, 4).We isolated and sequenced a clone encompassing the 5' proximal ORF of BCV mRNA2 (NS22) with two other overlapping cDNA clones (Fig 1). The 3' end of ORF2 had already been identified close to the start codon of the HE gene (3). The potential junction sequence ACTAAC of mRNA2 was found just upstream the first ATG codon which is in a strong context according to Kozack (10). These data indicated that BCV F15 genome contains an ORF homologous to the BCV Quebec ORF2 (8). Northern blot analysis of viral mRNA using NS22 and 4d67 as probes, suggested that ORF2 gene is transcribed by mRNA2. The nucleotide sequence of 4d67 and NS22 inserts upstream ORF2 shared 56% homology with the 3' end of the polymerase gene of MHV-A59 genome (11).

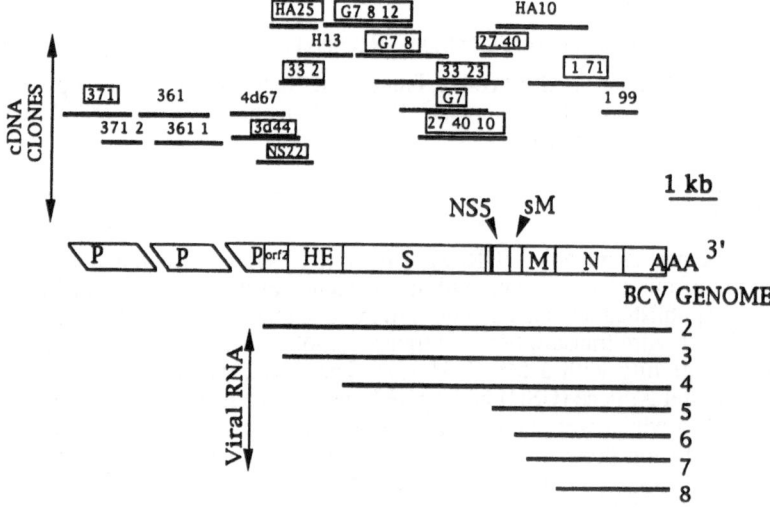

Figure 1. Cloning strategy for BCV F15 genome. Horizontal lines represent the relevant cDNA clones.The sequenced clones are boxed and the 7 subgenomic mRNA are shown under the BCV genome map.

EXPRESSION OF ORF2 GENE IN VITRO

The full length ORF2 was extracted from NS22 using *Pst*I and then submitted to a digestion with the Bal 31 nuclease (Stratagene) as described (13). The resulting insert was purified, cloned into pT3T7 plasmid (Pharmacia), thus yielding the p18U NS2 construct containing the ORF2 of BCV genome downstream the T7 promoter. We performed cell free transcription and translation of these transcripts in rabbit reticulocyte lysates (Promega). A unique 31kDa protein (Fig 2, lane5) was obtained, confirming the initiation at the first ATG codon.

EXPRESSION OF ORF2 GENE IN BACTERIA

We used two fusion protein vectors to express the full length ORF2 of BCV in *E. coli*. We first chose the pUEX vectors (14) to direct the synthesis of a *cro*-βgal-NS2 protein migrating with a Mr of 146 kDa.

The recombinant blotted protein (15, 16) was neither recognized by a serum of rabbit immunized with BCV purified virions nor by a serum of a convalescent young calf inoculated with BCV. As we failed to induce specific antibodies against NS2 protein with the cro-βgal-NS2 protein, we used the pGEX vector(17) which directs the synthesis of polypeptide in *E. coli* as fusion with the C-terminal part of glutathion S transferase. The resulting recombinant peptide can be purified by affinity chromatography on glutathion agarose beads. Factor Xa can cleave the GST from the recombinant fusion protein. The pGEX-ORF2 construct was transfected into *E. coli* strain NM522 where expression of the fusion protein was triggered by addition of 1mM IPTG in the cell culture. Following expression and cell lysis GST-NS2 fusion protein was examined by SDS-PAGE electrophoresis. The analysis revealed a major band at the expected Mr of 57 kDa which was not present in the control preparation. The fusion protein was synthesized to high level and accumulated in an insoluble form.

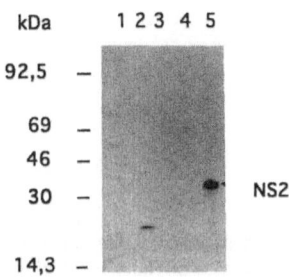

Figure 2. Cell free translation of mRNAs synthetized *in vitro* from p18UNS2 construct. Two ug of each of these mRNAs were used to proceed the translation in the presence of (35S) methionine and a rabbit reticulocyte lysate. The products were analysed on a 8-20% polyacrylamide gel .Lane1: molecular weight ; Lane 2: BMV m RNA ; Lane 3 : negatif control ; Lane 4 : internal control ; Lane 5 : NS2 mRNA .

SOLUBILISATION AND PURIFICATION OF NS2 EXPRESSED IN BACTERIA

Inclusion bodies containing GST-NS2 recombinant protein were washed with 2M urea in PBS in order to eliminate low Mr bacterial proteins. The GST-NS2 protein was then dissolved into 7.5M guanidine chloride (18) and dialysed against four buffer changes of progressively decreasing the molarity of urea in PBS. We obtained a soluble GST-NS2 (Fig 3) which was purified with glutathion agarose beads (SIGMA). The purified fusion protein was cleaved with the factor Xa into GST (26kDa) and NS2 (31kDa) (Fig. 3). The NS2 protein was easily recovered by centrifugation of the agarose beads.

OBTENTION OF SPECIFIC ANTI-NS2 POLYSERA

Mice were immunized by injecting intraperitoneally concentrated GST-NS2 emulsified in Freund's complete adjuvant. These antisera immunoprecipitated a 31 kDa protein from BCV F15 infected HRT 18 cells (Fig.4a) and recognized the purified recombinant protein by Western blot analysis (preimmune antisera failed to react with NS2). The NS2 protein expressed in Sf9 cells using a baculovirus vector was also immunoprecipitated, thus confirming the specificity of the anti-NS2 sera (Fig.4b). We were not able to immunoprecipitate NS2 protein from purified radiolabelled virions; this result together with the absence of reactivity of different anti-BCV polysera against NS2 indicated that this viral encoded protein is a nonstructural protein.

We determined the kinetics of synthesis of the NS2 protein in HRT 18 cells. NS2 appears as soon as 5h post infection in the same time as the other structural proteins. The synthesis of NS2 continued until the end of the viral cycle.

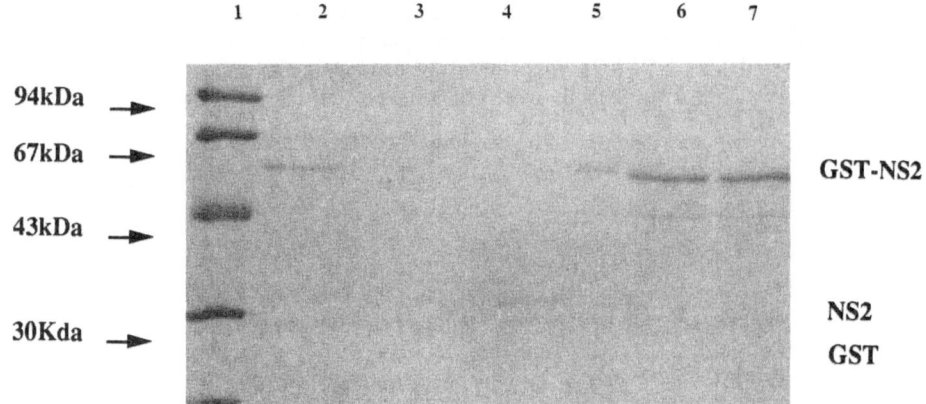

Figure 3. Cleavage of the GST-NS2 protein by Xa protease and purification of NS2. Lane 1: Molecular weight. Lane 2: GST-NS2 digested with Xa protease (1hour incubation at room temperature). Lanes 3 & 4: The GST-NS2 protein was purified by glutathion agarose beads after fixation of GST-NS2 and digestion with Xa protease.

DISCUSSION

These results enabled us to achieve the description of the genomic organisation of the BCV F15 strain, dowstream the polymerase gene. We can deduce the gene sequence of the BCV F15 genome from 5' to 3'end as: Pol-NS2-HE-S-NS5-sM-M-N. The study presented here confirmed that mRNA2 code for a protein of 30,5 kDa (NS2) (8). The predicted polypeptide of BCV F15 strain shares 97% homology with the corresponding protein of BCV Quebec strain (8). We could observe a difference in length at the C terminal part of the NS2 protein between the two BCV isolates, resulting from 3 additions (A 784 & 826, C 823) and 1 deletion (T 782, (8)) in BCV F15 strain. A striking feature is the existence of heterologous recombination in MHV A59 in the same part of the genome. A potential variability in this part of the BCV and MHV genomes with recombination, deletion/insertion, indicates another hot spot region in the coronavirus genomes.

No characteristics signatures was found in NS2 aa sequence except one potentiel glycosylation site and three potential phosphorylation sites. Comparison of our sequence data with those from the homologuous proteins of two MHV strains (19,20) did not reveal any known conserved motif.

We failed at detecting the BCV F15 mRNA2 product, a 31kDa Mr protein, as a structural protein embedded into virions.

The NS2 protein was synthesised in the same time as the other structural proteins (M, HE, S, N), but at a lower rate, possibly reflecting a lower concentration in mRNA2 in infected cells(2). The kinetics of synthesis of NS2 in infected HRT18 cell line appeared slightly different by comparison with the previously published kinetic of NS2 in infected MDBK cell line. The peak of NS2 synthesis occured quickly after the beginning of the viral protein synthesis (7-8 h p.i.). We observed a marked accumulation of NS2 protein 24h p.i.[as described (9)] but the increasing amount of NS2 reflected the occurence of a second viral cycle. So, as described for MHV (21), NS2 is expressed continuously during viral infection. We did not observe a NS2 processing similar to that of HE and S and the immunoreactivity of the recombinant protein expressed in insect cell was not modified.

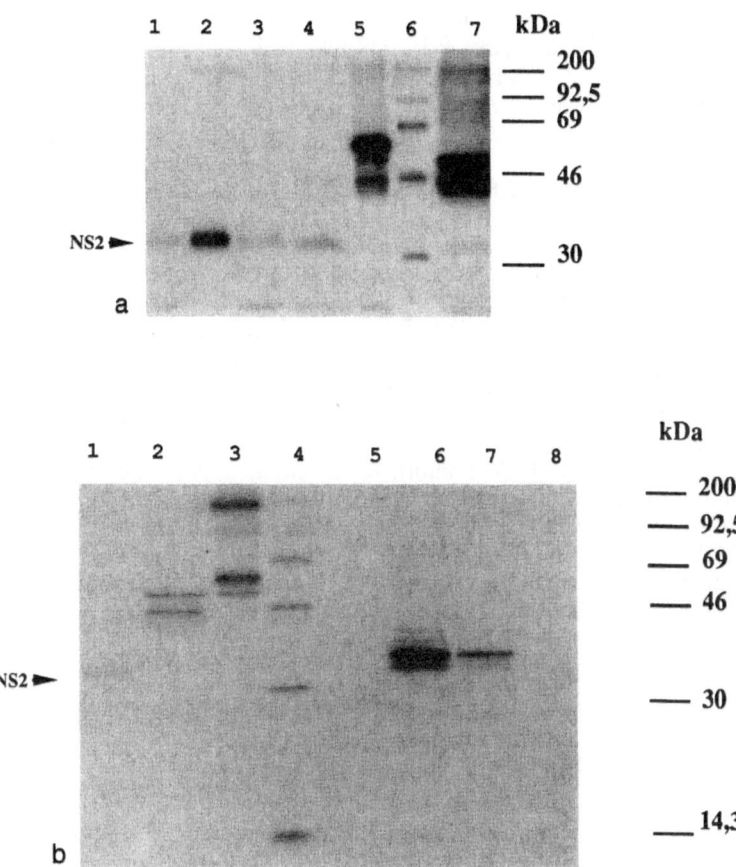

Figure 4a. Immunoprecipitations of the NS2 protein from radiolabelled BCV F15 infected cells lysates. Lanes 1, 2, 3, 4: Immunoprecipitations with four different polyclonal sera of mice immunised with GST-NS2 protein. Lane 5: Immunoprecipitation of HE protein with the monoclonal antibody J17b(16). Lane 6: Molecular weight markers. Lane 7: Negative control: immunoprecipitation with non immune mice serum.

4b. Immunoprecipitation of the the recombinant NS2 protein expressed in insect cells.
Lane 1: Immunoprecipitation of NS2 protein from BCV infected cell lysate. Lane 2: Negative control: immunoprecipitation of BCV infected cell lysate with non immune mice serum. Lane 3: Immunoprecipitation of HE protein with the monoclonal antibody J17b(16). Lanes 6 & 7: Immunoprecipitations of the recombinant NS2 protein expressed in insect cells. Lane 8: Negative control: immunoprecipitation of a recombinant sM baculovirus infected cells with anti-NS2 serum.

REFERENCES

1. Laporte, J. & Bobulesco, P., (1981), "Biochemistry and Biology of Coronaviruses",. V. ter Meulen, S. Siddel and H. Wege eds, pp 181-184, Plenum Publishing Corp., New-York
2. Keck, J.G, Hogue, B.G., Brian, D.A. & Lai, M.M.C. (1988).*Virus Res.*, 9, 343-356.
3. Boireau P., Woloszyn N., Crucière C., Savoysky E., Laporte J. (1990a). In: "Coronaviruses and their diseases", D. Cavanagh and D. Brown eds, pp 81-88.
4. Boireau P., Crucière C., Laporte J. (1990b). *J. Gen. Virol.*, 71, 487-492.
5. Savoysky E., Boireau P., Finance C., Laporte J. (1990). *Res. Virol.*, 141, 411-425.
6. Crucière, C. & Laporte, J. (1988). *Ann. Virol.* (Inst. Pasteur), 139, 123-138.
7. Woloszyn N., Boireau P., Laporte J. (1990).*Nucl. Ac. Res.* 18, 1303.
8. Cox, G.J., Parker, M.D. & Babiuk, L.A. (1989). *Nucleic Acids Res.*, 17(14), 5847.
9. Cox, G.J., Parker, M.D. & Babiuk, L.A. (1991). *Virology*, 185, 509-512.

10. Kozak, M. (1987). *J. Mol. Biol.*, 196, 947-950.
11. Pachuk, C.J., Bredenbeck, P.J., Zoltick, P.W., Spaan, W.J.M. & Weiss, S.R. (1989). *Virology*, 171, 141-148.
12. Cavanagh, D., Brian, D.A., Enjuanes, L., Holmes, K.V., Lai, M.M.C., Laude, H., Siddell, S.G., Spaan, W.J.M., Taguchi, F., & Talbot, P.J. (1990).*Virology*, 176, 306-307.
13. Sambrook, J., Fritsch, E.F. & Maniatis, T. (1989). Cold Spring Harbor Laboratory, Cold Spring Harbor, N.Y.
14. Bressan, G.M. & Stanley, K.K. (1987). *Nucleic Acids Res.*, 15(23), 10056.
15. Vautherot J.F., Laporte J., Boireau P.(1992). *J. Gen. Virol.*,.73, 1725-1737
16. Vautherot, J.F., Madelaine, M.F., Laporte, J. (1990). "Coronaviruses and their Diseases" (Cavanagh & Brown, Eds), pp 173-180.Plenum Publishing Corp., New-York.
17. Smith, D.B. & Johnson, K.S. (1988). *Gene*, 67, 31-40.
18. Lefèvre, F., L'Haridon, Borras-Cuesta, F. & La Bonnardière, C. (1990). *J. Gen. Virol.*, 71, 1057-1063.
19. Shieh, C-K., Lee, H-J., Yokomori, K., La Monica, N., Makino, S. & Lai, M.M.C. (1989). *J. Virol.*, 63, 3729-3736.
20. Luytjes, W., Bredenbeek, P.J., Noten, A.F.H., Horzinek, M.C. & Spaan, W.J.M. (1988). *Virology,* 166, 415-422.
21. Bredenbeek, P.J., Noten, A.F.H., Horzinek, M.C. & Spaan, W.J.M. (1989). *Virology*, 175, 303-306.

CHARACTERIZATION OF THE HUMAN CORONAVIRUS 229E (HCV 229E) GENE 1

Jens Herold, Thomas Raabe, and Stuart G. Siddell

Institut für Virologie
Universität Würzburg
Versbacherstr.7
8700 Würzburg

SUMMARY

The sequence of the HCV 229E gene 1 has been determined and compared with the homologous sequences of the murine hepatitis virus and the avian infectious bronchitis virus. The coding sequence of gene 1 is 20 273 nucleotides in length. Within this coding region are two large open reading frames, ORF 1a (4 086 codons) and ORF 1b (2 687 codons) which overlap by 40 nucleotides. In the overlapping region, the genomic RNA can be folded into a pseudoknot structure, an element which is known to mediate -1 ribosomal frame-shifting in other coronaviruses. Assuming that -1 frame-shifting occurs at the HCV sequence UUUAAAC (nucleotides 12 514 - 12 520), the ORF 1a - ORF 1b product is predicted to be 6 758 amino acids in length. Our sequence analysis of the HCV 229E gene 1 has revealed a high degree of similarity within the ORF 1b of HCV, MHV and IBV, whereas ORF 1a is much less conserved. Elements which are believed to be necessary for specific (e.g. frame-shifting) and general (e.g. NTP-binding/helicase) transcriptional functions have been identified. This study completes the genomic sequence of HCV 229E which is 27.27 kb long and one of the largest known RNA genomes.

INTRODUCTION

The coronavirus genome is a single stranded, polyadenylated RNA of positive polarity (1). It is 27.6 kb in length in the case of avian infectious bronchitis virus (IBV) and about 31 kb in the case of mouse hepatitis virus (MHV)(2, 3). Expression of the coronavirus genome is mediated by the genomic RNA (refered to as mRNA 1) and a set of 3′ coterminal subgenomic mRNAs. The mRNAs code for the coronavirus structural and nonstructural proteins. The known structural proteins are the S (spike), the M (membrane), the N (nucleocapsid), the SM (small membrane) and, for some coronaviruses, the HE (haemagglutinin esterase) proteins (4, 5, 6). The remaining mRNAs encode nonstructural proteins, most of which have not yet been detected in virus-infected cells.

Coronaviruses, Edited by H. Laude and J.F. Vautherot
Plenum Press, New York, 1994

To initiate infection, the genomic RNA is translated into a RNA-dependent RNA polymerase which subsequently uses its own mRNA as a template to create a negative stranded full-length copy of the viral genome and a set of positive and negative strand subgenomic RNAs (7). The RNA-dependent RNA polymerase has to be a multifunctional protein. It should contain the activities (i) to transcribe the genome into a negative stranded copy, (ii) to transcribe a set of subgenomic RNAs of both polarities, (iii) to transcribe the leader sequence and (iv) to produce progeny virion RNA. Furthermore, it is likely that it also possesses capping and polyadenylation activity. Here we present the sequencing strategy and a sequence analysis of the HCV 229E gene 1 and compare it with the homologous sequences of IBV and MHV.

MATERIALS AND METHODS

Virus and cells

HCV 229E virus was obtained from Dr. D.A.J. Tyrrell of the MRC Common Cold Unit, Salisbury, U.K. and cloned and propagated in C16 cells at 33°C as previously described (8). The cytoplasmic RNA from 10^8 cells which had been infected 48 h previously at an m.o.i. of 3 was extracted by standard procedures and the poly A-containing fraction selected by hybridization to poly U-Sepharose.

cDNA cloning

cDNA synthesis was performed by the method of Gubler and Hoffman (9) using either sequence specific oligonucleotides or random hexamers as reverse transcription primers. Double-stranded cDNA molecules were blunt-ended with T4 DNA polymerase and ligated into SmaI-linearized, phosphatase-treated pBluescript II KS+ (Stratagene) and transformed into competent *E.coli* TG-1 cells (10). The cDNA clones encompassing the gene 1 of HCV 229E were obtained by screening 3 independent cDNA libraries with specific, ^{32}P-labelled oligonucleotides.

DNA sequencing and PCR Amplification

PCR was performed using a GeneAmp / RNA PCR Kit (Perkin Elmer Cetus) according to the manufacturers procedures. A 5´-biotinylated, synthetic oligonucleotide (Oli B4: 5´-Biotin-GCCTATGAAAGTGCTGTTGTTAATGG-3´) was used as an upstream primer and a non-biotinylated oligonucleotide (Oli 36: 5´-TTAGATTTAAGAACAGCCTGTGACGC-3´) served as the downstream and reverse transcription primer. Oli B4 extends from nucleotide 5336 to 5371 of the genomic sequence, Oli 36 is complementary to nucleotides 6230 to 6255. The resulting cDNA strands were separated with streptavidin-coupled magnetic beads according to the manufacturers protocol (Dynal) and the nucleotide sequence of both strands was determined using a T7 DNA Polymerase Sequencing Kit (Pharmacia).

Double and single-stranded DNA sequencing was carried out by the dideoxynucleotide chain termination method (11). The sequence of all cDNA clones was determined on both strands.

Computer analysis of the sequence data

Sequence data were assembled with the programs of Staden (12) and analyzed using the UWGCG (University of Wisconsin, Genetic Computer Group) sequence analysis software package.

RESULTS

Selection of cDNA clones representing the gene 1 of HCV 229E

Most of the clones which have been used to obtain the sequence of gene 1 of HCV 229E were produced by a random priming method. Clone T5B5 was produced by priming with an 18-mer oligonucleotide which is complementary to a sequence at the 5′end of the HCV 229E S gene. Clone 35D5 was identified as coming from the 5′end of the genome by screening a random primed cDNA library with a leader-specific oligonucleotide. The other random-primed cDNA clones were found by probing the different cDNA libraries with ^{32}P-labelled oligonucleotides, walking either from the 5′end (clone 35D5) in a 3′direction or from the 3′end (clone T5B5) in a 5′direction. The sequence information of the region between clone 30F7 and J13D8, where no cDNA clone was found in the libraries, was obtained by PCR amplification and direct solid-phase sequencing of the PCR product. In this way, a set of overlapping cDNA clones representing the complete gene 1 of HCV 229E was obtained (Fig.1).

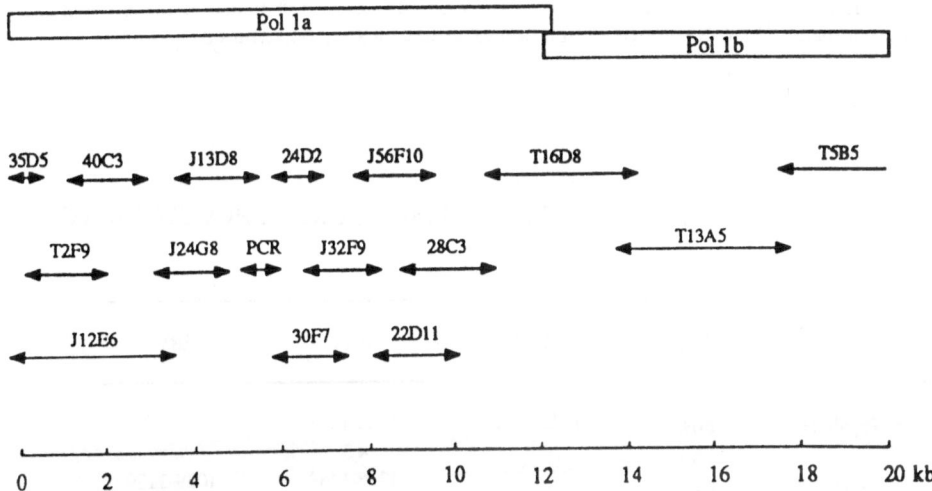

Figure 1. Diagram showing the positions of all cDNA clones used in obtaining the nucleotide sequence of the gene 1 of HCV 229E.

Analysis of the gene 1 nucleotide sequence

15 cDNA clones and a PCR derived DNA fragment were sequenced to obtain the complete sequence of HCV 229E gene 1. The coding sequence of gene 1 is 20 273 nucleotides in length. Analysis of the sequence revealed two large ORFs, 1a and 1b (Fig. 2). ORF 1a is 4 086 codons in length, ORF 1b, which overlaps ORF 1a by 40 nucleotides and is located in a -1 reading frame, is 2 687 codons in length.

In the ORF 1a - ORF 1b overlapping region the genomic RNA can be folded into a pseudoknot structure. This structure is analogous to elements known for IBV and MHV (3, 13, 14). It has been shown by mutational analysis that the pseudoknot structure and the 5′-located "slippery" sequence, UUUAAAC, are necessary for the coronavirus frame-shifting event *in vitro* and *in vivo* (15, 16).

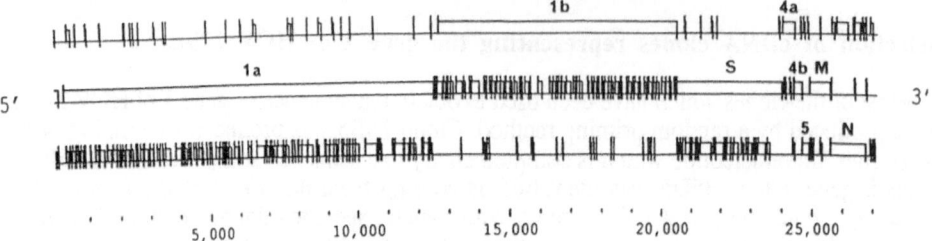

Figure 2. ORFs of the HCV 229E genomic RNA . GCG program (FRAMES).

Sequence similarities between ORF 1a and 1b of HCV 229E, MHV and IBV

After optimal alignment with the GCG program GAP, the overall similarity between the HCV 229E, MHV and IBV 1a ORFs is about 50% and for the 1b ORFs about 70%. However, when analogous domains in ORF 1a i.e. the 3C-like protease or the growth-factor-like domain (3) are compared, much higher similarities are determined.

Table 1. Similarity of particular domains in the 1a and 1b ORFs of HCV 229E, MHV and IBV

		HCV	MHV	IBV
Papain I[1]	Pos.[5]	1012-1272	1038-1319	
	%[7]		43	
X-domain[2]	Pos.[5]	1299-1396	1358-1459	1039-1139
	%		54	58
Papain II[3]	Pos.[5]	1611-1866	1635-1990	1378-1612
	%		45	42
3Cl pro[3]	Pos.[5]	2964-3267	3349-3653	2779-3086
	%		62	60
MP2[3]	Pos.[5]	3268-3546	3654-3941	3087-3379
	%		55	54
GFL-domain[3]	Pos.[5]	3934-4069	4338-4475	3784-3929
	%		64	64
Pol-module[4]	Pos.[6]	530-832	534-836	540-842
	%		78	77
Hel-motif	Pos.[6]	921-996	925-1000	931-1006
	%		84	79
MBD[1]	Pos.[6]	1199-1325	1201-1327	1208-1334
	%		74	79

[1] (17); [2] (18); [3] (3); [4] (19); [5] amino acid position in ORF 1a; [6] amino acid position in ORF 1b; [7] % similarity to the HCV 229E sequence

DISCUSSION

The sequence analysis of gene 1 completes the genomic sequence of HCV 229E which, excluding the poly(A) tail at the 3´end, is 27 270 nucleotides in length. Gene 1 is divided into two ORFs which can apparently be translated into a Pol 1a product of about 400 000 mol.wt.. or a Pol 1a - Pol 1b product of about 700 00 mol. wt. The structures necessary for -1 ribosomal frame shifting, i.e. the "slippery" sequence (UUUAAAC) and a potential pseudoknot structure have been identified in the polymerase gene of HCV 229E.

Gene 1 is believed to code for an RNA dependent RNA polymerase(s) which carries out replication and transcription functions. ORF 1b, which is highly conserved among the coronaviruses sequenced to date, contains polymerase, helicase and metal-binding motifs (3, 17), suggesting that this region may be directly involved into RNA synthesis.

It seems likely that the coronavirus gene 1 is translated as a polyprotein which is subsequently proteolytically cleaved into functionally distinct polypeptides, as is the case for alphaviruses (20) and picornaviruses (21). In this respect, Gorbalenya *et al.* (17) and Lee *et al.* (3) have identified several protease motifs in the ORF 1a sequences of IBV and MHV, respectively. Firstly a papain-like protease motif which is duplicated in MHV and which is thought to be responsible for the cleavage of the N-terminal p28, at least *in vitro* (22, 23). And secondly, a highly conserved 3C-like protease which is predicted to have a prefered specificity for QS or QA dipeptides. Both domains could also be identified in the HCV 229E genome.

In the long term, the availability of the cDNA clones and nucleotide sequence for the polymerase gene of HCV 229E will allow us to produce specific immunological reagents which can be used to elucidate the expression, processing and function of this complex locus.

REFERENCES

1. Wege, H., Muller, A., and ter Meulen, V., 1978, *J. Gen. Virol.*, 41:217
2. Boursnell, M.E.G., Brown, T.D.K., Foulds, I.J., Tomley, F.M., and Binns, M.M., 1987, *J. Gen. Virol.*, 68:57
3. Lee, H.-J., Shieh, C.-K., Gorbalenya, A.E., Koonin, E.V., La Monica, N., Tuler, J., Bagdzhadzhyan, A., and Lai, M.M.C., 1991, *Virology*, 180: 567
4. Cavanagh, D., Brian, D.A., Enjuanes, L., Holmes, K.V., Lai, M.M.C., Laude, H., Siddell, S.G., Spaan, W., Taguchi, F., and Talbot, P.J., 1990, *Virology*, 176:306
5. Liu, D.X., and Inglis, S. C., 1991, *Virology*, 185:911
6. Godet, M., L´Haridon, R., Vautherot, J-F., and Laude H., 1992, *Virology*, 188:666
7. Sawicki, S.G., and Sawicki, D.L., 1990, *J. Virol.*, 64:1050
8. Phillpotts, R.J., 1983, *J. Virol. Methods*, 6:267
9. Gubler, U., and Hoffman, B.J., 1983, *Gene*, 25:263
10. Hanahan, D., 1985, in:"DNA cloning Vol.I", D.M. Glover, ed., IRL Press, Oxford
11. Sanger, F., Nicklen, S., and Coulson, A.R., 1977, *Proc. Natl. Acad. Sci.*, 74:5463
12. Staden, R., 1982, *Nucleic Acids Res.*, 10:4731
13. Brierley, I., Boursnell, M.E.G., Binns, M.M., Bilimoria, B., Blok, V.C., Brown, T.D.K., and Inglis, S.C., 1987, *EMBO J.*, 6:3779
14. Bredenbeck, P.J., Pachuk, C.J., Noten, A.F.H., Charité, J., Luytjes, W., Weiss, S.R., and Spaan W.J.M., 1990, *Nucleic Acids Res.*, 18:1825
15. Brierley, I., Digard, P., and Inglis, S.C., 1989, *Cell*, 57:537
16. Brierley, I., Rolley, N.J., Jenner, A.J., and Inglis, S.C., 1991, *J. Mol. Biol.*, 220:889
17. Gorbalenya, A.E., Koonin, E.V., Donchencko, A.P., and Blinov, V.M., 1989, *Nucleic Acids Res.*, 17:4847
18. Gorbalenya, A.E., Koonin, E.V., and Lai, M.M.C, 1991, *FEBS Letters*, 288:201
19. Koonin, E.V., 1991, *J. Gen. Virol.*, 72:2197
20. Strauss, E.G., Rice, C.M., and Strauss, J.H., 1984, *Virology*, 133:92
21. Summer, D.F., and Maizel, J.V., 1968, *Proc. Natl. Acad. Sci.*, 59:966
22. Baker, S.C., Shieh, C.-K., Soe, L.H., Chang, M.-F, Vannier, D.M., and Lai, M.M.C., 1989, *J. Virol.*, 63:3693
23. Denison, M.R., Zoltick, P.W., Hughes, S.A., Giangreco, B., Olson, A.L., Perlman, S., Leibowitz, J.L., and Weiss, S.R., 1992, *Virology*, 189:274

IDENTIFICATION OF CORONAVIRAL CONSERVED SEQUENCES

AND APPLICATION TO VIRAL GENOME AMPLIFICATION

Anne Bridgen, Kurt Tobler and Mathias Ackermann

Institute for Virology
University of Zürich
Winterthurerstrasse 266a
CH-8057 Zürich

Our work with the porcine epidemic diarrhoea virus (PEDV) has led us to look more closely at sequences and sequence motifs which are conserved between different coronaviral genomes. This is possible, since sequence information from at least part of the genomes of twelve coronaviruses is now available. Many of these motifs are specific to coronaviruses while others are shared with other positive strand RNA viruses, for example with the torovirus Berne virus and the flavivirus equine arteritis virus[1]. Functions can be assigned to some of these conserved regions, for example the spike (S) and membrane (M) protein transmembrane regions and the S leucine-zipper motif[2]. Other well conserved domains, including regions within the M and nucleocapsid (N) protein genes, have as yet no defined role. The identification of such conserved sequences is important for the recognition of functional domains of the viral RNA and proteins and also of regions useful in the cloning of novel coronaviruses using techniques based on the polymerase chain reaction (pcr). They can also be used to assess the evolutionary relationships of different virus groups. We plan to discuss the nature and possible functions of the conserved sequence motifs in more detail elsewhere, and to concentrate on the application of these sequences to viral genome amplification in this article.

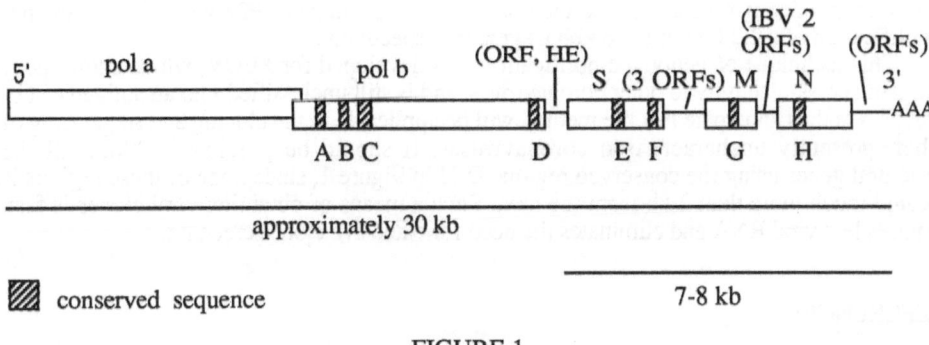

FIGURE 1

Figure 1 (not to scale) shows a representation of a standard coronavirus genome including the major open reading frames (ORFs) where HE, S, M and N represent the haemagglutinin-esterase, spike, membrane and nucleocapsid proteins, respectively. Pol a and b represent the two ORFs of the polymerase gene, of which the second is expressed through ribosomal frame shifting. IBV represents Infectious Bronchitis Virus. ORFs which are not found in all coronaviruses sequenced to date are shown in parentheses. Eight sequence regions which show a high level of homology beween coronaviral genomes and which are long enough to be used for pcr amplification of the viral genome are indicated by hatching. These sequences vary in length from twenty-four to several hundred bases and represent the following regions, where amino acids (aa) are displayed in the normal one letter code:

A. polymerase region I, GDD motif[1]
B. polymerase region II, cysteine rich region[1]
C. polymerase region III, helicase domain[1]
D. polymerase region IV, conserved domain[1]
E. S gene: region of nearly 300 amino acids showing high sequence conservation
F. S gene: consensus sequence YVKWPKYVWL
G. M gene: region including the consensus sequence SWWSFNPE
H. N gene: GYW -(4aa)- R -(4-7aa)- G -(10aa)- FYYLGTGP -(11aa)- DGV(V/F)WVA

We amplified the 2.2 kbp cDNA located nearest to the 3' end of the PEDV genome using primers based on the conserved sequences in regions G and H, as shown below. The primers were synthesised to contain restriction enzyme sites, but these were not used for product cloning. Rare amino acid variations from the consensus sequence are shown in parentheses.

```
                           (F)
                F    Y    Y    L    G    T    G    P      Amino acids (H, Fig. 1)
     P23   A AGC TTT  TAC  TAC  TTA  GGA  ACA  GGA  CC     Primer, mRNA sense
     5'end HindIII    T    G    C                          27mer, 18 fold degenerate
                           T    T


                D(E)  G   V(I) V/F   W    V    A           Amino acids (H, Fig. 1)
     P24        CTG  CCA  CAG  AAG  ACC  CAG  CGA GCT C     Primer, antisense
     3'end       T    T    C              XhoI              25mer, 8 fold degenerate


                S   W(F)  W    S   F(W)  N    P    E        Amino acids (G, Fig. 1)
     P25   G ACT AGT  TGG  TGG  AGA  TTT  AAC  CCA  GA     Primer, mRNA sense
     5'end   SpeI                T         T    T           27mer, 8 fold degenerate
```

Primers P24 and P25 were used to amplify a 680 bp fragment from single stranded cDNA primed with P24 or oligo dT. The 3' most sequences were amplified in a modification of the RACE (rapid amplification of cDNA ends) technique of Frohman et al.[3], which is described in more detail elswhere[4]. Briefly, the original Frohman protocol required the use of two related primers, of which one was used for cDNA synthesis from the poly A tail and the other for pcr amplification, in addition to a sequence specific primer. Instead, we created a dT containing primer for first strand cDNA synthesis by addition of thymidine residues to P25 with terminal deoxytransferase, and then used either P25 alone or P25 with P23 for the pcr amplification of a 2.2 kbp or a 1.6 kbp fragment, respectively.

This technique of genome amplification was developed for PEDV, which shows poor antigenic cross-reactivity to other coronaviruses and is still unclassified into an antigenic subgroup[5]. We thus anticipate that the method will be applicable to the cloning and sequencing of other, presently uncharacterised coronaviruses. It should be possible to clone all the structural genes using the conserved regions D-H in Figure 1, since none of these regions is located much more than 2 kb from the next. Such a means of obtaining viral clones is fast, requires less viral RNA and eliminates the need for extensive clone screening.

REFERENCES

1. Snijder, E.J., den Boon, J.A., Bredenbeek, P.J., Horzinek, M.C., Rijnbrand, R. and Spaan, W.J.M. (1990). *Nucleic Acids Res.* 18: 4535-4542.
2. Britton, P. (1991). *Nature* 353: 394.
3. Frohman, M.A., Dush, M.K. and Martin, G.R. (1988). *Proc. Natl. Acad. Sci.* 85: 8998-9002.
4. Tobler, K., Bridgen, A. and Ackermann, M. *Adv. Exp. Med. Biol.* This volume.
5. Pensaert, M.B., Debouck, P. and Reynolds, D.J. (1981). *Arch. Virol.* 68: 45-52.

Chapter 2

Transcription, Replication and Genome Engineering

STUDIES INTO THE MECHANISM FOR MHV TRANSCRIPTION

Mary C. Schaad, Wan Chen, Sheila A. Peel and
Ralph S. Baric

Program in Infectious Diseases, Department of Epidemiology, University of
North Carolina at Chapel Hill Chapel Hill, North Carolina 27599-7400

ABSTRACT

Previous studies have demonstrated that the MHV genome is divided into seven
transcriptional units which are transcribed from highly conserved intergenic start sites
(UCU/CAAAC) into mRNA containing a common leader RNA at the 5' end and a
coterminal 3' end. In this manuscript, we provide evidence that an additional transcriptional
unit is encoded at the 3' end of the MHV genome and is transcribed from a perfect
intergenic region into a leader-containing ~800 nt mRNA . This mRNA could potentially
encode a small 17-18 kDa protein which is identical to the C-terminal third of the
nucleocapsid gene.

INTRODUCTION

Mouse hepatitis virus, a member of the coronaviridae, contains a single-stranded
plus polarity RNA of about 31 kb in length. The genomic RNA is enclosed within a helical
nucleocapsid constructed from multiple copies of a 50 kDa nucleocapsid protein designated
N. The nucleocapsid is surrounded by a bi-lipid envelope derived from internal host cell
membranes and contains two or three virus specific glycoproteins. The S glycoprotein is
180/90 kDa in molecular weight and forms the distinct surface projections of the virus
particle. The M glycoprotein gene is about 23 kDa molecular weight and contains O-linked
glycosidic moieties. Some strains of MHV also contain a gene encoding a 65 kDa
hemagglutinin esterase protein[1].

The mechanism of MHV transcription is controversial and remains under study. The
majority of data suggest that the incoming genomic RNA is transcribed into a full-length
negative-stranded RNA which acts as template for the synthesis of 6 or 7 subgenomic
mRNAs. It is postulated that a free leader RNA of about 65-72 nucleotides in length is
transcribed from the full-length minus-stranded RNA and acts in trans to prime
transcription from highly conserved intergenic start sites (UCU/CAAAC) just 5' to each
mRNA. The subgenomic mRNA can then act as template for the synthesis of subgenomic
negative-stranded RNAs which function as templates for additional rounds of subgenomic
mRNA synthesis[2,3,4].

In this study, we provide evidence that a previously unrecognized mRNA is
transcribed from a highly conserved intergenic start site that is located within the N gene
sequence at the 3' end of the genome. The new mRNA could potentially encode a 154-155
amino acid protein which is in frame with the C-terminal third of the nucleocapsid gene of
MHV.

Coronaviruses, Edited by H. Laude and J.F. Vautherot
Plenum Press, New York, 1994

METHODS

Virus and Cells

Mouse hepatitis virus strains MHV-A59, MHV-S and MHV-1 were used throughout the course of this study. The virus was propagated on DBT cells maintained at 37°C in dMEM containing 10% Nu-serum, and 1% kanamycin/gentamicin.

Isolation of Viral mRNA and Replicative Form RNAs

Cultures of DBT cells were infected with MHV at a MOI of 10 and maintained at 37°C for 8-12 hrs postinfection. Medium was removed and the cells washed in ice cold PBS. To isolate RNA for replicative form (RF) analysis, intracellular RNA was isolated by lysing cells in 0.5 ml of Hypo TKM (10 mM Tris-Hcl, pH 7.5, 150 mM KCl, 1.5 mM $MgCl_2$) containing 100 ug/ml Proteinase K and 0.05% NP40. The nuclei were removed by slow speed centrifugation, supernatents extracted with multiple phenol, phenol/chloroform/isoamyl alcohol, and chloroform treatments then ethanol precipitated. To isolate intracellular RNA for mRNA analysis, the cells were lysed as previously described[2].

For RF analysis, the intracellular RNA was resuspended in 10 ul of water, adjusted to 1 X DNAse buffer (100 mM NaCl, 10 mM Tris, pH 7.8, 2 mM EDTA, 2 mM $MgCl_2$, 2 mM $CaCl_2$) and treated with 4 U of DNase 1 for 15 min at 30° C. Following DNase treatment, the sample was adjusted to 1 X RNAse buffer (700 mM NaCl, 10 mM Tris, pH 7.4, 30 mM EDTA) and treated with 50 ng/ml RNase A for 15 min at 30°C. The samples were loaded onto 1% agarose gels containing TBE (89 mM Tris, 89 mM Boric acid, 2 mM EDTA)[2].

PCR Cloning and Sequencing

Primers were designed from published sequences from the MHV-A59 N gene and leader RNA. The forward primer was 5'-TAAGAGTGATTGGCGTCCGTACG-3' which corresponded to nucleotides (nt) 3-25 in the leader RNA sequence. The reverse primer was derived from nt 920-944 in the N gene sequence and was 5'-GCAAGAATGGGGAACTGTGGATCAC-3'. Intracellular RNA was isolated from MHV-A59, MHV-S, and MHV-1-infected DBT cells. Two ug of RNA was reverse transcribed into cDNA in 1 X RT buffer (10 mM Tris, pH 8.8, 50 mM KCl, 0.1% Triton X-100), 5 uM oligo dT, 500 uM dNTPs, 0.5 U/ul RNASIN, 10 mM $MgCl_2$, 10 mM DTT and 8 U AMV reverse trans-criptase at 42°C for 1 hr. The reaction was stopped by the addition of 2 ul of 0.5 M EDTA, extracted with phenol/chloroform and then precipitated with ethanol. Approximately 30 ng of cDNA was amplified in a 50 ul reaction containing 1 x TAQ DNA polymerase buffer, 2 mM $MgCl_2$, 1.25 mM each dNTPs and 5 U of TAQ DNA polymerase (Promega). After 25 cycles, the PCR products were separated on 4.5% NuSieve (FMC) agarose gels and bands of the appropriate size were excised and electroeluted into 5 M potassium acetate. The eluted fragments were phenol/chloroform extracted and precipitated in ethanol. Pellets were washed 3X in 70% ethanol, 1X with 95% ethanol and vacuum dried.

PCR products were directly sequenced by the chain termination method using a sequenase kit (USB). Approximately 300 ng of purified PCR product was incubated with 20 pM primer and 10% DMSO and denatured at 97°C for 8.5 min. The reaction was cooled to room temperature for 1 min, and the extension reactions were performed on ice with a 1:10 dilution of the nucleotide mix following standard Sequenase version 2.0 protocols (USB).

RESULTS

Demonstration of a Small Eighth mRNA and RF RNA in MHV-Infected Cells

We have noted on occasion the presence of an additional small RNA less than 1,000 nt in length in MHV-infected cells. To determine if this small mRNA species was in fact a product of the MHV genome, cultures of DBT cells were infected with MHV-A59, treated with actinomycin D (10 ug/ml) at 2 hr postinfection, and radiolabeled with 100 uCi/ml [3]H-

86

Uridine from 3-7 hr postinfection. The RNA was isolated and separated on 1% agarose gels. As expected, the seven characteristic viral mRNAs previously reported in MHV-A59 infected cells were readily detected. In addition, a small <1000 nt RNA was also present in infected, but not uninfected cells (Figure 1).

If this small RNA is of viral origin, analysis of RF RNAs in infected cell lysates should demonstrate the presence of a corresponding double-stranded RF RNA. To test this hypothesis, cultures of DBT or 17CL1 cells were infected with MHV-A59, treated with actinomycin D, and radiolabeled with ^{3}H-Uridine. The intracellular RNA was digested with DNAse I and RNAse A and separated on 1% agarose gels. In agreement with

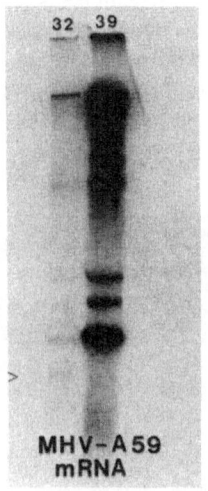

Figure 1. ^{3}H-Uridine-labeled MHV-A59 intracellular RNAs isolated at 10 hr postinfection from DBT cells labeled at 32°C and 39.5°C.

previous findings, seven RNase A resistant RF RNAs were present in MHV-infected cells[2]. In addition, an eighth subgenomic RF RNA was occasionally detected (data not shown). The size of this RF RNA was about 1600-1800 bp in length suggesting that it was involved in the transcription of an approximate 800-900 nt mRNA. The ability to detect the small mRNA and RF RNA by radiolabeling experiments was difficult, and at times their presence was not detectable suggesting that the mRNA is usually made in small quantities.

Table 1. Conservation of intergenic 7-8 in MHV

VIRUS	CONSENSUS INTERGENIC (Y=C or T)	PUTATIVE 7-8 INTERGENIC	PUTATIVE OPEN READING FRAME (AMINO ACIDS)
MHV-A59	TCYAAAC	TCCAAAC	154
MHV-JHM	TCYAAAC	TCCAAAC	155
MHV-1		TCCAAAC	155
MHV-3		TCCAAAC	154
MHV-S		TCCAAAC	154
			14

Identification of a Potential Intergenic Start Site for mRNA 8. Transcription.

A highly conserved intergenic sequence UCU/CAAAC is present in the MHV genome just 5' to the start site for the synthesis of each subgenomic mRNA[1,3]. If an additional ~800-900 nt mRNA is encoded in the MHV genome, it seems likely that such a conserved start site would be present within the 3' end of N gene coding sequences. Nucleotide sequence analysis of such a "start" site among the different MHV strains is shown in Table 1. A perfect intergenic start site was present at nt 828 in the MHV-A59 N gene[5]. In addition, sequence analysis also revealed "start sites" at similar locations at the 3' end of the genome among other group II coronavirus strains (BCV, OC43, data not shown).

To determine whether the small mRNA found in MHV-infected cells initiated from this putative start site, oligodeoxynucleotide primers were chosen which were located at the 5' end of the leader RNA sequence and downstream from the potential intergenic region in the MHV N gene coding sequences. Intracellular RNA was isolated from MHV-A59, MHV-S, and MHV-1 infected cultures and reverse transcribed into cDNA. Following 25 cycles of PCR amplification, the expected 160 base pair product was detected in all MHV strains tested, but not in uninfected controls. Surprisingly, two additional PCR products were also identified in MHV-1 (120 bp) and MHV-S (100 bp) infected cells (Figure 2).

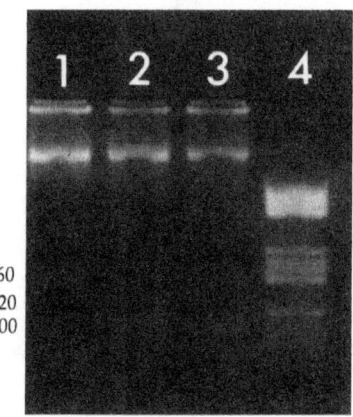

Figure 2. RNA extracted from MHV-1, MHV-S and MHV-A59-infected cells was reverse transcribed and amplified by 25 cycles of PCR using primers originating from the 5' end of the leader RNA sequence (nt 3-25) and within the N Gene (nt 920-944). Lanes 1,2 and 3: MHV-A59, MHV-1 and MHV-S. Lane 4: DNA Molecular Weight Marker V (Boehringer Mannheim).

Partial Sequence Analysis of the MHV mRNA 8 Gene

The partial sequence of the MHV-A59, MHV-S and MHV-1 "mRNA 8" and its start sites are shown in Figure 3. The sequence in this region was identical to those previously published[5]. As expected, all of the PCR-derived products contained the MHV leader RNA sequence at the 5' end of the mRNA consistent with the replication strategy of the virus. The perfect intergenic present at nt 828 was used as the start site for mRNA 8 synthesis from all MHV strains tested (Figure 3). In addition to the transcripts initiating from the perfect intergenic, MHV-1 and MHV-S had transcripts initiating from two "imperfect" intergenics. mRNAs initiating from either the perfect intergenic (UCCAAAC) in the MHV strains examined, or from the second start site in MHV-1 (CCAAUC), would encode a 154-155 amino acid protein, which was in frame with the carboxy terminus of the N protein[5] (Figure 4). mRNA initiated from the second start site in the MHV-S N gene (UAAAC) was, however, unique to MHV-S infected cells and started downstream from the ATG initiation codon for the 154-155 amino acid protein. This ORF would encode a putative 28 amino acid protein out of frame with respect to N. Unless additional start sites are located within the MHV-A59 and MHV-1 genomes to encode a mRNA that expresses the putative

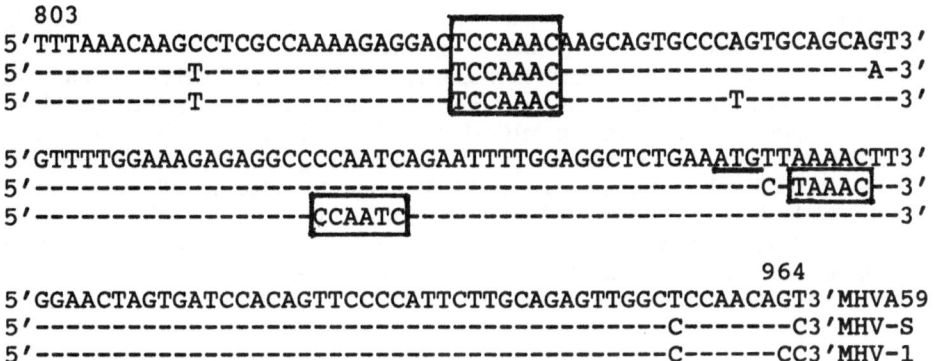

```
     803
5'TTTAAACAAGCCTCGCCAAAAGAGGACTCCAAACAAGCAGTGCCCAGTGCAGCAGT3'
5'----------T---------------TCCAAAC------------------A-3'
5'----------T---------------TCCAAAC----------T----------3'


5'GTTTTGGAAAGAGAGGCCCCAATCAGAATTTTGGAGGCTCTGAAATGTTAAAACTT3'
5'------------------------------------------------C-TAAAC-3'
5'-----------------CCAATC---------------------------------3'

                                                  964
5'GGAACTAGTGATCCACAGTTCCCCATTCTTGCAGAGTTGGCTCCAACAGT3'MHVA59
5'-------------------------------------------C------C3'MHV-S
5'-------------------------------------------C-----CC3'MHV-1
```

Figure 3. PCR amplified products spanning the putative mRNA 8 start site for MHV-A59, MHV-S and MHV-1 were sequenced. Boxed nucleotide sequences represent the start sites for mRNA 8 synthesis in each MHV strain. The initiation codon ATG is underlined and is in frame with the N gene sequence.

28 amino acid protein, it seems unlikely that this protein plays any significant role in virus replication.

DISCUSSION

The MHV genome is divided into seven transcriptional units which are synthesized from highly conserved intergenic start sites (UCU/CAAAC) into mRNA containing a common leader RNA sequence derived from the 5' end of the genome and a 3' co-terminal end[1]. In this manuscript, we provide evidence suggesting that a new transcriptional unit is encoded within the N gene at the 3' end of the MHV genome. The mRNA 8 transcript is initiated from a perfectly conserved intergenic sequence at nt 828 in the MHV-A59 sequence into a mRNA of about 900 nt in length. The ability to detect this mRNA and its corresponding RF RNA using radiolabeling experiments was highly variable suggesting that it was made in relatively small amounts or was difficult to detect because of its size and location in the gel.

Although we have not demonstrated the putative protein product from this mRNA, it should encode a 154-155 amino acid protein of about 17-18 kDa which is identical to the C-terminal third of the N gene of MHV. Interestingly, a variety of N "degradation" products have also been described in MHV infected cells[6]. It is unclear whether any of these products may represent the product of mRNA 8. While a variety of co-terminal truncated genes, which function very differently during infection, have previously been described in the SV40 and hepatitis delta virus genomes[7,8], this represents the first such demonstration of this type of genetic organization in the MHV genome. This transcriptional unit may also be expressed during other group II coronavirus infections. Highly conserved intergenic sequences are also present within the BCV and HCV-OC43 genomes which could be transcribed into mRNA encoding a truncated protein which is in frame with the C-terminus of the nucleocapsid protein. In addition, a small RNA of <1,000 nt and its corresponding RF RNA have been demonstrated in BCV-infected cells[4].

The N gene functions in nucleocapsid formation, transcription and pathogenicity. Sequence analysis among the different MHV strains indicate that the N protein can be divided into three highly conserved domains[5]. The putative mRNA 8 product would encode amino acids 300-454 containing a portion of the basic domain II and all of the acid domain III[5]. A second internal open reading frame (nt 65-688) is also present within the N gene which may encode a 207 amino acid protein[5] (Figure 4). A ts mutant, Alb4, has been isolated which contains a 28 amino acid in frame deletion in the N gene sequence between amino acids 380 and 408[9]. The finding of a new mRNA which potentially encodes a protein which is identical to amino acids 300-454 in the N gene raises questions about the actual role of the nucleocapsid gene in virus transcription and pathogenicity. It is also unclear whether the deletion in Alb4 preferentially alters the function of the N protein, the function of the truncated C-terminal product of mRNA 8, or both.

The finding of two "illegitimate" start sites in the MHV-S and MHV-1 genomes are puzzling, but support previous studies that imperfect start sites can be used to initiate mRNA transcription[10]. In the case of MHV-1, the CCAAUC "start" site is also conserved in the MHV-A59 and MHV-S genomes, but leader-containing transcripts are present only in MHV-1-infected cells. Interestingly, in MHV-1 and B1-infected cells, new mRNA species (2-2,3-1) have been identified which start at a very similar intergenic region (UC/UUAAUC) suggesting that CCAAUC may also represent a functional intergenic in the MHV-1 genome[10]. It was suggested that the presence of three UCUAA repeats encoded within the leader sequence at the 5' end of the MHV-1 genome may permit the use

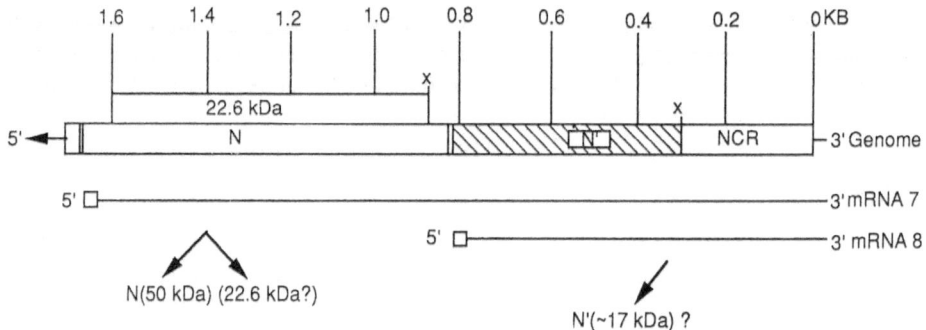

Figure 4. Tentative genetic map of the 3' end of the MHV genome. The boxed regions represent likely ORFs that are encoded in the 3' end of the genome in the N gene sequence. Solid double bars represent the location of perfect intergenic "start" sequences and the x denotes the termination codons for these ORFs. Hatched lines represent the putative 17-18 kDa protein that is identical to the C-terminus of the N gene and may be encoded by mRNA 8.

of these particular intergenic regions[10]. In the case of the "imperfect" start site in the MHV-S genome, a specific mutation is present within this intergenic region (UAAAC) that is not present in MHV-1 or MHV-A59 genomes (AAAAC), and this may account for initiation of mRNA synthesis. Since this mRNA is MHV-S specific and contains a 28 amino acid open reading frame that is probably not expressed in other strains of MHV, it seems likely that this gene is not critical for virus growth. Additional studies are needed to determine the role of any of these gene products in virus replication.

REFERENCES

1.   M.M.C. Lai. *Ann. Rev. Microbiol.* 44:303 (1990).
2.   S.G. Sawicki and D.L. Sawicki. *J. Virol.* 64:1050 (1990).
3.   S. Makino, M. Joo, and J.K. Makino. *J. Virol.* 65: 6031-6041 (1991).
4.   M.A. Hofmann, P.B. Sethna, and D.A. Brian. *J.Virol.* 64: 4108-4114 (1990).
5.   M.M.Parker and P. S. Masters. *Virology* 179:463 (1990).
6.   S.G. Robbins, M.F. Frana, J.J. McGowan, J.F. Boyle, and K.V. Holmes. *Virology* 150:402 (1986).
7.   J. Tooze DNA Tumor Viruses, Molecular Biology of Tumor Viruses, 2nd Ed., Cold Spring Harbor Laboratory, Cold Spring Harbor, NY (1981).
8.   F-L. Chang, P-J. Chen, S-J. Tu, C-J. Wang, and D-S. Chen. *PNAS USA* 88:8490 (1991).
9.   C.A. Koetzner, M.M. Parker, C.S. Ricard, L.S. Sturman, and P.S. Masters. *J. Virol.* 66:1841 (1992).
10.   N. La Monica, K. Yokomori, and M.M.C. Lai. *Virology* 8:402 (1992).

ANALYSIS OF THE CIS-ACTING ELEMENTS OF CORONAVIRUS TRANSCRIPTION

Myungsoo Joo and Shinji Makino

Department of Microbiology, The University of Texas at Austin, Austin, Texas 78712

INTRODUCTION

Mouse hepatitis virus (MHV), a coronavirus, is an enveloped virus containing a single-stranded, positive-sense RNA genome of approximately 31 kb (11, 13, 22). There are seven to eight species of virus-specific subgenomic mRNAs in MHV-infected cells. These subgenomic mRNAs comprise a 3'-coterminal nested-set (9, 14). In decreasing order of size, they are mRNAs 1 through 7 (9, 14). The 5' end of the MHV genomic RNA contains a 72- to 77-nucleotide-long leader sequence (8, 10, 26). Within the 3'-region of the leader sequence there is a pentanucleotide sequence, UCUAA. This sequence repeats two to four times in different MHV strains (19). The MHV-specific genes are downstream from the leader, and each gene is separated by a special short stretch of sequence, the intergenic sequence. The intergenic sequences include the unique consensus sequence UCUAAAC or a sequence very similar sequence (25). All MHV mRNA species have a sequence identical to the 5'-end genomic leader sequence. These leader sequences are fused to the intergenic consensus sequence, which marks the start of each gene (8, 10, 25, 26). In most MHV genes the degree of intergenic sequence nucleotide homology with the leader sequence correlates with the amount of mRNA transcribed (25). This correlation is not observed in infectious bronchitis virus mRNA transcription (4). The site where the leader fuses with the mRNA is somewhere within the repeated pentanucleotide (UCUAA). The number of repeats in each given mRNA varies (19). The pentanucleotide repeats at the genomic leader sequence and at the intergenic region are identical, making identification of the fusion site of these two sequences difficult.

Several models try to explain subgenomic RNA synthesis. One model is leader RNA-primed transcription; it proposes that a free leader RNA is transcribed from the 3'-end of the genomic-size, negative-strand template RNA, dissociates from the template, and then rejoins the template RNA at downstream intergenic regions to serve as the primer for mRNA transcription (3, 7). Another model, was put forth by Sawicki and Sawicki (23). This model suggests that subgenomic negative-strand RNAs are initially synthesized from the input genomic RNA followed by synthesis of the positive-strand subgenomic RNA on the subgenomic-sized, negative-stranded RNA. Another possible mechanism may be that leader RNA joins the subgenomic RNA body by a mechanism similar to RNA splicing (8, 26). None of these models has been exclusively proven. The mechanism of coronavirus transcription is unknown.

Defective interfering (DI) RNAs of MHV are vital for understanding the mechanisms of coronavirus mRNA transcription (16). An artificial subgenomic mRNA was made using the DI system; the intergenic region, derived from between genes 6 and 7 of the genome was inserted into a complete DI cDNA clone. This clone replicated and transcribed a subgenomic DI RNA after transfection of its RNA into MHV-infected cells.

In the present study, how nucleotide substitution in the consensus sequence influences mRNA transcription and the location of the leader-body fusion site were studied with mutant

DI cDNAs. The role of the sequences flanking the intergenic sequence in subgenomic DI RNA transcription was also examined.

MATERIALS AND METHODS

Viruses and cells. The plaque-cloned A59 strain of MHV (MHV-A59) (9) was used as a helper virus. Mouse DBT cells were used for growth of viruses.

DNA construction. A procedure based on recombinant polymerase chain reaction (PCR) was employed for site-directed mutagenesis (5). Construction of various plasmids will be described elsewhere.

RNA transcription and transfection. Plasmid DNAs were linearized by Xba I digestion and transcribed with T7 RNA polymerase as previously described (18). Lipofection was used for RNA transfection (16).

Preparation of virus-specific intracellular RNA and Northern (RNA) blotting. Virus-specific RNAs in virus-infected cells were extracted as previously described (20). Northern blotting of intracellular RNA species were described previously (6).

Direct sequencing of the PCR product. Direct PCR sequencing was performed according to the procedure established by Winship (27).

RESULTS

The effect of nucleotide substitutions within the consensus sequence on subgenomic DI RNA transcription

The effect of nucleotide substitutions within the conserved UCUAAAC consensus sequence on MHV subgenomic DI RNA synthesis was observed in a series of constructions. The "parent" construct, MJWT, had a TCTAAAC sequence in the middle of the intergenic region located between genes 6 and 7. MJWT also contained the wild-type regions flanking the intergenic consensus sequence: 0.1 kb upstream and 0.17 kb downstream. Twenty-one MJWT-derived mutants were made; each had a specific nucleotide substitution within the UCUAAAC sequence. These mutants were named according to the site of mutation at the UCUAAAC consensus sequence. For example, MJU3G contained a specific nucleotide substitution from U to G at the third nucleotide of the consensus sequence. In vitro-synthesized DI RNAs were transfected by lipofection into monolayers of DBT cells preinfected with MHV-A59 helper virus(18). Virus samples obtained after overnight incubation were further passaged to generate the passage 1 virus samples. The passage 1 virus sample harvested after 16 h of culture was used as the inoculum for the analysis of intracellular RNA species.

Virus-specific intracellular RNA was extracted at 7 h postinfection and analyzed by Northern blotting using a probe that specifically hybridizes with all MHV RNAs (Fig. 1). Northern blot analysis demonstrated that many of the single nucleotide substitutions in the consensus sequence did not abolish subgenomic DI RNA transcription. Furthermore, except for MJC2G and MJU3G, the mutants supported at least the same level of subgenomic DI RNA synthesis as MJWT. This observation was particularly unexpected for those mutants with G substitutions, because none of the naturally occurring transcriptionally functional intergenic consensus sequences contained G. During isolation of mutant DI cDNAs, a double mutant, MJA5GC7U, with nucleotide substitutions at both nucleotides 5 and 7 was isolated. MJA5G and MJC7U supported efficient subgenomic DI RNA synthesis, whereas no subgenomic DI RNA synthesis was observed in the MJA5GC7U-replicating cells (see Fig. 1). This would seem to indicate that although MHV transcription regulation is sufficiently flexible to recognize a one-nucleotide alteration in the consensus sequence, more than one mutation impedes such recognition.

MHV undergoes high-frequency RNA recombination (2, 15). Possibly, the increase in subgenomic RNA transcription efficiency of the mutants was due to recombination between the mutant DI RNA and the helper virus-derived intergenic sites. Sequences at the intergenic sites confirmed that there was subgenomic DI RNA synthesis from the mutated intergenic consensus sequences. No recombinants were detected in the mutant DIs (data not shown).

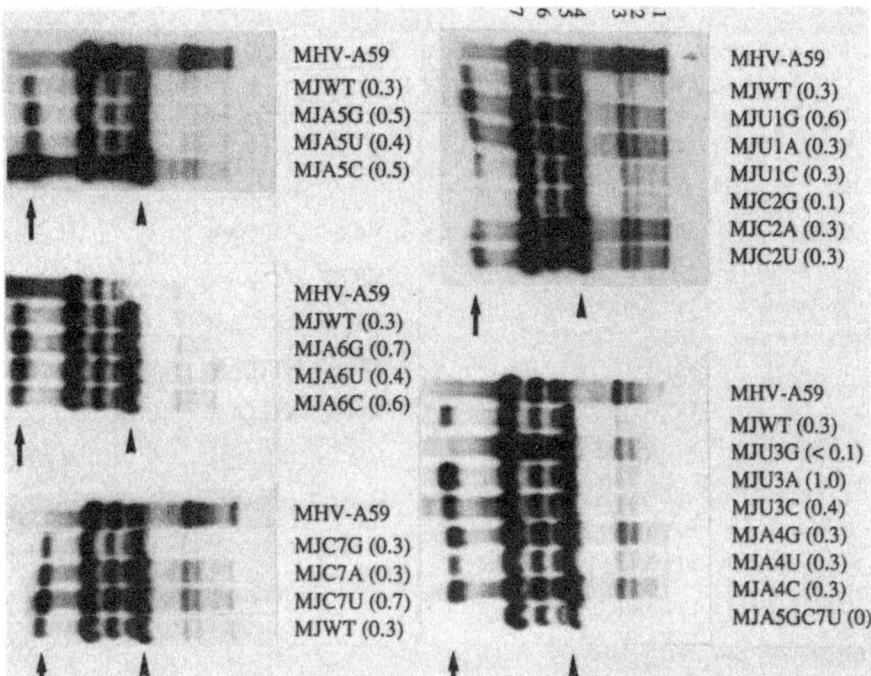

Fig. 1. Northern blot analysis of MJWT-derived mutant subgenomic DI RNAs. Passage 1 virus samples were used for virus inoculum. Intracellular RNAs were extracted 7 h postinfection, separated by 1% formaldehyde-agarose gel electrophoresis and transferred to nylon membrane. The probe was prepared by random-primed ^{32}P-labeling of the MHV-specific cDNA fragment corresponding to the 3'-region of MHV genomic RNA. Numbers 1 to 7 represent MHV-A59-specific mRNA species. Genomic DI RNAs and subgenomic DI RNAs are shown by arrowheads and arrows, respectively. The molar ratios of genomic DI RNA to subgenomic DI RNA are shown in parentheses.

Identification of the leader-body fusion site on subgenomic DI RNAs

The mutant DI RNAs created in the present study were used to locate the exact leader-body fusion site on subgenomic DI RNA. To analyze this region, a subgenomic DI RNA-specific PCR product was prepared. Leader-fusion site PCR products were examined first by direct sequencing. Those PCR products with ambiguous direct-sequencing results were further cloned into a plasmid vector for dideoxy sequencing. Figure 2 shows the results of these sequence analyses. All subgenomic DI RNAs were shown to contain two pentanucleotide sets at the leader-body fusion region. All of them had the UCUAA sequence as a first pentanucleotide. The sequence at the second set differed among the mutants. Mutants with nucleotide substitutions in intergenic site positions three through seven maintained the substituted nucleotides in the second pentanucleotide. Most of the subgenomic DI RNAs transcribed from the genomic DI RNAs with a substituted nucleotide at the first or second position of the consensus sequence demonstrated sequence heterogeneity in the second set. Some maintained the substituted nucleotide, while others contained UCUAA. The leader and body sequences appeared to fuse either at the first or the second nucleotide of the intergenic consensus sequence.

Besides the region examined above, two As upstream of the intergenic consensus sequence also represent an area of sequence homology where leader-body fusion may occur. The possibility that leader-fusion might occur at those upstream As was investigated by creating an additional four mutant DI cDNAs. Each mutant had the upstream two A nucleotides of the consensus sequence deleted and each contained a different substituted

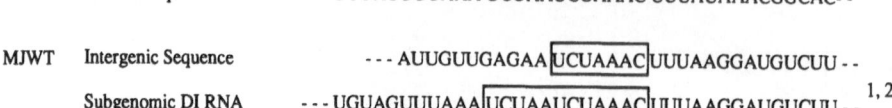

Leader Sequence   - - - UGUAGUUUAAA UCUAAUCUAAAC UUUAUAAACGGCAC- -

MJWT   Intergenic Sequence   - - - AUUGUUGAGAA [UCUAAAC] UUUAAGGAUGUCUU - -

        Subgenomic DI RNA   - - - UGUAGUUUAAA [UCUAAUCUAAAC] UUUAAGGAUGUCUU - - [1,2]

	Intergenic Sequence	Subgenomic DI RNA			Intergenic Sequence	Subgenomic DI RNA
MJU1G	GCUAAAC	UCUAAUCUAAAC [1,2]		MJU3A	UCAAAAC	UCUAAUCAAAAC [2]
		UCUAAGCUAAAC [1]		MJU3C	UCCAAAC	UCUAAUCCAAAC [1]
MJU1A	ACUAAAC	UCUAAUCUAAAC [2]		MJA4G	UCUGAAC	UCUAAUCUGAAC [2]
		UCUAAACUAAAC [2]		MJA4U	UCUUAAC	UCUAAUCUUAAC [1]
MJU1C	CCUAAAC	UCUAAUCUAAAC [1,2]		MJA4C	UCUCAAC	UCUAAUCUCAAC [2]
		UCUAACCUAAAC [1,2]		MJA5U	UCUAUAC	UCUAAUCUAUAC [1]
MJC2G	UGUAAAC	UCUAAUCUAAAC [2]		MJA5C	UCUACAC	UCUAAUCUACAC [2]
MJC2U	UUUAAAC	UCUAAUCUAAAC [1]		MJA6G	UCUAAGC	UCUAAUCUAAGC [2]
		UCUAAUUUAAAC [1,2]		MJA6U	UCUAAUC	UCUAAUCUAAUC [1]
MJC2A	UAUAAAC	UCUAAUCUAAAC [1]		MJC7G	UCUAAAG	UCUAAUCUAAAG [2]
		UCUAAUAUAAAC [1,2]		MJC7A	UCUAAAA	UCUAAUCUAAAA [2]

Fig. 2. Sequence comparison of the 5' end genomic leader sequence, intergenic regions and leader-body fusion sites of MJWT-derived mutants. The intergenic sequence and the leader fusion site on the subgenomic MJWT DI RNA sequences are surrounded by boxes and the corresponding regions of mutants are listed below. The subgenomic DI RNA sequences obtained by direct PCR sequencing of consistent and clear sequencing data are shown by a number 1, while those obtained from cloned PCR products are indicated by number 2. At least three cDNA clones were analyzed.

nucleotide at the first position of the consensus sequence. All the subgenomic DI RNAs had two repeats of the UCUAA sequence and demonstrated no sequence heterogeneity. When the intergenic region contained one consensus sequence and when most of its flanking sequences were conserved, the leader and body fused at the first or second nucleotide of the consensus sequence.

The role of the intergenic region flanking sequences in subgenomic DI RNA synthesis

Two series of mutants were constructed to test the effect of intergenic site flanking sequences on MHV transcription. One set of mutants was altered uptstream of the 18 nucleotide intergenic sequence and the other was altered downstream. Four mutants, each with a 435 nucleotide-long downstream sequence, were tested in the study of the upstream intergenic effect on transcription; the altered upstream flanking sequences contained 1440, 283, 89 and 0 nucleotides. An A nucelotide, two nucleotides upstream of the UCUAA repeats, is referred to as nucleotide 1. The sequence from nucleotides 1 through 18 represents the intergenic region sharing complete sequence homology with the 3'-region of the leader sequence (25). Analysis of the intracellular virus-specific RNA species from these mutants showed that the amount of subgenomic DI RNA was about 80% that of the genomic DI RNA (data not shown). For the downstream flanking sequence study, ten mutants with downstream intergenic flanking sequences ranging in size from 1671 nucleotides to zero nucleotides were examined. Each of these mutants lacked an upstream flanking sequence. Like the effect of the upstream flanking sequence, there seemed to be no appreciable differences in transcription levels of these mutants. These analyses suggested that upstream and downstream sequences flanking the intergenic sequence were not important in the regulation of subgenomic DI RNA transcription.

DISCUSSION

The present study showed that subgenomic DI RNAs were synthesized from almost all the intergenic consensus site mutants. The data indicated that sequence homology between the genomic RNA leader sequence and the intergenic consensus sequence was not the sole determinant of subgenomic DI RNA quantity. Instead, the amount of subgenomic DI RNA was affected by the site and species of the substituted nucleotide. It seemed that sequence integrity within the consensus sequence was more flexible for MHV transcription than was previously thought. A recent description, that sequences differing from the UCUAAAC consensus sequence are also used for MHV transcription (12, 17, 24, 28), is consistent with the data presented here. Therefore, not only the sequence but also the secondary or tertiary structure dictated by the intergenic consensus sequence regulated MHV transcription. Moreover, this structure interacted with a leader sequence structure that was also affected by its primary sequence. The intergenic sequence flanking sequences did not alter the regulation of subgenomic DI RNA synthesis. Although, there is the possibility that other genomic region(s) may regulate the efficiency of MHV subgenomic mRNA synthesis (see below).

If most of the mutants with single nucleotide substitutions within the consensus sequence supported subgenomic RNA transcription, then why is this experimentally "plastic" sequence so highly conserved in the virus? For reasons of simplicity the intergenic region used for this study contained only one UCUAAAC sequence and had flanking sequences derived from the gene 6-7 intergenic sequence. Regulation of transcription at this intergenic region, like any intergenic region, could be an interrelated process; the number of intergenic consensus sequence repeats, the type of flanking sequences and the leader sequence could be involved. It will be interesting to see how nucleotide substitutions at wild type virus-derived intergenic regions affect transcription and how flanking sequences influence the effect of those substitutions. It may be that the molar ratios of subgenomic mRNAs and those of the mRNA translation products are important for efficient MHV replication, and that alterations of these ratios may result in less efficient MHV replication. So far none of the MHV mutants demonstrated significantly altered ratios of any subgenomic RNA species encoding essential replication proteins (20). Therefore, it is possible that an MHV variant producing a different molar ratio of genomic RNA to subgenomic RNA may replicate less efficiently than wild type MHV. Such a variant would be easily eliminated from the virus population after several rounds of virus replication. Consequently, although MHV transcription regulation mechanisms are flexible enough to recognize single nucleotide substitutions in the consensus sequence, it is probably selectively advantageous for the wild type MHV to contain the UCUAAAC consensus sequence in vivo.

Another question raised from the present study is why sequences which are very similar to the UCUAAAC consensus sequence are not recognized for transcription. Analysis of entire MHV-JHM sequences demonstrated that MHV-JHM genomic RNA contains 19 regions where only one nucleotide is different from the UCUAAAC sequence and the UCUAAAC sequence is only present at the transcriptionally functional intergenic regions. The intergenic region between genes 1 and 2 is UCUAUAC, and that between genes 5 and 6 is UCCAAAC. Both of these different consensus sequences seemed to be at least equally as active as the UCUAAAC consensus sequence based on analysis of MJA5U and MJU3C (Fig. 1). The intergenic consensus sequence between genes 1 and 2 has two adjacent upstream and downstream nucleotides which are complementary to the leader sequence (24). The altered intergenic consensus sequence between genes 5 and 6 is also flanked by a number of nucleotides which are complementary with the leader sequence. None of the nonfunctional "intergenic regions" exhibit this flanking sequence homology. This may indicate that if a single nucleotide substitution is present in the consensus sequence, then the presence of adjacent leader-complimentary flanking sequences are necessary for RNA transcription. Alternatively, it is possible that transcription at the nonfunctional "intergenic regions" is prevented by the presence of flanking sequences which down-regulate transcription.

Sequence analysis of subgenomic DI RNA indicated that the leader-body fusion site on the mRNA was most probably the first or second nucleotide of the second consensus sequence repeat. This was consistent with the previous observation that the UCCAAAC sequence present at the intergenic region between genes 5 and 6 is conserved in mRNA 6 (19). All these data indicate that the site of the leader-body fusion was also affected by the nucleotide sequence or the structure of the intergenic sequence or both. This differs from the proposed MHV transcription model which hypothesizes that leader RNA is cleaved by viral

RNA polymerase at the mismatch site between the leader sequence and the intergenic sequence (1).

If MHV subgenomic mRNAs are synthesized by the leader-primed transcription mechanism then how does leader-body fusion take place? One possibility is that only small-sized free leader RNAs with one UCUAA sequence plus one or two additional U and C nucleotides at the 3' end are used for transcription. In this case, sequence complementarity between the free leader RNA and the intergenic sequence would be only two to three nucleotides and the leader RNAs would not need to undergo RNA cleavage. Alternatively, relatively long free leader RNA species may bind to the intergenic region along with virus-derived proteins and host cell factors. After RNA-protein complex formation at the intergenic site, one of the viral proteins might cleave the leader sequence with the resulting leader RNA binding to the intergenic region and priming RNA synthesis (7). If this model is correct, then the leader RNA cleavage site may be determined by nucleotide sequence or RNA structure, and not simply by a nucleotide mismatch between free leader RNA and the intergenic sequence. It is possible that a ribozyme activity could cleave leader RNA, although this activity has not been found in MHV RNA.

An unanswered question from these studies (16) is why the ratio of subgenomic DI RNA to the genomic DI RNA was significantly different from that of the mRNA 7 to mRNA 1 ratio. It was found that the amount of mRNA 1 is only 1.5% of that of mRNA 7 (14), whereas the amount of subgenomic DI RNA was about 80% that of the genomic DI RNA. One possibility is that the amount of helper virus-derived MHV transcription function in DI RNA-replicating cells is too low for efficient subgenomic DI RNA transcription. MHV DI RNA replication strongly inhibits helper MHV RNA replication and transcription (21) leading to the speculation that significantly decreased amounts of MHV-specific transcription factors are present in DI RNA-replicating cells. A limited amount of MHV transcriptional function might not be enough to drive recognition of the template DI RNA molecules and subsequent subgenomic DI RNA transcription. Another possibility is that host derived-factor(s) limit the efficiency of MHV RNA transcription. If MHV RNA transcription takes place at specific sites within the cells or requires host-derived factors(s) or both, then the maximum number of RNA molecules involved in MHV RNA transcription may be saturated after accumulation of large amounts of template RNA molecules. Although genomic DI RNA accumulates more efficiently than MHV genomic RNA, only a small fraction of the DI RNA would be used as a transcription template because of the limited availability of host functions. Alternatively, it is possible that an enhancer sequence(s) for subgenomic RNA synthesis is present somewhere within gene 1 to gene 4 of MHV genomic RNA. The lack of such a sequence(s) in the DI RNA might result in a lower efficiency of subgenomic DI RNA synthesis.

ACKNOWLEDGMENTS

This work was supported by Public Health Service grant AI29984 from the National Institutes of Health.

REFERENCES

1. Baker, S. C., and M. M. C. Lai. EMBO J. 9:4173-4179. (1990).
2. Baric, R. S., K. Fu, M. C. Schaad, and S. A. Stohlman. Virology 177:646-656. (1990).
3. Baric, R. S., S. A. Stohlman, and M. M. C. Lai. J. Virol. 48:633 (1983).
4. Brown, T. D. K., M. E. G. Boursnell, M. M. Binns, and F. M. Tomley. J. Gen. Virol. 67:221 (1986).
5. Higuchi, R. Recombinant PCR, *in* "PCR Protocols," M. A. Innis, D. H. Gelfand, J. J. Sninsky and T. J. White eds. Academic Press, San Diego (1990).
6. Jeong Y. S., and S. Makino. J. Virol. 66: 3339 (1992).
7. Lai, M. M. C. Annu. Rev. Microbiol. 44:303 (1990).
8. Lai, M. M. C., R. S. Baric, P. R. Brayton, and S. A. Stohlman. Proc. Natl. Acad. Sci. USA 81: 3626 (1984).
9. Lai, M. M. C., P. R. Brayton, R. C. Armen, C. D. Patton, C. Pugh, and S. A. Stohlman. J. Virol. 39: 823 (1981).

10. Lai, M. M. C., C. D. Patton, R. S. Baric, and S. A. Stohlman. J. Virol. 46: 1027 (1983).
11. Lai, M. M. C., and S. A. Stohlman. J. Virol. 26: 236 (1978).
12. La Monica, N., K. Yokomori, and M. M. C. Lai. Virology 188:402 (1992).
13. Lee, H.-J., C.-K. Shieh, A. E. Gorbalenya, E. V. Eugene, N. La Monica, J. Tuler, A. Bagdzhadzhyan, and M. M. C. Lai. Virology 180:567 (1991).
14. Leibowitz, J. L., K. C. Wilhelmsen, and C. W. Bond.Virology 114: 39 (1981).
15. Makino, S., J. G. Keck, S. A. Stohlman, and M. M. C. Lai. J. Virol. 57:729 (1986).
16. Makino, S., M. Joo, and J. K. Makino. J. Virol. 65.:6031 (1991).
17. Makino, S., and M. M. C. Lai. Virology 169:227 (1989).
18. Makino, S., and M. M. C. Lai. J. Virol. 63: 5285 (1989).
19. Makino, S., L. H. Soe, C.-K. Shieh, and M. M. C. Lai. J. Viol. 62:3870 (1988).
20. Makino, S., F. Taguchi, N. Hirano, and K. Fujiwara. Virology 139: 138 (1984).
21. Makino, S., K. Yokomori, and M. M. C. Lai. J. Virol. 64:6045 (1990).
22. Pachuk, C. J., P. J. Bredenbeek, P. W. Zoltick, W. J. M. Spaan, and S. R. Weiss. Virology 171:141(1989).
23. Sawicki, S. G., and D. L. Sawicki. J. Virol. 64:1050 (1990).
24. Shieh, C.-K., H.-J. Lee, K. Yokomori, N. La Monica, S. Makino, and M. M. C. Lai. J. Virol. 63:3729 (1989).
25. Shieh, C.-K., L. H. Soe, S. Makino, M.-F. Chang, S. A. Stohlman, and M. M. C. Lai. Virology 156: 321 (1987).
26. Spaan, W., H. Delius, M. Skinner, J. Armstrong, P. Rottier, S. Smeekens, B. A. M. van der Zeijst, and S. G. Siddell. EMBO J. 2: 1939 (1983).
27. Winship P. R. 1989. Nucleic Acids Res. 17:1266 (1989).
28. Yokomori, K., L. R. Banner, and M. M. C. Lai. Virology 183:647 (1991).

CONTROL OF TGEV mRNA TRANSCRIPTION

Julian Hiscox, David H. Pocock and Paul Britton

Division of Molecular Biology, A.F.R.C., Institute for Animal
Health, Compton, Newbury, Berkshire, RG16 0NN, United
Kingdom

ABSTRACT

Coronavirus proteins are translated from a nested set of subgenomic mRNAs which have common 3' termini with unique 5' extensions. Evidence suggests that coronavirus mRNAs are generated by a mechanism of leader primed transcription. Leader RNA binds to consensus sequences upstream of each gene on full length negative strand viral RNA and transcription proceeds to the 5' end of the negative strand to produce the nested set of mRNAs. Even though this gives rise to polycistronic mRNA species only the 5' extension of each mRNA is translated to give the viral proteins. The leader RNA for TGEV is about 90 nucleotides long and contains the sequence which recognises the leader binding sites on the negative strand RNA. Evidence suggests that the length of the leader binding sequence may be involved in transcriptional control of individual mRNAs. In order to investigate this a virus specific mRNA isolation method was developed to measure the relative amounts of mRNAs synthesised during an infection of LLC-PK1 cells with TGEV (strain FS772/70). Thus the relative quantity of each mRNA can be determined and correlated with the variation in size of the leader binding site.

INTRODUCTION

Transmissible gastroenteritis virus (TGEV) is an enteropathogenic virus of pigs, it causes severe diarrhoea and malabsorption which is fatal for piglets younger than two weeks old[1]. TGEV belongs to the family *Coronaviridae*, a group of pleomorphic enveloped viruses containing a single stranded RNA genome of positive polarity[2]. Coronaviruses naturally infect only closely related species, and in most cases virus replication is limited to the epithelial cells of respiratory or enteric tracts[3,4].

The size of the coronavirus genome is between 27 and 32kb making it the largest viral RNA genome so far detected. The single stranded genome is 5' capped and 3' polyadenylated which if introduced into eukaryotic cells acts as an infectious mRNA[5,6]. Genomic RNA is transcribed by the viral RNA polymerase to give a complementary, full length, negative strand RNA. Initiation of subgenomic mRNA transcription is primed by a common RNA leader sequence, derived from the 3'-end of the negative genomic RNA. The leader RNA sequence for TGEV is about 90 nucleotides long and sequence data indicated that the TGEV leader RNA binds to a heptameric sequence, ACUAAAC, upstream of each ORF[7,8]. An exception, is the leader binding sequence of TGEV mRNA 4 which contains a hexameric sequence CUAAAC. This mRNA is the least abundant of TGEV mRNA species. Analysis of the nucleotides upstream of the ACUAAAC sequences showed that the number of homologous residues varied depending on the gene, indicating the possibility that the leader binding sequences may play a role in the transcriptional control of TGEV mRNAs.

To understand transcriptional control of TGEV mRNA it is essential to determine the relationship of the leader binding motif to the amount of mRNA transcribed. In order to determine the amount of mRNA transcribed an assay system was developed to quantify the levels of mRNA produced during an infectious cycle.

METHODS

Virus and Cells

The FS772/70 strain of TGEV was grown in a porcine kidney cell line LLC-PK1. For the isolation of TGEV RNA cells were grown to confluence in 250cm^2 tissue culture flasks and infected at a m.o.i. 10. Two hours post infection the cells were washed, fresh media added, and at 13h pi pulse labelled for three hours with 200μCiml^{-1} of [5,6-^{3}H]uridine. Cellular mRNA synthesis was inhibited by the addition of 1μgml^{-1} of actinomycin D.

Preparation of Magnetic Beads

The Dynabeads were resuspended by shaking the vial and 1mg dispensed into a 1.5ml eppendorf tube, washed twice with 200μl of PBS, pH 7.5 containing 0.1% BSA (washing was facilitated by using the magnetic particle concentrator). For RNA work the beads were washed twice with 200μl of DEPC-treated 0.1M NaOH, 0.5M NaCl and once with DEPC treated 0.1M NaCL and resuspended in 20μl binding and washing buffer (BW; 10mM Tris-HCl pH 7.5, 1mM EDTA and 2M NaCl). Biotinylated oligonucleotide (600pmol) was added to the beads and incubated at 23°C for 30min with continuous agitation. Unbound oligonucleotide was removed by washing the beads four times with BW buffer.

RNA Isolation and Purification

Total cellular RNA from TGEV infected cells was isolated using guanidinium isothiocyanate and β-mercaptoethanol[9], resuspended in 50μl of DEPC-H$_2$O and 20μl aliquots used for purification. TGEV RNA was purified using Dynabeads® M-280 Streptavidin magnetic beads coupled to an antisense biotinylated oligonucleotide (Biotin-GTATATCACTATCAAAAGGAAAA) specific to the 3'-end of TGEV RNAs (see Fig.1).

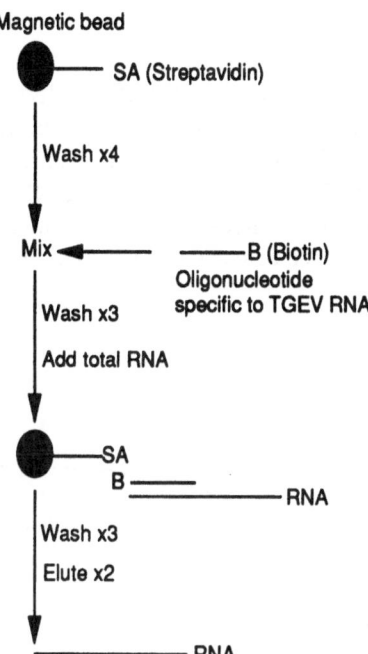

Figure 1.   Schematic representation of separating viral mRNA from cellular RNA. An antisense oligonucleotide specific to TGEV RNA was mixed with Dynal streptavidin magnetic beads. Unbound oligonucleotide was removed and total RNA added and hybridization allowed to proceed for 160min. Unbound RNA was removed during washing and bound virus RNA eluted using DEPC-H_2O.

Aliquots of total cellular RNA (40µg) from infected cells were heated at 65°C for 3min, cooled on ice and made up to 6xSSC and 0.1%SDS in 80µl.The mixture was added to the magnetic beads and the TGEV RNA hybridized to the bound oligonucleotide at 23°C for 3h. The beads were washed three times with 6xSSC and 0.1% SDS at 23°C and the RNA eluted by the addition of 100µl DEPC-H_2O and heating at 65°C for 3min. The RNA elution step was performed twice, the RNA precipitated with ethanol and resuspended in DEPC-H_2O. To determine when optimum binding of virus RNA to the biotinylated/oligonucleotide magnetic bead complex occurred, virus RNA was purified at various time points and resuspended in 200µl DEPC-H_2O, mixed with 3ml of Optiphase Safe (LKB) and counted in a Tri-CARB beta counter (Packard). The count obtained was taken to be directly proportional to the amount of virus RNA that had been purified. Thus optimum binding occurred after 160min (see Fig.2).

Analysis of RNA

Purified viral RNA was denatured with glyoxal and dimethylsulphoxide prior to separation by electrophoresis on a 1% agarose gel for 3hr at 85 volts. To enhance the radioactive signal, the gel was soaked in 0.8M sodium salicylate for 30min, dried onto Whatman 3MM paper and exposed to pre-flashed X-ray film at -70°C for 96hr.

RESULTS AND DISCUSSION

In order to calculate the amounts of each TGEV mRNA species produced during an infectious cycle an assay system was developed. The [³H]-uridine labelled mRNA species were separated by agarose gel electrophoresis after denaturation of the RNA by glyoxal. However, as can be seen from Fig. 3a analysis from total RNA is not possible due to the presence of the 28 and 18S ribosomal RNA (observed as unlabelled bulges) and potential breakdown products or the presence of incomplete transcripts.

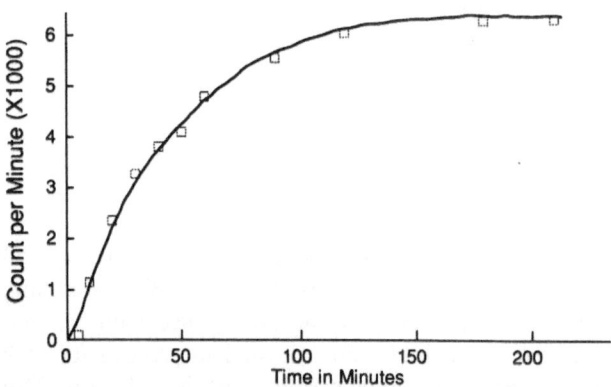

Figure 2.   Kinetics of labelled virus mRNA hybridization. Virus RNA was purified at various time points and binding measured using liquid scintillation counting. Optimum binding occurred after 160min.

Poly (A⁺) mRNA is often extracted from total cellular RNA with either poly-dU or oligo-dT bound to a solid matrix, the most common method is column chromatography though recently the use of oligo-dT bound to magnetic beads is proving to be popular[10,11]. A problem associated with oligo-dT columns is that ribosomal RNA can still contaminate the mRNA requiring multiple purification steps resulting in loss of mRNA . The use of magnetic beads allows repeated washing steps without loss of mRNA. A further problem associated with oligo-dT columns for the purification of mRNA is size selection in favour of smaller RNAs, either due to poorer hybridization of the larger species or that the larger species are eluted and lost during the washing steps[12]. Results using magnetic beads associated with oligo-dT have proved useful in the isolation of poly (A⁺) TGEV RNA. However in addition to the TGEV poly (A⁺) RNA, cellular poly (A⁺) RNA was also isolated which may lead to a decrease in the amount of TGEV RNA isolated or effect the isolation of the larger TGEV mRNA species. To overcome the problems associated with these methods it was decided to develop a viral specific purification method.

An advantage of the magnetic bead system is the use of streptavidin[13] bound magnetic beads as the solid support. Simply a biotinylated oligonucleotide, specific for TGEV RNA, can be bound to the beads and used to purify TGEV RNA. As can be seen from Fig. 3b only viral RNA is present following purification, the cellular ribosomal RNA has been removed. By using a specific oligonucleotide complementary to the 3' end of the viral genome any contamination associated with breakdown products and incomplete transcripts should be greatly reduced.

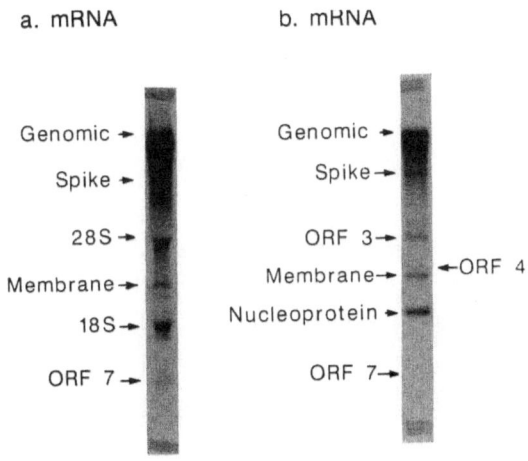

Figure 3.   Labelled TGEV mRNA after glyoxal/DMSO gel electrophoresis. (a) RNA from total cell preparations. The bulges are due to unlabelled cellular ribosomal RNA. (b) Sequence specific purification of virus RNA removes cellular RNA thus eliminating the bulges caused by ribosomal RNA.

In pulse labelling studies, cellular mRNA synthesis is greatly reduced by the addition of actinomycin D. This compound can affect virus replication and in some cases lead to reduced titres of progeny virus[14,15]. Using a sequence specific method to capture labelled and unlabelled virus RNA and not cellular RNA it would be possible to eliminate the use of actinomycin D during some pulse labelling studies.

This purification method can be used to investigate potential early and late gene activity by measuring the relative amounts of mRNA produced throughout infection and to characterise the relationship between the TGEV intergenic motif and the leader polymerase complex.

REFERENCES

1. L.P. Doyle and L.M. Hutchings, *J. Amer. Vet. Med. Assoc.* **108**:257-259 (1946).
2. D.J. Garwes, D.H. Pocock and B.V. Pike, *J. Gen. Virol.* **32**:283-294 (1976).

3. K. McIntosh, R.K. Chao, H.E. Krause, R. Wasil, H.E. Mocega and M.A. Mufson, *J. Infect. Dis.* **130**:502-510 (1974).

4. H. Wege, S. Siddell and V. ter Meulen, *Curr. Top. Microbiol. Immunol.* **99**:165-200 (1982).

5. B. Lomniczi, *J. Gen. Virol.* **36**:531-533 (1977).

6. R.S. Baric, S.A. Stohlman and M.M.C. Lai, *J. Virol.* **48**:633-640 (1983).

7. K.W. Page, P. Britton and M.E.G. Boursnell, *Virus Genes* **4**:289-301. (1990).

8. P.B. Sethna, S.L. Hung and D.A. Brian, *Proc. Natl. Acad. Sci. USA* **86**:5626-5630 (1989).

9. P. Chomczynski and N. Sacchi, *Anal. Biochem.* **162**:156-159 (1987).

10. K.S. Jacobsen, E. Breivold and E. Hornes, *N.A.R.* **18**:3669 (1990).

11. M. Uhlen, *Nature.* **340**:733-734 (1989).

12. S. Ekenberg, M. McKormick, L. Wu and C. Smith, *Promega Notes.* **39**:7-10 (1992).

13. A. Pahler, W.A. Hendrickson, M.A. Gawinowicz-Kolks, C.E. Argarana and C.R. Cantor, *J. Biol. Chem.* **262**:13933-13937 (1987).

14. M.C. Clarke, *J. Gen. Virol.* **3**:276-270 (1968).

15. D.A. Kennedy and C.M. Johnson-Lussenberg, *J. Virol.* **29**:401-404 (1979).

AN INTRALEADER OPEN READING FRAME IS SELECTED FROM A HYPERVARIABLE 5' TERMINUS DURING PERSISTENT INFECTION BY THE BOVINE CORONAVIRUS

Martin A. Hofmann, Savithra D. Senanayake, and David A. Brian

Department of Microbiology
University of Tennessee
Knoxville, TN 37996-0845

INTRODUCTION

In an effort to unambiguously establish the 5' terminal nucleotides on bovine coronavirus (BCV) mRNAs, we developed a method that employs the head-to-tail ligation of single-stranded cDNA from extended primers before PCR amplification, cloning and sequencing[1,2]. Only the mRNAs for the N, M, and S structural proteins were studied in this manner, and mRNAs isolated from cells at various times postinfection through 432 days were studied. We learned that the 5' terminal five nucleotides were hypervariable among four major terminus types, and that certain types predominated at various times postinfection. The type I terminus (GAUUGUG...) predominated within the first 4 days postinfection and was considered the wild type terminus. At 296 days postinfection and beyond, the type II terminus (GAUUAUG...) predominated. Types II, III and IV all possessed a G → A mutation that gave rise to an AUG codon and established an intraleader ORF for 11 amino acids. Although we have not yet been able to show the existence of the peptide in persistently infected cells, we have demonstrated that the intraleader ORF attenuates translation of downstream ORFs during translation *in vitro*.

We therefore hypothesize that the intraleader ORF serves to attenuate virus replication and that this may be a mechanism by which coronaviruses maintain a persistent infection.

MATERIALS AND METHODS

The infection of HRT-18 cells with the Mebus strain of BCV, establishment of persistent infection, and RNA extraction from cells were done as previously described[3].

Primer extensions (using oligodeoxynucleotides that bind within the body of each mRNA studied), head-to-tail ligations (with RNA ligase), PCR amplification (using primers that amplify across the head-to-tail junction), cloning, and DNA sequencing, have all been described[1,2].

Preparation of RNA transcripts of the cloned N mRNA having the 5' type I terminus (GAUUGUG...) and *in vitro* translation procedures were carried out as previously described[4]. The type II terminus (GAUU<u>AUG</u>...) was prepared from the N-containing clone (pLN, having the type I terminus) by oligodeoxynucleotide-directed *in vitro* mutagenesis using the Amersham mutagenesis kit.

RESULTS

A total of 337 clones of the N, M, and S mRNA termini were sequenced and the results of 307 clones are summarized in Fig. 1. Thirty of the 337 clones had 5'-ward extensions of 1 to 6 nucleotides (i.e., 5'-ward of base #1 in a type I, II or III terminus) and were not included in the data on Fig. 1 except to determine percentage values. The origin of the additional 5'-terminal bases is unknown to us at this time. Several of the clones had the 5'-most base (i.e., base #1) missing and, for this study, we concluded that these were the result of premature termination of first-strand cDNA synthesis during the cloning procedure. With the first base missing, the identity of the terminus was still apparent and the truncated clones were scored as one of the four terminus types.

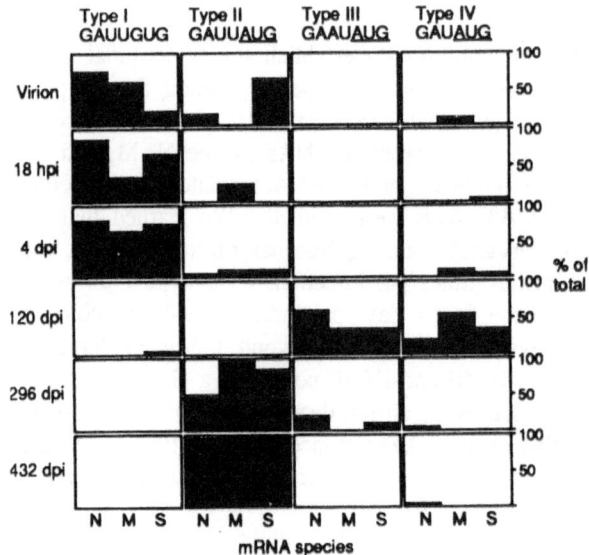

Fig. 1.   Occurrence of the four types of 5' termini at different times postinfection. Results for N, M and S mRNA species are depicted separately. Variant types resulting from 5'-ward extension (described in the text) are considered in the percentage calculations.

The G → A mutation in termini types II, III and IV caused the appearance of a methionine codon which begins an 11 amino acid ORF within the leader (Fig. 2). The potential peptide is neutral and hydrophilic, and its synthesis would be terminated by a weak UGA (umber) codon.

To (i) determine whether a peptide is translated from the ORF, and (ii) examine the effects of the intraleader ORF on the translation of a BCV mRNA, the cloned N mRNA (pLN) that yields T7 polymerase-generated transcripts with a type I (GAUUGUG...) terminus was mutated to yield transcripts with a type II (GAUUAUG...) terminus. To date we have not been able to identify an 11 amino acid (1.2 kDa) peptide by *in vitro* translation and radiolabeling with ^{35}S-met or ^{35}S-cys. To examine the effect of the intraleader ORF on translation of downstream ORFs,

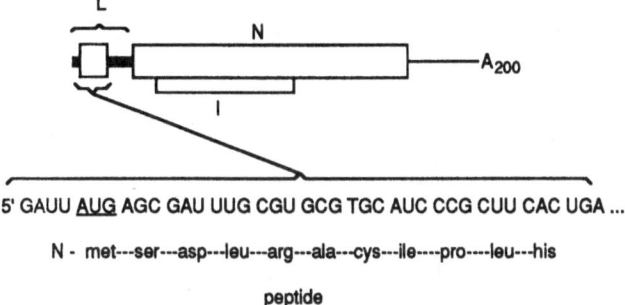

5' GAUU <u>AUG</u> AGC GAU UUG CGU GCG TGC AUC CCG CUU CAC UGA ...

N - met---ser---asp---leu---arg---ala---cys---ile----pro----leu---his

peptide

Fig. 2. Schematic diagram showing the nucleotide sequence and resulting amino acid sequence of the intraleader ORF on the BCV N mRNA. The I protein ORF (named for internal open reading frame) which exists on the bicistronic N mRNA[4] is also shown.

transcripts of pLN-type I and pLN-type II were generated, quantitated by Northern analysis, and equal amounts were translated in parallel. Quantities of N and I proteins, both products of downstream ORFs on the N mRNA[4], were determined by imaging on the Ambis Radioanalytic Imager (Fig. 3).

Fig. 3 illustrates that in the presence of the intraleader ORF, synthesis of N is decreased by 40% and I is decreased by 10%.

DISCUSSION

We do not know the mechanism by which the 5' sequence heterogeneity was generated. We do not believe it arose as a function of the analytical procedure (i.e., by the action of enzymes used in the copying, amplifying and cloning of the RNA) since a cloned transcript of known sequence (i.e, type type I terminus, GUAAGUG...) when prepared, transfected into cells, recovered and carried through the amplification and cloning procedure, yielded no variants (in 18 of 18 clones).

We think it is very likely that a variety of 5' termini could have been present in the virus inoculum and passed on by infectious virus. Despite the fact that the virus inoculum used had been sequentially plaque purified three times, it is conceivable that more than one leader-containing genome, mRNA, or defective-interfering RNA molecule serving as potential leader donors could have been packaged in a single particle and carried through the plaque purification process. Such molecules would have become the source of the variant 5' termini, possibly by a leader switching mechanism[5]. We have demonstrated that subgenomic mRNA and defective-interfering molecules of BCV do become packaged[3]. To determine whether 5' terminus hypervariability arises denovo awaits production of a clone of BCV derived from an infectious genomic molecule.

To our knowledge, a coronavirus intraleader open reading frame has not been previously reported. But then, published coronavirus leader sequences have been determined only on RNA obtained from cells immediately after an acute infection[6-11].

Although the mechanism(s) by which variant termini become distributed among the mRNA species and establish predominance at certain times during persistent infection are not known, we propose that the selection of leaders with an internal open reading frame is causally related to the persistent infection. Namely, we suggest that the

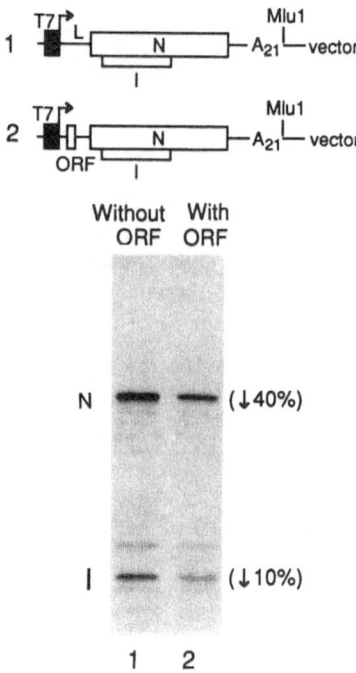

Fig. 3. Effect of the intraleader ORF on the translation of N and I proteins *in vitro*. Equal amounts (1 μg each as quantitated by RNA blot hybridization) of transcripts made *in vitro* by T7 RNA polymerase were translated in wheat germ lysate in the presence of ^{35}S-methionine and the radioactive products were analyzed by SDS-polyacrylamide gel electrophoresis and quantitated by the Ambis Radioanalytic Imager.

intraleader ORF serves to attenuate virus gene expression (translation), which in turn attenuates virus replication. Since cells infected with a less virulent (i.e., attenuated) virus are less likely to die, cells infected with the attenuated virus are more likely to survive as persistently infected cells. Thus, selection of the intraleader open reading frame may be a mechanism by which coronavirus maintain a persistent infection.

REFERENCES

1.   Hofmann, M. A., and D. A. Brian. 1991. *PCR Methods and Applications* **1**:43-45.
2.   Hofmann, M. A., and D. A. Brian. 1991. *BioTechniques* **11**:30-31.
3.   Hofmann, M. A., P. B. Sethna, and D. A. Brian. 1990. *J. Virol.* **64**:4108-4114.
4.   Senanayake, S. D., M. A. Hofmann, J. L. Maki, and D. A. Brian. 1992. *J. Virol.* **66**:5277-5283.
5.   Makino, S., and M. M. C. Lai. 1989. *J. Virol.* **63**:5285-5292.
6.   Brown, T. D. K., M. E. G. Boursnell, and M. M. Binns. 1984. *J. Gen. Virol.* **65**:1437-1442.
7.   Lai, M. M. M. C., R. S. Baric, P. R. Brayton, and S. A. Stohlman. 1984. *Proc. Natl. Acad. Sci. USA.* **81**:3626-3630.
8.   Kamahora, T., L. H. Soe, and M. M. C. Lai. 1989. *Virus Research* **12**:1-9.
9.   Schreiber, S. S., T. Kamahora, and M. M. C. Lai. 1989. *Virology* **169**:142-151.
10.   Sethna, P. B., M. A. Hofmann, and D. A. Brian. 1991. *J. Virol.* **65**:320-325.
11.   Shieh, C.-K., L. H. Soe, S. Makino, M.-F. Chang, S. A. Stohlman, and M. M. C. Lai. 1987. *Virology* **156**:321-330.

EFFECTS OF MOUSE HEPATITIS VIRUS INFECTION ON HOST CELL METABOLISM

Stanley Tahara[1], Cornelia Bergmann[12], Gary Nelson[12], Richard Anthony[12], Therese Dietlin[1], Shigeru Kyuwa[3], and Stephen Stohlman[12]

Departments of Microbiology[1] and Neurology[2], University of Southern California School of Medicine, Los Angeles, CA and the Department of Animal Pathology[3], University of Tokyo

ABSTRACT

A time dependent decrease in cell surface expression of major histocompatibility complex (MHC) class 1 proteins was found during JHMV infection of the mouse macrophage J774.1 cells line by radioimmunoassay. MHC class I, actin and CSF-1 receptor mRNA levels were also found to decrease during infection. Surprisingly, not all host cell mRNA were similarly affected, suggesting that the apparent MHV-induced translational shut off of host cell protein synthesis during infection was specific for only some host cell mRNAs. Interestingly, two mRNAs found to be refractory to JHMV infection encode monokines, suggesting a role in pathogenesis. To understand the mechanism(s) of this preferential mRNA stability and the apparent shut off of host cell mRNA, translation lysates were prepared from infected and uninfected cells. Translation of host mRNAs in these extracts showed no apparent loss of translational ability in the infected cells vs. the uninfected cells; however, viral mRNAs were preferentially translated in the lysates from the infected cells. Chimeric mRNAs containing the MHV leader upstream of a globin reporter gene showed that preferential translation was a property of the MHV leader RNA. Deletional analysis showed that the sequences responsible for this cis translational augmentation are in a 12 nucleotide (nt) tract at the 3' end of the leader. The previously reported interaction of the nucleocapsid protein with these nts suggest that it may play a role in translational augmentation of MHV mRNAs.

INTRODUCTION

Many cytolytic viruses divert host cell macromolecular synthesis for use by the virus. Indeed, the preferential shut off of host cell translation is a property common to many cytolytic viruses[1]. The mechanisms used by viruses include alterations in host translation factors, increased turnover of host mRNAs, and competition due to vastly increased viral mRNA levels. Analysis of protein synthesis during mouse hepatitis virus (MHV) infection of susceptible cells shows a conversion to viral protein synthesis at the expense of host protein synthesis[2]. MHV has been suggested to actively suppress host cell translation through a combination of effects manifested by a loss of polysomes and degradation of host mRNAs[3].

METHODS

Virus and cell lines: The DL isolate of JHMV and the A59 strain of mouse hepatitis virus (MHV) were propagated and plaque assayed as previously described[4]. Northern blot analyses were carried out using the J774.1 Balb/c monocytic cell line[5]. Translation lysates were prepared from either the J774.1 or DBT murine astrocytoma cell lines as previously described[6]. Lysates were prepared from uninfected DBT cells grown in suspension using Joklik's modified MEM while the lysates from infected cells were prepared from monolayers infected with MHV-A59 an m.o.i. = 5 to 10 at 4.5 - 5.0 hr post infection.

Northern Analysis: Infected and uninfected J774.1 cells were lysed in guanidine isothiocyanate and the RNA purified by centrifugation through CsCl. RNA (30µgm) was analyzed by electrophoresis through 1 to 1.2% agarose gels containing 0.66mM formaldehyde and transferred to High Bond N (Amersham) membranes. The membranes were hybridizided with probes specific for actin, Il-1a, Ilb, TNFa, tubulin, fms (the CSF receptor) 28S ribosomal RNA, interferon (IF)-a and MHC class 1 and class 2. Probes were labeled by random priming and hybriziation was carried out for 24-96 hrs at 42 C. RNA levels were determined by densitometry.

Plasmids: Chimeric plasmids containing the MHV mRNA #6 leader sequence 5' of the human globin sequence were cloned by PCR and inserted using standard techniques. All constructions were confirmed by sequence analysis.

In vitro translation: Cell-free extracts were treated with 125 U/ml micrococcal nuclease at 21 C for 0-12 min. In vitro translations were performed in a 15µl volume containing 9µl of nuclease treated extract and 10 µCi ^{35} S-methionine (Amersham, Arlington Heights, IL) 60 min at 37 C. Synthetic, capped mRNAs were added at a concentration of 16.7 µg/ml. Reactions were terminated by addition of 1 ml of cold acetone. Precipitates were collected, resuspended in Laemmli sample buffer and analyzed on 15% SDS-polyacrylamide gels (SDS-PAGE). Bands were quantitated by densitometery.

RESULTS

The mRNA levels in J774.1 cells infected with JHMV were determined by northern blot to insure the quality of the RNA and quantitated by densitometry and compared to the levels in uninfected cells. Figure 1a shows that there is little variation in the quantity of 28S ribosomal RNA. In contrast, decreased levels of actin mRNA were found consistent with previous data[3]. In addition, the levels of MHC class 1 mRNAs were also decreased, consistent with decreased cell surface expression as determined by radioimmunoassay[7]. Figure 1a also shows that the CSF-1 receptor mRNA decreased during infection. Three additional mRNAs were analyzed, MHC class 2, IL-Iα and IF-α. No mRNA for these two putative macrophage mRNAs could be detected, although MHC class 2 mRNA was detected after IFN-γ induction of uninfected cells (data not shown).

Not all host cell mRNAs decreased following JHMV infection. Figure 1b shows the analysis of tubulin, IL-1β and TNF mRNAs. All three of these mRNAs increased following infection. Neither the TNF nor IL-1β mRNAs were at significant levels in the uninfected cells. Treatment of the cells with 1 μg/ml actinomycin D prevented the accumulation of these mRNAs, suggesting that infection activated transcription and/or a

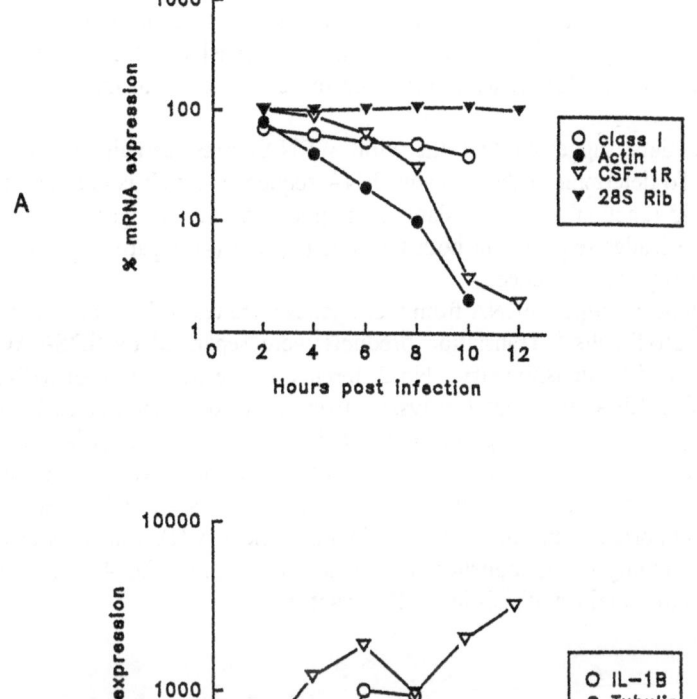

Figure 1. Relative mRNA levels determined from northern blots. Panel A: mRNA levels that decrease during infection. Panel B: mRNA levels that increased during infection. Levels are relative to those found in uninfected cells.

post transcription process. The most rapid and dramatic increase was seen in TNF mRNA which increased throughout infection. The induction and stability of these mRNAs suggested that the proteins were synthesized throughout the acute infection. Preliminary experiments indicate that TNF is indeed released from infected cells.

These data suggested that the conversion of protein synthesis from host to viral with time during infection may be abnormal in the J774.1 cells infected with JHMV. Analysis of protein synthesis in these cells infected with JHMV showed decreased host cell protein synthesis by 6 hr post infection accompanied by an increase in JHMV proteins. Later time points show only viral proteins being synthesized suggesting the possibility of translational regulation of protein synthesis in infected cells.

To determine if MHV altered host cell translation, lysates were prepared from either uninfected and JHMV-infected J774.1 cells or uninfected and A59-infected DBT cells for in vitro translation. Initial experiment examined the efficiency of translation of globin and CAT as model host genes and poly A selected MHV RNA from infected DBT cells. The globin mRNA was translated equally well in all lysates indicating that the apparent decrease in translation of host cell mRNAs is not due to diminished translation, and further indicating that there is no loss in host cell translation factors. By contrast, the poly A-selected viral mRNAs were translated more efficiently in the lysates of infected cells.

These data suggested that the viral mRNAs may contain a cis translational augmentor. To determine if the 5' leader RNA sequence, which is common to all MHV mRNAs, could function as the translational augmentor, chimeric genes were constructed containing the leader sequence in both the plus (+) and (-) negative orientations 5' of a human globin reporter sequence.

Synthetic 5' capped mRNA from these genes were tested in lysates from uninfected and infected DBT cells. Translation products were separated by SDS-PAGE and the globin quantitated by densitometry. No difference in the translation of native globin or synthetic CAT mRNAs was found in lysates from infected vs. uninfected DBT or J774.1 cells, therefore the quantity of globin translated was used for comparison to control for lysate differences. Translation of these mRNAs in the lysates from infected and uninfected cells showed that translation in the infected lysates was enhanced only by MHV leader in the (+) orientation. In addition, deletion of the 3' MHV leader sequence which is the N protein binding site 8, abolishes the enhanced translation (Fig. 4). Table 1 shows the relative levels of protein synthesis in the two lysates.

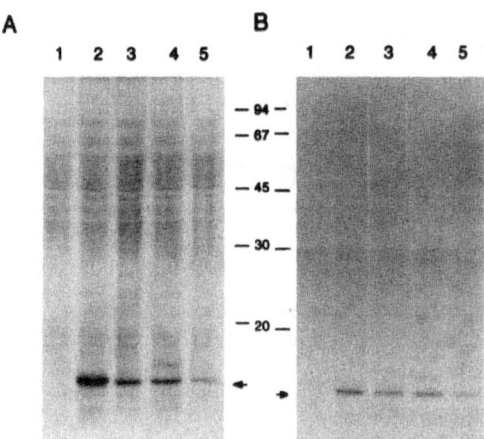

Figure 2. Translation of the MHV-leader globin chimeric mRNAs in lysates from infected (Panel A) and uninfected (Panel B) DBT cells using 250 ng of RNA per reaction. Lane 1: No mRNA added; Lane 2: haGL1 (Leader +); Lane 3: haGL12 (Leader -); Lane 4: haGL 62-74 (Leader deletion); Lane 4: haG (globin).

Table 1. Relative activity of MHV-leader globin chimeric RNAs.

mRNA	Uninfected Cells[1]	Infected Cells
haG	1.00	1.00
haGL1	2.70 ± 0.60[2]	10.31 ± 1.70
haGL12	2.38 ± 0.65	2.94 ± 0.68
haGL62-74	1.63 ± 0.35	3.39 ± 0.37

[1]Lysates of DBT cells. [2]Mean $\leq$ standard deviation

DISCUSSION

The data in this report contain a number of novel concepts relative to the biology and immunobiology of MHV infection. The analysis of host cell mRNA stability during infection indicated that not all host mRNAs are degraded, even though viral protein translation becomes predominant. These data suggest three fundamentally important concepts in the biology of MHV infection. First, in contrast to other cytolytic viral infections[1] MHV infection does not prevent the translation of host mRNAs. Secondly, the differential stability of the host mRNAs indicates that the degradation of host cell mRNA must be regulated since virus is able to persist without cytopathology, both in vitro and in vivo[8,9]. Third, the ability of the macrophage to upregulate mRNA levels for two cytokines following infection, one involved in enhancing antigen presentation (IL-1), and the other with antiviral activity (TNF) suggest that these cells are able to at least partially circumvent the deleterious effects of infection. TNF is cytolytic for oligodendroglial cells[11], suggesting that a vigorous immune response to infection within the CNS, particularly the release of TNF following infection of newly recruited macrophages, may contribute to JHMV-induced CNS demyelination.

Analysis of translational regulation in MHV-infected cells showed that there is no loss of the cells ability to translated host mRNAs. This is consistent with the 5' capped, poly adenylated nature of the MHV mRNAs and the ability of virus to persist in vitro and in vivo[2,8,9]. However, observations in a number of cell lines has shown that during infection there is decreased host protein translation with concommitant increases in viral translation[2]. Analysis of chimeric mRNAs clearly demonstrated that the 5' MHV leader RNA sequence functions as a cis augmentor and is at least partially responsible for the apparent increase in viral over host cell translation. The presence of this cis effect only in the lysates from the infected cells further suggested the presence of a transacting factor. Since this putative augmentor would bind leader RNA and such an interaction has been described[11], the N protein binding sequence (nts 62-74) was deleted. This deletion abolished the cis augmentation effect. Preliminary experiments replacing the 3' leader nts that comprise the N protein binding site within the 3' end of the native globin leader sequence show enhanced translation in infected vs. uninfected lysates. Although there is no direct proof that the N protein functions as the transacting protein, the kinetics of N protein synthesis and its subcellular distribution are both consistent with a major regulatory role in MHV infection, possibly controlling both transcription and translation as well as fulfilling its structural role in forming the helical nucleocapsid.

Acknowledgements

SMT was supported by grants from the NIH (GM38512, NS26991), the Life and Health Insurance Medical Research Fund and the Margaret E. Early Research Trust. RPA was supported by NIH grant NS07149. SAS was supported by NIH grant NS18146. CB was supported by grants from the NIH (NS26991) and the Universitywide AIDS Task Force.

References

1. Schneider, R.I., and T. Shenk. Ann. Rev. Biochem. 56:317-332, (1987).
2. Lai, M.M.-C. Ann. Rev. Microbiol. 44:303-333, (1990).
3. Hilton, A., Mizzen, L., Macintyre, G., Cheley S., and R. Anderson. J. Gen. Virol. 67:923-932, (1986).
4. Stohlman, S.A., Brayton, P.R., Fleming, J.O., Weiner,L.P., and M.M.-C. Lai. J. Gen. Virol. 63:265-275, (1982).
5. Stohlman, S.A., Kyuwa, S., Cohen, M., Bergmann, C., Polo, J.M., Yeh, J, Anthony, R., and J.G. Keck. Virol. 189:217-224, (1992).
6. Tahara, S.M., Dietlin, T.A., Dever, T.E., Merrick, W.C. and L.M. Worrilow. J. Biol. Chem. 266:3594-3601, (1991).
7. Williamson, J., Kyuwa, S., Wang, F.-I. and S.A. Stohlman. Adv. Exp. Biol. Med. 276:557-563, (1990).
8. Stohlman, S.A., Baric, R.S., Nelson, G.W., Soe, L.H., Welter, L.M. and R.J. Deans. J. Virol. 4288-4295, (1988).
9. Stohlman, S.A., and L.P. Weiner. Arch. Virol. 57:53-61, (1978).
10. Kyuwa, S. and S.A. Stohlman. Seminar Virol. 1:273-280, (1990).
11. Robbins, D., Shirazi, Y., Drysdale, B., Lieberman, A., Shin, H. and M. Shin. J. Immunol. 139:2593-2597, (1987).

THE EFFECT OF AMANTADINE ON MOUSE HEPATITIS VIRUS REPLICATION

Julian L. Leibowitz[1,2] and S. Jeffrey Reneker[1]

[1]Department of Pathology and Laboratory Medicine
[2]Department of Microbiology and Molecular Genetics
University of Texas Medical School
Houston, TX 77225

INTRODUCTION

Amantadine has been known to be a potent inhibitor of influenza A virus infection for many years (1). Although amantadine's antiviral effect on many other viruses is much less potent that observed for influenza A, it does have a fairly broad antiviral spectrum (2,3). The extreme sensitivity of influenza A virus to the drug has been shown by genetic and molecular biologic studies to reside with the M2 protein. Amantadine interacts with M2 and inhibits its ability to function as a pH-gated ion channel, and thus appears to interfere with a step in influenza virus uncoating and assembly (4-6). Amantadine exerts its more broad-spectrum antiviral effects by virtue of its lysomotropic properties. The drug accumulates in endocytic vesicles, raises their pH, and thus interferes with the ability of viruses requiring low pH to complete uncoating and penetration from entering the cytosol (2,3).

The effect of amantadine on coronavirus replication has not been fully investigated. It has been noted that the bovine coronavirus (BCV) is sensitive to amantadine, probably at a post-uncoating stage in the viral replicative cycle (7). In this work, we report that, like BCV, mouse hepatitis virus (MHV) is sensitive to amantadine at doses comparable to those reported to inhibit Semliki forest virus and vesicular stomatitis virus (VSV). However, unlike these viruses, amantadine exerts its antiviral effects on MHV at a late stage in viral replication.

MATERIALS AND METHODS

Cells and Virus

The origin and maintenance of the cell lines used in these studies has been reported previously (8). The origin, and growth of the MHV-A59, MHV-JHM, and

Coronaviruses, Edited by H. Laude and J.F. Vautherot
Plenum Press, New York, 1994

VSV stocks used in these studies has been described (8). Plaque reduction, assays were performed by diluting viral stocks to approximately 100 PFU/ml in medium containing various concentrations of amantadine hydrochloride (Sigma Chemicals), letting the virus adsorb to monolayers of L2 cells for 60 minutes, and then overlaying the cells with media containing 0.8% agarose and the same concentration of amantadine as the virus inoculum. Plaques were stained and enumerated at 2 days post infection.

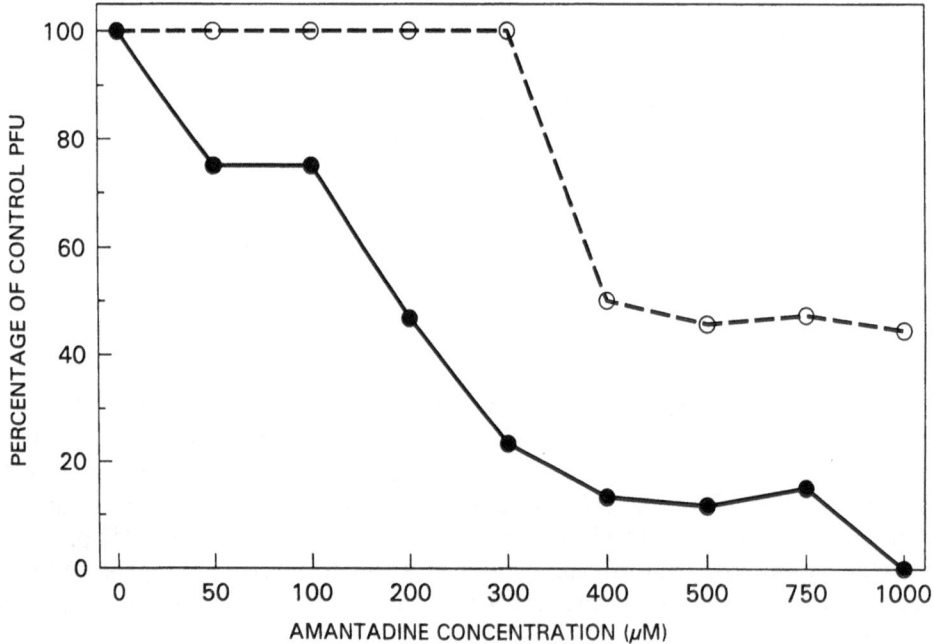

Figure 1. The effect of amantadine addition on plaque formation by MHV and VSV. Approximately 100 PFU of MHV (●) and VSV (o) were assayed in the presence of various concentrations of amantadine and in the absence of the drug. Results are expressed as a percentage of the plaques observed in the absence of amantadine.

RNA preparation and gel electrophoresis

Cells were incubated with ^{32}P-orthophosphate from 7-8 hours post infection in the presence of 5 μg/ml actinomycin D. RNA was extracted with guanidium-thiocyanate as described (9) and electrophoresed on 0.8% agarose gels containing formaldehyde as a denaturant.

Antibodies, immunoprecipitation and immunofluorescence

A monospecific polyclonal goat antisera directed against the MHV-A59 S protein was graciously provided by Dr. K.V. Holmes. The monoclonal antibodies J2.7 (anti-M, obtained from Dr. John Fleming) and 1.16.1 (anti-N) have been described previously (10,11). All other antibodies used in this work were purchased from Jackson Research.

The conditions of metabolic labeling of cells with ^{35}S-methionine, immunoprecipitation, and immunofluorescence microscopy have been described previously (12).

RESULTS

The effect of amantadine on MHV replication

To determine if amantadine inhibited MHV replication we performed a plaque reduction assay incorporating various amounts of amantadine in the virus inoculum and overlay medium. As shown in Figure 1, amantadine inhibited plaque formation with MHV-A59. Fifty percent inhibition was achieved at about 200 μM amantadine and complete inhibition was reached at 1 mM concentrations. Similar results were obtained with MHV-JHM (not shown), although slightly higher doses of amantadine were required to reach the same degree of inhibition. VSV, although reported in the literature to be sensitive to amantadine was considerably less so than MHV.

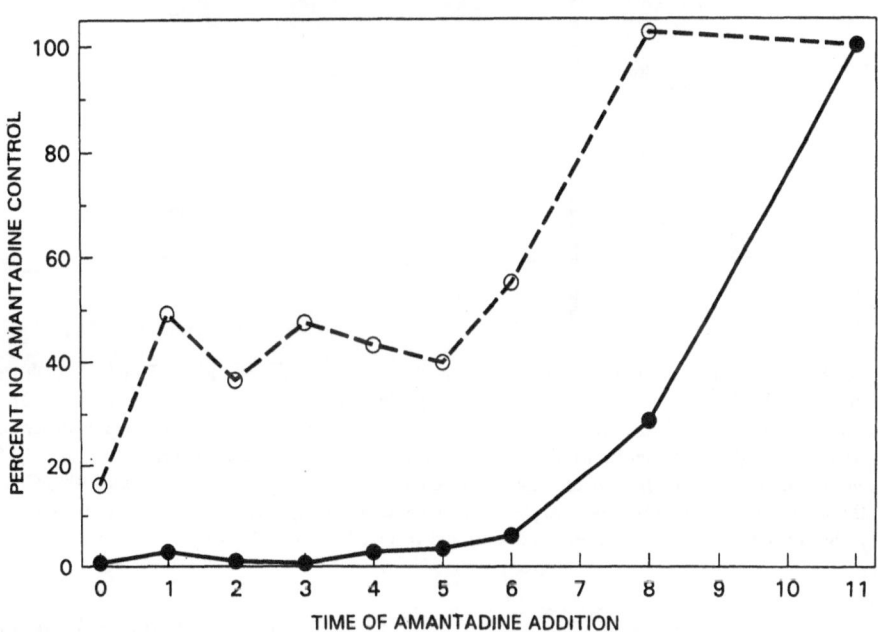

EFFECT OF TIME OF AMANTADINE ADDITION ON YIELD

Figure 2. The effect of time of addition of amantadine on MHV replication. Cells were infected with either MHV (●) or VSV (o) at a M.O.I. of 3 and amantadine added to 500 μM at the times indicated. Cultures were harvested at 11 hour post infection and virus yield determined by plaque assay in the absence of amantadine.

To investigate the step in viral replication which is sensitive to amantadine we added the drug at various times post infection and assessed the effect on viral yield. Replicate cultures were infected with either MHV-A59 or VSV in the presence or absence of amantadine. At various times post infection amantadine was added to cultures which had not contained the drug. All cultures were harvested at 11 hours post infection and the amount of virus determined by plaque assay in the absence of amantadine. As shown in Figure 2, amantadine inhibited MHV replication when added as late as six hours post infection. This did not to be due to general toxicity of the drug since the effect on VSV (Figure 2) replication was much less pronounced, the

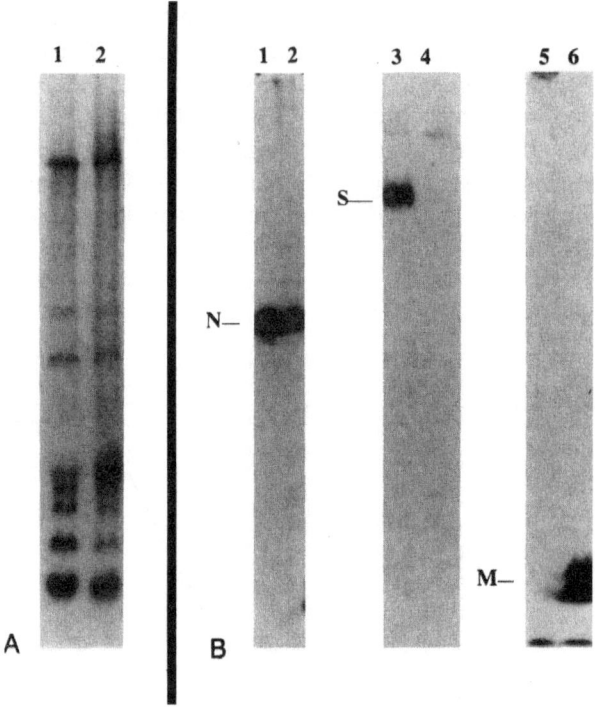

Figure 3. The effect of amantadine on MHV-specific RNA and protein synthesis. Panel A. Cultures were infected with MHV at a MOI=3 and incubated in the presence (lane 1) or absence (lane 2) of amantadine, labeled with ^{32}P-orthophosphate, and the RNA extracted and electrophoresed as described in Materials and Methods. Panel B. MHV-infected cultures were incubated in the presence (lanes 2,4,5) or absence (lanes 1,3,6) of 500 μM amantadine and labeled from 7-8 hours post infection with 100 μCi/ml ^{35}S- methionine. Cytoplasmic extracts were prepared and immunoprecipitated with antibodies to N (lanes 1and 2), S (lanes 3 and 4), and M (lanes 5 and 6) and resolved by SDS-PAGE..

cells were morphologically intact, and incorporation of ^{35}S-methionine into TCA-precipitable material was 75% of that in control cells when amantadine was added at time 0. The less potent inhibition of VSV replication by amantadine is consistent with published data (3).

This block in infectivity was also reflected in a decrease in the production of viral particles, as determined by banding in potassium tartrate gradients of ^{3}H-uridine labeled virus (not shown). However the inhibition in virus particle formation was considerably less than the inhibition in infectivity, suggesting that the particles produced in the presence of amantadine were less infectious.

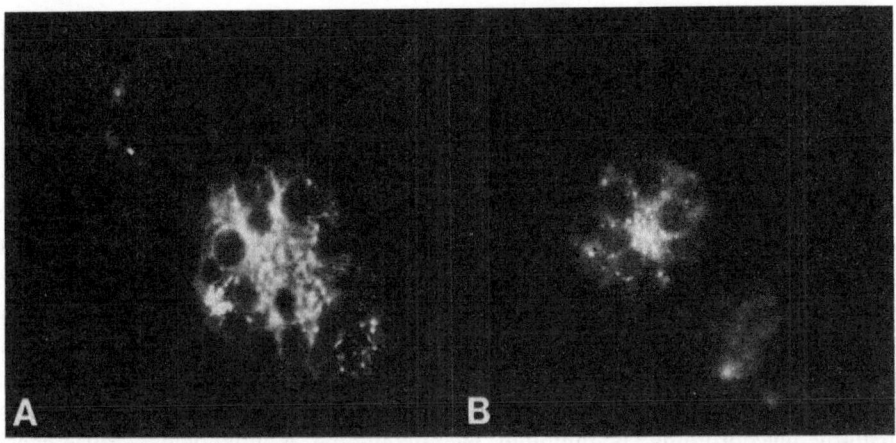

Figure 4. Immunofluorescent staining of amantadine treated cells. Cells were infected with MHV in the presence (Panel A) or absence (Panel B) of 500 μM amantadine, fixed at 7 hours post infection and stained with the anti-M monoclonal, J2.7.

Determination of the stage of replication effected by amantadine.

The above experiments suggested that amantadine exerted its effect on MHV replication at late times post infection. To further examine this, cells were infected with MHV-A59 and either treated with 500 μM amantadine from the time of infection or incubated in the absence of the drug. At seven hours post infection the cultures were labeled with ^{32}P-orthophosphate in the presence of actinomycin D. RNA was extracted and analyzed by gel electrophoresis. As shown in Figure 3A, there was no qualitative effect of amantadine on the species of MHV-specific RNAs synthesized, although the amount of virus-specific RNA was slightly decreased. Inhibition of MHV-specific RNA synthesis by amantadine was approximately fifty percent, as determined by actinomycin D resistant ^{3}H-uridine incorporation into TCA precipitatable material.

Since this degree of inhibition of MHV-specific RNA synthesis could not account for the observed effects on viral yield we then investigated the effect of the drug on virus-specific protein synthesis. Cells were infected with MHV-A59 and incubated for 7 hours in the presence or absence of 500 μM amantadine, labeled with ^{35}S-methionine for 60 minutes, and cytoplasmic extracts prepared. These extracts were analyzed by immunoprecipitation using antibodies to N, S, and M protein. As shown in Figure 3, Panel B, the accumulation of immunoprecipitable M and S glycoproteins was greatly diminished by amantadine. Pulse-chase experiments (data not shown) demonstrated that this decrease in MHV-specific glycoprotein accumulation was due to decreased synthesis, rather than enhanced turnover. In contrast to the large decrease in MHV glycoprotein synthesis in the presence of the amantadine, the amount of immunoprecipitatble nucleocapsid protein which accumulated was only moderately decreased (about 50%) by amantadine. The effect of amantadine on the accumulation of the M glycoprotein (Figure 4) and S protein (not shown) was confirmed by immunofluorescence microscopy.

DISCUSSION

We have demonstrated that amantadine inhibits MHV replication. The doses of amantadine required to exert this effect are one to two orders of magnitude higher than those required to inhibit influenza A replication. However, these doses approximate those reported to inhibit SFV and VSV by interfering with uncoating. For these viruses amantadine's mode of action appears to be due its accumulation in endosomes with a resultant increase in the pH of this cellular compartment. This increased pH interferes with a change in conformation of the VSV and SFV spike proteins, thereby interfering with the viruses' ability to escape the endosomes into the cytosol (2,3). Unlike the case for VSV or SFV, amantadine does not inhibit MHV uncoating since virus-specific RNA synthesis is not significantly inhibited by the drug.

The precise mechanism by which the accumulation of MHV-specific proteins, particularly glycoproteins, is inhibited by amantadine is unknown. Pulse-chase experiments have demonstrated that this represents a real decrease in synthesis of these proteins, not enhanced degradation in the presence of the drug. This decreased synthesis of the MHV glycoproteins is entirely out of proportion to the relatively minor, approximately 25%, inhibition of general protein synthesis which we observed with 500 μM amantadine. It cannot be accounted for by the 50% decrease in MHV-specific mRNA synthesis which we observed in the presence of the drug. It is possible that it is due to a decreased stability of membrane bound polysomes.

ACKNOWLEDGEMENTS

This work was supported in part by grants RG2203-A-5 from the National Multiple Sclerosis Society and by USPHS grant AI 31069.

REFERENCES

1. W.L. Davies, R.R. Grunert, R.F. Haff, J.W. McGahen, E.M. Neumayer, M. Paulshock, J.C. Watts, T.R. Woods, E.C. Hermann, C.E. Hoffmann. Science 144:862-863 (1964).
2. A. Helenius, J. Kartenbeck, K. Simons, and E. Fries. J. Cell Biol. 84:404-420 (1980).
3. F. Superti, L. Seganti, F.M. Ruggeri, A. Tinari, G. Donelli, and N. Orsi. J. Gen. Virol. 68:387-399 (1987).
4. A.J. Hay, A.J. Wolstenholme, J.J. Skehel, and M.H. Smith. EMBO J. 11:3021-3024 (1985).
5. R.J. Sugrue, and A.J. Hay. Virology 180:617-624 (1991).
6. L.H. Pinto, L.J. Holsinger, and R.A. Lamb. Cell 69:517-528 (1992).
7. H.R. Payne, J. Storz, and W.G. Henk. Arch. Virol. 114:175-189 (1990).
8. J.L. Leibowitz, K.C. Wilhelmsen, and C.W. Bond. Virology 114:39-51 (1981).
9. P. Chomczynski and N. Sacchi. Anal. Biochem. 162:156-159 (1987).
10. J.O. Fleming, S.A. Stohlman, R.C. Harmon, M.M.C. Lai, J.A. Frelinger, and L.P. Weiner. Virology 131:296-307 (1983).
11. J.L. Leiibowitz, J.R. DeVries, and M. Rodriguez. Adv. Exp. Med. Biol. 218:321-331 (1987).
12. E.L. Oleszak and J.L. Leibowitz. Virology 176:70-80 (1990).

ANALYSIS OF MESSENGER RNA WITHIN VIRIONS OF IBV

Dave Cavanagh, Kathy Shaw, and Zhao Xiaoyan

Division of Molecular Biology
AFRC Institute for Animal Health
Compton Laboratory
Compton, Newbury RG16 0NN, UK

ABSTRACT

The presence of subgenomic mRNAs (sgRNAs) in virions of infectious bronchitis virus was examined by probing Northern blots of RNA extracted from virions using as a probe a cDNA of the 3'-terminal nucleocapsid protein (N) gene. The sgRNAs were readily detected even after extensive purification of virions and after RNase A treatment of virions. The molar ratio of gRNA to each sgRNA was in the range 25 to 400 for IBV-M41 and 10 to 30 for IBV-Beaudette. After comparison with the molar ratios of genomic to intracellular viral sgRNAs it was estimated that the efficiency of incorporation of gRNA into virions was approximately 100 to 500-fold greater than for sgRNAs in the case of M41 and 20 to 100-fold for Beaudette, depending on the sgRNA species. It is concluded that sgRNAs can be present within IBV virions. Approximately 1 in 3 Beaudette virions and 1 in 20 M41 particles might contain a single copy of one sgRNA.

INTRODUCTION

Recently sgRNAs of transmissible gastroenteritis virus (TGEV) and bovine coronavirus (BCV) have been reported to be present in virions[1-3] although no very small defective-interfering (DI) RNAs were packaged into particles of murine hepatitis virus (MHV).[4] We have investigated the presence of sgRNAs in virions of IBV.

METHODS

Virus growth and RNA extraction

Most work was performed with M41 and IBV-Beaudette which were grown in Vero cells (Beaudette only), chick kidney (CK) cells and embryonated fowl eggs. IBV was

Coronaviruses, Edited by H. Laude and J.F. Vautherot
Plenum Press, New York, 1994

radiolabelled with ^{32}P-inorganic phosphate in CK cells and cell-associated RNA (CK cells and chorioallantoic membrane, CAM, from infected embryonated eggs) and RNA in pelleted virions was extracted using guanidinium isothiocyanate[5]. RNAs were separated in 1.2% agarose gels containing formaldehyde and the gel exposed to u.v. light (302 nm) for 2 min to nick the RNA sufficiently to improve the transfer of gRNA to nitrocellulose filters. The filters were probed with a cDNA of the N gene produced by the polymerase chain reaction (PCR). Radiolabelled (^{32}P) probes were made by the random hexanucleotide primer method and non-radioactive probe was made and used in accordance with the manufacturer's instructions for the ECL (enhanced chemiluminescence) direct nucleic acid labelling and detection system (Amersham International). The RNAs in extracts of ^{32}P-labelled infected cells were separated in agarose gels as described above. The gel was dried, an autoradiograph prepared and superimposed on the gel, bands were excised and the radioactivity determined in a scintillation counter in order to calculate the gRNA/sgRNA molar ratios of cell-associated viral RNA.

Differential purification of IBV virions

Virus was purified at 0-4°C and RNA extracted from some virions at each stage. Briefly, allantoic fluid from infected eggs was clarified, and then the virions centrifuged to produce primary pellets of virions. The resuspended virions were then pelleted through 25% (w/w) sucrose onto a 55% sucrose pad to give banded virus. This was then sedimented under isopycnic conditions through a linear 25-55% sucrose gradient.
After fractionation of the gradient and measurement of the A_{260} of each fraction, fractions 13 to 25 were pooled in pairs, diluted, and the virions pelleted (isopycnic gradient purified virus).

RNase A treatment of IBV virions

Briefly allantoic fluid from eggs infected with IBV-M41 was harvested, clarified and the virus pelleted, resuspended in NET buffer (100 mM NaCl, 1 mM EDTA, 10 mM tris-HCl, pH 7.4) and divided into 90 ul aliquots. Some aliquots were then incubated at 37°C for 5 or 30 min with RNase A (Sigma). Virions in other samples were first treated with 2% Nonidet P40 non-ionic detergent (BDH) to dissolve the virus envelope before addition of RNase. Controls were non-treated virus simply incubated at 37°C for 5 and 30 min and virus incubated with NP-40 after addition of RNAguard RNase inhibitor (Pharmacia). The RNA was extracted by addition of 2 ml of solution D,[5] 5 ug of tRNA added and the RNA purified. Half of each sample, equivalent to the virus from about 2.5 eggs, was used for electrophoresis.

RESULTS

Presence of IBV sgRNAs in preparations of virions

In addition to the gRNA and 5 sgRNAs (corresponding to mRNAs) previously described for IBV-Beaudette our analysis of many strain of IBV frequently revealed the presence of two additional sgRNAs, indicated by arrows in Figs. 1 and 2. The band between sgRNAs 2(encoding the spike,S, glycoprotein) and 3 hybridised to a cDNA probe corresponding to the 3' half of the S gene but not to any of six probes corresponding to regions throughout the polymerase (pol) gene, including the first 2 kb at the 5' terminus. The other band, between sgRNAs 4(M) and 5 did not bind any of the pol and S gene probes (data not shown).

RNA extracted from primary pellets of virions of Beaudette included sgRNAs (Fig. 1, lane e). Comparison with that extracted from infected Vero and CK cells (Fig. 1, lanes b and d) clearly showed that the gRNA/sgRNA ratios were greater for virions than for intracellular RNA, indicating that the sgRNAs had been incorporated at lower efficiency than gRNA. When Northern blots were probed with N gene-specific positive- and negative-sense radiolabelled oligonucleotides, only binding of the negative-sense oligonucleotide was detected, showing that the great majority of the molecules were plus-sense (mRNAs). The ratios for gRNA/sgRNA did not change with increased purification of virions (Fig. 2). Lanes (c) to (h) in Fig. 2 also show that after sucrose gradient sedimentation of M41 virions the fractions containing most sgRNA coincided with those containing most gRNA, which in turn corresponded to the peak fraction of virions as determined by A_{260} readings (not shown). This showed a close association of the sgRNAs with the virions.

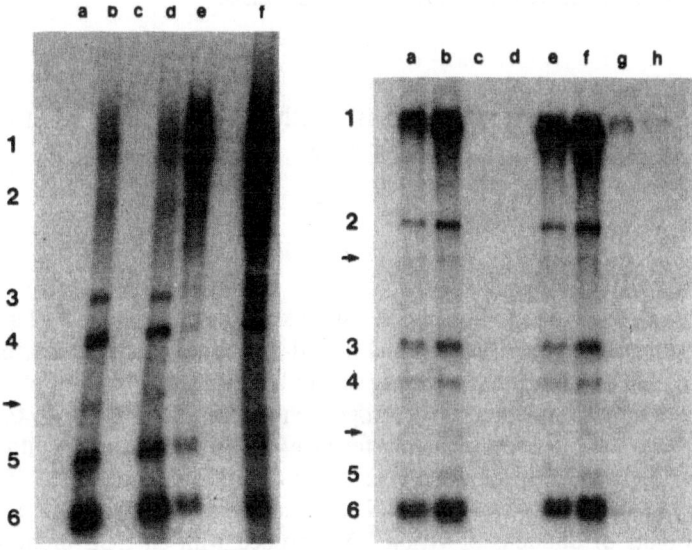

Figure 1. Northern blot of IBV RNAs probed with a ^{32}P-labelled N gene probe: RNA from mock-infected (a) Vero and (c) CK cells; cell-associated RNA from infected (b) Vero and (d) CK cells; virion-associated RNA from IBV strains (e) Beaudette and (f) UK/142/86. The arrow indicates an IBV RNA species which has not been identified as a functional mRNA.

Figure 2. Northern blot, probed with a ^{32}P-labelled N gene probe, showing the continued presence of sgRNAs during purification of M41 virions: (a) primary pelleted virions; (b) banded virions; (c) to (h) isopycnic sucrose density gradient fractions in the region of the virus peak. The arrow heads indicate two IBV RNA species which have not been identified as functional mRNAs.

Virion-associated sgRNAs were not susceptible to RNAase A

To determine if the sgRNAs were within virus particles, virions were incubated with RNase A, on the premise that the sgRNA would be digested if it were external to the virions. Fig. 3 shows that no viral RNA was destroyed at the highest concentration (10

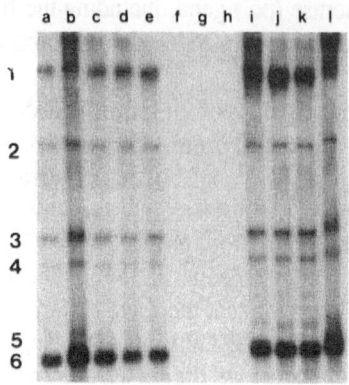

Figure 3. Northern blot, probed with a ^{32}P-labelled N gene probe, showing resistance of virion-associated sgRNAs to RNAase A. RNA was extracted from M41 virions after incubation at 37°C under the following conditions: (a) and (l) without either detergent Nonidet P-40 or RNAase; (b) with Np-40 but without RNAase; (c) to (e) five min incubation without NP-40 but with RNAase at (c) 0.1, (d) 1 and (e) 10 ug/ml; (f) to (h) as for (c) to (e), respectively, but with NP-40 to expose the RNAs previously within the virions; (i) to (k) as for (c) to (e), respectively, but longer (30 min) incubation.

ug/ml, 30 min, 37°C) of Rnase used whereas all RNA was destroyed when the virion membrane was dissociated with non-ionic detergent.

Low efficiency of incorporation of sgRNAs into virions

The molar ratios of the RNAs in virions were estimated from the amount of N gene probe bound to Northern blots. The 3' co-terminal nested set nature of the RNAs means that each molecule of RNA, irrespective of size, should bind the same amount of probe and hence the amount of probe bound is in proportion to the molar ratio of the RNAs. Account was taken of the finding that transfer of gRNA from agarose to nitrocellulose was only 63% efficient (mean of three separate transfers of radiolabelled IBV RNA) compared with almost 100% for the sgRNAs. The molar ratio of M41 RNAs was estimated from blots of dilutions of virion RNA probed with an ECL probe (data not shown). Comparison of band intensities indicated that the molar ratio gRNA/sgRNA ranged from about 25 for sgRNA 6 to 400 for mRNA 5. When account was taken of the molar ratios of gRNA/sgRNA within cells (not shown) i.e. in which most sgRNAs outnumber gRNA, the efficiency of incorporation of gRNA was estimated to be 100 to 500-fold greater for sgRNAs. The molar ratio for virion RNAs of Beaudette was calculated from blots probed with a ^{32}P-labelled probe, the gRNA/sgRNA ratio ranging from about 10 for sgRNA 6 to 30 for sgRNAs 3 and 4. The efficiency of incorporation of gRNA was 20 to 100-fold greater than for sgRNAs.

DISCUSSION

The presence of sgRNAs in virions is relevant to several aspects of coronavirus replication, including the requirement of signals for packaging of RNA into virions,[7,8] the capacity of coronavirus sgRNAs to function as replicons[1-3,9] and, following the experimental demonstration of recombination during MHV replication[10] and circumstantial evidence for recombination in the field for IBV,[11-13] the possibility that incorporation into virions of sgRNAs from both parents following mixed infection could increase the chance of recombination at subsequent replication cycles.

Estimates of the molar amounts of sgRNAs incorporated into coronavirus virions vary. Only minute amounts of very small DI RNAs were incorporated into MHV particles.[4] Sethna et al.[3] have reported that virions of TGEV contained 5 to 14-fold more gRNA than given sgRNAs, the N protein-encoding sgRNA being the most abundant. Bovine coronavirus (BCV) virions have been reported to contain more molecules of N and M protein-encoding sgRNAs than gRNA.[2] Our results for IBV fall somewhere in the middle. The amounts of sgRNAs in virions of IBV-Beaudette were less than but similar to those reported for TGEV, whereas the frequency of sgRNAs in M41 particles was about 4-fold less than in Beaudette. The majority of IBV virions probably do not contain any sgRNA species. The IBV sgRNAs were incorporated into virions broadly in proportion to the amount of each sgRNA in infected cells. An exception, with both Beaudette and M41, was sgRNA 2, encoding the spike protein, which was over represented.

The efficiency of incorporation of sgRNAs into IBV virions was some two orders of magnitude less than that of gRNA. This is not surprising in view of the finding that a sequence near the 3' end of the gene encoding open reading frame 1b of the 5'-most gene, that encoding the polymerase, is essential for packaging of RNA into virions of MHV.[6-8] None of the sgRNAs of IBV contain any part of gene 1. Our results support the view that specific sequences, present within gene 1, are necessary for efficient packaging i.e. for the formation of the ribonucleoprotein i.e. viral RNA surrounded by N protein molecules.

What role, if any, might intra-virion sgRNAs have in replication? It has been demonstrated for TGEV and BCV that during virus replication sgRNAs can function as replicons; infected cells contain negative sense versions of the mRNAs.[1,3,9] It is conceivable, therefore, that sgRNAs within virions might, upon their release from virions following penetration of host cells, be replicated. Indeed, it has been observed that virion-associated sgRNAs appeared to have served as templates for their own replication following infection of cells with BCV.[2] If this were to occur then some coronavirus particles would, in effect, have two copies of some genes which might result in over-production of the corresponding mRNA and, presumably, of the corresponding encoded protein. This could, conceivably, have an effect on replication by disturbing the balance of the various gene products. However, Makino et al.[14] have shown that when MHV-infected cells were transfected with sgRNA, the latter was not replicated. Input mRNA might be translated, but the effect of this would be expected to be minor compared with the translation of the ultimately much greater amount of mRNA produced de novo.

Another way in which virion-associated sgRNA might have a biological effect is in respect of recombination. During template-switching the polymerase might continue RNA synthesis on the input sgRNA and then switch back to synthesis on genomic RNA. (Recombination could also occur during production of plus-sense RNA if the input sgRNA were to be first replicated to produce negative-sense sgRNA). This could result in a recombinant if the input sgRNA was not homologous to the input gRNA. That is, during infection of a cell with two strains of virus, some progeny virions might have the gRNA from one parental strain and a sgRNA from the other. Of course, recombination could occur during the initial mixed infection. However, if virion-associated sgRNAs can be replicated i.e. serve as a template for the polymerase, then this extends the possibility of recombinants being produced when such virions infect other cells, including in other host individuals.

In conclusion, plus-sense sgRNAs were incorporated into virions of IBV, however the efficiency of incorporation was low and most particles would lack any sgRNA. Whether virion-associated sgRNAs play any significant role in coronavirus replication and recombination depends on whether input sgRNAs can serve as substrates for the virus polymerase.

REFERENCES

1. P.B. Sethna, S-L Hung and Brian, D.A. *Proc.Natl.Acad.Sci. USA.* 86:5626 (1989).
2. M.A. Hofmann, P.B. Sethna and D.A. Brian. *J.Virol.* 64:4108 (1990).
3. P.B. Sethna, M.A. Hofmann and D.A. Brian. *J.Virol.* 65:320 (1991).
4. S. Makino, F. Taguchi and K. Fujiwara. *Virology* 133:9 (1984).
5. P. Chomczynski and N. Sacchi. *Anal.Biochem.* 162:156 (1987).
6. S. Makino, K. Yokomori and M.C. Lai. *J. Virol.* 64:6045 (1990).
7. R.G. van der Most, P.J. Bredenbeek and W.J.M. Spaan. *J. Virol.* 65:3129 (1991).
8. J.A. Fosmire, K. Hwang and S. Makino. *J. Virol.* 66:3522 (1992).
9. S.G. Sawicki and D.L. Sawicki. *J. Virol.* 64:1050 (1990).
10. J.G. Keck, G.K. Matsushima, S. Makino, J.O. Fleming, D.M. Vannier, S.A. Stohlman and M.M.C. Lai. *J.Virol.* 62:1810 (1988).
11. D. Cavanagh and P.J. Davis. *J.Gen.Virol.* 69:621 (1988).
12. J.G. Kusters, E.J. Jager, H.G.M. Niesters and B.A.M. van der Zeijst. *Vaccine* 8:605 (1990).
13. D. Cavanagh, P.J. Davis and J.K.A. Cook. *Avian Pathol.* 21:413 (1992).
14. S. Makino, M. Joo and J.K. Makino. *J.Virol.* 65:6031 (1991).

INHIBITION OF MOUSE HEPATITIS VIRUS MULTIPLICATION BY ANTISENSE OLIGONUCLEOTIDE, ANTISENSE RNA, SENSE RNA AND RIBOZYME

Tetsuya Mizutani[1], Masanobu Hayashi[1], Akihiko Maeda[1], Nobuya Sasaki[1], Tadashi Yamashita[1], Noriyuki Kasai[2], and Shigeo Namioka[1]

Department of Laboratory Amimal Science, Faculty of Veterinary Medicine[1], Institute for Animal Experimentation, School of Medicine[2], Hokkaido University Sapporo 060, Japan

ABSTRACT

Antisense nucleic acids against specific sequences of mouse hepatitis virus (MHV)-RNAs were tested for their inhibitory effects on viral multiplication in mouse DBT cells. An antisense oligonucleotide containing a sequence complementary to leader RNA was synthesized and shown to induce a significant inhibitory effect on the multiplication of MHV-JHM. A vector which expressed the antisense or sense mRNA7 of MHV was transfected into DBT cells. A decreased multiplication of MHV was observed in both cell lines. The transfected cell line which expressed ribozyme against the 5'-end of the MHV genome was established. The rate of inhibition of MHV-multiplication and the quantity of synthesized virus-specific mRNAs in this transfected cell line were the same for both antisense and sense RNA. These results show that antisense nucleic acids might be eligible for use as antiviral agents against MHV multiplication.

INTRODUCTION

It has been reported that antisense nucleic acids inhibit the expression of their target genes (Reviewed in ref. 1). Three different classes of antisense nucleic acids (antisense oligonucleotide, antisense RNA and ribozyme) are currently in use. Inside the cell, antisense nucleic acids hybridize with target RNAs and inhibit the expression of the respective target genes. Successful demonstration of antisense mechanisms for inhibition of viral replication and multiplication have been described. However, no experimental data has been reported concerning the effects of antisense nucleic acids on positive strand RNA virus except retrovirus in infected cells [1]. To inhibit MHV multiplication *in vitro* and *in vivo*, three approaches were used in this study.

[1] A 14-mer antisense oligonucleotide containing the sequence complementary to the conserved pentanucleotide sequence, UCUAA, of the leader RNA, was tested for its inhibitory effects on MHV multiplication in DBT cells. The sequence UCUAA is conserved at the initiation sites for each of the six subgenomic mRNAs and is thought to be involved in the interaction between the leader RNA and negative-strand RNA. Thus, leader RNA may take part in leader-primed transcription [2].

[2] Vectors which express sense or antisense MHV mRNA7 were transfected into DBT cells. Nucleocapsid (N) protein encoded by mRNA7 might play important roles within the transcriptional complex, since the N protein associated with MHV-specific RNAs containing the leader RNA sequences *in vitro* [3] and anti-N monoclonal antibodies inhibit viral replication *in vivo* [4].

Furthermore, we examined the effect of sense RNA on viral multiplication because both positive- and negative-strand RNAs are synthesized in cells infected with MHV.

[3] A ribozyme designed to cleave sequences specific to MHV RNA might present a better form of antiviral agent. We used the ribozyme of the hammerhead motif[5] which was designed against the 5'-end of the MHV genome (gene1). Gene 1 is a gene important for viral transcription and replication because gene 1 is translated into proteins containing RNA-dependent RNA polymerases soon after infection[6].

MATERIALS AND METHODS

1. Cell line and MHV: Mouse astrocytoma DBT cells[7] were cultured in Eagle's minimum essential medium (MEM) supplemented with 5% calf serum (CS). The JHM strain of MHV[8] was used throughout this study.

2. Oligonucleotides, Plasmids and Transfection: The oligonucleotides were synthesized using the phosphoramidates method on a Beckman system Plus-1 DNA synthesizer and purified by HPLC. AL-oligo sequence is complementary to the consensus sequence, UCUAA, of the MHV leader RNA. SL-oligo is complementary to AL-oligo. GATA-oligo is a control oligonucleotide unlerated to the MHV leader RNA (Fig. 1a). A cDNA of MHV-mRNA7 was kindly provided by Dr. Siddell[9]. For RNA expression in cultured cells, cDNA of MHV were inserted into pZIP-Neo SV (x) 1 [10] or pEF321-T[11]. DBT cells were transfected with these plasmids using the standard calcium phosphate precipitation procedure [12] and selected in MEM containing 1 mg/ml G418.

3. Viral infection and Plaque assay: Transfected cells and parent DBT cells were infected with MHV-JHM at a m.o.i. of 0.1 or 1.0 for 1 h at 37°C under CS-free MEM conditions. After the incubation period, cells were washed twice with CS-free MEM and fresh CS-MEM was added. After the addition of MEM with CS, plaque assays were performed to titrate infectious progeny at various times post infection (p.i.). The infectivity was expressed as plaque forming units (PFU) /ml.

4. Northern blot hybridization: Cellular RNA was prepared according to the method of Silver *et al*.[13]. The RNA samples were electrophoresed in 1% agarose gels containing formaldehyde, blotted onto nitrocellulose membranes [14] and hybridized with the ^{32}P-labeled nick-translated cDNA of MHV mRNA7 as a probe [15]. Sense and antisense probes were expressed using T7 or SP6 polymerase [14] containing ^{32}P-UTP.

5. DNA sequencing: Sequencing was carried out according to Sanger's dideoxyribonucleotide chain termination method [16].

RESULTS AND DISCUSSION

1. Inhibition of MHV multiplication by antisense oligonucleotide

The oligonucleotide sequences are shown in Fig. 1a. To investigate the effects of antisense oligonucleotide (AL-oligo) on viral multiplication, DBT cells (5×10^4) were incubated with MHV-JHM (0.1 m.o.i.) in the presence of oligonucleotides at concentrations ranging from 1 to 25 μM for 1 h at 37°C. During the incubation time, approximately 1.7% of the oligonucleotides was incorporated into the cells (data not shown. ref. 17). This result is in agreement with reports by Loke *et al*.[18], and Kawamura *et al*.[19] Fig.1b shows that the inhibitory effect of AL-oligo on viral multiplication was specific for the antisense sequence of the leader RNA of MHV at 5 and 10 μM. When the cells were treated with 25 μM SL-oligo, significant inhibition (20% inhibition) of viral multiplication was observed. The interaction between sense oligonucleotide (SL-oligo) with the negative-strand RNA might interfere with transcription of mRNAs by the negative-strand RNA at high concentrations of SL-oligo. To investigate the effect of AL-oligo on the synthesis of specific viral mRNAs, cellular RNA was prepared from cells infected with MHV (1.0 m.o.i.) in the presence (10 μM at 4.5 h.p.i.) or absence of the oligonucleotides and analyzed by Northern blot hybridization (Fig. 1c). The synthesis of MHV-mRNAs in the cells treated with AL-oligo was reduced, whereas no inhibitory effect on the synthesis of viral mRNA was observed in the cells treated with SL-oligo and GATA-oligo. This result suggests that AL-oligo interferes with MHV-RNA transcription at the initial stages of viral infection.

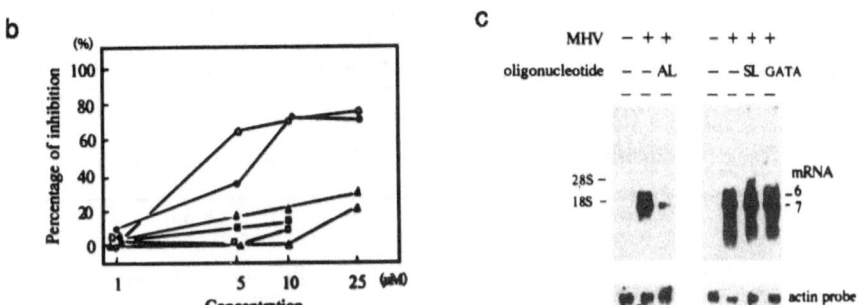

a   leader RNA
5'-AUCUAAUCUAAUCUAAACUUUA-3'

oligonucleotide	sequence 5'-3'
AL-oligo	GTTTAGATTAGATT
SL-oligo	AATCTAATCTAAAC
GATA-oligo	GATAGATAGATAGATA

Fig. 1. (a) Sequences of oligonucleotides. (b) Effects of oligonucleotides on MHV multiplication. DBT cells were infected with MHV-JHM (0.1 m.o.i.) in the presence of AL-oligo (○, ●), SL-oligo (△, ▲) and GATA-oligo (□, ■). Plaque assays were performed at 6 (○, △, □) and 12 (●, ▲, ■) h.p.i. (c) Effects of AL-oligo on the synthesis of viral RNA. Cellular RNAs, prepared from infected (0.1 m.o.i.) or mock-infected cells treated with oligonucleotide (10 μM) at 4.5 h.p.i., were analyzed by Northern blot hybridization.

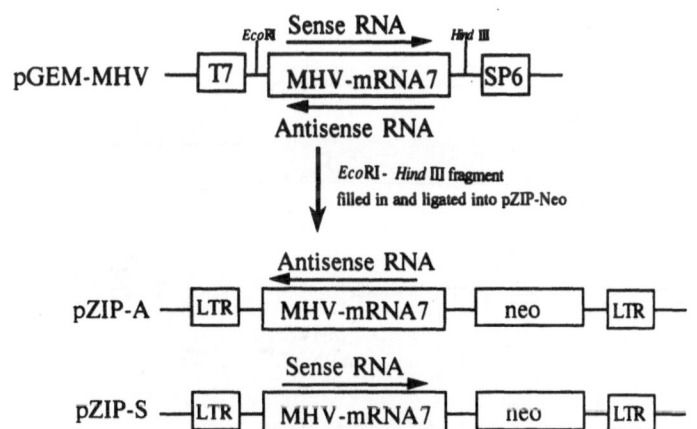

Fig. 2. Construction of plasmid pZIP-S and pZIP-A. The cDNA of MHV-mRNA7 was inserted into plasmid pGEM-1 at the *Pst* I site (pGEM-MHV). The pGEM-MHV sequence was cleaved with *Hind*III and *Eco*RI. The 1.8 kbp cDNA fragment of MHV-mRNA7 was isolated and the sticky-ends of this fragment were filled in by incubation with dNTPs in the presence of the DNA polymerase I Klenow fragments[14]. Plasmid pZIP-Neo SV (x) 1 was cleaved with *Bam*HI and filled in as explained above. The blunt-ends of both the MHV fragment and the pZIP-Neo were ligated in both sense and antisense orientation and designated pZIP-S and pZIP-A, respectively. T7 and SP6 : bacteriophage T7 and SP6 promoters , respectively ; LTR : Moloney murine leukemia virus-LTR ; neo : neomycin resistance gene.

2. Inhibition of MHV multiplication by sense or antisense RNA

To increase the level of inhibition of MHV-multiplication by antisense nucleic acids, we used sense and antisense RNA as inhibitory sequences and selected mRNA7 as a target. The treatment of infected cells with anti-N protein antibodies cause the inhibition of MHV RNA synthesis *in vitro*[20]. Therefore, the inhibition of synthesis of the N protein by antisense mRNA7 is thought to affect the MHV-transcription and/or viral multiplication. We used two vector constructions (Fig. 2), one construction being antisense RNA which is complementary to the mRNA7, and the 3'-portions of positive-strand genomic RNA and five subgenomic mRNAs. The other construction is the sense RNA which is complementary to the 5'-portions of negative-strand RNA and subgenomic negative-strand RNA. We selected negative-strand RNA as a target because some inhibitory effects could be observed at high concentrations of sense oligonucleotide (Fig. 1b).

Table I   Inhibition of MHV multiplication by antisense- or sense-RNA

	cell line	9 h.p.i.		12 h.p.i.	
		PFU/ml	inhibition (%)[a]	PFU/ml	inhibition (%)[a]
	parent DBT	2.47×10^5		1.47×10^6	
Antisense	A1	8.10×10^3	96.7	9.52×10^4	93.5
	A2	4.28×10^4	82.7	7.20×10^5	51.0
Sense	S1	1.13×10^4	95.4	1.90×10^4	98.7
	S2	3.35×10^5	——	1.60×10^6	——
	S3	7.65×10^4	69.0	3.47×10^5	76.4

[a] $$\text{Percentage of inhibition} = 1 - \left(\frac{\text{PFU obtained from transfected cells}}{\text{PFU obtained from parent DBT cells}} \right) \times 100 \ (\%)$$

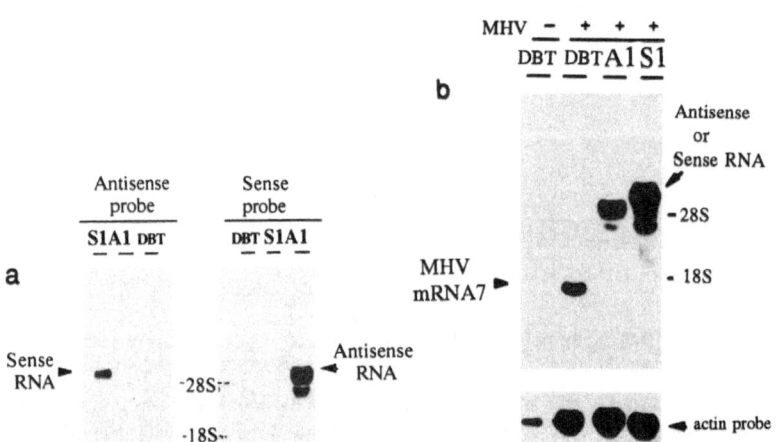

Fig. 3. (a) Single-strand RNA expression in the transfected cells. The RNAs were extracted from the S1 and A1 cells as described in the text. Samples (10 μg) were analyzed by Northern blot hybridization using single-strand ^{32}P-labeled probes. Sense or antisense probes was expressed from the pGEM-MHV sequence using T7 or SP6 polymerase, respectively (Fig. 2) and purified by Sephadex G-50 chromatography. (b) Effects of sense or antisense RNA on viral mRNA synthesis. The S1, A1 and DBT cells were infected with MHV-JHM (1.0 m.o.i.) and at 3.5 h.p.i., the cellular RNAs were analyzed by Northern blot hybridization using cDNA of MHV-mRNA7 as a probe.

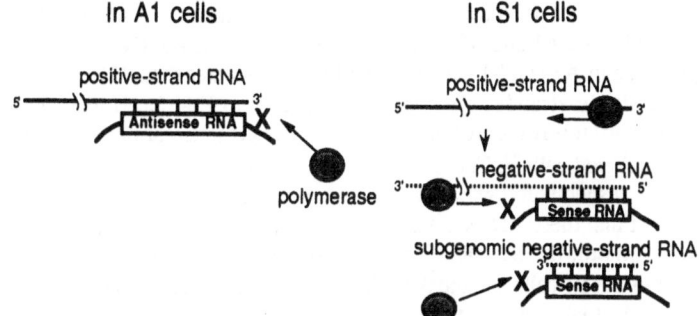

In A1 cells

In S1 cells

Fig. 4. Possible model for inhibition of MHV multiplication in the S1 and A1 cells.

Furthermore, in the case of Human Immunodeficiency Virus (HIV), sense RNA inhibited viral multiplication[21]. These vectors were introduced into DBT cells and the cells were selected by culture in medium containing G418. Several transfected cell lines which constitutively expressed sense or antisense RNA were established. To investigate the effect of sense or antisense RNA on viral multiplication, plaque assays were performed. The pZIP-Neo sequences and G418 did not directly

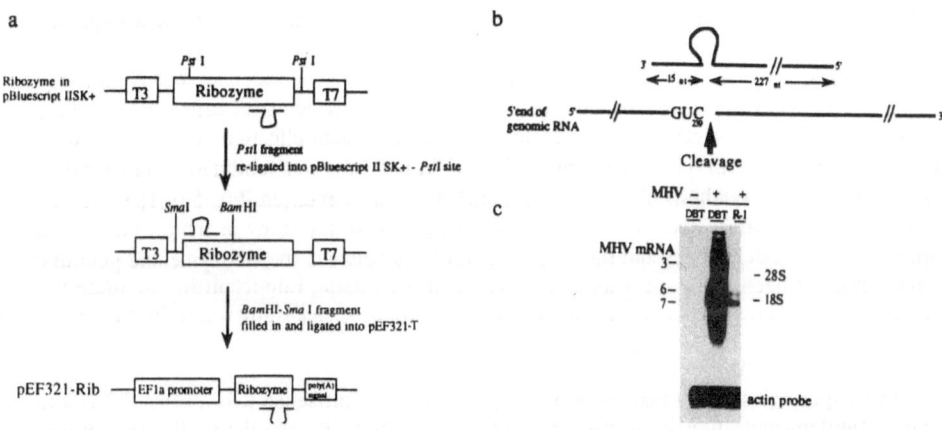

Fig. 5. (a) Construction of plasmid pEF321-Rib. Genomic RNA of MHV was extracted from MHV particles and the genomic RNA was reversed transcribed using the primer (5' GAATCCGTATACAGCATAGT 3') as forward primer and amplification by PCR was performed[14]. The temperature profile for the reaction was as follows : 95°C / 5 min then 50°C / 2 min, 72°C / 3 min, 93°C / 2 min for 30 cycles. The sequences of the forward and reverse PCR primers, (including ribozyme sequences, 5'ATGGGCAAATACGGTTTCGTCCTC ACGGACTCATCAGTCGGCTTCAAATGGG 3'), used in this reaction are based on the published sequences of the 5'-end genes of MHV-JHM[23]. The PCR reaction product was analyzed by gel electrophoresis, extracted from the gel and ligated into pBluescript II SK+ at the *Sma*I site and sequenced. This plasmid was digested with *Pst*I and re-ligated into pBluescript II SK+ at the *Pst*I site. The ribozyme fragment was digested with *Bam*HI and *Sma*I and after filling in, ligated into pEF321-T at the blunt-end (designated pEF321-Rib). This pEF321-Rib was re-sequenced for confirmation. (b) Schematic representation of ribozyme and the 5'-end of MHV genomic RNA. (c) Effect of ribozyme on MHV-RNA synthesis. DBT and transfected R-1 cells (5×10^4) were infected or mock-infected with MHV-JHM at 0.1 m.o.i. At 9 h.p.i., total RNA was extracted from the cells and Northern blot analysis was performed using mRNA7 cDNA as a probe.

affect the multiplication of MHV (data not shown). The multiplication of MHV in S1 and S3 cells (which express sense RNA), A1 and A2 cells (which express antisense RNA) was inhibited (Table I). MHV multiplication in A1 and S1 cells was inhibited by greater than 95% at 9 h.p.i. Northern blot analysis using single-stranded RNA probes is shown in Fig. 3a. It has been shown that interferon-mediated antiviral responses are induced by the presence of both sense and antisense RNA transcripts in infected cells[22]. The S1 and A1 cells expressed only the sense and antisense single-stranded RNA respectively, which RNAs were relatively stable in the cytoplasm (data not shown). It is thought that these RNAs are suitable candidates to hybridize with the target MHV-RNAs. On the other hand, the S2 and S3 cells expressed sense RNA at a low copy number compared with that of the S1 cells and the A2 cells expressed a longer antisense RNA as compared with that of the A1 cells (data not shown).

To investigate the effect of the sense and antisense RNA on the synthesis of viral mRNAs in the S1 and A1 cells, RNAs were prepared from cells infected with MHV at 1.0 m.o.i. and Northern blot analysis was performed. In the S1 and A1 cells, viral RNAs were hardly observed at 3.5 h.p.i. (Fig. 3b). This result suggests that the sense and antisense RNA hybridize with MHV-RNAs during the initial stages of infection. As shown in Fig. 4, it is thought that the antisense RNA of the A1 cells hybridizes with the genomic RNA during the initial stages of infection (3.5 h.p.i.). On the other hand, sense RNA which is expressed in the S1 cells, hybridizes with negative-and/or subgenomic negative-strand RNA. The MHV-RNA and antisense/sense RNA complex may delay replication of MHV in the A1/S1 cells.

3. Inhibition of MHV multiplication by ribozyme

Further investigation into the inhibition of MHV multiplication was performed using a ribozyme designed to cleave the 5'-end of the MHV genome between nucleotides (nt) 239-240 (Fig. 5b). The 5' flanking sequence of ribozyme contains 227 nt and the 3'-end contains 15 nt of a sequence complementary to MHV genomic RNA. The MHV gene 1 was cleaved at the target site with the ribozyme in cell free experiment (data not shown). This ribozyme gene sequence was cloned into pEF321-T vector and expressed in DBT cells (Fig. 5a). The so-transfected DBT cells were given the name R-1 cells and to investigate the effect of ribozyme on MHV multiplication, the R-1 cells were incubated with MHV (0.1m.o.i.). MHV-multiplication in R-1 cells was inhibited by greater than 98% at 12 h.p.i. and the synthesis of virus-specific mRNA was also reduced (Fig. 5c). These results suggest that genomic MHV RNA might be an adequate target for the study of inhibition of viral multiplication. The mechanisms and functional interactions between the ribozyme and genomic RNA have not yet been made clear and we consider this as being important to elucidate the mechanism of viral multiplication and its inhibition by ribozyme. Such a study is now in progress in our laboratory.

In this paper, we showed that antisense oligonucleotide, antisense RNA, sense RNA and ribozyme complementary to MHV-RNAs reduce viral multiplication and the synthesis of virus-specific mRNAs and that these antisense nucleic acids might find application as antiviral agents. We expect to produce transgenic mice which express antisense RNA against MHV mRNA7 to better-understand the effects of antisense RNA on viral multiplication *in vivo*.

ACKNOWLEDGEMENTS

We thank Dr. Taguchi, National Institute of Neuroscience, Kodaira, Japan, for providing DBT cells and MHV-JHM strain and for his useful suggestions.

REFERENCES

1. Colman, A. (1990). J.Cell Sci. 97: 399-409.
2. Shieh, C. K., Soe, L. H., Makino, S., Chang, M. F., Stohlman, S. A., and Lai, M. M. C. (1987). Virology. 156 : 321-330.
3. Baric, R. S., Nelson, G.W., Fleming, J.O., Deans, R. J., Keck, J. G., Casteel, N., and Stohlman, S. A. (1988). J. Virol. 62: 4280-4287.

4. Nakanaga, K., Yamanouchi, K., and Fujiwara, K. (1986). J. Virol. 59: 168-171.

5. Haseloff, J., and Guerlach, W. (1988). Nature. 334: 585-591.

6. Brayton, P. R., Lai, M. M. C., Patton, C. D., and Stohlman, S. A. (1982). J. Virol. 42: 847-853.

7. Hirano, N., Fujiwara, K., Hino, S., and Matumoto, M. (1974). Arch. Gesamte Virusforsch. 44: 298-302.

8. Makino, S., Taguchi, F., Hayami, M., and Fujiwara, K. (1983). Microbiol. Immunol. 27: 445-454.

9. Skinner, M. A., and Siddell, S. G. (1983). Nucleic Acids Res. 11: 5045-5054.

10. Constance, L. C., Bryan, E. R., and Richard, C. M. (1984). Cell. 37: 1053-1062.

11. Kim, D. W., Uetsuki, T., Kaziro, Y., Yamaguchi, N., and Sugano, S. (1990). Gene. 91: 217-223.

12. Graham, F. W., and Van Der EB, A. J. (1973). Virology. 52: 456-467.

13. Silver, S., Smith, M., and Nonoyama, M. (1979). J. Virol. 30: 84-89.

14. Maniatis, T., Frisch, E. F., and Samabrook, J. (1989). "Molecular Laboratory Manual." Cold Spring Harbor Laboratory, Cold Spring Harbor, N. Y.

15. Rigby, P. W. J., Dieckman, M., Rhodes, C., and Berg, P. (1977). J. Mol. Biol. 113: 237-251.

16. Sanger, F., Nicklen, S., and Coulson, A. R. (1977). Proc. Natl. Acad. Sci. U.S.A. 74: 5463-5467.

17. Mizutani, T., Hayashi, M., Maeda, A., Yamashita, T., Isogai, H., and Namioka, S. (1992). J. Vet. Med. Sci. 54: 456-472.

18. Loke, S. L., Stein, C.A., Zhang, X. H., Mori, K., Nakanishi, M., Subasinghe, C., Cohen, J. S., and Neckers, L. M. (1989). Proc Natl. Acad. Sci. U. S.A. 86: 3474-3478.

19. Kawamura, M., Hayashi, M., Furuichi, T., Nonoyama, M., Isogai, E., and Namioka, S. (1991). J. Gen. Virol. 72: 1105-1111.

20. Compton, S. R., Roger, D. B., Holmes, K. V., Fertsch, D., Remenick, J., and McGowan, J. J. (1987). J. Virol. 61: 1814-1820.

21. Joshi, S., Van Brunschot, A., Asad, S., Van Der Elst, I., Read, S.E., and Bernstein, A. (1991). J. virol. 65: 5524-5530.

22. Leiter, J. M. E., Krystal, M., and Palese, P. (1989). Virus. Res. 14: 141-160.

23. Soe, L., Shieh, C, K., Baker, S., Chang, M. F., and Lai, M. M. C. (1987). J. Virol. 61: 3968-3976.

SITE-SPECIFIC SEQUENCE REPAIR OF CORONAVIRUS DEFECTIVE INTERFERING RNA BY RNA RECOMBINATION AND EDITED RNA

Young-Nam Kim, [1] Michael M. C. Lai, [2] and Shinji Makino [1]

[1] Department of Microbiology, The University of Texas at Austin
Austin, Texas 78712
[2] Howard Hughes Medical Institute and Department of Microbiology
University of Southern California, School of Medicine
Los Angeles, California 90033

INTRODUCTION

Mouse hepatitis virus (MHV), a coronavirus, contains an approximately 31 kb-long genomic RNA (8, 9). In MHV-infected cells, seven to eight species of virus-specific subgenomic mRNAs with a 3'-coterminal nested-set structure (6, 10) are synthesized; these are numbered 1 to 7, in decreasing order of size (6, 10). The 5'-end of MHV genomic RNA contains a 72- to 77-nucleotide-long leader sequence (5, 7, 22). An identical sequence is found at the 5'-end of each MHV mRNA species; in each, the leader sequence is fused with the mRNA body sequence, which starts from a consensus sequence at the intergenic sites (17, 21).

When the JHM strain of MHV (MHV-JHM) was serially passaged in tissue culture at a high multiplicity of infection, a variety of defective-interfering (DI) RNAs of different sizes were detected (11). Two MHV DI RNAs, DIssE and DIssF were studied in a greter detail (15, 16, 19). DIssE is 2.3 kb in length and consists of three noncontiguous genomic regions, comprising the first 0.86 kb from the 5'-end, an internal 0.75 kb from gene 1, which presumably encodes the RNA polymerase of the virus, and 0.6 kb from the 3'-end of the parental MHV genome (16). The structure of the 3.6 kb-long DIssF consists of sequences derived from five noncontiguous regions of the genome of nondefective MHV (19). The first four domains (I to IV) from the 5' end are derived from gene 1 and the 3'-most domain (V) is derived from the 3'-end of genomic RNA. Both DI RNAs contain one large open reading frame (ORF), from which proteins are translated (16, 19). Significantly, MHV-A59-derived DI RNAs also contain a large ORF (23). Thus, it seems that the presence of a large ORF is a common feature of MHV DI RNAs, although the biological significance of the ORF is not known. A system has been established in which complete cDNA clones of DIssE and DIssF RNAs were placed downstream of T7 RNA polymerase promoters to generate DIssE and DIssF RNAs capable of efficient replication in the presence of a helper virus. This system has been used for studying MHV RNA replication (14), transcription (4, 13) and packaging (3, 19).

In the present study, we analyzed a DIssE-derived mutant DI RNA with a one-nucleotide deletion at position 376 from the 5'-end. This one-nucleotide deletion produces an ORF of one-tenth the size of the DIssE-specific ORF. The data obtained from this study demonstrated that DI RNA containing this small ORF can replicate at the same efficiency as DI RNA containing the large ORF. During RNA replication, however, DI RNA with the small ORF was replaced with novel DI RNAs containing the large ORF. Of this group, approximately half of the DI RNAs contained sequences created by RNA recombination and the remainder contained sequences in which a specific nucleotide was added at a specific site upstream of the deletion site.

Coronaviruses, Edited by H. Laude and J.F. Vautherot
Plenum Press, New York, 1994

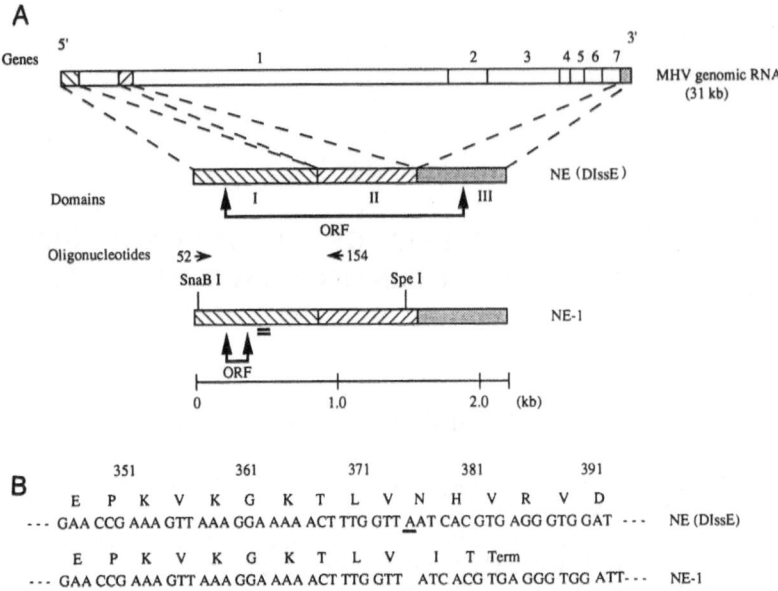

Fig. 1. Diagram of the structure of NE and NE-1. (A) Comparison of the structures of NE, NE-1 and standard MHV genomic RNA. Genes 1 through 7 represent the seven genes of MHV. The locations of the ORFs of NE and NE-1 are shown. The three domains of DIssE-derived DI RNAs (domains I through III) and the oligonucleotides used in the present study are indicated below the diagram of NE. The restriction enzyme sites are indicated above NE-1. The double-underlined region of NE-1 represents the region which is recognized by the anti-p28 antibody (1). (B) The nucleotide and deduced amino acid sequences of the 5'-regions of NE and NE-1. The numbers above these represent the nucleotide positions from the 5'-end of NE DI RNA (13). The A underlined in the NE sequence is missing from NE-1.

MATERIALS AND METHODS

Viruses and cells. The plaque-cloned A59 strain of MHV (MHV-A59) was used as a helper virus. Mouse DBT cells were used for RNA transfection and propagation of viruses.

RNA transcription and transfection. Plasmids were linearized by Xba I digestion and transcribed in vitro with T7 RNA polymerase as described previously (14). The lipofection procedure used for RNA transfection was described previously (13).

Preparation of virus-specific intracellular RNA. Intracellular virus-specific RNA was extracted as described previously (18).

Plasmid construction. Polymerase chain reaction (PCR) products corresponding to the 5' end 1.5 kb region of DIssF were obtained as previously described (19). Two cDNA clones, PCR-1 and PCR-2, were analyzed after the PCR products were cloned; the sequence of PCR-1 has been described (19). The sequence of PCR-2 was identical to that of PCR-1 except for a one-nucleotide deletion at position 376 from the 5'-end of PCR-2. Two complete DIssE-specific cDNA clones, NE and NE-1, were constructed by inserting the 1.5 kb SnaB I-Spe I fragments of PCR-1 and PCR-2, respectively, into the 2.9 kb SnaB I-Spe I fragment of the DIssE-specific cDNA clone DE5-w4 (14)(Fig. 1).

PCR. For amplification of MHV RNA species, MHV-specific cDNA was first synthesized from intracellular RNA as previously described (16), using as a primer oligonucleotide 154 (5'-CTGCTCCCTGGCAACGCC-3'), which binds to positive-stranded MHV DI RNA at nucleotides 952 to 966 from the 5'-end. MHV-specific cDNA was then incubated with oligonucleotide 52 (13), which binds to the leader sequence of negative-stranded MHV RNA, in PCR buffer (0.05 M KCl, 0.01 M Tris hydrochloride [pH 8.3], 0.0025 M $MgCl_2$, 0.01% gelatin, 0.17 mM each of dNTPs and 5U of Taq polymerase [Promega]) at 93°C for 30 s, 55°C for 30 s, and 72°C for 100 s for 25 cycles.

Radiolabeling of viral RNAs and agarose gel electrophoresis. Virus-specific RNAs in virus-infected cells were labeled with ^{32}Pi as previously described (18) and separated by electrophoresis on 1 % agarose gels after denaturation with 1 M glyoxal (20).

Labeling of intracellular proteins, immunoprecipitation, and SDS-PAGE. Labeling of intracellular proteins, immunoprecipitation, and SDS-polyacrylamide gel electrophoresis (PAGE) were performed as previously described (11, 19).

RESULTS

During sequence analysis of the cloned PCR products of DIssF DI RNA, we found that one of the cloned PCR products had a single A-nucleotide deletion at position 376 from the 5' end (Fig. 1). Due to this nucleotide deletion, the DI RNA-specific ORF was closed 6 nucleotides downstream of the deletion site. To test the ability of MHV DI RNA containing such a small ORF to replicate in MHV-infected cells, two complete DIssE-derived cDNA clones, NE and NE-1, were constructed. These clones contained identical sequences except that NE-1 had a single A deletion at nucleotide 376, producing a 57-amino-acid-long ORF. In contrast, the ORF of NE encoded 567 amino acids, similar to that of wild-type DIssE (16) (Fig. 1). In vitro-synthesized NE and NE-1 DI RNAs were translated in vitro, and the proteins synthesized were examined directly by SDS-PAGE. A 7.5-kDa protein was translated from NE-1 DI RNA; this molecular mass was close to the predicted molecular mass of 7,308. In contrast, an 88-kDa protein was translated from NE DI RNA, and was specifically immunoprecipitated by the anti-p28 antibody (data not shown). The 88-kDa size was consistent with the size of the protein of DIssE (16). These analyses demonstrated that the 7.5-kDa and the 88-kDa proteins were translated from the predicted small ORF of NE-1 DI RNA and the large ORF of NE DI RNA, respectively.

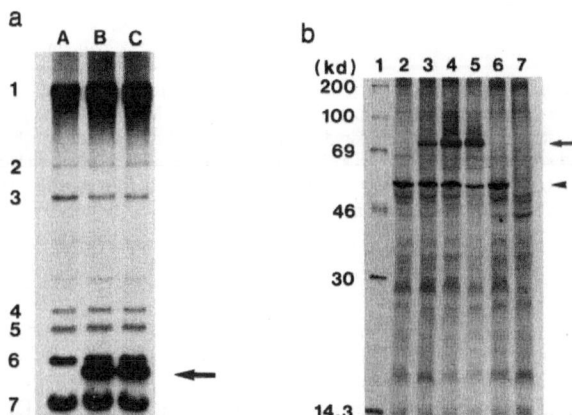

Fig. 2. Replication and protein synthesis by NE and NE-1 DI RNA in passage 0 virus-infected cells and RNA-transfected cells. (a) Agarose gel electrophoresis of MHV-specific intracellular RNAs which were obtained after infection of passage 0 virus samples. ^{32}P-labeled virus-specific RNA from DBT cells infected with MHV-A59 (lane A), transfected with NE (lane B) or transfected with NE-1 (lane C) was denatured with glyoxal and electrophoresed on a 1% agarose gel. Numbers 1 to 7 denote MHV-specific mRNA species. The arrow indicates DI RNAs. (b) SDS-PAGE of proteins from DI RNA-replicating cells. DBT cells were infected with MHV-A59 and transfected with in vitro-synthesized NE-1 DI RNA (lane 2), NE DI RNA (lane 3) or mock-transfected (lane 6). DBT cells were infected with passage 0 virus samples obtained from NE-1 DI RNA-transfected cells (lane 4), or from NE DI RNA-transfected cells (lane 5). Lane 7 represents mock-infection and mock-transfection. At 6.5 h p.i., cultures were labeled with [^{35}S]methionine for 20 min; total cell extracts were then prepared and immunoprecipitated with anti-p28 antibody. The 88-kDa DI-specific protein is indicated by the arrow. The band indicated by the arrowhead represents the nucleocapsid protein of MHV-A59. Lane 1, ^{14}C-labeled marker proteins.

To determine if NE-1 DI RNA replicates in MHV-infected cells, in vitro synthesized NE and NE-1 DI RNAs were independently transfected by lipofection into monolayers of DBT cells infected with MHV-A59 helper virus 1 h prior to transfection. The virus harvested at 16 hr p.i. is referred to as passage 0 virus sample. The passage 0 virus sample was used to infect DBT cells, and the ^{32}P-labeled intracellular RNA species were analyzed by agarose gel electrophoresis. Efficient replication of both NE- and NE-1-derived DI RNAs was observed (Fig. 2a). Next the replication efficiencies of NE DI RNA and NE-1 DI RNA were compared in DI RNA-transfected cells and it was found that there was no significant difference in the efficiency of replication and accumulation between the two DI RNAs (data not shown).

To examine whether NE-1-derived DI RNA still contained the small ORF, metabolic labeling of MHV-specific proteins and subsequent immunoprecipitation of MHV-specific proteins by the anti-p28 antibody were performed. As shown in Fig. 2b, synthesis of the 88-kDa protein was observed in NE DI RNA-transfected, MHV-infected cells as well as in cells infected with NE DI RNA-derived passage 0 virus sample. Surprisingly, the 88-kDa DI-specific protein was also

```
129 140 180    361          371        ↓   381 391 411  432

 A   A   G   AAGGAAAAACUUUGGUUAAUCAC    A   U   U      MHV-A59
 C   G   C   AAGGAAAAACUUUGGUUAAUCAC    G   C   C      MHV-JHM
 C   G   C   AAGGAAAAACUUUGGUUA UCAC    G   C   C      NE-1

                                                                        Exp. 1      Exp. 2      Total

 C   G   C   AAGGAAAAACUUUGGUUAAUCAC    A   U   U      Recombinant type 1   6 clones   12 clones   18 clones
 C   A   C   AAGGAAAAACUUUGGUUAAUCAC    A   U   U      Recombinant type 2   0 clone    1 clone     1 clone
 C   G   C   AAGGAAAAAACUUUGGUUAUCAC    G   C   C      Edited type          4 clones   13 clones   17 clones

                                                              Total   10 clones   26 clones   36 clones
```

Fig. 3. Sequences of cloned PCR products of DI RNAs derived from cells infected with passage 0 virus sample obtained from NE-1 DI RNA-transfected, MHV-infected cells. MHV-A59, MHV-JHM and NE-1 sequences are also shown for comparison. Nucleotide positions numbered from the 5'-end of the genome are shown above the MHV-A59 sequence. Underlined nucleotides indicate MHV-A59-specific nucleotides. The A nucleotide inserted into the edited DI RNA is indicated by double underlining. The number of clones used for sequence analysis in each of the two independent experiments are listed under Exp.1 and Exp.2.

detected in cells infected with passage 0 virus sample obtained from NE-1 DI RNA-transfected cells, although it was not detected in the NE-1 RNA-transfected and MHV-infected cells. This result suggested that the small ORF present in the original NE-1 DI RNA was changed to the large ORF during DI RNA replication and amplification.

To confirm the change of the NE-1 ORF during RNA replication, the sequence of the DI RNA present in the passage 0 virus-infected cells was examined. Intracellular RNA was extracted at 7 h p.i. from cells infected with passage 0 virus samples obtained from NE-1 DI RNA-transfected, MHV-infected cells. DI RNA-specific PCR product with the predicted size of 0.96 kb was prepared. The gel-purifed PCR product was cloned into a plasmid vector and sequenced. This experiment was conducted twice independently, under identical conditions, with virtually no difference in sequencing results (Fig. 3). Sequence analysis of the 36 clones obtained demonstrated that the sequences of DI RNAs in passage 0 virus-infected cells could be classified into two types. Nineteen of the 36 clones had a sequence in which the A nucleotide deletion at position 376 was repaired. As it is known that MHV undergoes high-frequency RNA recombination (2, 12), it seems likely that this sequence alteration was probably caused by RNA recombination between NE-1 DI RNA and MHV-A59 helper virus genomic RNA. The remaining 17 clones had a sequence which differed from the recombinant type: the single A nucleotide deletion at position 376 was not repaired and no RNA recombination was observed. However, each of these clones contained an additional A insertion 9 nucleotides upstream of the deletion site; this insertion produced a stretch of 6 consecutive A's. This nucleotide addition converted the NE-1-specific small ORF into a long ORF encoding 567 amino acids. These sequence analyses clearly demonstrated that the NE-1 DI RNA sequence was altered by RNA recombination to produce a large ORF. Furthermore, DI RNA species in which an A nucleotide was inserted 9 nucleotides upstream of the deletion site, thus producing a large ORF, also accumulated during replication of DI RNAs.

DISCUSSION

The present study demonstrated that DIssE-derived NE-1 DI RNA, which had a small ORF as compared to wild-type DIssE, replicated efficiently in MHV-infected cells, suggesting that a large ORF is not necessary for the replication of DI RNA. Analysis of DI RNAs obtained from passage 0 virus-infected cells demonstrated that most of these DI RNAs contained a large ORF similar to wild-type DI RNA. Of this group, approximately half of the DI RNAs contained sequences created by RNA recombination and the remainder contained sequences in which an A nucleotide was inserted 9 nucleotides upstream of the deletion site. It is not known whether the addition of the A nucleotide at this specific site on DI RNA occurs during in vtro transcription of NE-1 RNA by T7 RNA polymerase or during replication of NE-1 DI RNA.

Although a large ORF was not necessary for efficient MHV DI RNA replication, our study showed that DI RNAs with the large ORF accumulated during DI RNA replication. Although the mechanism of the accumulation of DI RNAs with large ORF is not clear, several possibilities may be considered. Due to the one-nucleotide deletion, the secondary or tertiary structures of NE-1 DI RNA might differ slightly from those of DI RNAs with the large ORF. It is possible that the MHV RNA replication machinery has a slightly better affinity for DI RNAs with the large ORF than for NE-1 DI RNA, resulting in the preferential replication of DI RNAs with the large ORF. The possible structural differences between these DI RNAs may be caused by the association of DI RNAs with ribosomes. It is possible that nascent positive-stranded DI RNA associates with ribosomes immediately after transcription and the DI-specific protein is synthesized cotranscriptionally. If this is the case, nascent strands of NE-1 DI RNA will dissociate from the ribosome 0.38 kb from the 5'-end, while nascent strands of DI RNAs with the large ORF are associated with the ribosome to 1.9 kb from the 5'-end. This difference may result in structural differences between different DI RNAs, and the RNA structure made by the longer association with ribosomes may have a slightly better affinity for the limited amount of MHV RNA replication component. Further studies are needed to determine the role of the DI-specific ORF in MHV DI RNA replication.

ACKNOWLEDGEMENTS

We thank Susan Baker for the anti-p28 antibody and Sima Vafaee for technical assistance. We thank Jennifer Fosmire for critical reading of the manuscript. This work was supported by Public Health Service grants AI29984 (to SM) and AI16144 (to MMCL) from the National Institutes of Health. MMCL is an investigator of Howard Hughes Medical Institute.

REFERENCES

1. Baker, S. C., C.-K. Shieh, L. H. Soe, M.-F. Chang, D. M. Vannier, and M. M. C. Lai. J. Virol. 63:3693 (1989).
2. Baric, R. S., K. Fu, M. C. Schaad, and S. A. Stohlman. Virology 177:646 (1990).
3. Fosmire, J. A., K. Hwang, and S. Makino. J. Virol. 66: 3522 (1992).

4. Jeong Y. S., and S. Makino. J. Virol. 66: 3339 (1992).
5. Lai, M. M. C., R. S. Baric, P. R. Brayton, and S. A. Stohlman. Proc. Natl. Acad. Sci. USA 81: 3626 (1984).
6. Lai, M. M. C., P. R. Brayton, R. C. Armen, C. D. Patton, C. Pugh, and S. A. Stohlman. J. Virol. 39: 823 (1981).
7. Lai, M. M. C., C. D. Patton, R. S. Baric, and S. A. Stohlman. J. Virol. 46: 1027 (1983).
8. Lai, M. M. C., and S. A. Stohlman. J. Virol. 26: 236 (1978).
9. Lee, H.-J., C.-K. Shieh, A. E. Gorbalenya, E. V. Eugene, N. La Monica, J. Tuler, A. Bagdzhadzhyan, and M. M. C. Lai. Virology 180:567 (1991).
10. Leibowitz, J. L., K. C. Wilhelmsen, and C. W. Bond. Virology 114: 39 (1981).
11. Makino, S., N. Fujioka, and K. Fujiwara. J. Virol. 54:329 (1985).
12. Makino, S., J. G. Keck, S. A. Stohlman, and M. M. C. Lai. J. Virol. 57:729 (1986).
13. Makino, S., M. Joo, and J. K. Makino. J. Virol. 65.:6031 (1991).
14. Makino, S., and M. M. C. Lai. J. Virol. 63: 5285 (1989).
15. Makino, S., C.-K. Shieh, J. G. Keck, and M. M. C. Lai. Virology 163: 104 (1988).
16. Makino, S., C.-K. Shieh, L. H. Soe, S. C. Baker, and M. M. C. Lai. Virology 166:550 (1988).
17. Makino, S., L. H. Soe, C.-K. Shieh, and M. M. C. Lai. J. Viol. 62:3870 (1988).
18. Makino, S., F. Taguchi, N. Hirano, and K. Fujiwara. Virology 139: 138 (1984).
19. Makino, S., K. Yokomori, and M. M. C. Lai. J. Virol. 64:6045 (1990).
20. McMaster, G. K., and G. G. Carmichael. Proc. Natl. Acad. Sci. USA 74:4835 (1977).
21. Shieh, C.-K., L. H. Soe, S. Makino, M.-F. Chang, S. A. Stohlman, and M. M. C. Lai. Virology 156: 321 (1987).
22. Spaan, W., H. Delius, M. Skinner, J. Armstrong, P. Rottier, S. Smeekens, B. A. M. van der Zeijst, and S. G. Siddell. EMBO J. 2: 1939 (1983).
23. van der Most, R. G., P. J. Bredenbeek, and W. J. M. Spaan. J. Virol. 65:3219 (1991).

SITE-DIRECTED MUTAGENESIS OF THE GENOME
OF MOUSE HEPATITIS VIRUS BY TARGETED
RNA RECOMBINATION

Paul S. Masters, Cheri A. Koetzner, Ding Peng, Monica M. Parker, Cynthia S. Ricard, and Lawrence S. Sturman

Wadsworth Center for Laboratories and Research
New York State Department of Health
Albany, New York 12201-0509 U.S.A.

ABSTRACT

We have genetically characterized a nucleocapsid (N) protein mutant of the coronavirus mouse hepatitis virus (MHV). This mutant, designated Alb4, is both temperature-sensitive and thermolabile, and its N protein is smaller than wild-type N. Sequence analysis of the Alb4 N gene revealed that it contains an internal deletion of 87 nucleotides, producing an in-frame deletion of 29 amino acids. All of these properties of Alb4 made it ideal for use as a recipient in a targeted RNA recombination experiment in which the deletion in Alb4 was repaired by recombination with synthetic RNA7, the smallest MHV subgenomic mRNA. Progeny from a cotransfection of Alb4 genomic RNA and synthetic RNA7 were selected for thermal stability. PCR analysis of candidate recombinants showed that they had regained the material that is deleted in the Alb4 mutant. They also had acquired a five nucleotide insertion in the 3' untranslated region, which had been incorporated into the synthetic RNA7 as a molecular tag. The presence of the tag was directly verified, as well, by sequencing the genomic RNA of purified recombinant viruses. This provided a clear genetic proof that the Alb4 phenotype was due to the observed deletion in the N gene. In addition, these results demonstrated that it is possible to obtain stable, independently replicating progeny from recombination between coronaviral genomic RNA and a tailored, synthetic RNA species. To date, we have constructed three additional mutants by this procedure. For two of these, a second-site point mutation that reverts the Alb4 phenotype has been transduced into a wild type background, which does not contain the Alb4 deletion. In the third, a portion of the region deleted in Alb4 has been replaced by its counterpart from the N protein of bovine coronavirus (BCV).

INTRODUCTION

The full application of molecular genetic techniques to the study of RNA viruses has been hindered by the unique genomic composition and replication strategies of these viruses. Positive-sense, single-stranded RNA viruses offer the advantage that their genomes, free of other viral structural components, are infectious when introduced into host cells. Nevertheless, RNA genomes present inherent obstacles to genetic engineering. RNA is chemically much more labile than DNA. Moreover, in contrast to the panel of techniques widely used to manipulate DNA, there are currently few tools available to directly construct recombinant RNA molecules in vitro.

A milestone in RNA virology was reached when a full-length cDNA clone of poliovirus was assembled and was shown to be infectious.[1] This advance enabled site-specific mutagenesis of the polio genome and allowed investigators to answer questions that previously could only be addressed indirectly, if at all. A number of positive-stranded RNA viruses have yielded to this approach, the largest to date being Sindbis virus, with a genome of 11.7 kb.[2] For these various viruses, infectious RNA has been synthesized in vitro from full-length cDNA clones by use of bacteriophage RNA polymerases, or, alternatively, it has been generated in vivo from eukaryotic expression vectors. In principle, coronaviruses ought to be amenable to this methodology. However, owing to both the enormous size of coronavirus genomes (26 to 31 kb) and the intrinsically high mutation rate of RNA-dependent RNA polymerases, the production of a full-length, error-free coronavirus cDNA will be technically difficult. In addition, it is not clear how such a cDNA, once constructed, will be easily manipulated.

As an alternative, we[3] and another group[4] have taken advantage of the high RNA recombination rate of coronaviruses[5-7] in order to target site-specific mutations to particular regions of the genome of mouse hepatitis virus (MHV). In our case, this was accomplished with a powerful selection provided by an N protein mutant designated Albany-4 (Alb4).

METHODS

The methods used in these studies have been described in detail elsewhere.[3] Briefly, a mixture of purified Alb4 genomic RNA and synthetic subgenomic RNA7 were cotransfected into mouse L2 or 17 clone 1 cells by either DEAE-dextran or electroporation. A cotransfection (rather than infection followed by transfection) protocol was used to ensure that the subgenomic RNA was delivered to the same subset of cells that was infected by genomic RNA. Progeny virus were harvested at 24-30 h and were incubated for 24 h at 39-40°C in 50 mM Tris-maleate, 100 mM NaCl, 1 mM EDTA, 10% fetal bovine serum. Surviving virus capable of forming large plaques on mouse L2 cells were chosen for further analysis.

RESULTS AND DISCUSSION

Characterization of the Alb4 Mutant

The MHV-A59 mutant Alb4 has two significant phenotypic features: temperature-sensitivity and thermolability. At the nonpermissive temperature (39° C), Alb4 forms plaques that are very small by comparison with those of its wild-type parent. Alb4 virions, upon incubation at 39-40°C (pH 6.5), undergo a dramatic drop in infectious titer (ca. 3 log), whereas wild-type virus titer is relatively unaffected by the same treatment.

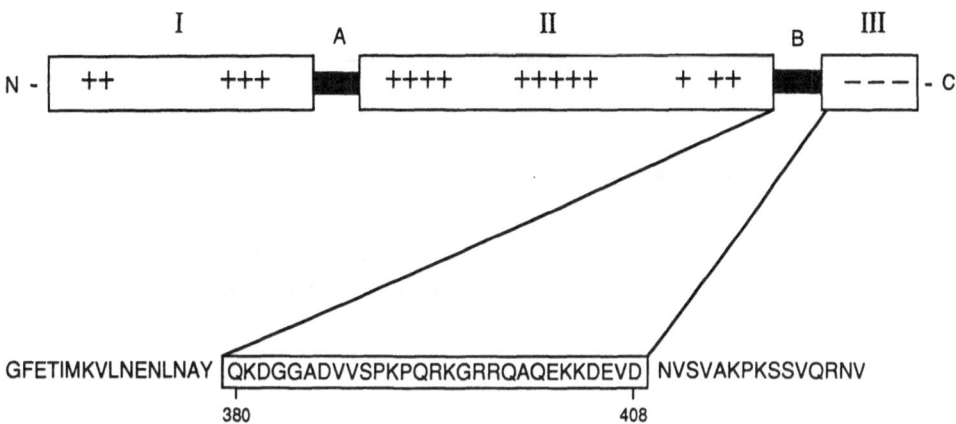

Figure 1. Region of the MHV-A59 N protein sequence that is deleted in Alb4 (boxed) shown in the context of a three-domain model of the N protein.

The N protein of Alb4 migrates markedly faster than wild-type N protein on SDS-PAGE, and cloning and sequence analysis has shown that this is due to the deletion of 87 nucleotides from the distal third of the gene.[3] Elsewhere, the Alb4 and wild-type N genes are identical. The Alb4 mutation results in an in-frame deletion of 29 amino acids from the MHV N protein, almost exactly excising a region (B) that we have previously proposed to be a spacer between two structural domains (II and III)[8] (Fig. 1). Clearly, spacer B cannot be essential for N protein function, since Alb4 is fully viable at the permissive temperature (33°C) and is even partially viable at the nonpermissive temperature.

Use of Alb4 as a Recipient for Targeted RNA Recombination

The characteristics of Alb4 suggested that it might be an ideal recipient for targeted RNA recombination. The reasons for attempting this were two-fold. First, it would establish genetically that the deletion in Alb4 was, indeed, the mutation responsible for the phenotype of that mutant. Second, it might provide a means for conducting site-directed mutagenesis in the 3' portion of the MHV genome.

A nearly authentic copy of wild-type MHV subgenomic RNA7 (the smallest of the MHV RNAs, which serves as the mRNA for N protein) was transcribed in vitro from a T7 vector (Fig. 2). The synthetic RNA7 differed from authentic RNA7 at only two loci: (i) two additional G residues at its 5' terminus, which were added to increase the efficiency of in vitro transcription, and (ii) a 5 nucleotide insertion, GTAAC, disrupting a BstEII site in the 3 UTR, which was added as a molecular tag to label recombinants (Fig. 2).

Alb4 genomic RNA and this synthetic RNA 7 were cotransfected into cells, heat-stable progeny were selected, and candidate recombinants were picked from among progeny capable of forming large (wild-type sized) plaques at the nonpermissive temperature. These candidates were first examined by PCR analysis of genomic RNA using primers that produced a product encompassing both the region deleted in Alb4 as well as the region of the 3' UTR containing the 5 nucleotide insertion. Both the size of the PCR fragment and the presence of an AccI site (Fig. 2) showed that the Alb4 deletion had been repaired in the putative recombinants. In addition, the recombinants had lost the BstEII site (Fig. 2) where the tagging insertion had been made. Finally, genomic RNA from recombinant viruses was directly sequenced to demonstrate the presence of the 5 nucleotide insertion. These results showed that it was possible to use targeted recombination to repair a substantial deletion and to cotransduce a site-directed mutation into the MHV genome and obtain stable, independently replicating recombinant viruses.

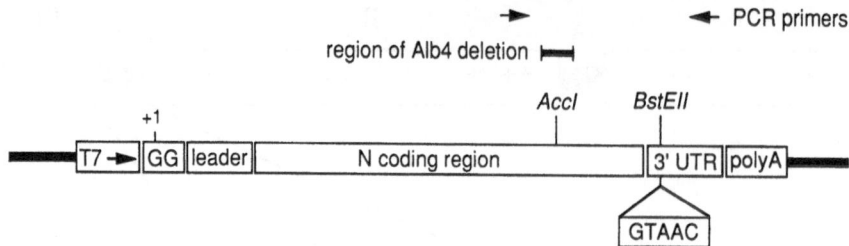

Figure 2. Transcription vector used for the synthesis of subgenomic RNA7 of MHV tagged with a 5 nt insertion. The start of the encoded transcript is denoted by +1.

We have shown further that this technique can be applied to the coding region of the N gene to produce specified point mutations as well as a large substitution. In the first case, two point mutations, each of which causes second-site reversion of the Alb4 lesion (Fig. 3), were separately transduced into a wild-type background to examine their effect in the *absence* of the Alb4 deletion. The presence of each of the mutations in recombinant progeny was verified, as before, by PCR analysis followed by direct sequencing of genomic RNA. The reverting mutations, in the absence of the original Alb4 deletion, are not deleterious to N protein function, although they may have more subtle effects on virus growth, which we are currently examining.

In the second case, a portion of the N gene encoding 15 amino acids within the spacer B region was replaced by its aligned counterpart from bovine coronavirus (BCV)[9] (Fig. 4). (This stretch of sequence is identical between BCV and human coronavirus OC-43.) Eight of the 15 positions in the exchanged sequences differ between MHV and BCV. The presence of the heterologous material in recombinant MHV viruses was verified by PCR analyses similar to those of the previous experiments (and including the demonstration of the creation of a new restriction site) and was further confirmed by sequencing of genomic RNA. Moreover, we could also demonstrate that the chimeric N protein from labeled, purified recombinant virus had a variant electrophoretic mobility compared to N

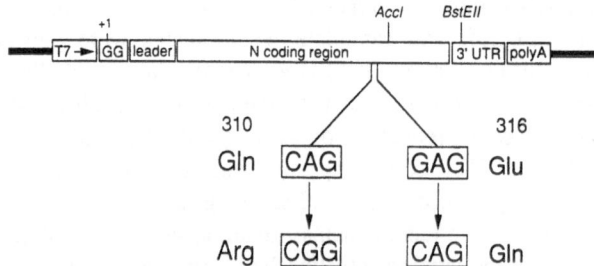

Figure 3. Transcription vectors used for the synthesis of subgenomic RNA7 containing Alb4 second-site reverting mutations in the absence of the Alb4 deletion.

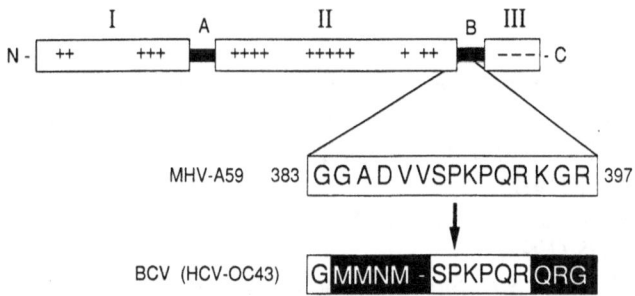

Figure 4. Schematic of the MHV N protein showing the wild-type sequence of a portion of spacer B aligned with the corresponding portion of the BCV (and HCV-OC43) N protein.

protein of wild-type MHV, presumably due to size and charge differences. The recombinant virus is phenotypically indistinguishable from wild-type MHV, showing that the composition of spacer B can vary considerably without having much effect on N protein function. We are presently attempting to construct more extensive BCV/MHV N protein chimeras in order to establish what portions of these two N proteins are functionally interchangeable.

Future Prospects

The studies described here have been limited to the neighborhood of the N gene, but in principle, the same methods could be applied to the other structural genes of MHV. This would require that suitable recipient mutants be found that map further upstream from Alb4 and have sufficiently low rates of reversion. As well, the technique could be used to insert addition genes into the coronavirus genome and to map conditional lethal mutants by marker rescue.

We have roughly estimated the efficiency of recovery of recombinants in our system to be on the order of 10^{-5}, which requires the use of the powerful counterselection afforded by Alb4. Recently, however, van der Most et al.[4] have reported the transfer of marker mutations between a synthetic MHV defective interfering (DI) RNA and the MHV genome with an efficiency on the order of 10^{-2}. This was high enough to allow successful screening of recombinants *without* selection. We have very recently confirmed this result by inserting into our transcription vector an additional 380 nucleotides corresponding to the sequence following the leader at the 5' extreme of the MHV genome. Such a system will potentially allow the construction of a broader range of recombinants, including mutants with impaired viability and mutants requiring two crossover events in order to be generated. We are currently seeking to use this system to construct mutants that will help elucidate coronavirus structural protein structure and function.

ACKNOWLEDGMENTS

This work was supported in part by Public Health Service grant AI31622 from the National Institutes of Health.

REFERENCES

1. V.R. Racaniello and D. Baltimore, *Science* 214:916 (1981).
2. C.M. Rice, R. Levis, J.H. Strauss, and H.V. Huang, *J. Virol.* 61:3809 (1987).
3. C.A. Koetzner, M.M. Parker, C.S. Ricard, L.S. Sturman, and P.S. Masters, *J. Virol.* 66:1841 (1992).
4. R.G. van der Most, L. Heijnen, W.J.M. Spaan, and R.J. de Groot, *Nucl. Acids Res.* 20:3375 (1992).
5. S. Makino, J.G. Keck, S.A. Stohlman, and M.M.C. Lai, *J. Virol.* 57:729 (1986).
6. J.G. Keck, L.H. Soe, S. Makino, S.A. Stohlman, and M.M.C. Lai, *J. Virol.* 62:1989 (1988).
7. R.S. Baric, K. Fu, M.C. Schaad, and S.A. Stohlman, *Virology* 177:646 (1990).
8. M.M. Parker and P.S. Masters, *Virology* 179:463 (1990).
9. W. Lapps, B.G. Hogue, and D.A. Brian, *Virology* 157:47 (1987).

HOMOLOGOUS RNA RECOMBINATION ALLOWS EFFICIENT INTRODUCTION OF SITE-SPECIFIC MUTATIONS INTO THE GENOME OF CORONAVIRUS MHV-A59 VIA SYNTHETIC CO-REPLICATING RNAs

Robbert van der Most, Leo Heijnen, Willy Spaan and Raoul de Groot

Department of Virology, Institute of Medical Microbiology
Faculty of Medicine, Leiden University
Postbus 320, 2300 AH Leiden, The Netherlands

ABSTRACT

We describe a novel strategy to site-specifically mutagenize the genome of an RNA virus by exploiting homologous RNA recombination between synthetic defective interfering (DI) RNA and the viral RNA. Marker mutations introduced in the DI RNA were replaced by the wild-type residues during replication. More importantly, however, these genetic markers were introduced into the viral genome: even in the absence of positive selection MHV recombinants could be isolated. This finding provides new prospects for the study of coronavirus replication using recombinant DNA techniques. As a first application, we describe the rescue of the temperature sensitive mutant MHV Albany-4 using DI-directed mutagenesis. Possibilities and limitations of this strategy are discussed.

INTRODUCTION

During mixed infection with different MHV strains, RNA recombination occurs at a remarkably high frequency, both in tissue culture and in infected mice (1-5). Although the mechanism is still unknown, homologous RNA recombination in coronavirus-infected cells presumably occurs via template switching ('copy-choice'). As proposed for picornavirus RNA recombination (6,7), polymerase complexes containing nascent RNA are thought to dissociate from their original template and anneal to another, after which RNA synthesis proceeds (2).

We are interested in genetic manipulation of coronaviruses. For several other RNA viruses full-length cDNA clones have been constructed from which infectious RNA can be transcribed *in vitro* (8-11). However, the construction of a full-length cDNA clone of a coronavirus has been hampered by the extreme length of the coronavirus genome. Here, we describe an alternative strategy to site-specifically introduce mutations into the MHV genome. This strategy exploits the high-frequency RNA recombination that occurs in MHV-infected cells (1,2) and involves the use of synthetic defective interfering (DI) RNAs. We provide evidence for homologous RNA recombination between DI RNA and the standard virus genome. Marker mutations introduced into the synthetic DI RNA were replaced by the wild-type residues during replication in MHV-infected cells. More importantly however, these marker mutations were incorporated into the genome of MHV-A59. DI-directed mutagenesis provides exciting new prospects for molecular genetic studies on coronaviruses, as was demonstrated by the rescue of the temperature-sensitive mutant Albany-4. This work was previously published in Nucleic Acids Research (vol. 20, p.3375-3388, 1992).

RESULTS

Construction of pMIDI-C

We previously reported the construction of pMIDI, a full-length cDNA clone of an MHV DI RNA (12). pMIDI consists of three non-contiguous regions of the MHV genome, namely the 5' most 3889 nucleotides, the 3' most 806 nucleotides and in between, 799 nucleotides derived from ORF1b of the polymerase gene. The three segments are joined in-frame, although in pMIDI the reading frame is interrupted by a UAA termination codon at position 3357 (12). To study recombination between MHV DI RNAs and the MHV genome, we constructed a pMIDI-derivative, pMIDI-C, in which we introduced three silent point mutations at positions 1778, 2297, 3572 (mutations A, B and C, respectively; Fig. 1). Marker mutations A (G→A) and B (T→C) were introduced in pMIDI by PCR-mutagenesis (13). The termination codon at position 3357 was replaced by the wild-type CAA codon by exchanging the *Xho*II-*Spe*I fragment (nucleotides 3237-3689) of pMIDI for the corresponding fragment derived from an independent, DI-derived cDNA clone, pDI02 (12). This fragment contained the wild-type CAA codon at position 3357 and a serendipitous T→C mutation at position 3572 (Marker C), presumably acquired during cDNA synthesis.

Point mutations in MIDI-C are replaced by the wild-type sequence during replication in MHV-A59 infected cells

RNA transcribed from pMIDI-C was used for transfection of MHV-infected mouse L-cells as described previously (12). At 12 hrs after transfection the tissue-culture supernatant (passage 0 virus), containing virus and DI particles, was harvested and passaged twice on fresh L-cells, yielding passage 1 and 2 virus. Total intracellular RNA was isolated from passages 0, 1 and 2 and analyzed by hybridization using (-)-sense oligonucleotide probe VI (Fig. 1). MIDI-C RNA was replicated in MHV-infected cells and strongly interfered with viral mRNA synthesis in cells infected with passage 1 (p1) virus/DI mixture (not shown). To determine whether recombination had occurred between the MHV genome and the synthetic DI RNA, the sequence of p2 MIDI-C RNA was examined. For this purpose, the DI RNA was subjected to cDNA synthesis followed by PCR amplification (RT-PCR) using the oligonucleotide primers I and III (Fig. 1). To prevent *in vitro* recombination of genomic and DI sequences during RT-PCR, the MIDI-C RNA used as a template was gel-purified. DI-specific PCR-DNA was digested with *Hind*III (positions 1985 and 3782 in pMIDI-C) and cloned into pUC20. Sequence analysis of 64 independent clones showed that mutations B and C had been replaced by the wild-type sequence in 14 (22%) and 4 clones (6%) respectively.

Since (i) the mutations were exclusively replaced by the wild-type sequence and (ii) the frequency with which the mutations were replaced depended on their distance from the artificial junction of the 1a and 1b segment, reversion by point mutation seems very unlikely. In fact, these results are best explained by homologous RNA recombination between MIDI-C RNA and the genome of the standard virus.

Isolation of MHV recombinants

RNA recombination between MIDI-C RNA and the viral genome could, in principle, also yield recombinant viruses carrying the marker mutations. To study this possibility, p2 virus was plaqued and 150 well-isolated plaques were used to inoculate mouse L-cells. Total cellular RNA, isolated from the infected cells, was subjected to RT-PCR using oligonucleotides I and II (Fig. 1). In all cases a DNA fragment with the expected length of 0.6 kb was generated. Evidently, the 0.6 kb fragment would also be produced with MIDI-C

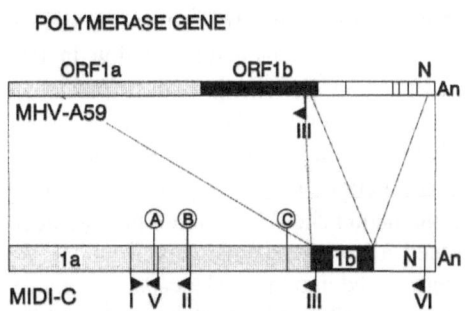

Figure 1. Schematic representation of the structure of MIDI-C RNA and the MHV-A59 genome. The different parts of MIDI-C, derived from the 5'end (ORF1a), ORF1b and 3' end of the MHV-A59 genome (N) are indicated. The locations are shown of marker mutations A, B and C, and of the oligonucleotides I, II, III, V and VI. The orientations of these oligonucleotides are indicated by arrowheads.

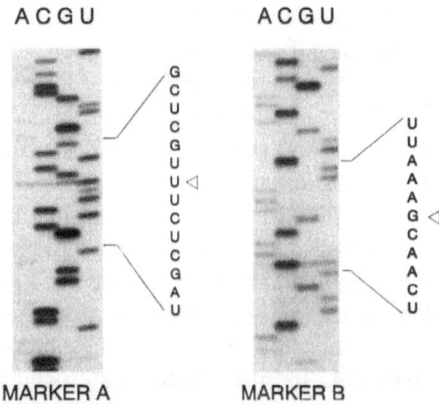

Figure 2. Sequence analysis of RNA isolated from cells infected with MHV recombinant 4, according to Fichot and Girard (15). Mutations A and B were analyzed using oligonucleotides V and II, respectively. Sequences are presented as (-) strand RNA. Arrowheads indicate the introduced mutations. Identical results were obtained for recombinant 138 (not shown).

RNA as template (Fig. 1). However, MHV DIs are rapidly lost by infecting at a low m.o.i. (12,14) and the obtained PCR fragments were therefore expected to be genome-specific.

The PCR-DNAs were screened for the presence of marker mutation B by differential hybridization using the 15-mer oligonucleotide probe IV (5' TGTCAACGAAATTCT 3', the guanine residue complementary to the introduced cytidine is underlined). PCR-DNA derived from two plaques, 4 and 138, specifically hybridized with this probe (not shown). The RNA preparations that had been used for RT-PCR were confirmed to be devoid of MIDI-C RNA by Northern blot analysis (not shown).

To determine whether viruses 4 and 138 were true recombinants, four consecutive plaque purifications were performed. Of each plaque generation, three to five well-isolated plaques were analyzed by differential hybridization of PCR-amplified cDNA. Wild-type MHV, treated identically, served as a negative control. In all cases, the progeny of viruses 4 and 138 contained mutation B (not shown), confirming that this genetic marker had been introduced into the viral genome and strongly arguing against MIDI-C contamination.

As a final control, we performed direct RNA sequencing. Stocks of viruses 4 and 138, that had been plaque-purified twice, were used to infect L-cells at an m.o.i. of 10. Intracellular RNAs were harvested and used for RNA sequence analysis using the oligonucleotide primers II and V (Figs. 1, 2). In addition to mutation B, the RNAs of viruses 4 and 138 contained mutation A (Fig. 2). Analysis of virus 4 RNA showed that mutation C was absent (not shown).

On the basis of our combined data, we concluded viruses 4 and 138 to be MHV mutants, generated by homologous RNA recombination between the synthetic MIDI-C RNA and the MHV-A59 genome.

Rescue of the MHV Albany-4 *ts*-mutant by homologous recombination

Having demonstrated that sequences of synthetic DI RNAs can be introduced into the viral genome via RNA recombination, we explored 'DI-directed mutagenesis' as a means to identify and localize mutations that result in a conditionally lethal phenotype. For this purpose, we used MHV strain Albany-4, a temperature-sensitive (*ts*-) mutant of MHV-A59. The *ts*-phenotype of this virus is the result of an 87 nucleotides deletion in the nucleocapsid (N) gene (nt 1138-1224 of the N-ORF). At 39°C, virus growth is impaired. Also, incubation of Albany-4 virions at 39°C abolishes infectivity (16).

pMIDI-C contains the 3' terminal 510 nucleotides of the wild-type N-ORF including the 87 nucleotides that are deleted in Albany-4. Incorporation of the N-sequences of MIDI-C RNA into the Albany-4 genome via RNA recombination should eliminate the *ts*-defect and generate wild-type MHV. To distinguish rescued Albany-4 recombinants from MHV-A59 contaminants, a silent T→A substitution was introduced into the nucleocapsid sequence of pMIDI-C (position 5030, corresponding to position 1200 of the N gene) by PCR-mutagenesis, yielding plasmid pMIDI-C*.

In this set of experiments, we used DNA transfection and *in vivo* transcription by vaccinia virus-expressed T7 polymerase as an alternative to RNA transfection to introduce MIDI-C* into MHV-infected cells. First, L-cells were infected with recombinant vaccinia virus vTF7-3 (17) at an m.o.i of 5 for 45 min, followed by DNA or mock-transfection at 90 min post infection. Two hours after transfection the cells were super-infected with Albany-4 at an m.o.i. of 10. After a 25 hr incubation at 33°C, the tissue culture supernatants were harvested and passaged once at 33°C in fresh L-cells, followed by a single passage at 39°C. Vaccinia virus remains cell-associated, and is therefore lost during passage. The 39°C/p2 virus stocks were used to infect monolayers of L-cells at 37°C. Intracellular RNAs were separated in formaldehyde-agarose gels and hybridized to 5'-end labelled oligonucleotide VI. This probe binds to the sequence, that has been deleted in Albany-4 (Fig. 3) and is therefore specific for the wild-type N-gene. Strikingly, probe VI not only detected MIDI-C* RNA but also the nested set of MHV mRNAs (Fig. 3, lanes 'MOCK' and 'MIDI-C*'), indicating that during passage, viruses had accumulated that had incorporated MIDI-C* sequences into their genomes.

To obtain direct evidence for this, the p2 virus/DI mixture was plaqued at 39°C and virus from 4 plaques was propagated in L-cell monolayers. After hybridization of the intracellular RNAs to oligonucleotide VI, the nested set of MHV RNAs was found; MIDI-

C* RNA was not detected (Fig. 3, lanes 1-4). Sequence analysis of these RNA preparations showed that the wild-type N-sequence had been restored and that the U→A marker mutation in the N-ORF was present (Fig. 4), providing formal evidence that the isolated viruses were generated by recombination between MIDI-C* RNA and the Albany-4 genomic RNA.

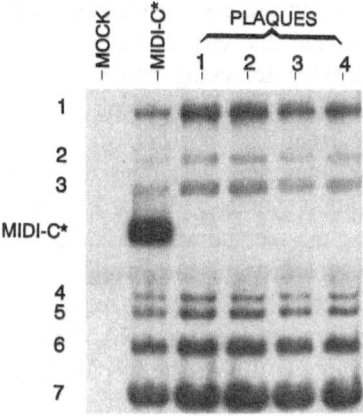

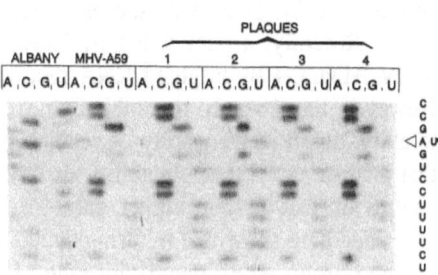

Fig. 3. Hybridization analysis of intracellular RNAs of MHV Albany-4 recombinants. Stocks of MHV Albany-4 (p2, grown at 39°C) obtained after initial transfection with pMIDI-C* or mock-transfection were either used to infect fresh L-cells at 37°C to isolate intracellular RNAs (lanes 'Mock' and 'MIDI-C*') or plaqued at 39°C. Virus from four randomly-chosen plaques was propagated and intracellular RNAs were isolated (lanes 1-4).

Fig. 4. RNA sequence analysis of recombinant Albany-4 viruses. Intracellular RNA of four plaque-purified recombinant Albany-4 viruses, described in Fig. 3, was subjected to sequence analysis. Intracellular RNA from Albany-4- and MHV-A59-infected cells served as controls. Sequences are presented as (-)strand RNA. An arrowhead indicates the introduced marker mutation.

DISCUSSION

For the coronavirus MHV homologous recombination of genomic RNAs has been well documented (1-4,18). Here, we extend these observations by providing evidence for recombination between DI RNAs and the MHV-A59 genome. We have shown that silent mutations introduced into the ORF1a sequence of MIDI-C were exclusively replaced by the wild-type residues during DI replication in MHV-infected cells. These marker mutations, however, were not replaced at an equal rate: markers B and C, located 1500 and 300 nucleotides upstream of the artificial ORF1a/1b border in MIDI-C, were replaced in 22% and 6% of the passage 2 DI RNAs, respectively. Apparently, the frequency of replacement correlated with the distance between the mutation and the ORF1a/1b border, i.e. the region in which template-switching has to occur in order to remove the mutation while maintaining the original MIDI-C structure. These finding are in accordance with the current model for coronavirus RNA recombination.

Most convincingly, recombination between MIDI-C RNA and the MHV genomic RNA was demonstrated by the identification of viruses that had incorporated the genetic markers A and B into their genomes. It should be emphasized that these viruses were isolated in the absence of selection. These findings lead to the important conclusion that synthetic DI RNAs can be used to site-specifically alter the MHV genome by exploiting RNA recombination. The potential of DI-directed mutagenesis is illustrated by the rescue of the conditionally lethal mutant Albany-4. The *ts*-phenotype of this mutant is caused by a deletion at the 3' end of the N-gene (16). Viruses that had restored the N-gene by recombination with MIDI-C* RNA, were selected for by passaging the virus/DI mixture at the restrictive temperature.

Clearly, there will be limitations in this system: thus far, we have introduced mutations only in the 5' and 3' terminal regions of the MHV genome. It remains to be determined whether mutations can be introduced efficiently into more internal regions, since this would require double recombination events. Also, in the case of mutations causing reduced replication *in vitro*, the screening for recombinant viruses will be difficult. Presumably, such problems can be solved by improving screening procedures and by applying selection, e.g. via rescue of *ts*-markers (this paper) or by using neutralizing antibodies (18). Studies to address these issues are currently in progress.

Recently, a publication by Koetzner et al. (19) appeared describing the rescue of MHV Albany-4 by targeted RNA recombination using a synthetic mRNA 7 homologue. However, the observed recombination frequency using this method was considered too low to allow direct identification of recombinants without selection. In fact, it was anticipated that a more general applicability of targeted RNA recombination would require finding conditions that favour higher rates of recombination of exogenous RNA. The results described here show that by using synthetic co-replicating DI RNAs site specific mutations can be introduced efficiently into the MHV genome.

ACKNOWLEDGMENTS

We thank Dr. Paul Masters for generously providing MHV Albany-4 and Dr. Willem Luytjes for stimulating discussions. R.G.M. was supported by grant 331-020 from the Dutch Foundation for Chemical Research (SON). We gratefully acknowledge Oxford University Press for permission for the use of copyright material.

REFERENCES

1. M.M.C. Lai, R.S. Baric, S. Makino, J.G. Keck, J. Egbert, J.L. Leibowitz and S.A. Stohlman, *J. Virol.*, 56:449-456 (1985).
2. S. Makino, J.G. Keck, S.A. Stohlman and M.M.C. Lai, *J. Virol.* 57:729-737 (1986).
3. J.G. Keck, S.A. Stohlman, L.H. Soe, S. Makino and M.M.C. Lai, *Virology* 156:331-341 (1987).
4. J.G. Keck, L.H. Soe, S. Makino, S.A. Stohlman and M.M.C. Lai, *J. Virol.* 62:1989-1998 (1988).
5. J.G. Keck, G.K. Matsushima, S. Makino, J.O. Fleming, D.M. Vannier, S.A. Stohlman and M.M.C. Lai, *J. Virol.* 62:1810-1813 (1988).
6. K. Kirkegaard and D. Baltimore, *Cell* 47:433-443 (1986).
7. T.C. Jarvis and K. Kirkegaard, *EMBO J.* 11:3135-3145 (1992).
8. P. Ahlquist, R. French, M. Janda and L.S. Loesch-Fries, *Proc.Natl.Acad.Sci. USA* 81:7066-7070 (1984).
9. C.M. Rice, R. Levis, J.H. Strauss, H.V. and Huang, *J. Virol.* 61:3809-3819 (1987).
10. C.M. Rice, A. Grakoui, R. Galler and T.J. Chambers, *New Biol.* 1:285-296 (1989).
11. C.-J. Lai, B. Zhao, H. Hori and M. Bray, *Proc.Natl.Acad.Sci. USA* 88:5139-5143 (1991).
12. R.G. van der Most, P.J. Bredenbeek and W.J.M. Spaan, *J. Virol.* 65:3219-3226 (1991).
13. J. Sambrook, E.F. Fritsch and T. Maniatis, Molecular Cloning. Cold Spring Harbor Laboratory Press, Cold Spring Harbor (1989).
14. S. Makino, F. Taguchi and K. Fujiwara, *Virology* 133:9-17 (1984).
15. O. Fichot and M. Girard, *Nucleic Acids Res.* 18:6162 (1990).
16. P.S. Masters and L.S. Sturman, *Adv. Exp. Med. Biol.* 276:235-238 (1990).
17. T.R. Fuerst, E.G. Niles, F.W. Studier and B. Moss, *Proc.Natl.Acad.Sci. USA* 83:8122-8126 (1986).
18. S. Makino, J.O. Fleming, J.G. Keck, S.A. Stohlman and M.M.C. Lai, *Proc.Natl.Acad.Sci. USA* 84:6567-6571 (1987)
19. C.A. Koetzner, M.M. Parker, C.S. Ricard, L.S. Sturman and P.S. Masters, *J. Virol.* 66:1841-1848 (1992).

Chapter 3

Characterization and Functions of Viral Proteins

IDENTIFICATION OF PEPLOMER CLEAVAGE SITE MUTATIONS ARISING DURING PERSISTENCE OF MHV-A59

James L. Gombold, Susan T. Hingley, and Susan R. Weiss

Department of Microbiology
University of Pennsylvania School of Medicine
Philadelphia, PA 19104-6076

ABSTRACT

Primary mouse glial cell cultures were infected with mouse hepatitis virus strain A59 (MHV-A59) and maintained over an 18 week period. Viruses isolated from these cultures 16-18 weeks postinfection produce small plaques on fibroblasts and cause only minimal levels of cell-to-cell fusion at times when wild type causes nearly complete cell fusion. However, when mutant-infected cultures were examined 24-36 hours postinfection approximately 90% of the cells were in syncytia showing that the fusion defect is not absolute but rather delayed. Addition of trypsin to mutant-infected cultures enhanced cell fusion a small (2- to 5-fold) but significant degree. Sequencing of portions of the spike genes of six fusion-defective mutants revealed that all contained the same single nucleotide mutation resulting in a substitution of aspartic acid for histidine in the spike cleavage signal. Mutant virions contained only the 180 kDa form of spike protein suggesting that this mutation prevented the normal proteolytic cleavage of the 180 kDa protein into the 90 kDa subunits. Examination of revertants of the mutants supports this hypothesis. Replacement of the negatively-charged aspartic acid with either the wild type histidine or a non-polar amino acid was associated with the restoration of spike protein cleavage and cell fusion.

INTRODUCTION

Mouse hepatitis virus strain A59 (MHV-A59) is a positive stranded enveloped virus with an approximately 31 kb RNA genome. Three structural proteins have been identified in MHV-A59 (1). One of these proteins, the peplomer or spike glycoprotein (S), is present on the surface of the virion and is responsible for binding to the cellular receptor. Antibodies directed at specific epitopes on S are capable of neutralizing viral infectivity. In infected cells, S is transported by the secretory system to the cell surface where it is free to interact with adjacent cells resulting in cell fusion (syncytia). Expression of S (via vaccinia virus vectors) in the absence of infection or other viral proteins is sufficient for the induction of efficient cell-to-cell fusion (2,3).

A portion of the spike in virions is cleaved into two 90 kDa fragments termed S1 (N terminal) and S2 (C terminal). In the case of MHV, the level of fusion observed correlates with the amount of cleaved S; thus, the cleaved protein rather than the precursor is believed to be fusogenic (4). Proteolytic cleavage of coronavirus spike proteins occurs adjacent to a sequence of basic amino acids on the C terminus of S1 (5). This motif is conserved among many MHV strains and in avian infectious bronchitis virus (IBV). Similar stretches of basic amino acids occur adjacent to the glycoprotein cleavage site in paramyxoviruses, many retroviruses, and influenza A viruses (6,7,8).

In this report, we present data showing that small plaque, fusion-defective mutants arise during persistence in primary murine glial cell cultures. Sequencing of the spike protein gene revealed that six mutants derived from two independent cultures contained the same histidine to aspartic acid substitution in the cleavage signal region. Analysis of virion structural proteins suggests that this mutation prevents proper proteolytic cleavage of the 180 kDa spike precursor. Evidence that the defect in cleavage is responsible for the fusion deficiency seen in infected fibroblasts comes from studies of revertants in which fusion, cleavage of the spike, and loss of the acidic aspartic acid residue coincide. These data confirm and extend previous reports concluding cleavage of S is required for efficient cell-cell fusion by MHV.

MATERIALS and METHODS

Virus and Cells

MHV strain A59, obtained from Dr. Lawrence Sturman (Albany, NY), was propagated in mouse L2 cells. The virus stock had been grown previously at 40C to eliminate any temperature-sensitive (ts) mutants that might be present.

Primary mixed glial cell cultures were made from dissociated brains of newborn C57BL/6 mice essentially as described (9) and were used 10-15 days after plating. These cultures were 90-95% astrocytes as determined by positive immunostaining for glial fibrillary acidic protein (data not shown). Primary glial cultures, L2 cells, and 17 Cl-1 cells were grown in Dulbecco's modified Eagles medium (DMEM) containing 10% fetal bovine serum (FBS).

Infection of Glial Cells and Isolation of Virus

Three independent glial cell cultures (A,B, & C) were infected with MHV-A59 at a multiplicity of infection (MOI) of 5 and maintained at 37C. The medium was removed twice weekly and the cells refed with fresh medium. Virus was isolated from the medium by three plaque-to-plaque purifications and then grown to high titer on L2 cells at 32C. Three mutants were isolated from culture B (B10, B11, B12) after 18 weeks of infection, and three from culture C (C10, C11, C12) after 16 weeks of infection. Fusion-competent revertants were isolated following serial, low-multiplicity passages of mutants on L2 cells by plaque-purification as above.

Virus Titrations

Plaque titration of virus was done on L2 cells and at 37C in DMEM containing 2% FBS. In some cases titrations were done at 39C & 32C to identify and quantitate ts mutants. Titration of virus in the presence of 10-100 ug/ml trypsin (Sigma) was done as described above except that cell monolayers were washed twice in Tris-buffered saline prior to infection and FBS was omitted from the medium.

Viral Proteins

For analysis of the spike glycoproteins in virions, L2 or 17Cl-1 cells were infected at an MOI of 5. At 5 hpi, the cells were incubated for 30 min in DMEM lacking serum and methionine and then labelled with 50 uCi/ml of [35S] TransLabel for 6 hours. Virus in the medium was collected by centrifugation and the pellets were lysed in RIPA buffer (50 mM Tris, pH 8, 100 mM NaCl, 1% NP-40, 1% deoxycholate, 0.1% SDS, 100 ug/ml PMSF). The spike protein was immunoprecipitated from virion lysates using a polyclonal goat serum (AO4) (kindly provided by Dr. K. Holmes, Bethesda, MD). Lysates were incubated with antiserum for 60 min at 4C and the antigen-antibody complexes were then precipitated with *S. aureus* (Pansorbin, Calbiochem). The pellets were washed three times, resuspended in Laemmli sample buffer, and analyzed by SDS-PAGE.

Fusion Assay

L2 cells were infected with virus at an MOI of 5. The proportion of nuclei contained in syncytia was determined as a function of time after infection by counting three random fields using ethanol-fixed crystal violet-stained cells.

To assess the effect of trypsin on cell fusion, the cells were infected as above except that serum was omitted from the growth medium. At selected times, the cells were washed three times with PBS and then treated for 30 min at 37C with 2.5 ug or 5 ug of trypsin per ml in PBS. Afterwards, the trypsin was removed, and the cells washed once and then fed with DMEM containing 10% FBS. The degree of fusion was determined after a 90 min incubation at 37C.

RNA Sequencing

For preparation of RNA templates, L2 cell monolayers in 50 mm plates were infected with A59 or the fusion defective mutants at an MOI of 5. At 8 hpi, the cells were washed three times on ice with ice-cold PBS and lysed in 50 mM Tris-HCl (pH 7.5), 100 mM NaCl, 1% NP-40. The lysates were digested with 200 ug of Proteinase K (Boehringer Mannheim) per ml, extracted with phenol and phenol-chloroform, and precipitated with ethanol. RNA was quantitated by measuring adsorbance at 260 nm.

Sequence information was obtained both by direct sequencing of viral RNA and by sequencing DNA by PCR using cDNAs transcribed from the viral RNA. For direct RNA sequencing, cytoplasmic RNA (50-70ug) was sequenced using reverse transcriptase and dideoxynucleotides (BMB). Primer CSN (5'-CATCGGAGTGTATGGCTC-3'), complementary to sequence approximately 30 nt downstream of the cleavage signal region (nucleotides 2185-2202 in the sequence of Luytjes et al. (5)), was purchased from Operon Technologies. Sequencing products were treated with 10-20 U of terminal transferase (BMB) to avoid ambiguities due to "strong stops".

Sequencing of the S genes of wild type and mutant viruses was determined also from DNA fragments obtained by reverse transcription-polymerase chain reaction (RT-PCR). Cytoplasmic RNA was reverse transcribed using random hexamers (BMB) and Muloney MLV reverse transcriptase (Gibco BRL). PCR amplification of the cDNA templates with Taq polymerase (BMB) and six pairs of primers (obtained from Dr. K. Holmes) yielded 6 overlapping fragments approximately 700-800 nucleotides in length spanning the entire S gene. PCR products were purified using Promega's Magic PCR Prep, then sequenced using the Promega fmol DNA sequencing kit.

RESULTS AND DISCUSSION

We have reported previously (9,10) that while infection of murine fibroblasts with MHV-A59 causes extensive fusion of the cell monolayer, infection of primary glial cell cultures causes minimal if any cytopathic effect (cpe). To better understand virus-glial cell interactions, we have begun to examine the evolution of MHV during long-term infection in these cultures. We infected at high multiplicity three separate cultures of glial cells with A59. At weekly intervals the medium was removed from the cultures and stored at -80C, and the cells refed with fresh medium.

Infectious virus in the medium was titrated by plaque assay. All three cultures produced virus continuously over the 16 to 18 week period with levels of virus varying from 10^6 to 10^7 pfu per culture (data not shown). These titrations also showed that virus present at the latest times postinfection made small, non-lytic plaques compared to wild type (data not shown). For further studies, we plaque purified viruses from the culture supernates as described in Materials and Methods. Infection of L2 cells with the plaque purified mutants caused a productive infection but the cytopathic effect normally observed with A59 was markedly reduced. At 10 hpi, a time at which A59 typically causes extensive syncytia, cultures infected with the mutants displayed 5-10 percent cpe (Fig. 1). With the exception of C10 and C11, fusion in mutant-infected cells increased over the next 12-18 hours to include 90% of the cells. Syncytia in C10- and C11-infected cells involved less than 50% of the cells at 24 hpi and this lower level of fusion may be related in part to the temperature-sensitive phenotype of these mutants (data not shown; see below). It is important to note that the mutants are not absolutely defective in their ability to induce fusion, but rather, induce fusion with slower kinetics than wild type A59.

The small plaque morphology and the delayed appearance of fusion was not due to poor growth kinetics of the clones, which, except for C10 and C11, were indistinguishable

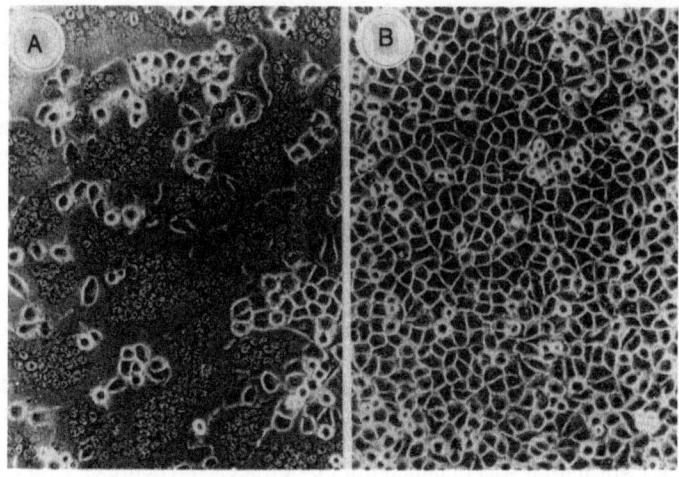

Figure 1. Mutants from persistently-infected glial cell cultures are fusion defective. Wild type A59 (A) or mutant B10 (B) were used to infect L2 cells and photographed at 10 hpi. Unlike wild type, the mutant induces minimal cytopathic effect (syncytia).

from wild type (data not shown). Furthermore, immunofluorescence staining of mutant infected cells suggested that the delayed fusion was not a result of a delay in the appearance of cell-surface spike.

Since fusion of cells during infection appears to correlate with the extent of cleavage of the spike protein (4), we addressed the possibilty that the mutants were defective in cleavage of the 180 kDa spike precursor to the 90 kDa S1 and S2 subunits. Cells were infected with either wild type or mutants at high multiplicity, and at various times after infection, the cells were incubated for 30 min in the presence of 2.5 or 5.0 ug/ml trypsin. All mutants except C10 and C11 induced higher levels of fusion in the presence of trypsin than in control cultures (data not shown). However, the increase in fusion observed in cells treated with 5 ug of trypsin per ml, though statistically significant, was generally 5-fold or less and never more than 30% of the cells fused. In addition, a wild type plaque morphology is observed when the mutants are assayed in the presence of 30 ug/ml trypsin.

Together, these data showed that trypsin was able to at least partially restore the fusion phenotype of the mutants, suggesting that the mutant spike protein was not being proteolytically cleaved into S1 and S2 subunits. To examine this directly, cells were infected with wild type or the mutants and labeled with [35S] TransLabel from 6 to 12 hpi. Virions were pelleted from the medium and the spike protein was immunoprecipitated from these virions and analyzed by SDS-PAGE. The spike protein in wild type A59 was present in both the 180 kDa and the 90 kDa forms (Fig. 2). In contrast, the mutant virions contained only the uncleaved 180 kDa spike precursor.

To better understand the nature of the defect in these mutants, we directly sequenced the RNA encoding the S1/S2 junction in several of the clones. The sequence obtained for wild type A59 was identical to that previously published by Luytjes et al. (5). Shown in Table 1 is the highly basic amino acid sequence directly upstream from the S2 amino terminus that is thought to act as a signal for proteolytic cleavage in wild type MHV. In all six mutants, a single nucleotide change at nucleotide 2146 was observed that caused a substitution of aspartic acid for histidine. The introduction of a negatively-charged amino acid into this highly basic region may mask the signal or prevent or otherwise interfere with its usage and thereby prevents cleavage of the spike.

To identify additional mutations in the spike genes of the mutants that may effect the fusion phenotype, we sequenced in entirety gene 3 from wild type and mutants B11 and C12 as described in Materials & Methods. Wild type S varied by two nucleotides and one amino acid from the published sequence. One mutation, a A to G transition at nucleotide 3045 was silent; the other was a G to A transition at nucleotide 293 which substitutes a serine for an asparagine in S1 (S.T.H., unpublished observation; C. Ricard and L.

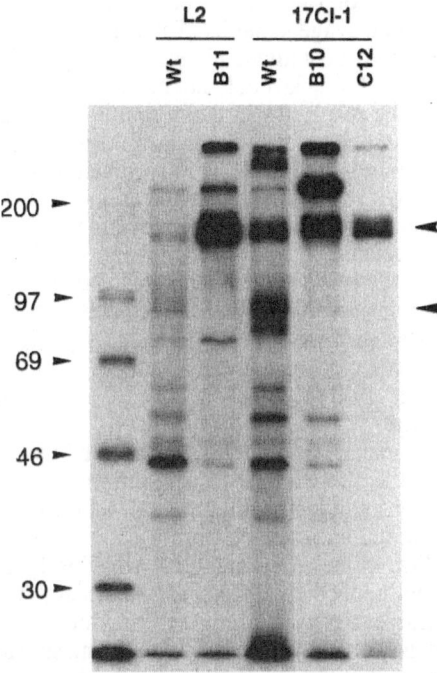

Figure 2. The 180 kDa spike protein in fusion defective mutants is not cleaved. L2 or 17Cl-1 cells were infected with wild type A59 or the fusion mutants and labeled with [35S] TransLabel. Radiolabelled virions in the medium were pelleted and the spike protein immunoprecipitated. The 90 kDa spike protein subunits are detectable only in wild type infected cells. Arrows on the right indicate the position of the 180 kDa spike precursor (upper arrow) and the 90 kDa cleavage products (lower arrow).

Table 1. Predicted amino acid sequence of the spike cleavage signal in wild type A59 and the fusion mutants.

Virus	Cleavage Signal	Fusion(a)	Cleavage of S
Wild type	RRAHR	+	+
B10	RRADR	-	-
B11	RRADR	-	-
B12	RRADR	-	-
C10	RRADR	-	-
C11	RRADR	-	-
C12	RRADR	-	-
B11R1	RRADR	-	-
C11R1	RRAAR	+	+
C12R1	RRAHR	+	+
C12R3	RRAGR	+	+

(a) Cell fusion measured at 10 hpi

Sturman, personal commmunication). The B11 and C12 mutant spike genes were identical to wild type with the exception of the cleavage signal mutations described above and an additional mutation at nucleotide 476, a Gln to Leu substitution in S1. As described below, this latter mutation does not appear to be related to the fusion phenotype of the mutants.

To test the correlation between fusion competence, cleavage of S and the sequence in the putative cleavage signal, fusion-competent revertants of mutants C11 and C12 were isolated based on their wild type plaque morphology. Infection of cells with these viruses induces similar cytopathic effect when compared to wild type infected cells (data not shown). Immunoprecipitation of the spike from revertant virions detected both the uncleaved precursor and the cleavage subunits (Table 1). In contrast, one clone (B11R1) that was isolated but retained the mutant plaque phenotype did not revert in fusion phenotype. Cleavage of the B11R1 spike could not be detected by immunoprecipitation.

Sequence analysis of the revertants in the region of the cleavage site revealed that restoration of cleavage was associated with the loss of the aspartic acid residue seen in the mutants (Table 1). One clone reverted to the wild type histidine residue, while the others substituted the aspartic acid with a small non-charged amino acid. The Gln to Leu mutation in subunit S1 of the mutants S1 was retained in the revertants (data not shown) suggesting that this mutation is not associated with the fusion phenotype. We conclude that efficient fusion requires cleavage of the spike protein, and that alterations in the cleavage signal, such as the incorporation of negatively-charged amino acids, can prevent its recognition by the (cellular) protease. These results support a previously published report that efficient fusion by MHV-A59 requires cleavage of S (4) and suggest that the fusion-defective phenotype of the mutants is a direct result of the inablility of the spike protein to be processed into its mature form.

We report here the characterization of viruses isolated from persistent infections of murine glial cells. In general, these viruses display normal growth kinetics in murine fibroblasts and achieve titers similar to the parental A59 virus from which they were derived. However, fusion in mutant-infected fibroblasts is markedly delayed and plaques are small and non-lytic. Two clones, C10 and C11, are slightly different. These mutants are temperature-sensitive and replicate with slower kinetics at 37C. However, even though they achieve titers comparable to the other mutants, they do not induce the same level of fusion in cells, even if the assays are done at 32C (the permissive temperature for C10 and C11). Although this suggests that the fusion defect is not due directly to the ts defect, the slow growth kinetics of C10 and C11 may indirectly affect the level of fusion in infected cells.

All six mutants that were examined contain a histidine to aspartic acid substitution within the basic amino acid cleavage signal thought to be responsible for the proper proteolytic processing of the peplomer. Analysis of the spike protein present in virions shows that these fusion mutants do not contain significant amounts of cleaved spike, but rather incorporate the uncleave 180 kDa precursor. Restoration of fusion-competence in revertant viruses is associated with the replacement of the acidic aspartic acid residue with either the wild type histidine or a non-charged amino acid. We suggest that incorporation of aspartic acid within the dibasic signal prevents cleavage either by preventing its recognition by or its interaction with the protease responsible for cleavage. Sawicki (11) isolated from persistently-infected 17Cl-1 cells a small plaque mutant of MHV-A59 that is similar to the mutants described here. Purified virions of this mutant contained only the uncleaved 180 kDa spike protein; furthermore, 100-fold more trypsin was required to convert the 180 kDa protein to the 90 kDa form. Interestingly, this mutant contains a substitution of the histidine residue in the cleavage signal for an asparagine (S. Sawicki, personal communication).

Two other classes of fusion defective mutants of MHV have been described previously. Gallagher et al. (2) showed that a mutants of MHV-4 recovered from a persistently-infected neural cell line contained mutations within the heptad repeats of the S2 subunit of the spike protein. The heptad repeat regions are believed to be important in the oligomerization of protein monomers (12). These mutations caused a change in the pH-dependence of fusion such that an acid pH was required by the mutant to induce syncytia in infected cells. Gallagher et al. (13) also reported that neutralization-resistant mutants of MHV-4 selected with a monoclonal antibody to S were defective in their ability to induce syncytia. These mutants contain deletions in the S1 subunit in the "hypervariable" region (13,14). A JHM isolate from the spinal cord of a demyelinated rat (15) was also found to contain a deletion in S1 (16). While the precise mechanism by which any of these mutations effect fusion in these MHV-4 variants in unknown, the identification of these variants clearly demonstrates that different regions in the protein are crucial for proper function of the protein. Here, we have identified a third region of S, the cleavage signal region, that affects efficient cell-to-cell fusion by the spike.

Proteolytic processing of envelope proteins is common in many viruses, and for some is a requirement for cell-to cell-fusion as well as virus-cell fusion or infectivity. The fusion proteins of paramyxoviruses, orthomyxoviruses and retroviruses are cleaved into two subunits by cellular proteases (6,7,8) and, like MHV, the site of cleavage is preceded by a group of basic amino acids. In paramyxoviruses, cleavage of the fusion protein, F0, apears to be most efficient when at least 4 of the 5 amino acids in the cleavage signal are basic (7). The importance of cleavage of F0 stems from the observation that processing of the fusion protein is necessary for infectivity and virulence, showing that mutations within the signal sequence can be attenuating. Cleavage of the hemagglutinin of influenza virus also is influenced by the content of basic amino acids near the cleavage site (17) and is associated with virulence (6). The coronaviruses differ from these other classes of viruses in some ways: (1) Cleavage of the spike protein of porcine transmissible gastroenteritis virus and feline infectious peritonitis virus does not occur. Since these viruses cause fusion of infected cells, it is clear that cleavage of S is not a strict requirement for fusion for all coronaviruses. However, while the mutations within the MHV cleavage signal do not prevent fusion, they do appear to reduce the efficiency of fusion. (2) Cleavage of the paramyxovirus, myxovirus and retrovirus fusion envelope glycoproteins exposes a hydrophobic fusion peptide at the newly generated amino terminus; this is not true of the cleaved coronavirus spike protein. A hydrophobic fusion peptide has not yet been conclusively identified for any of the coronaviruses. Furthermore, the mutants described here do not cleave the spike protein, but are still infectious in cultured cells and cause a significant amount of cell-to-cell fusion at late times after infection. The phenotype of the mutants described here suggest that cleavage of S is a prerequisite for efficient cell-to-cell fusion. The fusion observed following prolonged incubations could be due to low and undetectable levels of cleaved S that accumulate over time on the cell surface. Alternatively, fusion could be due to high levels of uncleaved spike that inefficiently induces cell fusion.

ACKNOWLEDGMENTS

We thank Ms Marianne Allessio and Ms Xiurong Wang for technical assistance, Dr. Ehud Lavi for helpful discussions and Dr. K. Holmes for the AO4 antiserum. This work was supported by Public Health Service Grant NS-21954. JLG and STH were supported in part by training grant NS-07180 and NS-11037.

REFERENCES

1. Spaan, W. J. M., Cavanagh, D., Horzinek, M. C. J. Gen. Virol. 69:2939-2952, 1988.
2. Gallagher, T. M., Escarmis, C., Buchmeier, M. J. J. Virol. 65:1916-1928, 1991.
3. Vennema, H., Heijne, L., Zijderveld, A, Horzinek, M. C., Spaan, W. J. M. J. Virol. 64:339-346, 1990.
4. Frana, M. F., Behnke, J. N., Sturman, S., Holmes, K. V. J. Virol. 56:912-920, 1985.
5. Luytjes, W., Sturman, L., Bredenbeck, P. J., Charite, J., van der Zeijst, B. A. M., Horzinek, M. C., Spaan, W. J. M. Virology 161:479-487, 1987.
6. Bosch, F. X., Garten, W., Klenk, H. D., Rott, R. Virology 113:725-735, 1981.
7. Glickman, R. L., Syddall, R. J., Iorio, R. M., Sheehan, J. P., Bratt, M. A. J. Virol. 62:354-356, 1988.
8. Perez, L. G., Hunter, E. J. Virol. 61:1609-1614, 1987.
9. Gombold, J.L., Weiss, S.R. Microb. Path. In Press.
10.Lavi, E., Suzumura, A., Hirayama, M., Highkin, M. K., Dambach, D. M., Silberberg, D. H.,Weiss, S. R. Microb. Path. 3:79-86, 1987.
11. Sawicki, S. G. Adv. Exp. Med. Biol. 218:169-174, 1987.
12. DeGroot, R. J., Luytjes, W., Horzinek, M. C., van der Zeijst, B. A. M., Spaan, W. J. M., Lenstra, J. A. J. Mol. Biol. 196:963-966, 1987.
13. Gallagher, T. M., Parker, S. E., Buchmeier, M. J. J. Virol. 64:731-741, 1990.
14. Parker, S.E., Gallagher, T. M., Buchmeier, M. J. Virology 173:664-673, 1989.
15. Morris, V.L., Tieszer, C., MacKinnon, J., Percy, D. Virology 169:127-136, 1989.
16. LaMonica, N., Banner, L., Morris, V. L., Lai, M. M. C. Virology 182:883-888, 1991.
17. Kawaoka, Y.,Webster, R. G. Proc. Natl. Acad. Sci. 85:324-328, 1988.

PROTEOLYTIC CLEAVAGE OF THE MURINE CORONAVIRUS SURFACE GLYCOPROTEIN IS NOT REQUIRED FOR ITS FUSION ACTIVITY

Roland Stauber, Michael Pfleiderer and Stuart Siddell

Institut für Virologie
Universität Würzburg
Versbacherstr.7
8700 Würzburg

SUMMARY

The surface glycoprotein (S) of the murine hepatitis coronavirus MHV normally undergoes proteolytic cleavage during transport to the cell surface. To determine whether the cleavage of the MHV-JHM S glycoprotein is required to activate its ability to fuse cellular membranes, the protease recognition sequence in a cDNA copy of the S gene was altered from Arg-Arg-Ala-Arg-Arg into Ser-Val-Ser-Gly-Gly by site directed mutagenesis. The mutated and wild type S genes were expressed by means of recombinant vaccinia viruses and it could be shown that the mutated S protein was not cleaved when it was expressed in mouse DBT cells, in contrast to the wild type S protein. Nevertheless, the non-cleaved S protein induced extensive syncytium formation in mouse DBT cells. These results clearly indicate that the non-cleaved form of the MHV S protein is able to mediate cell membrane fusion. Thus, proteolytic cleavage is not an absolute requirement for its fusion function.

INTRODUCTION

The coronaviruses are a group of pleomorphic, enveloped, positive stranded RNA viruses associated with diseases of economic importance in both animals and humans (1). The murine coronavirus (murine hepatitis virus [MHV]) has a genomic RNA of 32 kb (2) which encodes four major structural proteins: the nucleocapsid protein N, the membrane glycoprotein M, the hemagglutinin-esterase HE, and the surface or spike glycoprotein S (3). As the name suggests, the MHV S protein forms the projecting spikes or peplomers on the surface of the virus particle. The S protein is synthesized as a cotranslationally glycosylated precursor which then undergoes oligomerization and carbohydrate processing. The polypeptide is also proteolytically cleaved by a host cell enzyme which recognizes a basic sequence, Arg-Arg-Ala-Arg-Arg, located approximately in the middle of the molecule (4). Thus, the mature spike structure consists of two S protein molecules, each of which is a heterodimer comprised of two non-covalently bound subunits. The S protein is able to mediate attachment of the virus to the cell surface and the fusion of cellular and viral membranes (5).

The entry of the enveloped virus particle is thought to be initiated via an interaction between a virus glycoprotein and the cellular receptor, which is followed by fusion of the viral envelope with the host cell plasma membrane or a host endosomal membrane. A common characteristic of fusion proteins is cleavage, which is thought to expose an internal hydrohobic domain that can interact with the lipid bilayer. With the exception of rhabdoviruses, the majority of viral fusion proteins must be cleaved to be functional. This has been demonstrated for the influenza hemagglutinin (6) and human immunodeficiency virus type 1 (HIV-1) gp 160 (7). The lines of evidence which argue that also for the MHV-S protein cleavage is required for the fusion activity are as follows. First, the addition of exogenous protease to the growth medium can enhance plaque and syncytium formation by some strains of MHV (8). Second, MHV virions grown in 17Cl-1 cells contain approximately equal amounts of cleaved and uncleaved S protein and are unable to mediate fusion from without. After treatment with trypsin in vitro (which converts the majority of the S protein to the cleaved form), they acquire this ability (9). Third, the addition of protease inhibitors to the medium of MHV-infected cells causes a delay in the onset of cell fusion (10). The fact however that a number of coronavirus S proteins do not undergo post-translational cleavage and are nevertheless able to initiate infection and induce syncytium formation (11) prompted us to reinvestigate the relationship of MHV S protein cleavage to fusion. The experiments reported here lead us to the conclusion that proteolytic cleavage is not an absolute requirement for fusion activity.

MATERIAL AND METHODS

Cells and viruses

HeLa cells, DBT cells, Sac(-) cells and TK⁻ 143B cells were grown in monolayers in minimal essential medium containing 10% heat-inactivated foetal calf serum, glutamine, antibiotics and non-essential amino acids. The MHV used in this study is the MHV Wb1 isolate described in Schwarz et al. (12). Vaccinia virus (WR strain) and the recombinant vaccinia virus, vTF7-3 (13) were plaque purified twice and grown to stocks of approximately 2×10^9 p.f.u./ml in HeLa cells as described by Mackett et al. (14).

cDNA cloning and sequence analysis

Poly(A)-containing RNA was isolated from MHV infected Sac(-) cells and cDNA synthesis was essentially done by the method of Gubler & Hoffman (15) using MHV-specific oligonucleotides. Recombinant clones were identified by colony hybridisation with two MHV-S specific oligonucleotides. The cDNA of one clone, pBS+/S, which gave a positive signal with both oligonucleotides in the colony hybridization was sequenced. The cDNA insert of pBS+/S was cloned into the vaccinia transfer vector, pTF7-5 (16). The correct clone, pTF7-5/S, was identified by colony hybridization, restriction endonuclease digestion and sequence analysis.

Oligonucleotide-directed mutagenesis

Mutagenesis was performed using the Amersham system which is based on the method of Nakamaye. Briefly, the cDNA insert of pBS+/S was cloned into the M13mp19 RF DNA which had been cut with the same enzymes. The resulting construct M13mp19/S, was used to transform competent E. coli TG-1 cells and single-strand DNA was isolated. Site directed mutagenesis was performed on the single stranded M13mp19/S-DNA according to the manufacturer.

Cloning of the mutated MHV-S gene

After performing the oligonucleotide-directed mutagenesis a 632 bp DNA fragment containing the mutated cleavage site, was cut out of double stranded M13mp19/S-Mut. DNA and used to replace the wild type cleavage site sequence in the pBS+/S DNA. The resulting pBS+/S-Mut. DNA was cloned into the vaccinia transfer vector pTF7.5 (16).

Isolation of vaccinia virus recombinants

Recombinant vaccinia virus were constructed by established procedures. Briefly, HeLa cells were infected at a low multiplicity of infection with wild type vaccinia virus strain WR and transfected with different recombinant transfer plasmids by using the calcium phosphate method. Progeny virus was plaqued on human TK⁻-cells in the presence of 5-bromodeoxyuridine (25 µg/ml). Recombinant vaccinia viruses were identified by DNA hybridization using specific probes and indirect immunofluorescence. Plaque purification and screening were repeated three times before stocks of recombinant viruses were grown in HeLa cells.

SDS-PAGE and Immunoblotting

SDS-PAGE was done under reducing conditions on 7.5% polyacrylamide gels according to the method of Laemmli. Western immunoblots were stained with undiluted hybridoma tissue culture supernatant from the MHV-S protein specific hybridoma 11F (17), followed by peroxidase-linked rabbit anti-mouse immunoglobulin and the substrate 4-chloronaphthol.

Indirect immunofluorescence

DBT cell monolayers on glass coverslips were infected with recombinant vaccinia viruses at an m.o.i. of 1 p.f.u./cell. After 2 h, the virus containing medium was replaced with medium or medium containing 0.5% ascites fluid. The ascites fluid contained monoclonal antibody 11F (17) or a monoclonal antibody specific for the MHV 5b protein. At 14 h post-infection, the cells were washed twice with ice-cold PBS, fixed with acetone-methanol (1:1), washed again with ice-cold PBS and incubated for 16 h at 4°C with tissue culture supernatant containing a mixture of 12 monoclonal antibodies which recognize linear or discontinuous determinants on the MHV S protein (18). The cells were then washed extensively at room temperature with PBS and incubated for 2 h with fluorescein-conjugated goat anti-mouse immunoglobulin (0.5% in PBS containing 5% goat serum). After extensive washing with PBS, the coverslips were mounted in 90% glycerol (pH 8.0) and fluorescence micrographs were taken on a Leitz Aristoplan microscope.

RESULTS

The cDNA insert of pBS+/S was sequenced and found to extend from a position 29 nucleotides downstream of the S protein gene, to a position 63 nucleotides upstream of the initiation codon. The sequence upstream of the initiation codon corresponds to the MHV leader sequence showing that the cDNA was copied from mRNA 3 (4). In order to demonstrate the fusogenic activity of the S protein encoded by this cDNA, we performed a transient expression assay using a system based on a recombinant vaccinia virus that synthesizes T7 RNA polymerase (16). DBT cells which have been infected with vTF7-3 and transfected with pTF7.5/S DNA show extensive syncytium formation. In contrast, cells which have been infected with vTF7-3 and transfected with pTF7-5 DNA show no syncytium formation. This result demonstrates that the S protein gene used in these studies is functional and is able to induce fusion from within in the absence of other viral proteins.

Having confirmed that the recombinant MHV-JHM S protein possesses fusion activity we addressed the question if cleavage of the S protein is necessary to cause syncytia formation. For this purpose we changed by means of site directed mutagenesis the basic recognition sequence Arg-Arg-Ala-Arg-Arg (amino acids 624-628), required for the cleavage by a host cell protease, into Ser-Val-Ser-Gly-Gly. In the resulting plasmid pTF7.5/S-Mut. the changed sequence was confirmed by sequence analysis and the DNA used for the generation of recombinant Vaccinia virus.

Using the Vaccinia virus system it was possible to monitor the expression of the wild type and the mutant MHV-JHM S gene. For this purpose lysates of DBT cells infected with the recombinant vaccinia virus vTF7-3 and the vT7.5/S or vTF7.5/S-Mut. at a m.o.i. of 10 respectively were analyzed on 7.5% polyacrylamide gels and immunoblotting was performed. In Fig. 1 the uncleaved and the cleaved form of the S protein can be seen for the expressed wild type MHV-JHM S gene (lane 3), whereas no cleavage products are visible for the mutant S protein (lane 2).

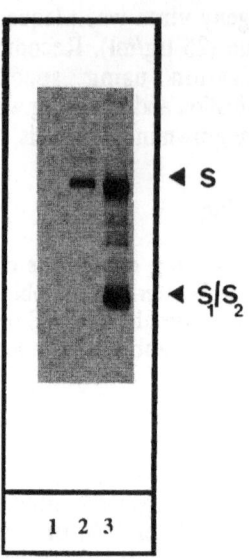

Figure 1 Western blot analysis of the wild type and mutated MHV S protein expressed by means of recombinant vaccinia virus. Lane 1 vTF7-3 infected, lane 2 vTF7-3 and vTF7-5/S-Mut. infected, lane 3 vTF7-3 and vTF7-5/S infected.

Cleavage of viral fusion glycoproteins is known to be prerequisite to activate their biological function. To determine wether this is also the case for the MHV-JHM S glycoprotein, we tested the wild type and the mutated S protein for their ability to cause syncytia formation.

In Fig. 2 DBT cells which had been infected with vTF7-3 as well as vTF7-5/S or vTF7-5/S-Mut were examined by indirect immunofluorescence 14 h post infection. Figure 2 (panel A, B) demonstrates that the wild type and the mutated S protein are able to elicit syncytia formation in DBT cells to the same extent.

The fact that an anti S MoAb, which inhibits MHV-JHM mediated fusion, is able to block the fusion activity of the wild type as well as the mutant S protein (panel C, D), proves that this cell-to-cell fusion is specifically caused by the expression of the recombinant S proteins. As expected the control MoAb against the MHV 5b has no inhibitory effect on the fusion activity (panel E, F).

DISCUSSION

Although virus mediated cell-fusion has been studied quite extensively, little is known about the mechanistic details of the initial interaction between the viral fusion protein and the lipid bilayer. In many enveloped virus families cleavage of the fusion glycoprotein is essential for its function. Besides the cleavage event another important factor is the pH dependence of fusion, which often reflects the route of virus entry. The alphaviruses and the orthomyxoviruses bind to plasma membrane receptors and are internalized via the endocytic pathway which has an acidic pH. The low pH causes conformational changes in the fusion glycoprotein and triggers fusion of the viral envelope with the endosomal membrane to release the viral RNA into the cytoplasm. On the other hand viruses, which enter the cell via direct fusion between the cellular plasma membrane and the viral membrane such as retroviruses or coronaviruses (19), possess fusion glycoproteins which are functional at neutral pH. To render the situation even more complex physical parameters such as the lipid composition of the host cell membrane seems to play a role in virus induced fusion (20).

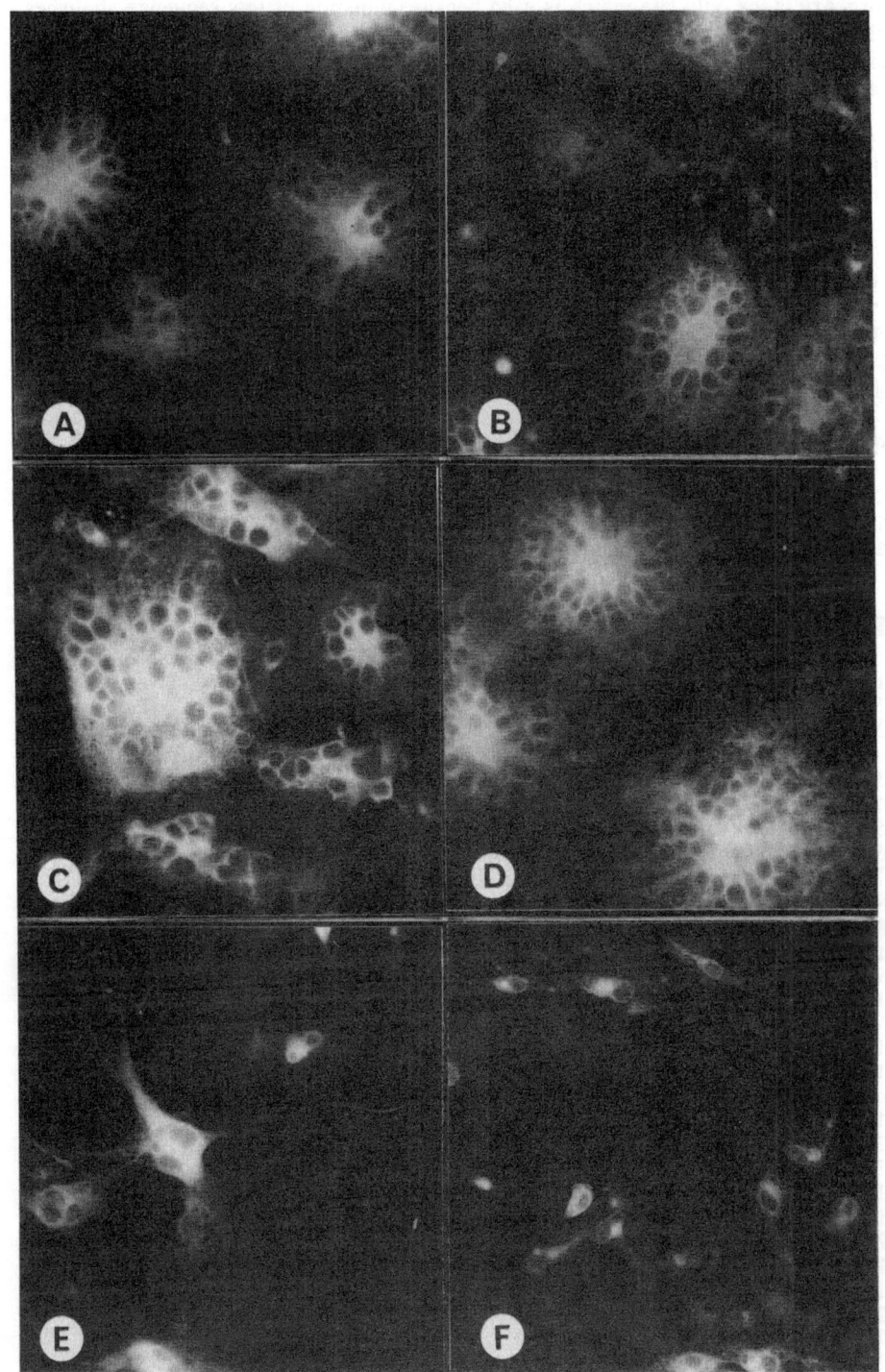

Figure 2 Fusion activity of the wild type and mutated S proteins.
Panel A vTF7-5/S infected, medium; panel B vTF7-5/S-Mut., medium
Panel C vTF7-5/S infected, medium with Mab 5b; panel D vTF7-5/S-Mut., medium with Mab 5b
Panel E vTF7-5/S infected, medium with Mab 11F; panel F vTF7-5/S-Mut., medium with Mab 11F

The results reported in this paper show that the putative cleavage site Arg-Arg-Ala-Arg-Arg is the only site which is used because mutation of this sequence abolishes cleavage. Therefore no other potential protease recognition sites, e.g. the recognition site for thermolysin which was suggested by Frana, et al. (10) play a role in the cleavage of the MHV-JHM S protein in DBT cells.

Finally it could be demonstrated that the MHV-JHM S protein does not need to be cleaved to activate its fusion function. Although we conclude that proteolytic cleavage of the MHV S protein is not required for its fusion activity, we do not wish to imply that this processing event is without significance. The cleavage site motif is well conserved amongst MHV strains (4) and processing of the S protein precursor occurs in a wide variety of murine cell lines. One possibility is that proteolytic processing "enhances" the fusion activity of the MHV S protein. The selective advantage of an enhanced fusogenic potential may be manifest in the extensive development of syncytia in vivo and the spread of virus from cell to cell without exposure to the humoral immune response. It is possible however that cleavage or the extent of cleavage is also an essential factor governing tropism and pathogenesis. For example the S protein of the neurotropic Coronavirus MHV-4 shows a different cleavage pattern compared to the S protein of MHV-JHM (21).

In our opinion, however, answers to questions such as these require the development of an infectious MHV cDNA clone or a system of targeted recombination (22). With such a system, it would be possible to introduce specific changes into the MHV genome and analyse their effect on the virus phenotype in vitro and in vivo.

REFERENCES

1. Spaan, W.J.M., Cavanagh, D. & Horzinek, M.C. (1988). Journal of General Virology 69, 2939-2952.
2. Lee, H.-J., Shieh, C.-K., Gorbalenya, A.E., Koonin, E.V., LaMonica, N., Tuler, J., Bagdzhadzhyan, A. & Lai, M.M.C. (1991). Virology 180, 567-582.
3. Siddell, S.G. (1982). Journal of General Virology 62, 259-269.
4. Schmidt, I., Skinner, M. & Siddell, S. (1987). Journal of General Virology 68, 47-56.
5. Collins, A.R., Knobler, R.L., Powell, H. & Buchmeier, M.J. (1982). Virology 119, 358-371.
6. Bosch, F.X., Orlich, M., Klenk, H.-D. & Rott, R. (1979). Virology 95, 197-207.
7. Bosch, V.,& Rawlita, M. (1990). Journal of Virology 64, 2337-2344.
8. Yoshikura, H. & Tejima, S. (1981). Virology 113, 503-511.
9. Sturman, L.S., Ricard, C.S. & Holmes, K.V. (1985). Journal of Virology. 56, 904-911.
10. Frana, M.F., Behnke, J.N., Sturman, L.S. & Holmes, K.V. (1985). Journal of Virology 56, 912-920.
11. de Groot, R.J., van Leen, R.W., Dalderup, M.J.M., Vennema, H., Horzinek, M.C. & Spaan, W.J.M. (1989). Virology, 171 493-502.
12. Schwarz, B., Routledge, E. & Siddell, S.G. (1990). Journal of Virology 64, 4784-4791.
13. Fuerst, T.R., Niles, E.G., Studier, F.W. & Moss, B. (1986). Proceedings of the National Academy of Sciences, USA 83, 8122-8126.
14. Mackett, M., Smith, G.L. & Moss, B. (1985). In "DNA Cloning: A Practical Approach", pp191-211. Edited by D.M. Glover, Oxford, IRL Press.
15. Gubler, U. & Hoffman, B.J. (1983). Gene 25, 263-269.
16. Fuerst, T.R., Earl, P.L. & Moss, B. (1987). Molecular and Cellular Biology 7, 2538-2544.
17. Routledge, E., Stauber, R., Pfleiderer, M. & Siddell, S.G. (1991). Journal of Virology 65, 254-262.
18. Wege, H., Dörries, R. & Wege, H. (1984). Journal of General Virology 65, 1931-1942.
19. Kooi, C., Cervin, M. & Anderson, R. (1991). Virology 180, 108-119.
20. Mizzen, L. Cheley, S., Rao, M., Wolf, R. & Anderson, R. (1983). Virology 128, 407-417.
21. Gallagher, T.M., Parker, S.E. & Buchmeier, M.J. (1990). Journal of Virology 64, 731-744.
22. Koetzner, C.A., Parker, M.M., Ricard, C.S., Sturman, L.S. & Masters, P.S. (1992). Journal of Virology 66, 1841-1848.

FUSOGENIC PROPERTIES OF UNCLEAVED SPIKE PROTEIN OF MURINE CORONAVIRUS JHMV

Fumihiro Taguchi, Toshio Ikeda, Keiichi Saeki, Hideyuki Kubo and Tateki Kikuchi

National Institute of Neuroscience, NCNP, 4-1-1 Ogawahigashi, Kodaira Tokyo 187, Japan

ABSTRACT

We have tested the fusogenic properties of cleaved and uncleaved spike (S) protein of murine coronavirus (MCV) JHMV variant cl-2 by expressing the S protein by recombinant vaccinia viruses (RVVs). The amino acid sequence of the putative cleavage site of cl-2 S protein, Arg-Arg-Ala-Arg-Arg, was replaced by Arg-Thr-Ala-Leu-Glu by *in vitro* mutagenesis of cl-2 S gene. The RVVs having cl-2 S gene [RVV t(+)] or mutated cl-2 S gene [RVV t(-)] were tested for their ability to induce fusion as well as cleavability in DBT cells. After inoculation with RVV t(+) onto DBT cells, the fusion formation was first observed at 8 h postinoculation (p.i.) and spread throughout the whole culture by 24 h. In cells infected with RVV t(-), fusion appeared by 2 h and most of cells were fused by 30 h p.i. The S protein and its cleavage products were detected in DBT cells expressing wild type S protein. However, no cleavage products of the S protein were detected in RVV t(-) infected cells producing mutated S protein, even though fusion was clearly visible. These results suggest that the cleavage event of JHMV-S protein of MCV is not a prerequisite for fusion formation, but that it enhances fusion.

INTRODUCTION

The S protein of MCV comprising the peplomer on the virion surface is 150 to 180 k daltons, depending upon the virus [1, 2]. This protein is multifunctional [2, 3]. It is known that the S protein is involved in the binding of virions to the receptor on susceptible cells [4]. Fusion of infected cells into polykaryocytes is also shown to be dependent upon S protein [2, 5]. In addition. the major neutralizing epitopes exist on the S protein. Furthermore, it is speculated that the S protein is a major determinant of the virulence of the virus [6, 7].

It has been reported that a cleavage event is important for the fusogenic properties of S protein [8, 9] as in the cases of orthomyxoviruses and paramyxoviruses [10]. The cleavage of S protein is supposed to be carried out by a host cell-derived trypsin like proteolytic enzyme. In the case of JHMV, the putative cleavage site is composed of the basic amino acid cluster of Arg-Arg-Ala-Arg-Arg [11, 12]. On the other hand, it has been reported that the membrane anchoring subunit of cleavage product has no hydrophobic amino acid cluster that might function as a putative fusion peptide [2, 11, 12]. This differs from the situation of other viruses with fusion activity [10]. This finding suggests that mechanisms of coronavirus fusion formation may be different from that of the ortho- and paramyxoviruses. To delineate the mechanism of fusion formation by coronaviruses, it is of great importance to determine whether the cleavage of S protein is requisite for fusionability by the S protein.

Coronaviruses, Edited by H. Laude and J.F. Vautherot
Plenum Press, New York, 1994

RESULTS

Removal of the endoproteolytic cleavage site of S protein

The amino acid sequence 765 Arg-Arg-Ala-Arg-Arg 769, is predicted to be involved in the cleavage by host cell-derived proteolytic enzyme and this cleavage could occur between residues 769 and 770 (Fig. 1). We replaced this amino acid sequence by the Arg-Thr-Ala-Leu-Glu sequence, predicted to be resistant to such an enzyme. Therefore, a mutant with this uncleavable peptide was created to test the effect of S protein cleavability on cell fusion. The site-directed mutation was carried out by making use of polymerase chain reaction (PCR). A pair of oligonucleotides were used to insert the mutation, one of which is a JHM-Try(-), complementary sequence encoding a mutated amino acid sequence which can be obtained by three nucleotides substitution as shown in Fig. 1. The other oligonucleotide is genomic sense and corresponds to the sequence ca. 650 nucleotide upstream from JHM-Try(-). After amplification by PCR, a 443 nucleotide fragment was obtained by cutting with *Cla* I and *Bal* 1. This was inserted in the wild type S gene where the corresponding wild type *Bal* I-*Cla* I 443 nucleotide fragment had been removed. The mutated S gene expected to encode the uncleavable S protein was then inserted into the transfer vector pSF for obtaining the RVV. Finally, RVVs with the wild type S gene {RVV t(+)} and mutated S gene {RVV t(-)} were obtained.

Fusion formation with wild type and mutated S proteins expressed by RVVs

DBT cells were infected with RVV t(+) to produce wild type S protein or RVV t(-) to produce mutated S protein at an m.o.i of 0.5 and observed for the presence or absence of fusion. Fused cells first appeared at about 8 h p.i. with RVV t(+). By 24 h, more than 90 per cent of cells were involved in the fusion. In DBT cells infected with RVV t(-) under the same condition, the fusion was also first observed at 12 h p.i. Thereafter, the fusion continued as that by RVV t(+) (Fig. 2), except for a 2 to 4 h delay in appearance and development. The fusion produced by RVV t(+) and RVV t(-) was due to the expressed S protein as fusion formation was prevented by monoclonal antibodies (MAbs) specific for cl-2 S protein (data not shown). These data clearly show that the wild type S protein as well as the mutated S protein containing Arg-Thr-Ala-Leu-Glu in place of Arg-Arg-Ala-Arg-Arg induced fusion in DBT cells.

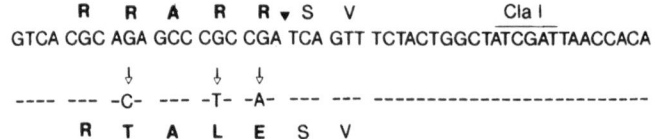

Figure 1. The amino acid and nucleotide sequences in the vicinity of the putative proteolytic cleavage site as well as the oligonucleotide (JHM-Try(-)) used to replace the cleavage site on wild type S protein. Black arrow shows the cleavage site.

Analysis by immunoprecipitation and Western blotting of wild type and mutated S proteins expressed by RVVs

RVV t(+) and RVV t(-) as well as control vaccinia virus were used to infect DBT cells at an m.o.i. of 0.5. At 14 h after the infection, when about 50% of RVV t(+) infected cells and 20% of RVV t(-) infected cells were fused, the cells were pulse-labeled with ^{35}S methionine for 30 min to detect the precursor S protein of an approximate molecular weight of 170 K. Thereafter, the labeled proteins were chased for 1 to 3 h to detect S protein cleavage products. As shown in Fig. 3, in the pulse-labeled samples, only one band corresponding to about 170 k daltons was precipitated by the MAbs specific for cl-2 S protein in the lysates prepared from DBT cells infected either with RVV t(+) or with RVV t(-). In the chase samples, Sl and S2 cleavage products were detected from RVV t(+) infected cells, but no such bands were found in the chased samples from RVV t(-) infected cells. Since the infected cells were labeled for 30 min only , it may have been that the incorporation of ^{35}S methionine was insufficient for the detection of trace amounts of S protein cleavage products by immunoprecipitation. Therefore, the presence of Sl and S2 was examined by more sensitive Western blotting with enhanced chemiluminescence. Here again only the 170 K S protein and no cleavage products were detected in RVV t(-) infected DBT cells, while both the 170 K and its cleavage products were observed in cells infected RVV t(+) infected cells (data not shown).

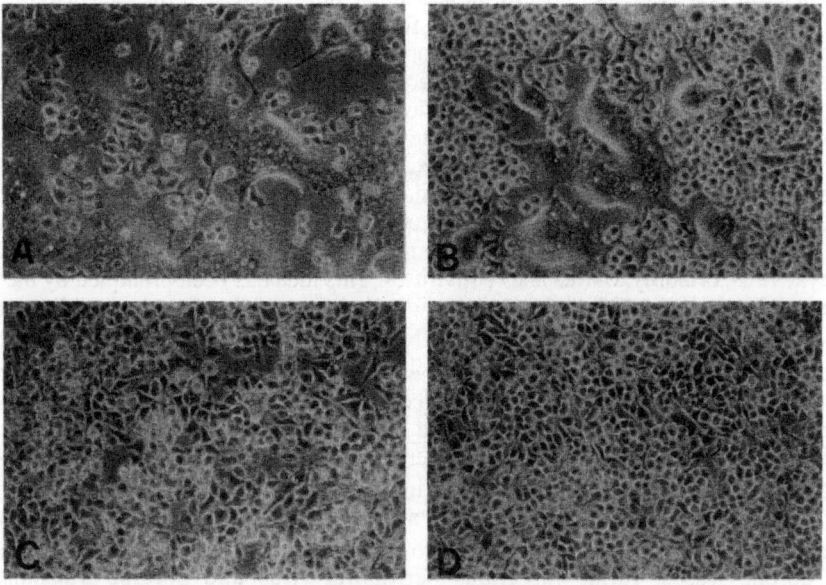

Figure 2. The fusion formation in DBT cells at 12 h after the infection with RVV t(+) (A), RVV t(-) (B) vaccinia virus (C) or mock infected DBT cells (D).

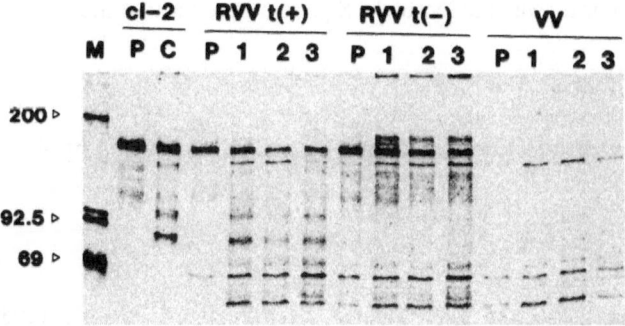

Figure 3. Immunoprecipitation of wild type and mutated S proteins expressed in DBT cells by the infection with RVV t(+), RVV t(-) or vaccinia virus (VV). Infected cells were 30 min pulse-labeled at 14 h p.i. (P) and chased for 1 (1), 2 (2) and 3 (3) h. The S protein produced in DBT cells infected with cl-2 (cl-2) was also precipitated.

DISCUSSION

Fusion formation has been reported to be induced by orthomyxoviruses, paramyxoviruses and retroviruses [10]. The proteins involved in fusion formation by these viruses are envelope proteins which in most cases comprise the peplomer or spike of the virions [10]. The cleavage of the envelope protein is an absolute requirement for membrane fusion of cells infected with these viruses, which exposes the hydrophobic amino acid cluster at the N terminus of the membrane anchoring subunit of cleavage product [10]. This hydrophobic amino acid cluster designated as fusion peptide is believed to be a fusion active site.

The cleavage event was suggested to be an important step for fusion formation by coronaviruses from the facts that MCV with a fusion negative phenotype was rendered fusionable after treatment with proteolytic enzymes [9, 13]. Also it has been reported that a correlation exists between the degree of cleavage or protein and the extent of fusion in different virus-cell systems [8]. All of these data are suggestive that the cleavage of the S protein was a prerequisite for the fusion formation by S protein of MCVs. This is analogous to similar mechanisms of fusion formation in other fusogenic RNA viruses. The results presented in this paper, however, indicate that fusion formation was caused by the uncleaved S protein which contained a mutation in the putative cleavage site, implying that the cleavage of S protein is not a prerequisite for the fusogenic activity of S protein. The possibility, however, cannot completely be excluded that undetectable trace amounts of the cleaved S protein were responsible for the fusion activity.

The data shown in the present paper are compatible with the idea that the mechanisms of fusion formation by coronavirus S protein may differ from that of other viruses such as paramyxovirus, orthomyxovirus and retroviruses. This idea has been prompted by the facts that no hydrophobic amino acid cluster, homologuous to the fusion peptide of other fusogenic RNA viruses [10], was found at the N terminus of the membrane anchoring chain S2, after cleavage of coronavirus S protein [2, 12, 14, 15]. Semliki forest virus and Sindbis virus are capable of inducing fusion in infected cells after cleavage of the envelope protein and are known to have a stretch composed of 16 to 17 hydrophobic amino acids that constitute a putative fusion peptide in an internal position but not at the N terminus of the membrane anchoring chain [10]. However, in the S protein of JHMV cl-2 variant [12] as well as the S proteins of other coronaviruses such a large stretch of hydrophobic amino acid has not been detected [2, 11, 1 1, 15]. It has been also documented by sequence analysis that some other coronaviruses, feline infectious peritonitis virus and porcine transmissible gastroenteritis virus, do not have the predicted proteolytic cleavage site on the S protein [2]. Nevertheless, fusion was found in cells infected with such viruses [16].

Little is known about the mechanisms of fusion formation by coronavirus S protein and there are only a few reports analyzing the relationship between the structure and fusion of the S protein in murine coronaviruses. To understand fusion mechanisms of coronaviruses, it is essential to locate the S protein fusion active site.

Acknowledgments

This research was partly supported by the grant provided by the Ichiro Kanehara Foundation.

REFERENCES

1. S. Siddell, H. Wege, and V. ter Meulen, *J.gen.Virol.* 64: 761-776 (1983)
2. W. Spaan, D. Cavanagh, and M.C. Horzinek, *J.gen.Virol.* 69: 2939-2952 (1988)
3. K.V. Holmes, E.W. Doller, and J.N. Behnke, *Adv.Exp.Med.Biol.* 142: 1331-142 (1981)
4. R.K. Williams, C. Jiang, and K.V. Holmes, *Proc.Natl.Acad.Sci*, USA, 88: 5533-5536 (1991)

5. A.R. Collins, R.L. Knobler, H. Powell, and M. Buchmeier, *Virology*, 119: 358-371 (1982)
6. R.G. Dalziel, P.W. Lampert, J. Talbot, and M.J. Buchmeier, *J.Virol.* 59: 463-471 (1986)
7. Y. Matsubara, R. Watanabe, and F. Taguchi, *Virus Res.* 20: 45-58 (1991)
8. M.F. Frana, J.N. Behnke, L.S. Sturman, and K.V. Holmes, *J.Virol.* 56: 912-920 (1985)
9. L.S. Sturman, C.S. Ricard, and K.V. Holmes, *J.Virol.* 56: 904-911 (1985)
10. J.M. White, *Annu.Rev.Physiol.* 52: 675-697 (1990)
11. I. Schmidt, M. Skinner, and S. Siddell, *J.gen.Virol.* 68: 47-56 (1987)
12. F. Taguchi, T. Ikeda, and H. Shida, *J.gen.Virol.* 73: 1065-1072 (1992)
13. H. Yoshikura, and S. Tejima, *Virology* 113: 503-511 (1981)
14. T.M. Gallagher, C. Escarmis, and M.J. Buchmeier, *J.Virol .* 65: 1916-1928 (1991)
15. S.E. Parker, T.M. Gallagher, and M.J. Buchmeier, *Virology* 173: 664-673 (1989)
16. R.J. De Groot, R.W. Vanleen,M.J.M. Dalderup, H. Vennema, M.C. Horzinek, and W.J.M. Spaan,*Virology* 171: 493-502 (1989)

CHARACTERIZATION OF A MONOCLONAL ANTIBODY RESISTANT VARIANT OF MHV

Bettina Grosse and Stuart G. Siddell

Institut für Virologie
Universität Würzburg
Versbacher Str.7
8700 Würzburg

SUMMARY

A monoclonal antibody resistant (MAR) variant of MHV was isolated after infection of hybridoma cells secreting the neutralizing and fusion-inhibiting monoclonal antibody, mAb 11F. The isolated variant was able to mediate syncytia formation even in the presence of high concentrations of mAb 11F. The S gene of the variant was cloned and sequenced. There were three nucleotide exchanges in comparison to the wild-type S gene, resulting in two amino acid alterations. However, both amino acid substitutions (at positions 255 and 1116) were located outside the binding site of mAb 11F.

INTRODUCTION

The murine coronaviruses are enveloped, positive strand RNA virus. The length of the genome is about 32 kb (1, 2) and it encodes four major structural proteins: the nucleocapsid protein N (50-60K Mr), the membrane protein M (23-26K Mr), the haemagglutinin-esterase protein HE (65K Mr) and the surface protein S (180K Mr) (3). The S protein forms peplomers on the surface of the virus which are composed of two or three S protein molecules, each of which is cleaved into two subunits, S1 and S2 (4,5). The S protein plays a crucial role in MHV infections. It is able to mediate the attachment of the virus to the cell surface and the fusion of cell and virus membranes (6,7). However, the relationship between the structure and function of the S protein is poorly understood. One approach to study this aspect of S protein biology is to isolate and characterize MAR variants which escape the effects of monoclonal antibodies capable of inhibiting biological functions such as virus infectivity and S protein mediated membrane fusion.

Coronaviruses, Edited by H. Laude and J.F. Vautherot
Plenum Press, New York, 1994

MATERIALS AND METHODS

Cells and viruses

Hybridoma cells were grown in suspension in RPMI 1640 medium containing 25mM HEPES, L-glutamine, 20% heat inactivated FCS and antibiotics (8). L cells and DBT cells (9) were grown in monolayers in MEM containing 5% heat inactivated FCS and antibiotics. The virus used in this study is the MHV Wb1 isolate (10). DBT cells were infected with MHV at an m.o.i. of 5 p.f.u./cell. After 16h, RNA was isolated according to standard procedures. To determine virus titers, dilutions of 10^{-1} to 10^{-8} of the stock virus were used to infect L cells cultivated in microtiter plates. After 48 h of incubation at 37°C the titer was determined on the basis of cytopathic effect as described of Reed and Muench.

Recombinant DNA

Poly A^+ RNA was isolated from MHV infected DBT cells and cDNA synthesis was done by the method of Gubler and Hoffman (12) using sequence specific oligonucleotides. DNA sequencing was done on single strand DNA templates (11). Sequence data were assembled with the programs of Staden (14) and analyzed using the UWGCG (University of Wisconsin, Genetic Computer Group) sequence analysis software package. Oligonucleotides were synthesized using phosphoamidite chemistry on a Cyclone DNA Synthesizer.

Fusion inhibition test

L-cells were grown as monolayers in microtiter plates. Each well was infected with 40 p.f.u. of MHV. After 45 minutes of virus adsorption at 37°C, the virus containing medium was replaced by mAb 11F containing medium. After 16h infection, the cells were stained with Giemsa solution. The ELISA test was performed as described (13).

RESULTS

Generation of MAR 11F variants

To obtain MAR 11F variants, the mAb 11F secreting hybridoma cells were infected with MHV Wb 1. For this purpose, most of the neutralizing mAb 11F was washed away before infection (Fig.1a). During the infection, the production and secretion of mAb 11F remained constant and consequently a permanent and strong selection pressure was maintained. At days 0 and 1 post infection, the culture supernatants contained virus titers of 10^3 p.f.u./ml. These virus particles were predominantly wild type MHV Wb1 (data not shown). At days 2 and 3 no infectious virus was detectable in the cultures. From day 4 onwards virus titers of up to 10^5 p.f.u./ml were present in the culture supernatant (Fig.1b).

Isolation of the MAR 11F/1 variant

The MAR 11F/1 variant was isolated by plaque purification from the supernatant of the infected hybridoma culture. The plaque purifications were done in the presence of mAb 11F containing media. To confirm that the isolated virus was indeed an escape variant, we performed a fusion inhibition test. The fusion activity of the isolated virus could not be inhibited even in the presence of high concentrations of the mAb 11F (Fig.2).

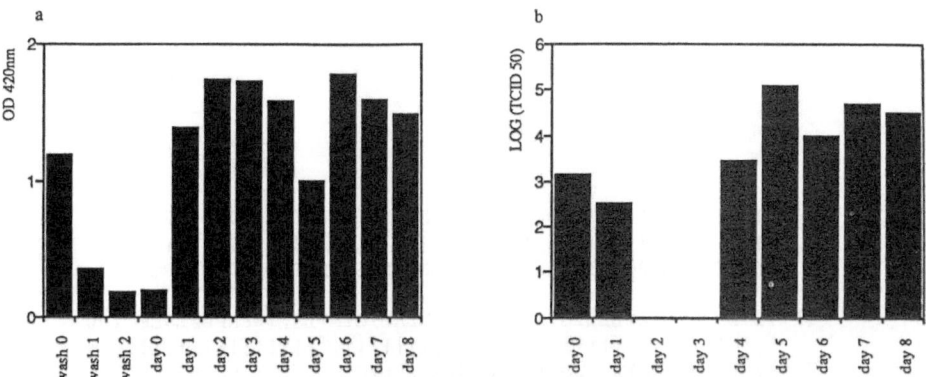

Figure 1. a) The titers of the mAb 11F present in the supernatants collected during the infection period. b) The virus titers present in the collected supernatants

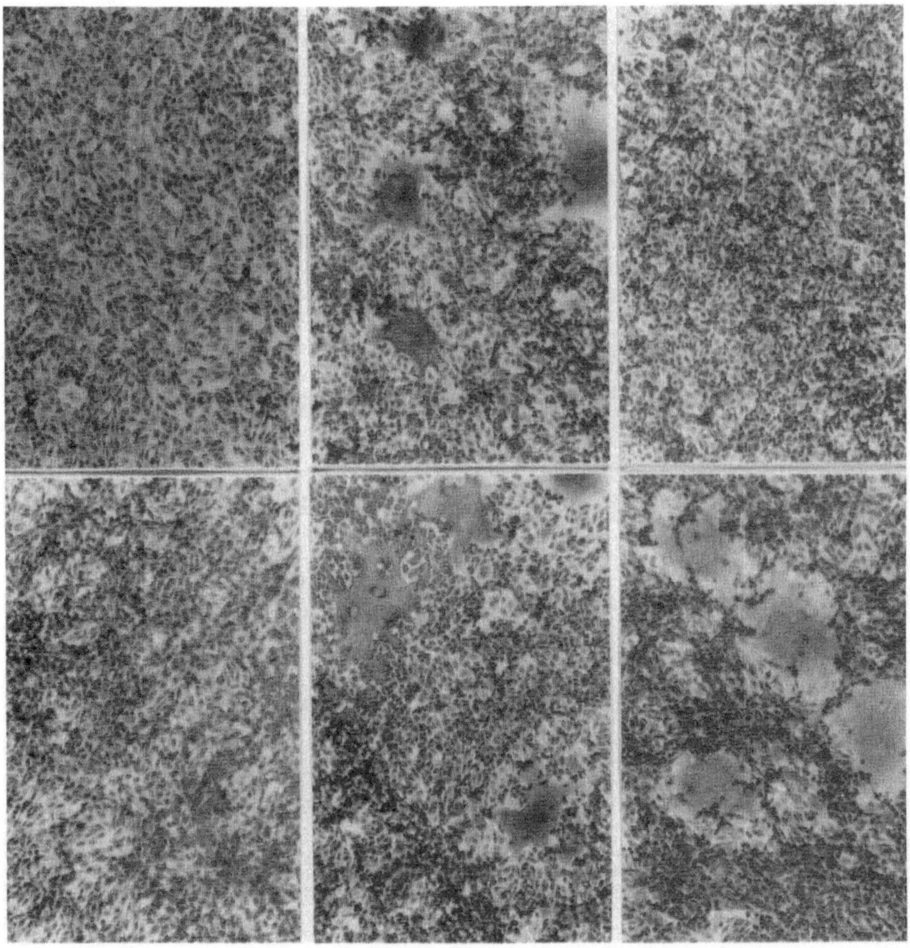

Figure 2. Upper panels: mock infected L cells, left panel ; wt MHV infected L cells treated with mAb 11F (right panel) and without mAb 11F (central panel). Lower panels: mock infected L cells, left panel ; MAR 11F/1 infected L cells treated with mAb 11F (right panel) and without mAb 11F (central panel).

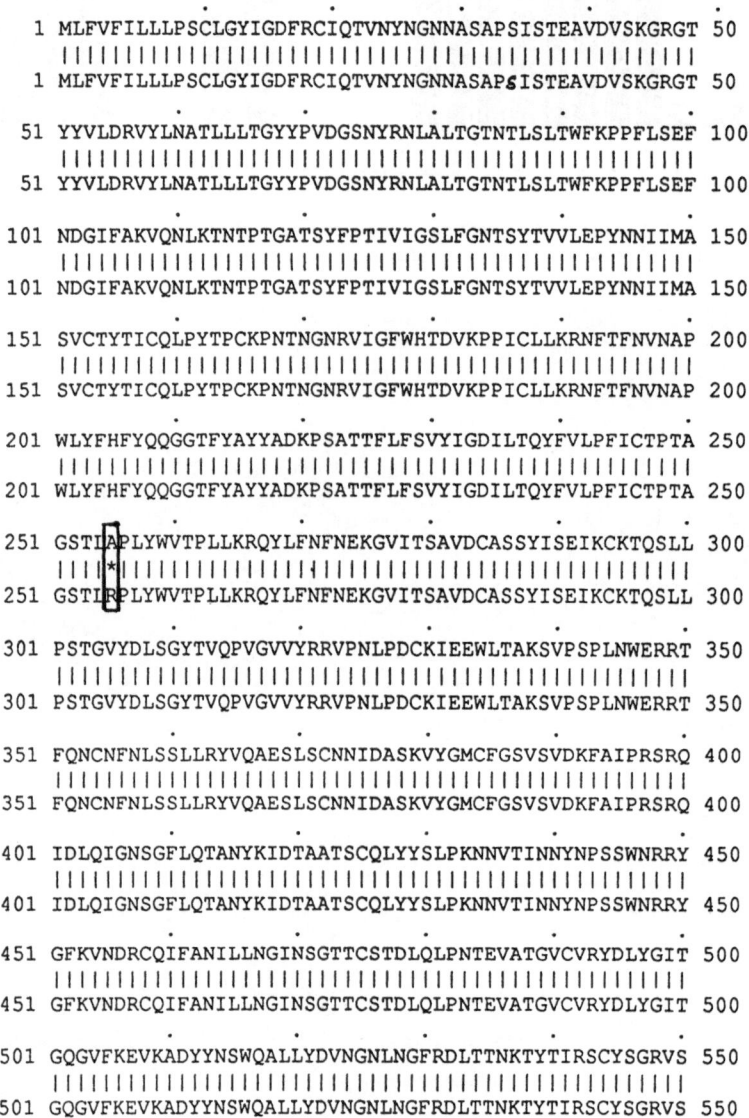

```
  1 MLFVFILLLPSCLGYIGDFRCIQTVNYNGNNASAPSISTEAVDVSKGRGT  50
    ||||||||||||||||||||||||||||||||||||||||||||||||||
  1 MLFVFILLLPSCLGYIGDFRCIQTVNYNGNNASAPSISTEAVDVSKGRGT  50

 51 YYVLDRVYLNATLLLTGYYPVDGSNYRNLALTGTNTLSLTWFKPPFLSEF 100
    ||||||||||||||||||||||||||||||||||||||||||||||||||
 51 YYVLDRVYLNATLLLTGYYPVDGSNYRNLALTGTNTLSLTWFKPPFLSEF 100

101 NDGIFAKVQNLKTNTPTGATSYFPTIVIGSLFGNTSYTVVLEPYNNIIMA 150
    ||||||||||||||||||||||||||||||||||||||||||||||||||
101 NDGIFAKVQNLKTNTPTGATSYFPTIVIGSLFGNTSYTVVLEPYNNIIMA 150

151 SVCTYTICQLPYTPCKPNTNGNRVIGFWHTDVKPPICLLKRNFTFNVNAP 200
    ||||||||||||||||||||||||||||||||||||||||||||||||||
151 SVCTYTICQLPYTPCKPNTNGNRVIGFWHTDVKPPICLLKRNFTFNVNAP 200

201 WLYFHFYQQGGTFYAYYADKPSATTFLFSVYIGDILTQYFVLPFICTPTA 250
    ||||||||||||||||||||||||||||||||||||||||||||||||||
201 WLYFHFYQQGGTFYAYYADKPSATTFLFSVYIGDILTQYFVLPFICTPTA 250

251 GSTIAPLYWVTPLLKRQYLFNFNEKGVITSAVDCASSYISEIKCKTQSLL 300
    |||||*||||||||||||||||||||||||||||||||||||||||||||
251 GSTIRPLYWVTPLLKRQYLFNFNEKGVITSAVDCASSYISEIKCKTQSLL 300

301 PSTGVYDLSGYTVQPVGVVYRRVPNLPDCKIEEWLTAKSVPSPLNWERRT 350
    ||||||||||||||||||||||||||||||||||||||||||||||||||
301 PSTGVYDLSGYTVQPVGVVYRRVPNLPDCKIEEWLTAKSVPSPLNWERRT 350

351 FQNCNFNLSSLLRYVQAESLSCNNIDASKVYGMCFGSVSVDKFAIPRSRQ 400
    ||||||||||||||||||||||||||||||||||||||||||||||||||
351 FQNCNFNLSSLLRYVQAESLSCNNIDASKVYGMCFGSVSVDKFAIPRSRQ 400

401 IDLQIGNSGFLQTANYKIDTAATSCQLYYSLPKNNVTINNYNPSSWNRRY 450
    ||||||||||||||||||||||||||||||||||||||||||||||||||
401 IDLQIGNSGFLQTANYKIDTAATSCQLYYSLPKNNVTINNYNPSSWNRRY 450

451 GFKVNDRCQIFANILLNGINSGTTCSTDLQLPNTEVATGVCVRYDLYGIT 500
    ||||||||||||||||||||||||||||||||||||||||||||||||||
451 GFKVNDRCQIFANILLNGINSGTTCSTDLQLPNTEVATGVCVRYDLYGIT 500

501 GQGVFKEVKADYYNSWQALLYDVNGNLNGFRDLTTNKTYTIRSCYSGRVS 550
    ||||||||||||||||||||||||||||||||||||||||||||||||||
501 GQGVFKEVKADYYNSWQALLYDVNGNLNGFRDLTTNKTYTIRSCYSGRVS 550
```

Figure 3. Comparison of the amino acid sequence between wt S-protein (lower sequence) and MAR 11F/1 S-protein (upper sequence).

```
 551 AAYHKEAPEPALLYRNINCSYVFTNNISREENPLNYFDSYLGCVVNADNR  600
     ||||||||||||||||||||||||||||||||||||||||||||||||||
 551 AAYHKEAPEPALLYRNINCSYVFTNNISREENPLNYFDSYLGCVVNADNR  600

 601 TDEALPNCDLRMGAGLCVDYSKSRRARRSVSTGYRLTTFEPYMPMLVNDS  650
     ||||||||||||||||||||||||||||||||||||||||||||||||||
 601 TDEALPNCDLRMGAGLCVDYSKSRRARRSVSTGYRLTTFEPYMPMLVNDS  650

 651 VQSVGGLYEMQIPTNFTIGHHEEFIQIRAPKVTIDCAAFVCGDNAACRQQ  700
     ||||||||||||||||||||||||||||||||||||||||||||||||||
 651 VQSVGGLYEMQIPTNFTIGHHEEFIQIRAPKVTIDCAAFVCGDNAACRQQ  700

 701 LVEYGSFCDNVNAILNEVNNLLDNMQLQVASALMQGVTISSRLPDGISGP  750
     ||||||||||||||||||||||||||||||||||||||||||||||||||
 701 LVEYGSFCDNVNAILNEVNNLLDNMQLQVASALMQGVTISSRLPDGISGP  750

 751 IDDINFSPLLGCIGSTCAEDGNGPSAIRGRSAIEDLLFDKVKLSDVGFVE  800
     ||||||||||||||||||||||||||||||||||||||||||||||||||
 751 IDDINFSPLLGCIGSTCAEDGNGPSAIRGRSAIEDLLFDKVKLSDVGFVE  800

 801 AYNNCTGGQEVRDLLCVQSFNGIKVLPPVLSESQISGYTAGATAAAMFPP  850
     ||||||||||||||||||||||||||||||||||||||||||||||||||
 801 AYNNCTGGQEVRDLLCVQSFNGIKVLPPVLSESQISGYTAGATAAAMFPP  850

 851 WTAAAGVPFSLNVQYRINGLGVTMNVLSENQKMIASAFNNALGAIQEGFD  900
     ||||||||||||||||||||||||||||||||||||||||||||||||||
 851 WTAAAGVPFSLNVQYRINGLGVTMNVLSENQKMIASAFNNALGAIQEGFD  900

 901 ATNSALGKIQSVVNANAEALNNLLNQLSNRFGAISASLQEILTRLDAVEA  950
     ||||||||||||||||||||||||||||||||||||||||||||||||||
 901 ATNSALGKIQSVVNANAEALNNLLNQLSNRFGAISASLQEILTRLDAVEA  950

 951 KAQIDRLINGRLTALNAYISKQLSDSTLIKFSAAQAIEKVNECVKSQTTR 1000
     ||||||||||||||||||||||||||||||||||||||||||||||||||
 951 KAQIDRLINGRLTALNAYISKQLSDSTLIKFSAAQAIEKVNECVKSQTTR 1000

1001 INFCGNGNHILSLVQNAPYGLCFIHFSYVPTSFKTANVSPGLCISGDRGL 1050
     ||||||||||||||||||||||||||||||||||||||||||||||||||
1001 INFCGNGNHILSLVQNAPYGLCFIHFSYVPTSFKTANVSPGLCISGDRGL 1050

1051 APKAGYFVQDNGEWKFTGSNYYYPEPITDKNSVVMISCAVNYTKAPEVFL 1100
     ||||||||||||||||||||||||||||||||||||||||||||||||||
1051 APKAGYFVQDNGEWKFTGSNYYYPEPITDKNSVVMISCAVNYTKAPEVFL 1100

1101 NNSIPNLPDFKEELDNWFKNQTSIAPDLSLDFEKLNVTFLDLTYEMNRIQ 1150
     ||||||||||||||||||*|||||||||||||||||||||||||||||||
1101 NNSIPNLPDFKEELDKWFKNQTSIAPDLSLDFEKLNVTFLDLTYEMNRIQ 1150

1151 DAIKKLNESYINLKEVGTYEMYVKWPWYVWLLIGLAGVAVCVLLFFICCC 1200
     ||||||||||||||||||||||||||||||||||||||||||||||||||
1151 DAIKKLNESYINLKEVGTYEMYVKWPWYVWLLIGLAGVAVCVLLFFICCC 1200

1201 TGCGSCCFRKCGSCCDEYGGHQDSIVIHNISAHED 1236
     |||||||||||||||||||||||||||||||||||
1201 TGCGSCCFRKCGSCCDEYGGHQDSIVIHNISAHED 1236
```

Sequence analysis of the MAR 11F/1 S protein gene

Poly A+ RNA was isolated from MAR 11F/1 infected DBT cells and a full length cDNA copy of the S gene was synthesized. The sequence of the cloned S gene was determined and compared with the wild-type MHV sequence. This analysis showed that the S gene of the variant has three mutated bases causing two amino acid changes (Fig.3). The amino acid alanine at position 255 is substituted for the amino acid arginine and at position 1116 a lysine is replaced by an asparagine. The location of both substitutions was unexpected because the binding site of the mAb 11F lies between the amino acids 33 to 40 (8).

DISCUSSION

The results presented here were unexpected and at the present time are not easy to explain. Both mutations of the MAR 11F/1 variant S protein are located outside the binding site of mAb 11F. There are at least two questions we want to answer: 1) which of the amino acid substitutions in the variant S protein is responsible for the phenotype change? and 2) does the S-protein contain multiple protein domains which are involved in, or necessary for its fusion activity ? We hope to address these questions by expression of the variant S-proteins in a vaccinia virus system following manipulation by site directed mutagenesis.

REFERENCES

1. Lee, H.-J., Shieh, C.-K., Gorbalenya,A.E., Koonin, E.V., LaMonica, N., Tuler, J., Bagdzhadzhyan, A., and Lai, M.M.C., 1991, *Virology*, 180:567
2. Pachuk, C.J., Bredenbeek, P.J., Zoltick, P.W., Spaan, W.J.M., and Weiss,S.R., 1989,*Virology*, 171:141
3. Siddell, S.G., 1982, *J. Gen. Virol.*, 62:259
4. Luytjes, W., Sturman, L.S., Bredenbeek, P.J., Charic, J., van der Zeijst, B.A.M., Horzinek,M.C., and Spaan, W.J.M., 1987,*Virology*, 161:479
5. Schmidt, I., Skinner, M., and Siddell, S., 1987, *J. Gen. Virol.*, 68:47
6. Collins, A.R., Knobler, R.L., Powell, H., and Buchmeier, M.J., 1982, *Virology*, 119:358
7. Vennema, H., Heijen, L., Zijderfeld, A., Horzinek, M.C., and Spaan, W.J.M., 1990, *J. Virol.*, 64:339
8. Routledge, E., Stauber, R., Pfleiderer, M., and Siddell, S.G., 1991, *J. Virol.*, 65:254
9. Kumanishi, T., 1967, *Jap. J. Exp. Med.*, 37:461
10. Schwarz, B., Routledge, E., and Siddell, S.G., 1990, *J. Virol.*, 64:4784
11. Sanger, F., Nicklen, S. and Coulson, A.R., 1977, *Proc. Natl. Acad. Sci., USA*, 74: 5463
12. Gubler, U., and Hoffman, B.J., 1983, *Gene*, 25:263
13. Wege, H., Dörries, R., and Wege, H., 1984, *J. Gen. Virol.*, 65:1931
14. Staden, R., 1982, *Nucleic Acids Res.*, 10:4731

MOLECULAR MIMICRY BETWEEN S PEPLOMER PROTEINS OF CORONAVIRUSES (MHV, BCV, TGEV AND IBV) AND Fc RECEPTOR

Emilia L. Oleszak[1], Stanley Perlman[2], Rebecca Parr[3], Ellen W. Collisson[3] and Julian L. Leibowitz[1]

[1]Department of Pathology and Laboratory Medicine, The University of Texas Health Science Center Medical School at Houston, Houston, TX; [2]Department of Pediatrics, The University of Iowa, Iowa City, IA and [3]Department of Veterinary Microbiology and Parasitology, College of Veterinary Medicine, Texas A & M University, College Station, TX

ABSTRACT

In previous studies we have demonstrated molecular mimicry between the S peplomer protein of Mouse Hepatitis Virus (MHV) and Fcγ Receptor (FcγR) of IgG. Rabbit IgG, but not its F(ab')$_2$ fragments, monoclonal rat and mouse IgG and the rat 2.4G2 anti-mouse FcγR monoclonal antibody (mab) immunoprecipitated natural and recombinant MHV S protein. On the basis of a number of criteria, MHV S peplomer protein exhibits Fc IgG binding ability. We report here a molecular mimicry between the S peplomer protein of Bovine Coronavirus (BCV) and FcγR. BCV S peplomer protein which belongs to the same antigenic subgroup as MHV also binds Fc portion of rabbit IgG and is immunoprecipitated by the 2.4G2 anti-FcγR mab. In contrast, Transmissible Gastroenteritis Coronavirus (TGEV) and Infectious Bronchitis Virus (IBV) S peplomer proteins which represent two distinct antigenic subgroups of Coronaviridae do not bind rabbit IgG and do not react with anti-FcγR mab. However, homologous swine IgG, but not its F(ab')$_2$ fragments, immunoprecipitated from TGEV-infected cells a polypeptide chain with molecular mass of 195 kDa, identical to that immunoprecipitated by the T36 mab anti-TGEV S peplomer protein.

INTRODUCTION

We have previously demonstrated molecular mimicry between S peplomer protein of Mouse Hepatitis Virus (MHV) and Fcγ Receptor (FcγR) of IgG[1-3]. The rat 2.4G2 anti-mouse FcγR monoclonal antibody (mab), irrelevant rat and mouse IgGs and rabbit IgG, but not its F(ab')$_2$ fragments, immunoprecipitated the S peplomer protein from cells infected with several strains of MHV, namely JHM, A-59 and MHV-3, but not from uninfected cells. In order to demonstrate that the S protein exhibits FcγR binding ability we have expressed the S gene using recombinant vaccinia virus. Both irrelevant rabbit IgG, but not its F(ab')$_2$ fragments and the 2.4G2 anti-

FcγR mab[4] immunoprecipitated recombinant MHV-JHM S peplomer protein from cells of murine, rabbit and human origin. These results unequivocally proved that an FcγR binding site resides on MHV S peplomer protein[1-3]. The FcγR binding site is expressed on the cell surface, since MHV-JHM infected cells but not uninfected cells formed rosettes with anti-erythrocyte antibody coated sheep erythrocytes[3]. The 2.4G2 anti-FcγR mab inhibits MHV-S mediated fusion "from within" and also neutralizes the virus[3]. Using Dayhoff Align program we have identified two regions of sequence similarity between MHV S peplomer protein and FcγR. The purpose of this study is to determine whether there is a molecular mimicry between FcγR and S peplomer protein of representative viruses of three separate antigenic subgroups within the family of Coronaviridae. We have chosen: (a) Bovine Coronavirus (BCV) as another virus that belongs to the same antigenic subgroups as MHV; (b) Trans-missible Gastroenteritis Virus (TGEV) as the representative of the second subgroup, that includes also Feline Infectious Peritonitis Virus (FIPV) and Human Coronavirus 229 E (HCV-229E) (c) Infectious Bronchitis Virus (IBV), which is the only member of the third subgroup. Viruses within each subgroups show antigenic crossreactivity but they infect different species and induce different clinical symptoms.

MATERIALS AND METHODS

Viruses and cells. BCV (MEBUS strain) was provided by Dr. B. Hogue (Baylor College of Medicine, Houston, TX); TGEV was obtained from Dr. C. Bond (Montana State Univ., Bozeman, MT). Tissue culture adapted IBV was provided by Dr. B. Sefton, Salk Inst., San Diego, CA. Bovine HRT-18 cell line, ST cell line and chicken embryo kidney (CEK) cells were previously described[5-7].

Antibodies. Gnotobiotic calf anti-BCV serum and bovine sera to either Breda virus or Cryptosporidium were obtained from Dr. G. Woode (Texas A & M Univ.). Mab to TGEV (clone T36) was provided by Dr. Bond. A mouse anti-IBV serum and the 951A5 anti-IBV S peplomer protein mab were developed by R. Parr & E.Collisson and will be described elsewhere. Purified whole rabbit IgG specific for Micrococcus lysodeicticus (M. lysodeicticus) as well as its F(ab')$_2$ fragments were a generous gift of Dr. S.Rodkey (Univ. Texas Medical Sch., Houston). The hybridoma producing the 2.4G2 anti-mouse FcγR mab was purchased from the ATCC (Rockville, MD). Affinity purified goat anti-mouse IgG, goat anti-rat IgG, goat anti-bovine IgG, goat anti-rabbit IgG and affinity purified bovine, swine and chicken IgG and their F(ab')$_2$ fragments were purchased from Jackson Immunoresearch Lab. (West Grove, PA).

Metabolic labeling of cells and immunoprecipitation. Confluent ST cells were infected with TGEV at m.o.i.=5 and labeled with ^{35}S Methionine (Met) at 8 hrs post infection (p.i.). Cytoplasmic lysates were prepared 12 hrs p.i. as described[1]. Confluent CEK cells were infected with tissue culture adapted IBV and labeled with ^{35}S Met 14 hrs p.i. Cytoplasmic lysates were prepared 15.5 hrs p.i. Confluent HRT-18 cells were infected with BCV at m.o.i.=1 and labeled with ^{35}S Met 9 hrs later. Cytoplasmic lysates were prepared 12 hrs p.i. as described. Control cells were not infected and labeled with ^{35}S Met as described[1]. Aliquots of 50-70 μl (approximately 5 x 10^5 cells) were immunoprecipitated with appropriate antibodies using Protein-A Sepharose beads (Pharmacia) and the immune complexes were analyzed by SDS-PAGE[1,2].

RESULTS

Molecular mimicry of BCV S glycoprotein and FcγR. There is remarkable sequence homology between S peplomer proteins of MHV-JHM and BCV.To examine

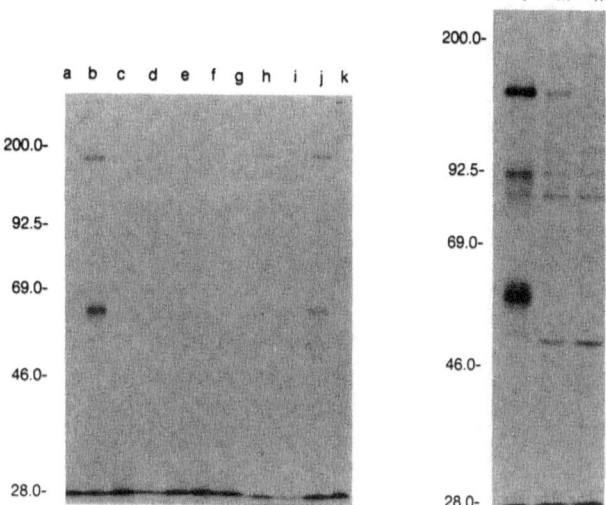

Figure 1. Immunoprecipitation of BCV-S peplomer protein by purified rabbit IgG: Lysates were prepared from [35]S Met labeled BCV- or mock-infected cells. BCV-infected lysates were immunoprecipitated with goat anti-bovine IgG (lane a); gnotobiotic bovine anti-BCV (lane b and l); bovine anti-Breda virus anti-bodies (lane c); bovine anti-Cryptosporidium antibodies (land d); 2.4G2 anti-FcγR mab (lane h); goat anti-rat IgG (lane i); goat anti-MHV S serum (lane j);purified rabbit IgG (100µg, specific for M. lysodeicticus) (lane m). Mock-infected lysates were immunoprecipitated with specific polyclonal antibodies to BCV (lane e); bovine anti-Breda virus (lane f); bovine anti-Cryptosporidium (lane g) and purified rabbit IgG (100µg, specific for M. lysodeicticus)(lane n). The immunoprecipitates were analyzed by SDS-PAGE[1,2].

whether BCV S peplomer protein exhibits the ability to bind the Fc portion of IgG, we immunoprecipitated BCV-infected cell lysates with either the 2.4G2 anti-FcγR mab or purified rabbit IgG or its F(ab')₂ fragments and analysed the immune complexes by SDS-PAGE (Fig. 1). Both the 2.4G2 anti-FcγR mab and the purified rabbit IgG, but not its F(ab')₂ fragments, immunoprecipitated a polypeptide chain with molecular mass of 190 kDa, identical to that of the BCV S peplomer protein, which was immunoprecipitated by specific bovine serum to BCV. This anti-BCV specific serum also immunoprecipitated the BCV nucleocapsid protein (52 kDa)(Fig. 1). The same two proteins were immunoprecipitated by polyclonal anti-serum specific for MHV. In addition, the same two proteins of 190 and 52 kDa were immunoprecipitated by "irrelevent" commercially available affinity purified bovine IgG, suggesting that these commercially available preparations of bovine IgG were from animals exposed to the virus (data not shown). This is not unexpected in view of the high incidence of BCV infection[8,9]. Gnotobiotic homologous control antibodies specific for Breda virus or specific for Cryptosporidium, used at concentrations significantly lower than that of the affinity purified bovine IgG, as well as secondary antibodies, did not immunoprecipitate any protein from BCV-infected cells or control cells (Fig. 1). All these results suggest that BCV S peplomer protein also binds the Fc region of IgG.

Reactivity of TGEV and IBV S peplomer proteins with Fc IgG. BCV, TGEV and IBV belong to three distinct antigenic subgroups of Coronaviridae[10]. We examined the ability of TGEV and IBV S peplomer proteins to bind rabbit IgG and to react with the 2.4G2 anti-FcγR mab (Fig. 2 and 3, respectively). Neither purified rabbit IgG

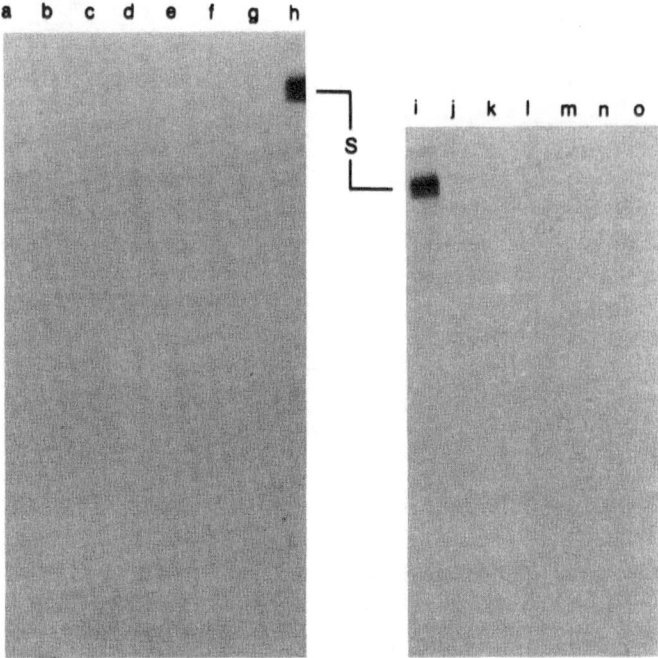

Figure 2. Immunoprecipitation of TGEV-S peplomer protein: Lysates from ^{35}S Met labeled TGEV-infected cells were immunoprecipitated with goat anti-rat IgG (lane c); 2.4G2 anti-FcγR mab (lane d); goat anti-mouse IgG (lane f); mouse anti-TGEV S peplomer protein T36 mab (lane h & i); goat anti-rabbit IgG (lane k); purified rabbit IgG (lane m); F(ab')$_2$ of rabbit IgG (lane n); goat anti-MHV-JHM (lane o) and analyzed by SDS-PAGE[1,2]. Mock-infected lysates were immunoprecipitated with goat anti-rat IgG (lane a); goat anti-mouse IgG (lane e); T36 mab (lane g); goat anti-rabbit IgG (lane j); purified rabbit IgG (lane l) and analyzed by SDS-PAGE[1,2].

nor the 2.4G2 anti-FcγR mab immunoprecipitated any proteins from TGEV-infected cells. The T36 mab specific for TGEV S peplomer protein immunoprecipitated a 195 kDa protein from TGEV-infected cells (Fig. 2; lane h and i), but not from control, uninfected cells (Fig. 2, lane g). Secondary antibodies did not immunoprecipitate any proteins from TGEV-infected or control uninfected cells. Anti-MHV S protein antibody did not immunoprecipitate TGEV S protein (Fig. 2, lane o), in agreement with the fact that TGEV belongs to a distinct antigenic group than MHV and BCV. However, homologous swine IgG (50 μg), but not its F(ab')$_2$ fragments, immunoprecipitated from TGEV-infected cells a protein of 195 kDa, identical to that immunoprecipitated by the T36 mab anti-TGEV S peplomer protein (data not shown). This homologous swine IgG did not immunoprecipitate any other TGEV proteins, in agreement with the findings that the prevalence of anti-TGEV antibodies is low in these animals[11].

Polyclonal antibodies to IBV, the only member of the third antigenic subgroup of Coronaviridae, immunoprecipitated several structural IBV proteins: S peplomer protein (180 kDa) nucleocapsid (51 kDa) and matrix protein (Fig. 3, lane d). S peplomer protein was also immunoprecipitated by the 951A5 anti-IBV mab. None of these anti-bodies immunoprecipitated any polypeptide chains from control uninfected cells. The 2.4G2 anti-FcγR mab or purified rabbit IgG did not immunoprecipitate any protein from IBV-infected cells (Fig. 3, lane h and f). We have also carried out studies using "irrelevant" commercially available purified chicken IgG. Homologous, chicken IgG immunoprecipitated several stuctural IBV proteins including S peplomer protein,

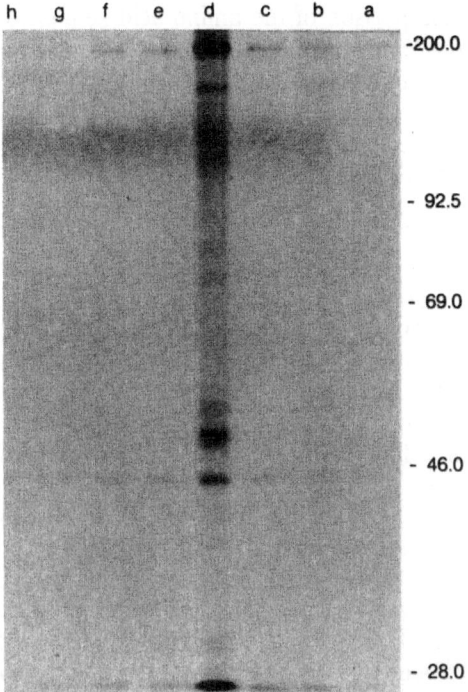

Figure 3. Immunoprecipitation analysis of IBV S peplomer protein: [35]S Met labeled IBV infected cell lysates were immunoprecipitated with the 951A5 mab specific for IBV peplomer protein (lane b); mouse polyclonal antibodies specific for IBV (lane d); purified rabbit IgG (lane f); F(ab')$_2$ fragments of rabbit IgG (lane g); the 2.4G2 anti-FcγR mab (lane h). Mock-infected cell lysates were immunoprecipitated with 951A5 mab (lane a); mouse anti-IBV antibodies (lane c); and purified rabbit IgG (lane e). The immunoprecipitates were analyzed by SDS-PAGE[1,2].

nucleocapsid and matrix proteins (data not shown). In view of the high incidence of IBV infection[12], this is not surprising and suggests that these commercially available IgG preparations were from chicken exposed to IBV. Experiments using truly "irrelevant" chicken IgG (without anti-IBV contaminating antibodies) are needed to determine whether IBV S peplomer protein binds homologous IgG and are being carried out.

DISCUSSION

We have previously shown molecular mimicry between S peplomer protein of MHV and FcγR[1-3]. Using Dayhoff Align Program and Monte Carlo Analysis we have identified two regions of sequence similarity between S peplomer protein (of MHV-JHM or A-59) and mouse FcγR (Domains #1 & #2)[1]. These two regions of sequence similarity are located within the N-terminal subunit (S1) of the S peplomer protein. In this report, we have examined the binding of immunoglobulin to S peplomer proteins of several coronaviruses representing three distinct antigenic subgroups of Coronaviridae. The 2.4G2 anti-FcγR mab and purified rabbit IgG were employed as two representative FcγR recognizing reagents. Both immunoprecipitated BCV S peplomer protein, but not S peplomer proteins of TGEV or IBV (Fig. 2 and 3). F(ab')$_2$ fragments of rabbit IgG did not immunoprecipitate BCV S peplomer protein. Therefore, there is molecular mimicry between BCV S peplomer protein and FcγR.

BCV and MHV belong to the same antigenic subgroup and show remarkable sequence homology[13]. Direct comparison of the amino acid sequence of the S1 subunits of JHM and BCV showed 62% identity. The S1 subunits of A-59 and BCV showed 60% identity. Also, there are high levels of homology in the S2 subunits of JHM, A-59 and BCV (75% & 74% respectively). It has been previously demonstrated, that the S peplomer protein of MHV and BCV antigenically crossreact[14]. We report here another common property of S peplomer proteins of these two viruses: both are able to bind the Fc portion of rabbit IgG and both are immunoprecipitated by the 2.4G2 anti-FcγR mab. The region of sequence similarity between MHV S peplomer protein (Domain #1) and FcγR is conserved in BCV S peplomer protein, which is in agreement with our findings that both peplomer proteins have an Fc binding site. However, it remains to be established whether this domain #1 contain the Fc binding epitope.

TGEV and IBV represent two distinct antigenic subgroups of Coronaviridae[10]. Computer analysis of the sequences of the structural proteins of MHV, TGEV and IBV did not reveal any homologies[15]. Bovine, swine and chicken Fc receptors have not been sequenced and therefore their sequences can not be compared to those of the corresponding S peplomer proteins.

The results that purified rabbit IgG or the 2.4G2 anti-FcγR mab did not immunoprecipitate any polypeptide chains from TGEV- or IBV-infected cells do not imply that there is no molecular mimicry between the TGEV and IBV peplomer proteins and the FcγR. To the contrary, homologous swine IgG, but not its $F(ab')_2$ fragments, immunoprecipitated from TGEV-infected cells the TGEV S peplomer protein (data not shown). This homologous swine IgG did not contain anti-TGEV antibodies and did not immunoprecipitate any other TGEV proteins, in agreement with the findings that the prevalence of anti-TGEV antibodies is low in swine[11]. Studies to determine whether "irrelevant" bovine IgG or chicken IgG immunoprecipitated the S peplomer proteins from BCV- or IBV-infected cells, respectively, were hindered by the fact that the "irrelevant", homologous IgG preparation available to us were very likely contaminated with anti-viral antibodies, because they immunoprecipitated several structural viral proteins. Immunoprecipitations experiments using bovine and chicken IgG preparations that do not contain anti-viral antibodies are needed and are being carried out.

REFERENCES

1. Oleszak E. and Leibowitz J. *Virology* 176:70(1990).
2. Oleszak E., Perlman S. and Leibowitz J. *Virology* 186:122(1992).
3. Oleszak E. and Leibowitz J. *Adv. Exper. Biol. Med.* 276:51(1991).
4. Unkeless J.C. *J. Exp. Med.* 150:580(1979).
5. Lapps W., Hogue B.G. and Brian D.A. *Virology* 157:47(1987).
6. Collisson E.W., Li J.Z., Sneed L.W., Peters M.L. and Wang L. *Vet. Microbiol.* 24: 261(1990).
7. Choi C.S., Murtaugh M.P. and Molitor T.W. *Arch. Virol.* 115:227(1990).
8. Alenius S., Niskanen R., Juntti N. and Larsson B. *Acta Vet. Scand.* 32:163(1991).
9. Heckert R.A., Saif L.J., Myers G.W. and Agnes A.G. *Am. J. Vet. Res.* 52:845(1991).
10. Siddell S., Wege H. and Ter Meulen V. *J. Gen. Virol.* 64:761(1983).
11. Brown I.H. and Paton D.J. *Vet. Rec.* 128:500(1991).
12. Gelb J. Jr., Leary J.H. and Rosenberger J.K. *Avian Dis.* 27:667(1983).
13. Hogue B., King B. and Brian D. *J. Virol.* 51:384(1984).
14. Boireau P., Cruciere C. and Laporte J. *J. Gen. Virol.* 71:487(1990).
15. Jacobs L., deGroot R., van der Zeijst B.A.M., Horzinek M. and Spaan W. *Virus Res.* 8:363(1987).

COMPLEX FORMATION BETWEEN THE SPIKE
PROTEIN AND THE MEMBRANE PROTEIN DURING
MOUSE HEPATITIS VIRUS ASSEMBLY

Dirk-Jan E. Opstelten, Marian C. Horzinek, and Peter J. M. Rottier

Institute of Virology
Department of Infectious Diseases and Immunology
Faculty of Veterinary Medicine
University of Utrecht
The Netherlands

SUMMARY

Using different approaches we have demonstrated the formation of a complex between the S protein and the M protein in the process of mouse hepatitis virus (MHV) assembly. Preservation of the M/S heterocomplexes was critically dependent on the solubilization conditions. Pulse-chase labeling of MHV-infected cells followed by a co-immunoprecipitation assay revealed that newly synthesized S and M engage in complex formation with different kinetics, the S protein reacting much slower. Sedimentation experiments showed the M/S heteromultimer complexes to be very large. A model for the role of the complex formation in MHV assembly is presented.

INTRODUCTION

Coronaviruses mature by budding into intracytoplasmic pre-Golgi membranes (1). In the case of mouse hepatitis virus (MHV) electronmicroscopic observations have shown helical nucleocapsids to attach to regions of these membranes (2). Assuming that the viral envelope proteins localize the budding one would expect these proteins to accumulate here. However, when expressed independently both the M and the S proteins appear to localize differently: M accumulates in the Golgi apparatus (3, 4) while S is transported to the plasma membrane (H.Vennema and P.Rottier, unpublished observations). This suggests that in infected cells an interaction between the two viral membrane proteins may be required to direct them to and retain them at the site of budding. This interaction is probably very specific since virions contain only trace amounts of host proteins. Using different approaches we describe here the detection of complexes between the S protein and the M protein and their initial characterization.

MATERIALS AND METHODS

Virus, cells and antisera

MHV strain A59 (MHV-A59) was propagated in Sac(-) cells as described by Spaan et al. (5). Sac(-) and 17Cl1 cells were maintained in Dulbecco's minimal essential medium containing 5% fetal calf serum, penicillin, and streptomycin (DMEM/5%FCS). The production of the polyclonal antiserum to MHV-A59 has been described (6). The monoclonal antibodies J7.6 (anti-S; 7) and J1.3 (anti-M; 8) were kindly provided by Dr. J. Fleming.

Infection and metabolic labeling

Subconfluent monolayers of Sac(-) or 17Cl1 cells in 35 mm-dishes were inoculated for 60 min at a multiplicity of infection of 10-50. From 5.5 hr post inoculation cells were incubated for 30 min in MEM (GIBCO) without methionine and then pulse-labeled with 20-200 μCi ExpreSS[^{35}S]-label (Dupont). Cells were washed once with PBS and either lysed directly or chased for various times in DMEM/5% FCS supplemented with 2 mM L-methionine. The cells were lysed in 600 μl detergent solution (50 mM Tris-Cl, [pH 8.0], 62.5 mM EDTA, 0.5% sodiumdeoxycholate, 0.5% Nonidet P-40 [NP40]) containing 2mM phenylmethylsulfonyl fluoride and 40 μg/ml aprotinin (Sigma). In the experiment of Fig. 1 detergent solutions were varied as indicated. The composition of the MNT detergent solution has been described before (9). For the analysis of complexes from virions (Fig. 3A) detergent solution was added from a 5 times concentrated stock solution directly to the cleared culture medium.

Sucrose gradient centrifugation

Samples of solubilized ^{35}S-labeled cells or virions were analyzed in sucrose gradients (15-30% [wt/wt] sucrose in detergent solution). The gradients were centrifuged for 30 min at 50,000 rpm in an SW50.1 rotor (Beckman Instruments Inc., Palo Alto, CA). They were fractionated from the bottom into 15 fractions of approx. 330 μl. Any material pelleted to the bottom of the tube was dissolved in 333 μl detergent solution.

Immunoprecipitation

Samples of cell lysates or gradient fractions were brought to 600 μl with detergent solution and antibodies were added (2 μl anti-MHV; 10 μl anti-S; 2 μl anti-M). The samples were incubated at 4°C for at least 3 hr and immune complexes were then bound to Staph. A (Pansorbin Cells, Calbiochem) for at least 30 min at 4°C. Staph. A was pelleted by centrifugation and washed three times with detergent solution. The final pellets were resuspended in Laemmli sample buffer containing 20 mM dithiothreitol and the samples heated for 2 min at 95°C. They were analyzed in 10 or 15% SDS poly-acrylamide gels.

RESULTS

Detection of M/S complexes

Sofar, specific complexes between the S and M protein of MHV-A59 have not been clearly observed. We reasoned that such complexes may have escaped detection due to the analytical conditions used. Assuming that these conditions were not suitable to preserve the intermolecular interactions in the complexes we studied the effects of different detergents and buffers used to solubilize the membrane proteins from infected cells. Fig. 1 shows some results. Parallel cultures of MHV-A59 infected cells were labeled for 60 min with ^{35}S-methionine and lysed with a panel of detergent solutions containing ionic as well as nonionic detergents. Each lysate was split in two parts and the viral proteins were immunoprecipitated with a polyclonal anti-virion serum or with a monospecific anti-S serum. It should be noted that the same buffer/detergent conditions were maintained throughout the analytical procedure. Clearly, under all conditions shown in Fig.1 the anti-S serum exclusively precipitated the S protein except in one case. When a combination of NP40 and NaDOC was used, significant co-precipitation of M with S was observed. Thus, in infected cells a major fraction of M is present in a physical complex with S and these complexes are well preserved when suitable lysis conditions are used. No co-immunoprecipitation of the nucleocapsid protein was observed. It is interesting to note that the detergent solution MNT/1% Triton X-100 which we have been using to detect viral spike oligomers (9) apparently disrupted the interactions between S and M.

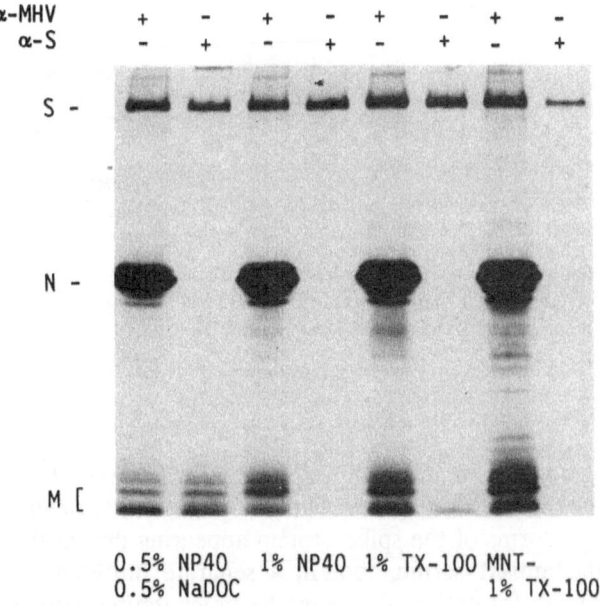

Figure 1. MHV-infected Sac(-) cells were labeled for 1 hr and solubilized with a set of different detergent solutions as indicated. Each lysate was split in two parts and the viral proteins were precipitated with the polyclonal anti-MHV serum and with the monospecific anti-S serum. Structural proteins are indicated (S, N, and M).

Kinetics of co-immunoprecipitation

To further confirm the existence of the M/S complexes and to determine the kinetics with which they are formed we performed a pulse-chase labeling of MHV-infected cells. The cells were labeled for 5 min and chased for various time periods. Using the detergent solution containing 0.5% NP40 and 0.5% NaDOC the cells were lysed and the viral proteins were precipitated with the polyclonal anti-MHV serum and with a monospecific anti-M serum. As Fig. 2 demonstrates the anti-M serum exclusively precipitated the M protein from the pulse-labeled sample. However, after 10-20 min of chase a detectable amount of the spike precursor gp150 started to become co-precipitated. Using the amount of labeled S protein precipitated with the polyclonal serum as a reference, the fraction of S that co-pre-

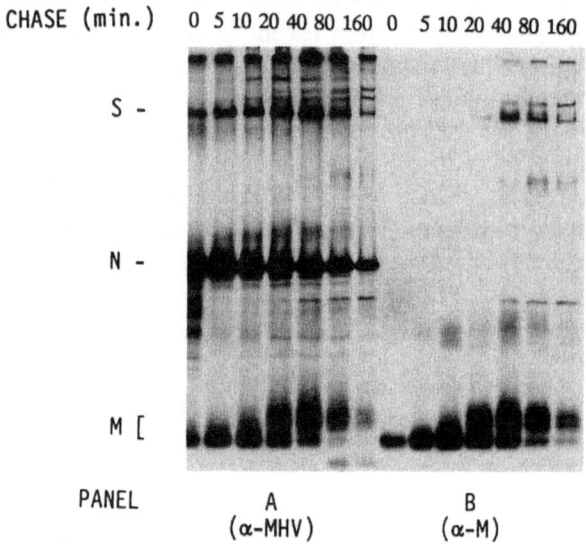

Figure 2. MHV-infected Sac(-) cells were pulse labeled for 5 min and chased for various time periods as indicated. Viral proteins in one half of each cell lysate was precipitated with anti-MHV serum (panel A) and in the other half with the anti-M serum (panel B).

cipitated with M gradually increased during longer chase times. This result indicates that S becomes involved in complexes with M post-translationally. Note that the S/gp180 and S/gp90 forms of the spike protein appearing during the chase were co-precipitated by the anti-M serum, too. In a separate pulse-chase experiment we examined the complex formation vice versa by precipitating with the monospecific anti-S serum. The results of this experiment indicated that the amount of co-immun-oprecipitated M reached its maximum already after about 10-20 min of chase (data not shown). We conclude that the envelope proteins engage in heterocomplex formation with different kinetics M appearing in complexes earlier after its synthesis than S.

Our second approach to study the complex formation between S and M was sedimentation analysis of the viral proteins using the same NP40/DOC detergent conditions both for solubilization and in the sucrose gradients. Therefore, MHV-infected 17Cl1 cells were labeled for 10 min and chased for 60 min and 180 min. To examine complexes derived from virions, the culture medium from the cells chased for 180 min was treated with the detergents and analyzed in 15-30% sucrose gradient. Fig. 3A shows that under the conditions described the major fraction of M and S co-sedimented to fractions 7-9. Given the composition of the gradient and the relatively short run time this position in the gradient implies that the complexes must be very large. The nucleocapsid protein appeared to sediment much slower confirming the conclusion from the co-immunoprecipitation assay that N was not engaged in M/S complexes. A fraction of S and M stayed at the top of the gradient. The nature of these molecules is unclear; they may have been derived from disrupted cells or from dissociation of the complexes.

In parallel with the virion material the gradient analyses of the labeled cell samples were carried out. Most viral protein labeled during the 10 min pulse was recovered from the top of the gradient (Fig. 3B). A significant fraction of M, however, sedimented heterogenously from the top to deep positions in the gradient and was apparently present in complexes. Although no labeled S protein was detected in these fractions the M protein was shown in a separate experiment to be precipitated by the anti-S monoclonal antibody (data not shown). After the 60 min chase a significant fraction of the labeled S protein also appeared in the M/S complexes of heterogenous sizes (Fig. 3C). At this time point most of the M protein was present in these complexes. Comparison of the different profiles in Fig. 3 shows that the intracellular complexes are more heterogeneous than those derived from virions. Part of the nucleocapsid protein from the cell samples co-sedimented with the M/S complexes. We do not believe that this N protein was physically associated with the M/S complexes as the protein did not co-immunoprecipitate with these complexes.

DISCUSSION

We here describe the detection of complexes between the spike protein and the membrane protein of MHV-A59 in infected cells. Specific detergent conditions appeared to be required to preserve the intermolecular interactions in the complexes. Pulse-chase labeling of MHV-infected cells followed by co-immunoprecipitation and sedimentation analyses revealed that M after its synthesis readily engages in complex formation with S. In contrast, S appears to associate with M with much slower kinetics. The data imply that M interacts with S molecules that were already synthesized. Apparently, S needs more time to become interaction-competent. Preliminary experiments show that this behaviour correlates well with the slow folding characteristics of the protein.

We do not know yet where in the cell the M/S complexes are formed. An important observation, however, is that already after a 10 min pulse labeling a significant fraction of unglycosylated M protein was found in a complex with S. This suggests that the complexes are formed before or at the site of budding.

Our sedimentation analyses demonstrated that the M/S complexes isolated from infected cells are rather heterogeneous, in contrast to the complexes obtained from

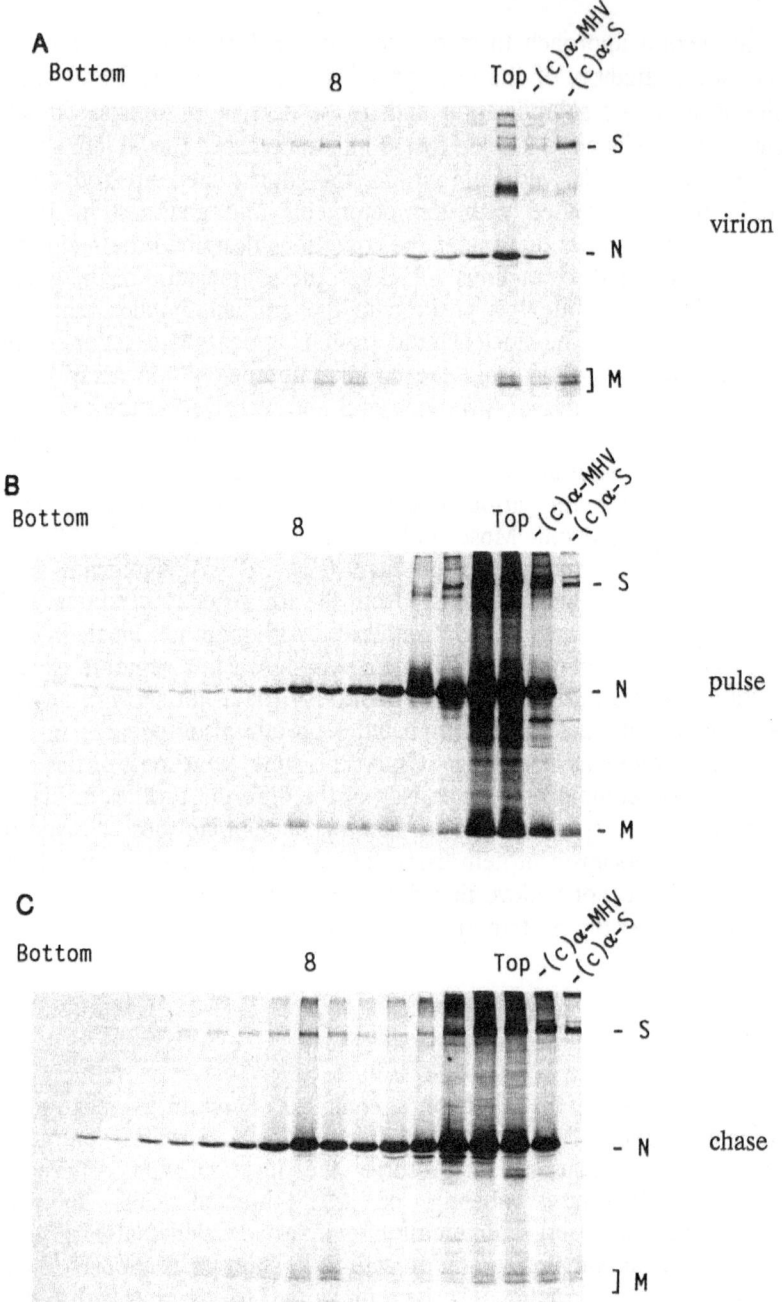

Figure 3. MHV-infected 17Cl1 cells were labeled for 10 min and chased for 60 min and 180 min. Detergent treated culture medium from the 180 min chase (A) and lysates of 10 min pulse labeled cells (B) or the 60 min chase (C) were run for 30 min at 50,000 rpm into a 15-30 % (wt/wt) sucrose gradient. Viral proteins in the fractions from bottom to top were precipitated with anti-MHV serum. Viral proteins in the control lanes (c) were precipitated with anti-MHV and with anti-S serum from aliquots taken from the samples before loading them onto the gradient.

virions. A significant fraction of the intracellular complexes sedimented slower than those derived from virions, while also faster sedimenting complexes were observed. The homogeneous nature of the virion M/S complexes suggests that the structures we are observing are real and do not represent procedural artefacts. Though no size determinations have been done yet it is clear that the complexes are very large.

Collectively, the data lead us to hypothesize that S and M by virtue of their interaction congegrate at the site of viral budding thereby forming a matrix into which the nucleocapsids can bud. Implicit to this hypothesis is the idea that the formation of the complexes between S and M causes the proteins to be retained at the budding site instead of being transported beyond this site when expressed independently. Finally, since coronaviruses like most other viruses are very selective in incorporating only virus-specific proteins, we postulate that the interaction between S and M is very specific. Accordingly, our results sofar demonstrate that the viral M/S complexes are not contaminated with non-viral proteins.

ACKNOWLEDGEMENTS

We thank Harry Vennema for helpful discussions and Pieter de Groote for technical assistance. We are grateful to Dr. J. Fleming for providing the monoclonal antibodies J7.6 and J1.3.

REFERENCES

1. J. Tooze, S. Tooze and G. Warren. (1984) *Eur. J. Cell Biol.* **33**:281-293.
2. M.K. Dubois-Dalcq, E.W. Doller, M.V. Haspel, and K.V. Holmes. (1982) *Virology* **119**:317-331.
3. P.J.M. Rottier and J.K. Rose. (1987) *J. Virol.* **61**:2042-2045.
4. J. Krijnse-Locker, G. Griffiths, M.C. Horzinek, and P.J.M. Rottier. (1992) *J. Biol. Chem.* 267:14094-14101.
5. W.J.M. Spaan, P.J.M. Rottier, M.C. Horzinek, and B.A.M. van der Zeijst. (1981) *Virology* **108**:424-434.
6. P.J.M. Rottier, M.C. Horzinek, and B.A.M. van der Zeijst. (1981) *J. Virol.* **40**:350-357.
7. D.G. Weismiller, L.S. Sturman, M.J. Buchmeier, J.O. Flemming, and K.V. Holmes. (1990) *J. Virol.* **64**:3051-3055.
8. J.O. Fleming, R.A. Shubin, M.A. Sussman, N. Casteel, and S.A. Stohlman. (1989) *Virology* **168**:162-167.
9. H. Vennema, P.J.M. Rottier, L. Heijnen, G.J. Godeke, M.C. Horzinek, and W.J.M. Spaan. (1990) *In* Coronaviruses and their diseases, ed. D. Cavanagh and T.D.K. Brown, Plenum Press, New York.

PRELIMINARY CHARACTERIZATION OF A MONO-CLONAL ANTIBODY SPECIFIC FOR A VIRAL 27 kD GLYCOPROTEIN FAMILY SYNTHESIZED IN PORCINE EPIDEMIC DIARRHOEA VIRUS INFECTED CELLS

Anna Utiger[1], Max Rosskopf[1], Franco Guscetti[2], and Mathias Ackermann[1]

[1]Institute for Virology, Vet.-med. Faculty, University of Zürich
[2]Institute of Veterinary Pathology, University of Zürich

ABSTRACT

We describe a new monoclonal antibody No. 204 (mcAb 204) which recognized a family of four polypeptides, consisting of a 27kD, a 24/23kD double band and a 19kD protein present within PEDV infected cell lysates. These proteins were identified by immunoprecipitation as well as by staining of immunoblots. In infected Vero cell cultures, the synthesis of the 27kD protein was initiated between 6 and 8 hours post inoculation. The 24/23kD double band and the 19kD protein were only detectable later. At least the 27 and the 24/23kD proteins were apparently glycosylated and present in purified virions.

Pulse-chase as well as solubilization experiments indicated that the faster migrating bands represented processed products of the 27kD glycoprotein. The nature of the processing is not known at present.

We suggest that the 27kD protein family may represent the integral membrane protein M of PEDV. Since this protein is highly abundant in virions as well as in infected cells, and since mcAb 204 is able to react with its antigen under various conditions, this monoclonal antibody may be useful to further studies of the M-protein of PEDV. In addition, it may provide a useful tool for routine diagnosis.

INTRODUCTION

Porcine epidemic diarrhoea virus (PEDV), a causative agent of severe diarrhoea in pigs, was identified in 1978 by Pensaert and Debouck[1] but only in 1988 could the virus be propagated in cell cultures using Vero cells and medium containing trypsin[2]. Consequently, little is known about the structural PEDV proteins synthesized in infected cell cultures. Using *in vitro* cultivated virus, the viral surface protein S and the nucleocapsid protein N have been unambiguously identified and partially characterized[3].

Coronaviruses, Edited by H. Laude and J.F. Vautherot
Plenum Press, New York, 1994

Egberink et al.,[4] who, in contrast to the above study, used purified PEDV from intestinal perfusates of infected pigs, were able to show by sodium dodecyl sulphate polyacrylamide gel electrophoresis (SDS-PAGE) three virion structural proteins, N, S, and a cluster of proteins with molecular weights ranging from 20 to 32kD which was believed to represent the integral membrane protein M.

In this study, we describe a monoclonal antibody which identified a family of PEDV glycoproteins with molecular weights similar to those designated M-protein by Egberink et al.[4] The different forms apparently originate from a single precursor. Based on evidence provided in this paper as well as on previous reports referring to other coronaviruses (reviewed in [5]), we suggest that the 27kD protein family which is highly abundant in infected Vero cells may represent the integral membrane protein M of PEDV.

METHODS

PEDV strain CV777 was propagated in Vero cells essentially as described[2,3]. Viral proteins were metabolically labelled by supplementing the medium with 0,1 MBq ^{35}S-methionine (Amersham) per ml. Hybridomas secreting monoclonal antibody No. 204 (mcAb 204) were obtained as reported previously[6]. Standard methods were used for polyacrylamide gel electrophoresis, transfer to nitrocellulose (BA85, Schleicher&Schüll), and radioimmunoprecipitation[3]. Prior to immunostaining, the blots were treated with Triton X-100 (1%, 10 min.) and trypsin (80 μg/ml, 3 min.)[7]. The reactions were visualized using rabbit-anti-mouse antibodies and protein-A peroxidase as conjugate and H_2O_2 and chloronaphtol as substrate[3]. The solubilization experiments were done in 30 mM phosphate buffer with pH values as indicated in the text. OBG (n-octyl-beta-d-glucopyranoside, Sigma) 1.7% and 1.4% Triton X-100 (Boehringer) were added as detergents. The Glycan Detection Kit (Boehringer) was used according to protocol A of the manufacturer. Immunohistochemical staining with mcAb 204 was done on formaldehyde fixed and paraffin embedded gut samples of PEDV infected pigs (kindly supplied by A. Pospischil). The sections were pretreated with protease type XXVII (0.1%, Sigma), the binding reaction was visualized using a DAKO-PAP-KIT[8].

RESULTS AND DISCUSSION

mcAb 204 reacts with a 27kD structural glycoprotein of PEDV

In order to identify and characterize the viral protein corresponding to mcAb 204, individual Vero cell cultures were infected with PEDV and mock infected. The newly synthesized proteins were metabolically pulse labelled for consecutive two hour intervals before harvesting. The cellular lysates were separated on SDS polyacrylamide gels, transferred to nitrocellulose, and immunologically stained with mcAb 204. Starting from between 6 and 8 hours post infection (hpi), a ^{35}S-methionine labelled 27kD protein, which was synthesized exclusively in the PEDV infected cells, could be observed by autoradiography. As shown in Fig. 1, the corresponding polypeptide could be immune stained with mcAb 204 (lane 2). During the following intervals, faster migrating bands were detected. Between 8 and 10 hpi a double band of 24 and 23kD (24/23, lane 3) and between 10 to 12 hpi a 19kD band became detectable (lane 4). When crudely purified virions were tested with mcAb 204 on Western blots, the 27 and the 24/23kD bands were also visible, whereas the 19kD band remained undetectable (lane 5, arrow). In order to test wether these polypeptides represented glycoproteins, purified virion preparations were oxidized in the course of the glycan detection procedure prior to immunoprecipitation. Following polyacrylamide gel electrophoresis and transfer of these pretreated immunoprecipitates to nitrocellulose, the 27 and 24/23kD bands were identified as glycoproteins by the use of the glycan detection kit (lane 8, dots). In contrast to the autoradiograms (lane 7), the glycan detection revealed two additional bands, one of which

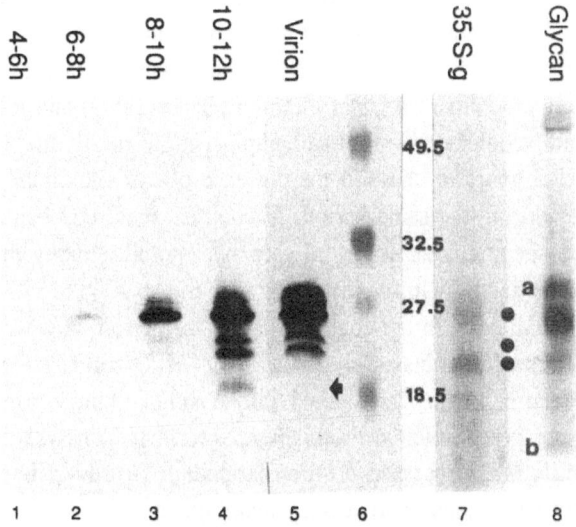

Figure 1. Western blot stainings (with mcAb 204 or with Glycan Detection Kit) and autoradiographic images are shown. Lanes 1 to 4: Immunostained PEDV infected cell lysates harvested at the time indicated at the top of each lane. Lane 5: Immunostained virion preparation. Arrow: lacking 19kD protein in virion preparation. Lane 6: Molecular weight (MW) marker. The numbers on the right indicate MW in Kilodaltons (kD). Lane 7: Autoradiographic image of ^{35}S-methionine labelled and with mcAb 204 immunoprecipitated PEDV antigen. Lane 8: Glycan detection of the material in lane 7. Dots indicate the position of the viral glycoproteins.

migrated more slowly than 27kD (a) and the other one faster than 19kD (b). Judging from mock infected controls (results not shown), band (a) apparently represented a fraction of the immunoglobulins. The nature of band (b) remained unresolved. Results obtained with immunoprecipitations of tritium-glucosamine labelled, infected cell lysates (data not shown) indicated in addition that all four members of the 27kD protein family were glycosilated.

In accordance with previous studies on PEDV structural proteins[4] as well as with reports concerning the structural proteins of coronaviruses in general[5], we conclude that the 27kD glycoprotein family may represent the integral membrane protein M of PEDV.

The 24/23 and the 19kD bands are processed products of the 27kD glycoprotein

It was not clear from the above described experiments if the faster migrating proteins were synthesized independently or if they represented processed products of the 27kD polypeptide. In order to study these faster migrating reaction partners of mcAb 204 more precisely, two sets of experiments i) pulse-chase and ii) solubilization studies were done.

i) Pulse-chase experiments. PEDV infected cells were pulse labelled for one hour intervals with ^{35}S-methionine. The radioactive precursor was then removed from the medium and, in order to prevent further translation of mRNA into protein without interfering with processive events, the cells were chased in the presence of 170 µg/ml cycloheximide for 30, 60 or 180 minutes. Finally, the cellular lysates were electrophoresed, blotted onto nitrocellulose, probed with mcAb 204 and exposed to autoradiography. The results (not shown) indicated that at least the 27 and the 23kD protein were radioactively labelled during the pulse period. In the course of the chase periods, increasing amounts of a faster migrating band (22kD) and less of the 27 and the 23kD bands could be detected by autoradiography. Interestingly, the 27kD band remained visible even after 3 hours of chase in the presence of cycloheximide, whereas the 23kD band disappeared during the same period. In contrast, the 24kD band was barely visible and the 19kD polypeptide was not at all seen by autoradiography throughout the chase period. On the other hand, by immunostaining with mcAb 204 the 27, 24, and 23kD bands could be seen immediately. In the course of the chase period, however, the 22 as well as the 19kD bands appeared, whereas the 27 and the 24kD bands grew increasingly faint. Finally, the 23 and 22kD bands merged into a single, blurry band. Such bands are typically seen with glycoproteins. From these experiments, it was clear that the 27kD glycoprotein family was processed posttranslationally. However, it was not possible to elucidate the ways and means of processing unambiguously.

ii) Solubilization studies. Using either OBG or Triton X-100 as detergents, infected cells were lysed and solubilized under variing pH conditions. The soluble and the insoluble fractions were separated by high speed centrifugation. The supernates containing the soluble proteins were removed, dialyzed, and adjusted to SDS buffer before electrophoresis. The pellets were resuspended and solubilized directly in SDS buffer. Samples of each fraction were electrophoresed and probed with mcAb 204 on Western blots before the bands were quantitated using a computing densitometer. The results of these experiments are summarized in Fig. 2. Under mildly acidic conditions (pH 5.5-6.0), the 27kD protein family appeared to be poorly soluble. With increasing pH (pH 6.6-8), however, higher amounts of 27kD polypeptide were present in the soluble supernatant fraction, until the pellet disappeared completely. Although in each quantifiable fraction the 27kD protein as well as faster migrating bands were observed, the amounts of the individual members of the protein family varied considerably with the conditions used for extraction. At pH 8, with OBG as the detergent, the 27kD polypeptide represented more than 90% of the immunostained protein. In contrast, when the cellular lysates were solubilized with Triton X-100 under otherwise identical conditions, only minor amounts of the 27kD protein could be detected, whereas the faster migrating 24/23kD double band represented more than 80% of the protein identified by mcAb 204.

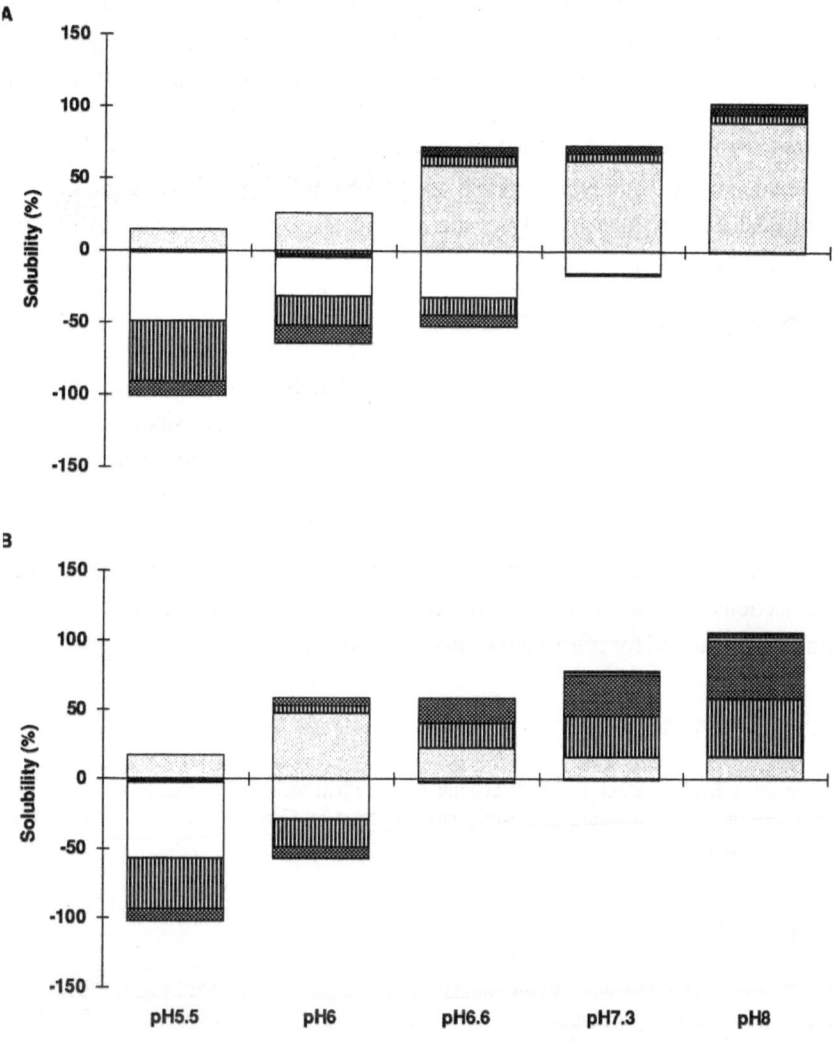

Figure 2. Solubility of 27kD protein family with OBG (A) and Triton X-100 (B). PEDV infected cell lysates were treated as described in the text. The pH values used for extraction are indicated on the x-axis. The mcAb 204 stained bands were quantitated by computing densitometry and the values were expressed in % of maximal staining (y-axis). Positive values represent proteins of the soluble fractions; negative values represent proteins contained in the insoluble pellets. 27kD protein (soluble fraction=grey; insoluble fraction=empty). 24/23kD double band (striped). 19kD protein (hatched).

We conclude that the 27kD protein family was processed to faster migrating forms, by both posttranslational processing and chemical breakdown during sample preparation. Although the various forms of the 27kD protein family were observed in abundant amounts when the infected cells were extracted directly with SDS buffer, we have insufficient evidence to suggest unambiguously that the faster migrating proteins represent naturally occuring polypeptides which are processed within infected cells from the 27kD precursor. Multiple forms of the integral membrane protein of other coronaviruses have been reported by others[5]. It has been suggested that dispersely migrating integral membrane glycoproteins may arise from different degrees of glycosylation or heterogeneity of the attached oligosaccharide chains[4]. On the other hand, the influence of sample treatment on producing various polypeptide patterns has been reported for a number of coronavirus M-proteins (reviewed in [5]). Thus, it is possible that the extraction at pH 8 with OBG as detergent somehow protected the 27kD protein from being artificially processed. In any event, the means of processing are not known at present. Therefore both post translational processing and chemical breakdown have to be considered. Studies employing specific glycohydrolases, proteases and protease inhibitors as well nucleotide sequence analyses are in progress to provide further insight into these unanswered questions.

Potential applications for mcAb 204

The monoclonal antibody described in this study may provide a versatile tool to further study the integral membrane protein of PEDV, since it was able to immunoprecipitate the corresponding proteins as well as to stain its antigen on immunoblots. It should be mentioned that the latter technique required pretreatment of the transferred proteins with proteinase. Methanol or formaldehyde fixation as well as paraffin embedding did not interfere with binding of mcAb 204 to PEDV antigens in tissue sections (not shown). This monoclonal antibody may therefore be applied for routine diagnosis and for retrospective investigations[8].

ACKNOWLEDGMENTS

We thank Evelyne Staub and Beat Scheier for excellent technical assistance and Anne Bridgen for carefully reading the manuskript. This work was supported by Grant No. 012.91.7 of the Swiss Federal Veterinary Services.

REFERENCES

1. M.B. Pensaert and P. Debouck, A new coronavirus-like ..., *Arch. Virol.* 58: 243-247 (1978).
2. M. Hofmann and R. Wyler, Propagation of the virus ..., *J. Clin. Microbiol.* 26: 2235-2239 (1988).
3. M. Knuchel, M. Ackrmann, H.K. Müller, and U. Kihm, An ELISA for detection ..., *Vet. Microbiol.* 32(2): 117-134 (1992).
4. H. Egberink, J. Ederveen, and M.C. Horzinek, Characterization of the structural proteins ..., *Am. J. Vet. Res.* 49: 1320-1324 (1988).
5. L. Sturman and K.V. Holmes, The molecular biology of Coronaviruses, *Adv. Virus Res.* 28: 35-112 (1983).
6. L. Jöhr, Induktion und Charakterisierung ..., Thesis, Vet.-med. Faculty, University of Zurich (1989).
7. J. Pan and N. Auersperg, Protease treatment ..., *J. Immunol. Meth.* 141: 139-144 (1991).
8. F. Guscetti, M. Rosskopf, A. Pospischil, and M. Ackermann, Monoclonal antibodies for detection of PEDV ..., Manuscript in preparation.

INVOLVEMENT OF LIPIDS IN MEMBRANE BINDING OF MOUSE HEPATITIS VIRUS NUCLEOCAPSID PROTEIN

Fred Wong and Robert Anderson

Department of Microbiology and Infectious Diseases
University of Calgary
Calgary, Alberta T2N 4N1

ABSTRACT

Evidence is presented which indicates that membrane binding of the MHV nucleocapsid (N) protein is influenced by membrane lipid composition. Binding of N protein to membranes of mouse fibroblast L-2 cells is very specific and occurs under conditions in which no other viral or cellular proteins show detectable binding. Binding occurs rapidly and does not require the presence of divalent cations such as Ca^{++} or Mg^{++}. Purified phospholipid liposomes compete against N protein binding to membranes. Phospholipids consisting of cardiolipin are the most effective in inhibiting membrane binding. Because of certain structural similarities between phospholipids and nucleic acids, we speculate that membrane lipid association of the N protein may compete for RNA binding sites on the N protein. Such a mechanism may be important for processes such as nucleocapsid uncoating and nucleocapsid assembly.

INTRODUCTION

Shortly after its synthesis the nucleocapsid protein becomes associated with intracellular membranes (1). The biological significance of this event is uncertain although it may serve to ensure coalescence of the various subviral components which are known to assemble at a specific cell membrane site (2,3). There is evidence that post-translational phosphorylation of the N protein is necessary for membrane association to occur (1). The nature of this association is unknown but may involve interaction between N and the integral membrane protein M which is synthesized on ribosomes present on the endoplasmic reticulum (4).

Membrane association of viral macromolecular synthesis and assembly appears to be a hallmark of coronavirus replication. Viral structural proteins are either translated on membrane-bound polysomes or become quickly associated with membranes shortly after synthesis. Virion assembly also occurs via budding through intracytoplasmic membranes, specifically those of transitional elements between the endoplasmic reticulum and the Golgi apparatus (2,3). In addition to virion assembly and synthesis of viral proteins, viral RNA is also synthesized on cytoplasmic membranes, possibly distinct with regard to the synthesis of (+) and (-) RNA species (5,6).

We report here further studies on the membrane-binding properties of the MHV N protein. We provide evidence that membrane association of N protein may involve specific phospholipids, thus raising the intriguing possibility that N protein is binding to a host cell membrane lipid "receptor" which may then be required for some step in virus replication.

Coronaviruses, Edited by H. Laude and J.F. Vautherot
Plenum Press, New York, 1994

MATERIALS AND METHODS

Preparation of cell extracts

Monolayer cultures of mouse fibroblast L-2 cells (7) were mock-infected or inoculated with MHV (A59 strain) at a multiplicity of infection of five, adsorbed for 1h at 4° and then incubated at 37°. At 7h post-infection, cultures were radiolabeled for 1h with ^{35}S-methionine. The monolayers were washed with phosphate buffered saline (PBS) and extracted (15 min on ice) with 1% Triton X-100 in 10 mM tris, pH 7.4, 100 mM NaCl, 0.2 mM PMSF. Extracts were freed of detergent by overnight shaking with SM-2 Biobeads (BioRad) and clarified by centrifugation at 100,000xg. The resultant supernatants were used in binding assays.

Binding assays

Monolayer cultures of mouse fibroblast L-2, LM (8) and LM-K (9) cells were incubated for various times at 4° with radiolabeled cell extract supernatants prepared as above. Monolayers were then washed several times with PBS and the cells solubilized in dissociation buffer for analysis on SDS-PAGE and autoradiography (10).

Lipids and preparation of liposomes

Individual phospholipids, cardiolipin (CDL), phosphatidylcholine (PC), phosphatidylethanolamine (PE), phosphatidylserine (PS) and phosphatidylinositol (PI) were obtained from Sigma. Liposomes were prepared by sonication of phospholipids in PBS (11).

RESULTS

Cell membrane binding of MHV N protein

In addition to binding to intracellular membranes during its synthesis in MHV-infected cells (1), the nucleocapsid N protein was surprisingly found to bind to the outer surface of intact mouse fibroblast cells (Fig. 1). Extracts from ^{35}S-methionine-labeled mock- or MHV-infected L-2 cells were exposed to monolayer cultures of various sublines of mouse fibroblast cells for 1h at 4°. The cultures were then washed and solubilized for analysis on an SDS-polyacrylamide gel which was visualized by autoradiography. As shown in Fig. 1, all cells bound the N protein under conditions in which no other protein was observed to bind. This result suggested that the cellular component responsible for membrane binding of the N protein is present not only on intracellular membranes but also on the outer surface of the cell plasma membrane. Moreover, the results indicate that membrane-association of N protein can occur in the absence of other viral proteins, including M.

It should be mentioned that no binding of S protein was observed in the assay employed, presumably because the method of cell extract preparation disrupted higher order structure of the oligomeric form (12,13) which may be required for receptor binding.

Kinetics and ion requirements for membrane binding of N protein

Binding of N protein to L-2 cells occured rapidly as shown by a time course study in which an extract from ^{35}S-methionine-labeled MHV-infected L-2 cells was added to L-2 cell monolayers, adsorbed for various times at 4° and the monolayers then washed to remove unbound material. SDS-PAGE analysis of the cell-bound radiolabeled proteins showed the presence of N protein as early as 5 min, with increasing amounts observed over time (Fig. 2a). Binding of N protein to cell membranes did not require the presence of cations such as Mg^{++} or Ca^{++}, although binding was slightly enhanced in response to Ca^{++} concentrations of 5 or 10 mM (Fig. 2b).

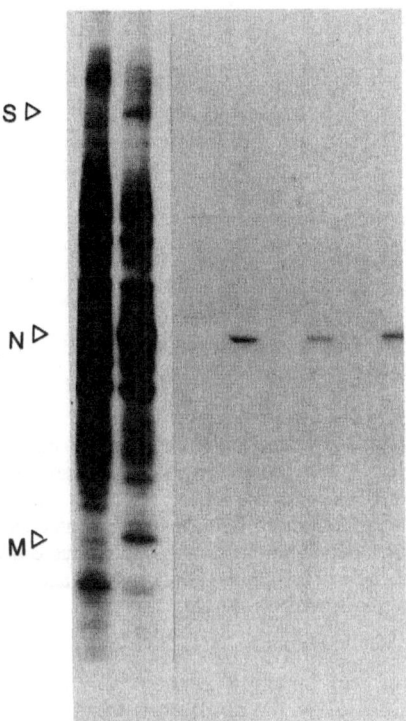

Figure 1

Autoradiographic SDS-PAGE analysis of binding of radiolabeled N protein to intact cells. Triton - X-100-solubilized extracts from ^{35}S-methionine labeled, mock- or MHV-infected L-2 cells were freed of detergent by treatment with SM-2 Biobeads (BioRad) and subsequently ultracentrifuged (1h at 100,000 x g) to yield supernatant extracts shown in lanes 1 and 2, respectively. Extracts were allowed to adsorb for 1h at 4° to monolayer cultures of mouse fibroblast cells: L-2, lanes 3 and 4, LM, lanes 5 and 6, and LM-K, lanes 7 and 8. Monolayers were washed to remove unbound material and then solubilized for analysis on SDS-PAGE. Lanes 3, 5 and 7 show cell-bound radiolabeled proteins from mock-infected L-2 cells, while lanes 4, 6 and 8 show cell-bound radiolabeled proteins from MHV infected L-2 cells.

Blocking of membrane binding of N protein by phospholipid liposomes

In examining the possible nature of the cell membrane component(s) responsible for binding of the N protein, we considered the involvement of both proteins and lipids. Treatment of L-2 cells with the proteases trypsin or proteinase K did not impair the ability of these cells to bind N protein, arguing against the possibility of a membrane protein "receptor" for N. In contrast, the addition of phospholipid liposomes to L-2 cells was found to block cell-binding of N protein, specifically when cardiolipin (CDL) was the phospholipid constituent (Fig. 3). The inhibition of cell-binding was highly dependent on the phospholipid composition of the liposomes, as phosphatidylcholine (PC), phosphatidylethanolamine (PE), phosphatidylserine (PS) and phosphatidylinositol (PI) were relatively ineffective in blocking binding of N to L-2 cells.

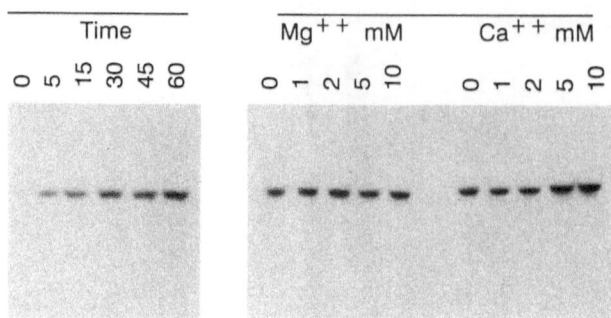

	a						b										

Kinetics and effects of divalent cations on membrane binding of N protein. A total extract of ^{35}S-methionine-labeled MHV-infected cells, prepared as described in the Materials and Methods, was added to monolayer cultures of L-2 cells and incubated for various times at 4° (A) or for 1h at 4° in the presence of various concentrations of MgCl$_2$ or CaCl$_2$ (B). Monolayers were washed and the cells solubilized for autoradiographic SDS-PAGE.

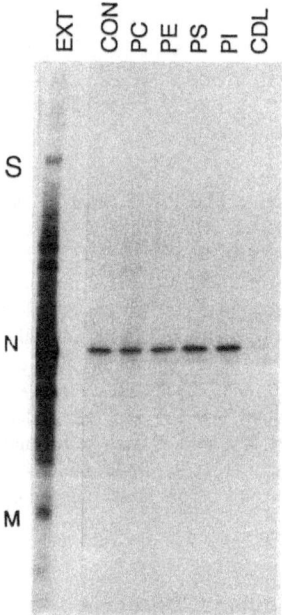

Figure 3

Blocking of membrane binding of N by phospholipid liposomes. An extract (EXT) of ^{35}S-methionine-labeled MHV-infected cells, prepared as described in the Materials and Methods, was added to monolayer cultures of L-2 cells in the absence (CON) or presence of phospholipid liposomes (PC, PE, PS, PI, CDL) at a final concentration of 250 μg/ml. After incubation for 1h at 4° the monolayers were washed and the cells solubilized for SDS-PAGE and autoradiography.

DISCUSSION

The nucleocapsid N protein of coronaviruses, including MHV, is a protein of approx MW 43-50k. Basic amino acids are localized in the amino-terminal two-thirds of the protein; in contrast, the carboxy terminal one-third is acidic. The N protein sequence contains a considerable proportion of serines (7-10%) some of which are phosphorylated. Phosphorylation occurs soon after biosynthesis of the N protein at which time N becomes associated with internal cell membranes (1). This observation has prompted speculation that phosphorylation is the trigger for membrane association of N. The biological significance of this event is uncertain although it may serve to ensure coalescence of the various subviral components which are known to assemble at a specific cell membrane site (2,3). Nucleocapsid N protein incorporated into virions is also phosphorylated; the recent finding of a cellular N protein-reactive phosphatase localized in the endosomal compartment may suggest a role for N protein dephosphorylation during the entry process (14). A potentially crucial observation is that cultured glial cells, induced to differentiate in vitro become resistant to MHV (strain JHM) and less capable of dephosphorylating the N protein (15).

MHV N protein binds RNA in a manner (depending on *in vitro* conditions) which suggests sequence specificity (16,17). In particular a short sequence present within the 5'-terminal leader region of both genomic and messenger RNAs has been reported to bind N protein (18). The finding of N protein bound to mRNAs is somewhat surprising, since mRNAs are not encapsidated during assembly. It has been speculated that binding of N may act as a translational regulator of viral mRNAs (19). The N protein is apparently required for viral RNA transcription, since anti-N monoclonal antibody has been shown to inhibit MHV RNA synthesis *in vitro* (20). Evidence that discreet regions of the N protein participate in RNA binding comes from varied sources. For example, a 40k processing product of the N protein has been reported which no longer binds RNA (21). Furthermore, carboxy-truncated forms of the N protein show much reduced RNA-binding ability (22).

The findings from the present study suggest that, in addition to RNA, the MHV nucleocapsid protein has the capacity to bind membrane phospholipids. Such dual affinity might conceivably have a role in facilitating either assembly or disassembly, should RNA and membrane phospholipid compete for the same binding sites on N. Such considerations are perhaps similar to those described for the bacterial dnaA (23) and recA (24) proteins which also bind both nucleic acid and phospholipid. Interestingly, recA and dnaA proteins show strong affinity for cardiolipin (23,24), the phospholipid which was found to be the most effective in blocking membrane association of the MHV N protein.

Although cardiolipin is a major component of bacteria, it is a relatively minor component of animal cells in which it is mainly associated with mitochondria (25). We suggest that the MHV nucleocapsid protein may associate either with non-mitochondrial cardiolipin or a phospholipid with similar structural characteristics. A possible candidate is bis(monoacylglycero)phosphate a lysosomal phospholipid, the synthesis of which is enhanced in certain virus infections, eg. vaccinia (26,27) and mengo (28). Studies are in progress to identify possible alterations in cellular lipid metabolism upon infection with MHV.

REFERENCES

1. Stohlman, S.A., Fleming, J.O., Patton, C.D., and Lai, M.M.C. (1983) *Virology* **130**, 527-532.
2. Rottier, P.J.M. and Rose, J.K. (1987) *J. Virol.* **61**, 2042-2045.
3. Tooze, S.A., Tooze, J., and Warren, G. (1988) *J. Cell Biol.* **106**, 1475-1487.
4. Niemann, H., Boschek, B., Evans, D., Rosing, M., Tamura, T., and Klenk, H.D. (1982) *EMBO J.* **1**, 1499-1504.
5. Brayton, P.R., Lai, M.M.C., Patton, C.D.and, and Stohlman, S.A. (1982) *J. Virol.* **42**, 847-853.
6. Brayton, P.R., Stohlman, S.A.and, and Lai, M.M.C. (1984) *Virology* **133**, 197-201.

7.   Rothfels, K.H., Axelrad, A.A., Siminovitch, L., McCulloch, E.A., and Parker, R.C. (1959) *Canad. Cancer Conf.* 3, 189-214.
8.   Merchant, D.J. and Hellman, K.B. (1962) *Proc. Soc. Exp. Biol. Med.* 110, 194-198.
9.   Kit, S., Dubbs, D.R., Piekarski, L.J., and Hsu, T.C. (1963) *Exp. Cell Res.* 31, 297-312.
10.   Laemmli, U.K. (1970) *Nature* 227, 680-685.
11.   Reeves, J.P. and Dowben, R.M. (1969) *J. Cell. Physiol.* 73, 49-60.
12.   Delmas, B. and Laude, H. (1990) *J. Virol.* 64, 5367-5375.
13.   Vennema, H., Rottier, P.J.M., Heijnen, L., Godeke, G.J., Horzinek, M.C., and Spaan, W.J.M. (1990) *Adv. Exp. Med. Biol.* 276, 9-19.
14.   Mohandas, D.V. and Dales, S. (1991) *FEBS Lett.* 282, 419-424.
15.   Beushausen, S., Narindrasorasak, S., Sanwal, B.D.and, and Dales, S. (1987) *J. Virol.* 61, 3795-3803.
16.   Robbins, S.G., Frana, M.F., McGowan, J.J., Boyle, J.F., and Holmes, K.V. (1986) *Virology* 150, 402-410.
17.   Stohlman, S.A., Baric, R.S., Nelson, G.N., Soe, L.H., Welter, L.M., and Deans, R.J. (1988) *J. Virol.* 62, 4288-4295.
18.   Baric, R.S., Nelson, G.W., Fleming, J.O., Deans, R.J., Keck, J.G., Casteel, N., and Stohlman, S.A. (1988) *J. Virol.* 62, 4280-4287.
19.   Masters, P.S. and Sturman, L.S. (1990) *Adv. Exp. Med. Biol.* 276, 235-238.
20.   Compton, S.R., Rogers, D.B., Holmes, K.V., Fertsch, D., Remenick, J., and McGowan, J.J. (1987) *J. Virol.* 61, 1814-1820.
21.   Beushausen, S. and Dales, S. (1987) *Adv. Exp. Med. Biol.* 218, 239-253.
22.   Masters, P.S., Parker, M.M., Ricard, C.S., Duchala, C., Frana, M.F., Holmes, K.V., and Sturman, L.S. (1990) *Adv. Exp. Med. Biol.* 276, 239-246.
23.   Sekimizu, K. and Kornberg, A. (1988) *J. Biol. Chem.* 263, 7131-7135.
24.   Krishna, P. and van de Sande, J.H. (1990) *J. Bacteriol.* 172, 6452-6458.
25.   White, D.A. (1973) in *BBA Library 3. Form and Function of Phospholipids* (pp. 441-482, Elsevier, New York.
26.   Hiller, G., Eibl, H., and Weber, K. (1981) *Virology* 113, 761-764.
27.   Stern, W. and Dales, S. (1974) *Virology* 62, 293-306.
28.   Schimmel, H. and Traub, P. (1987) *Lipids* 22, 95-103.

A NOVEL GLYCOPROTEIN OF FELINE INFECTIOUS PERITONITIS CORONAVIRUS CONTAINS A KDEL-LIKE ENDOPLASMIC RETICULUM RETENTION SIGNAL

H. Vennema[1], L. Heijnen[2], P.J.M. Rottier[1], M.C. Horzinek[1] and W.J.M. Spaan[2]

[1]Department of Virology, Faculty of Veterinary Medicine, University of Utrecht, Yalelaan 1, P.O. Box 80.165, 3508 TD Utrecht, The Netherlands
[2]Department of Virology, Faculty of Medicine, University of Leiden, The Netherlands

ABSTRACT

A new protein of the feline infectious peritonitis virus (FIPV) was discovered in lysates of infected cells. Expression of the gene encoding open reading frame (ORF) 6b of FIPV in recombinant vaccinia virus infected cells was used to identify it as the 6b protein. It is a novel type of viral glycoprotein whose function is not clear. It is a soluble protein contained in microsomes; its slow export from the cell is caused by the presence of an ER-retention signal at the C-terminus. This amino acid sequence, KTEL, closely resembles the consensus KDEL-signal of soluble resident ER proteins. A mutant 6b protein with the C-terminal sequence KTEV became resistant to digestion by endo-ß-N-acetylglucosaminidase H with a half-time that was reduced threefold. In contrast, a mutant with the sequence KDEL was completely retained in the ER. The FIPV 6b protein is the first example of a viral protein with a functional KDEL-like ER-retention signal.

INTRODUCTION

The feline infectious peritonitis virus (FIPV) and transmissible gastroenteritis virus (TGEV) of swine are genetically closely related. However, FIPV contains a complete open reading frame (ORF-2) in the 3' region of the genome, which is lacking in TGEV (1). ORF-2 is located 3' from ORF-1 of mRNA 6 and we suggested that it is produced from a bicistronic mRNA. According to the recently proposed nomenclature (2) the first and second ORF of mRNA 6 are now called ORF 6a and ORF 6b, respectively. The FIPV ORF 6b potentially encodes a polypeptide of 24-kilodalton (kDa), it contains a short N-terminal hydrophobic region (1), which may function as a signal sequence, and one consensus N-glycosylation site. In this report we show that the protein encoded by ORF 6b is produced in FIPV-infected cells and that it is a novel type of viral glycoprotein.

MATERIALS AND METHODS

Cells and viruses

FIPV strain 79-1146 (3) was grown in Crandell feline kidney (CrFK) cells. For vaccinia virus infections, HeLa cells were used. Cells were maintained in Dulbecco modified Eagle medium (GIBCO Laboratories) containing 5% fetal bovine serum.

Radio immunoprecipitation assays (RIPA)

FIPV or vaccinia virus infected cells were labeled with L-[^{35}S]methionine or L-[^{35}S]cysteine (>1,000 Ci/mmol, Amersham Corp.). Lysis, radio immunoprecipitation assay (RIPA) and endo-ß-N-acetylglucosaminidase H (endo H, Boehringer Mannheim Biochemicals) treatment were carried out as described (4). Analysis by sodium dodecyl sulphate polyacrylamide gel electrophoresis (SDS-PAGE) was as described (5).

Cloning and expression of ORF 6b

A cDNA fragment extending from a *Spe*I-site, located 84 base pairs upstream of the initiation codon of ORF 6b to a *Sal*I-site in the polylinker of cDNA clone E7 (1) was recloned in a bacteriophage T7 expression vector. The resulting construct, was designated pTF6b.

RESULTS

Identification of the FIPV 6b protein

ORF 6b of FIPV contains one methionine codon and seven cysteine codons. The electrophoretic patterns of immunoprecipitated lysates from [^{35}S]methionine and [^{35}S]cysteine labeled FIPV-infected cells were compared to identify ORF 6b derived polypeptides. After [^{35}S]methionine labeling the FIPV membrane protein (M) appeared as a strong band of 29 kDa and a minor band of 26 kDa (Fig. 1), corresponding to the glycosylated and the unglycosylated forms of the M protein, respectively (4). In [^{35}S]cysteine labeled immunoprecipitates a double band was observed of 26 to 26.5 kDa, at the position of the unglycosylated M protein band. When the samples were digested with endo H, which cleaves asparagine-linked high mannose oligosaccharide side chains, a 24 kDa protein was detected exclusively after [^{35}S]cysteine labeling (Fig. 1). To test the hypothesis that ORF 6b encodes this glycoprotein, the 6b gene was recloned into a bacteriophage T7 expression vector. The resulting construct, designated pTF6b, was used in the transient T7 expression system with recombinant vaccinia virus vTF7-3 producing the T7 RNA polymerase (6). The protein detected after immunoprecipitation comigrated with the 26.5 kDa protein from FIPV-infected cells (Fig. 1). Digestion with endo H yielded a protein which comigrated with the 24 kDa protein mentioned above.

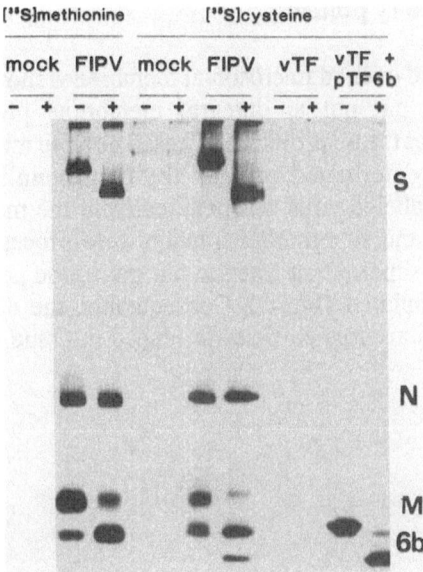

Figure 1. Identification of the 6b protein. RIPA and SDS-PAGE analysis of lysates from infected cells. Immunoprecipitates were split and one half was treated with endo H, the other serving as a control (indicated by + and -, respectively). FIPV- and mock-infected cells were labeled at 5 h p.i. with [^{35}S]methionine or [^{35}S]cysteine. Recombinant vaccinia virus vTF7-3-infected and transfected cells (lanes marked vTF and vTF + pTF6b, respectively) were labeled with [^{35}S]cysteine.

To prove that the recombinant 6b protein and the 26 to 26.5 kDa protein in FIPV-infected cells are the same, we set up a competition RIPA. Unlabeled recombinant 6b protein producing cells was added to lysate from [^{35}S]cysteine labeled FIPV-infected cells. The samples were then immunoprecipitated. This resulted in a decreased intensity of the 26 to 26.5 kDa doublet, but not of the FIPV M protein band (Fig. 2A). The same was done with recombinant M protein. This resulted in a decrease of M protein band intensity (Fig. 2B).

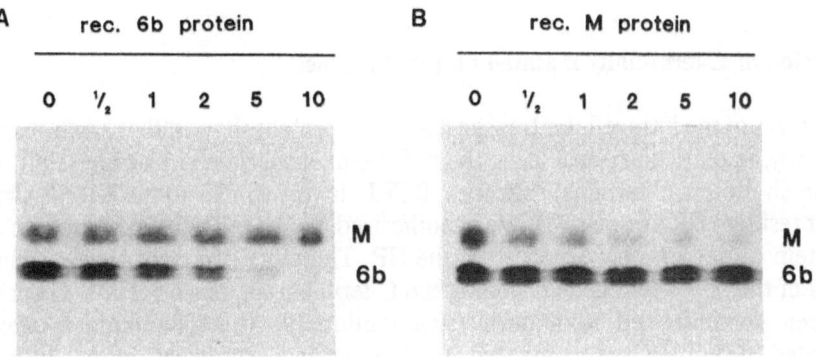

Figure 2. Competition of immunoprecipitation with recombinant 6b and M proteins. Lysate from recombinant 6b (panel A) or M (panel B) protein producing cells were added to samples of [^{35}S]cysteine labeled lysate of FIPV-infected cells and processed for RIPA. The relative amount of unlabeled cell lysate is indicated above each lane.

The 6b protein is a secretory protein

Isolation of closed and opened microsomal membranes showed that the 6b protein is a membrane protein but not an integral membrane protein (7). Preliminary observations indicated that the 6b protein is released into the medium of FIPV-infected cells. It remained to be determined whether the 6b protein is a structural protein. Therefore, [³⁵S]cysteine labeled virus was pelleted from the medium of infected cells by centrifugation; pellet and supernatant fraction were processed for RIPA. The 6b protein was found in the supernatant fraction but not in the pellet fraction where the structural proteins accumulated (Fig. 3). Consequently, the 6b protein is not stably associated with extracellular virus particles produced in tissue culture.

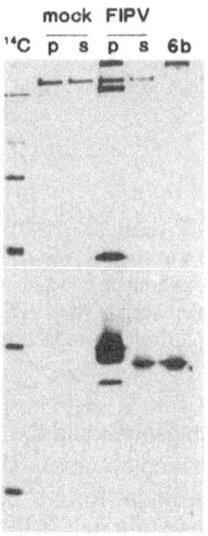

Figure 3. Virus purification. Radioactively labeled virus was purified by centrifugation. Pellet and supernatant (marked p and s, respectively) of mock- and FIPV-infected cells were processed for RIPA. Secreted recombinant 6b protein and ¹⁴C-labeled marker proteins were run in parallel for comparison (lanes marked 6b and ¹⁴C, respectively).

Expression of C-terminally mutated 6b protein genes

Analysis of the intracellular transport of the 6b protein showed that it is transported very slowly, in particular when using the recombinant vaccinia virus expression system (7). The 6b protein C-terminal sequence, KTEL, is very similar to the KDEL-signal of cellular resident ER proteins (8). We hypothesized that the KDEL-like sequence of the 6b protein caused its slow export from the ER. Therefore, we constructed a mutated version of the 6b protein gene encoding the C-terminal sequence KTEV. This change has been demonstrated to abolish ER-retention (9, 10). The mutant construct, designated pTF6bV, was compared to the original construct in a pulse-chase experiment (Fig. 4). The 6bV protein was secreted into the medium faster than the wild-type expression product. It became endo H resistant with a half-time that was reduced threefold. Next, we changed the C-terminal sequence to KDEL, to see if this would confer complete ER retention. The 6bD protein was analyzed in a pulse-chase experiment (Fig 4C). It was completely retained in the ER during a 3 h chase period; it could not be detected in the medium and remained entirely endo H sensitive. These

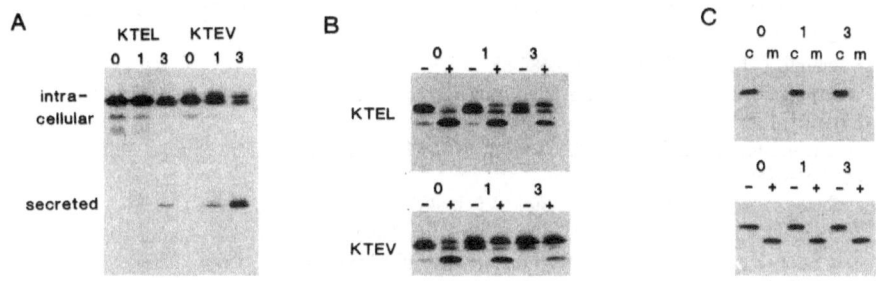

Figure 4. Pulse-chase analysis of wild-type and mutated 6b proteins. Cells infected with vTF7-3 were transfected with pTF6b or pTF6bV (indicated by KTEL and KTEV, respectively), pulse-labeled for 1 h and analyzed immediately or after chase times of 1 and 3 h, as indicated. Analysis was performed as described in the legends to Fig. 1. (A.) Intracellular and secreted polypeptides were analyzed separately. (B.) Lysates including the culture media were analyzed by endo H digestion. (C.) Similar experiments were performed with pTF6bD. Cells and medium were processed separately (indicated by c and m, respectively). Intracellular material was subsequently analyzed after treatment with endo H or mock treatment (indicated by + and -, respectively).

experiments demonstrate that the original C-terminal sequence of the FIPV 6b protein confers partial ER-retention by the same mechanism as used for cellular resident ER proteins.

DISCUSSION

In this paper we report the identification and characterization of a new virus specific protein in FIPV-infected cells. A glycoprotein of 26.5 kDa was detected in [^{35}S]cysteine but not in [^{35}S]methionine labeled cell lysates, which explains why it had not been observed before (4, 11). The protein was identified by recombinant gene expression of the second ORF of mRNA 6 (7), currently designated ORF 6b. Further studies showed that the 6b protein is a soluble protein contained in microsomes, that it is secreted from infected cells, and that it is not stably associated with virus particles in tissue culture medium.

Detailed analysis of its intracellular transport showed that the 6b protein was released slowly from the ER, particularly in recombinant vaccinia virus infected cells. The 6b protein C-terminal sequence, KTEL, is almost identical to the KDEL-signal of resident ER proteins. This signal or a closely related sequence at the C-terminus of a protein with an N-terminal signal sequence are strong indications that it is a resident lumenal ER-protein (8). Proteins with a KDEL-signal are recognized by a receptor located in a compartment between the ER and the Golgi apparatus and recycled back to the ER. Site-directed mutagenesis of the signal to KTEV abolished retention. The wild-type 6b protein was not completely retained in the ER, while mutagenesis of its C-terminus to KDEL resulted in complete ER retention.

The FIPV 6b protein is the first example of a viral protein with a functional KDEL-like ER-retention signal.

REFERENCES

1. Groot, R.J. de, A.C. Andeweg, M.C. Horzinek, and W.J.M. Spaan. (1988) *Virology* 167:370-376.
2. Cavanagh, D., D.A. Brian, L. Enjuanes, K.V. Holmes, M.M.C. Lai, H. Laude, S.G. Siddell, W. Spaan, Taguchi, and P.J. Talbot. (1990) *Virology* 176:306-307.
3. McKeirnan, A.J., J.F. Evermann, A. Hargis, L.M. Miller, and R.L. Ott. (1981) *Feline Pract.* 11:16-20.

4. Vennema, H., L. Heijnen, A. Zijderveld, M.C. Horzinek, and W.J.M. Spaan. (1990) *J. Virol.* 64:339-346.
5. Laemmli, U.K. (1970) *Nature* 227:680-685.
6. Fuerst, T.R., E.G. Niles, F.W. Studier, and B. Moss. (1986) *Proc. Natl. Acad. Sci. USA* 83:8122-8126.
7. Vennema, H., L. Heijnen, P.J.M. Rottier, M.C. Horzinek, and W.J.M. Spaan. (1992) *J. Virol.* 66:4951-4956.
8. Pelham, H.R.B. (1990) *Trends in Biol. Sci.* 15:483-486.
9. Andres, D.A., J.D. Rhodes, R.L. Meisel, and J.E. Dixon. (1991) *J. Biol. Chem.* 266: 14277-14282.
10. Zagouras, P., and J.K. Rose. (1989) *J. Cell Biol.* 109:2633-2640.
11. Groot, R.J. de, R.J. ter Haar, M.C. Horzinek, and B.A.M. van der Zeijst. (1987) *J. Gen. Virol.* 68:995-1002.

ALTERED PROTEOLYTIC PROCESSING OF THE POLYMERASE POLYPROTEIN IN RNA(-) TEMPERATURE SENSITIVE MUTANTS OF MURINE CORONAVIRUS

Susan C. Baker,[1] HongQiang Gao,[1] and Ralph S. Baric[2]

[1]Department of Microbiology and Immunology
  Loyola University Medical Center, Maywood, IL 60153
[2]Department of Epidemiology, University of North Carolina
  at Chapel Hill, Chapel Hill, NC 27599

ABSTRACT

We examined the synthesis and processing of the polymerase polyprotein in RNA(-) temperature sensitive mutant of murine coronavirus strain A59. These temperature sensitive mutants of MHV-A59 synthesize viral RNA at the permissive temperature (33.0°C), but are unable to synthesize viral RNA at the nonpermissive temperature (39.5°C). The ts mutants have been mapped to five different complementation groups in the polymerase gene. The 5'-most complementation groups, Groups A and B, map to a region encoding an autoproteinase responsible for the cleavage of p28, the amino-terminal product of the polymerase polyprotein. We screened six temperature sensitive mutants to determine if there was an alteration in the proteolytic processing of the polymerase polyprotein, particularly in the cleavage of the p28 protein. Two mutants, tsNC9 and tsLA16, had altered proteolytic products at both the permissive and nonpermissive temperatures. One Group B temperature sensitive mutant, designated tsNC11, was defective in the production of p28 protein at the nonpermissive temperature. To further localize the site of the mutation in tsNC11, RNA representing the 5'-most 5.3 kb region of the polymerase gene was transfected into tsNC11-infected cells and virus production monitored. The transfected RNA was able to complement the defect in tsNC11, resulting in viral RNA synthesis and production of viral particles at the nonpermissive temperature. These results indicate that a gene product from the 5.3 kb region of gene 1 is required for coronavirus RNA synthesis.

INTRODUCTION

The unique transcription and recombination mechanisms of coronaviruses implicate a polymerase with interesting properties. The polymerase may be able to disassociate and re-associate with template RNA to generate recombinant RNA molecules and seems to be able to identify relatively small intergenic sequences interspersed in the 31 kb RNA genome and initiate leader primed transcription (1). In addition, functional domains such as a helicase, polymerase, zinc-finger motif and proteinases have been predicted from the primary sequence of the polymerase gene (2). Our understanding of the complex functions of the MHV polymerase gene have been hampered because of the large size of the gene (22 kb) and its gene product (800 kDa) (2, 3). It has been previously shown that the gene 1 protein product is a polyprotein which is cleaved to yield subunits (4-7), but little is known about the

Coronaviruses, Edited by H. Laude and J.F. Vautherot
Plenum Press, New York, 1994

functional domains of either the polyprotein or the subunits. We have initiated a genetic approach to investigate the functional domains of the polymerase gene. Temperature sensitive mutants of MHV-A59 which are defective for RNA synthesis at the nonpermissive temperature and have been mapped to the 5'-end of the polymerase gene (8, 9) were analyzed for alteration in expression of the 5'-most proteolytic product, p28. To further localize the lesion in the ts mutant, RNA encoding the 5'-most open reading frame was transfected into ts virus infected cells and analyzed its ability to complement the ts defect. This approach allowed us to more precisely localize a mutation in one temperature sensitive mutant of coronavirus, tsNC11, and provides evidence that p28 may function in coronavirus RNA synthesis.

MATERIALS AND METHODS

Viruses and Cells. Temperature sensitive mutants of MHV-A59 designated tsNC8, tsNC9, tsNC11, tsNC13, tsLA6 and tsLA16 were derived by chemical mutagenesis and mapped to gene 1 by complementation and recombination analysis as previously described (8, 9). Virus isolates were plaque purified three times and stocks were prepared in DBT cells. Plaque assays were performed on monolayers of DBT cells in MEM containing 3% FCS and 0.4% Noble agar. Plates were incubated at 33.0ºC or 39.5ºC for 48 hr, and plaques visualized with neutral red.

Labeling of infected cells and immunoprecipitation. Duplicate monolayers of DBT cells were infected at a m.o.i. of 5 PFU/cell for 1 hr at 32.0ºC. At 1.0 hr postinfection the innoculum was replaced with MEM containing 3% FCS and Actinomycin D (2 ug/ml). At 5.5 hr postinfection, medium was replaced with methionine-free MEM and half the cultures were shifted to 39.5ºC. At 6.25 hr postinfection monolayers were labeled for 45 min with medium containing 200 uCi/ml (^{35}S)-trans-label (ICN), after which the cells were lysed and prepared for immunoprecipitation. Immunoprecipitations were carried out using an antipeptide antiserum specific for p28 (5). Products were analyzed on 10% SDS-polyacrylamide gels followed by autoradiography.

in vitro transcription and transfection of RNA. Plasmid pT7-NBgl consists of MHV-JHM gene 1 sequence from nucleotide 187 to nucleotide 5273 inserted downstream from the T7 polymerase promoter (5). Plasmid DNA was linearized by digestion with BamHI (nucleotide 467), Xho I (nucleotide 1129) or EcoRI (which cuts in the polylinker following the MHV-JHM sequences) and capped RNA was synthesized in vitro with T7 RNA polymerase as previously described (5). The resulting RNA was transfected into tsNC11 infected DBT cells using lipofectin (Gibco/BRL) following the procedure of Makino (10).

RESULTS AND DISCUSSION

As an initial approach to investigate the functions of the coronavirus polymerase gene products, we screened temperature sensitive mutants of MHV-A59 for alterations in expression or processing of the 5'-most polyprotein cleavage product, p28. These temperature sensitive mutants of MHV-A59 are unable to synthesize viral RNA at the nonpermissive temperature (39.5ºC). A large panel of ts mutants were mapped by complementation and recombination studies into 5 groups (8, 9). Group A and B mutants map to the 5'-most 10 kb of gene, suggesting that the ts defect may reside in the p28 protein (encoded in the first 1 kb of the RNA genome), or in the proteinase domain (encoded in the 3.5 to 4.5 kb region of the RNA genome). The temperature sensitive mutants were screened by labeling duplicate dishes of infected cells with (^{35}-S)-methionine at the permissive and nonpermissive temperatures, followed by immunoprecipitation of the cell lysate with antisera specific for p28 (5). Two ts mutants, tsNC9 and tsLA16, displayed an additional, higher molecular weight band that was co-precipitated with p28 at both the permissive (32.0ºC) and nonpermissive (39.5ºC) temperature (Figure 1A). A third ts mutant, tsNC11, produced p28 at levels comparable to wild type MHV-A59 at the permissive temperature. However, little or no p28 was detected in tsNC11 infected cell lysates labeled at the nonpermissive temperature (Fig. 1B). This reduction in the level of p28 in the tsNC11 infected cells at the

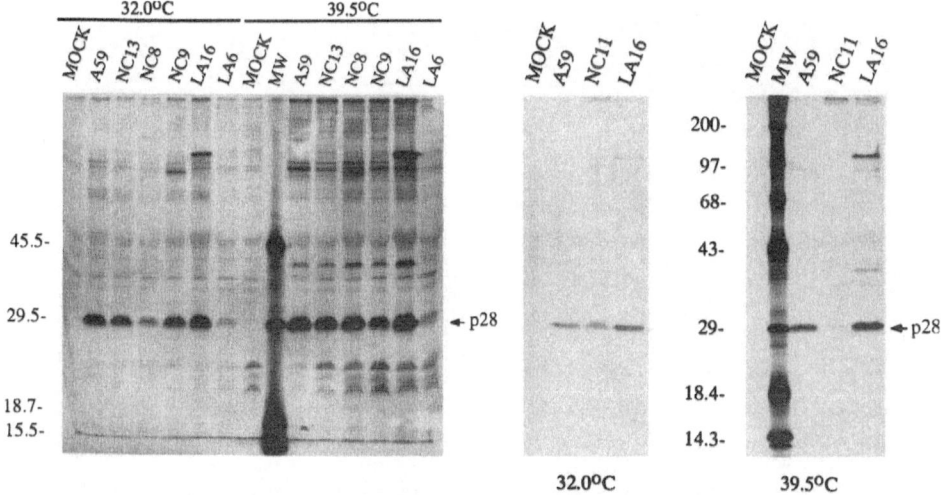

Figure 1. Immunoprecipitation of p28 from MHV-A59 and temperature sensitive mutant infected DBT cells labeled with (^{35}-S)-methionine at the permissive and nonpermissive temperature.

nonpermissive temperature suggested that tsNC11 may have a defect in either the autoproteinase domain responsible for cleavage of p28, the p28 cleavage site or in the p28 coding region.

To further test whether the loss of the p28 protein product was responsible for the ts phenotype of tsNC11, we established a protocol to provide translatable RNAs in trans to determine if the gene products produced from these RNAs would act in trans to complement the ts phenotype. The RNAs used for these experiments were synthesized from the pT7-NBgl plasmid DNA (5). As shown in Figure 2, this plasmid DNA contains the 5' open reading frame of MHV strain JHM downstream of a T7 RNA polymerase promoter. Plasmid DNA was linearized by digestion with Eco RI and capped RNA synthesized via T7 RNA polymerase. The capped RNA was then transfected in tsNC11 infected DBT cells using lipofectin (10). At four hours after infection and transfection, the cells were shifted to the nonpermissive temperature and incubated for an additional 16 hours. Supernatants were harvested from the infected/transfected cultures and plaque assays were performed at both the permissive and nonpermissive temperatures.

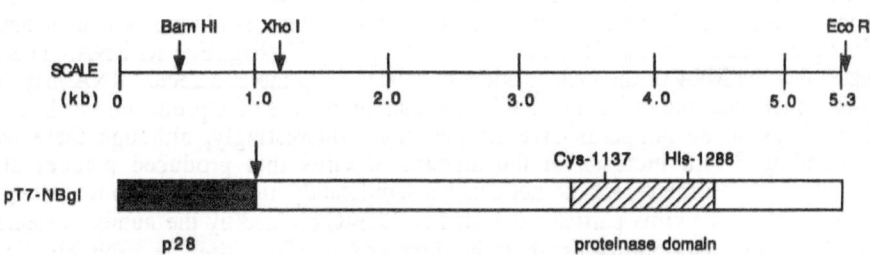

Figure 2. Schematic diagram of pT7-NBgl plasmid DNA. The open reading frame encodes the p28 protein and the proteinase domain responsible for autoproteolytic cleavage of p28. Restriction enzyme sites used to linearize plasmid DNA before in vitro transcription with T7 RNA polymerase are indicated.

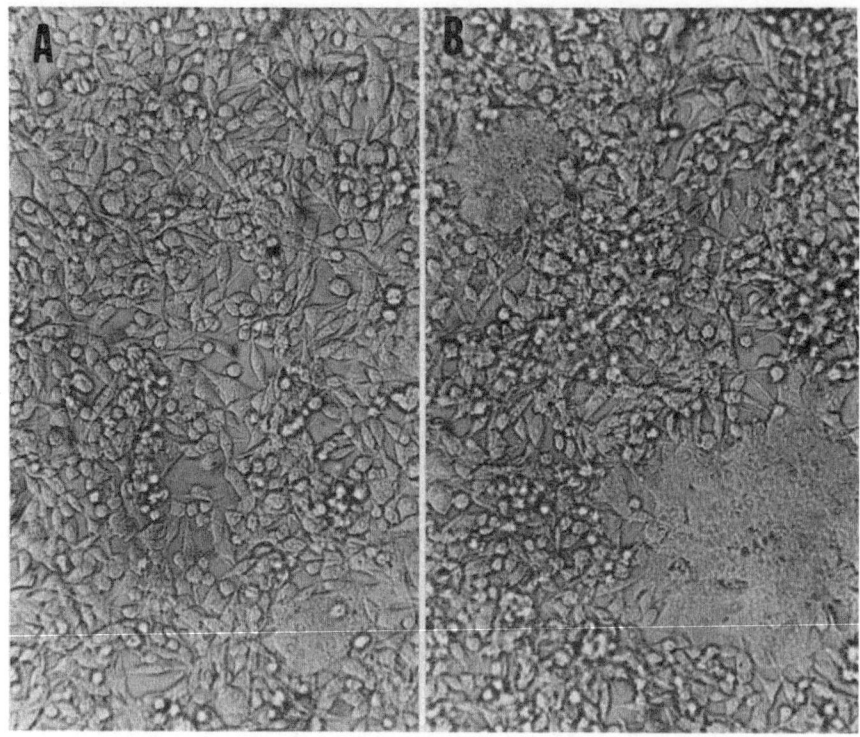

Figure 3. Fusion of tsNC11 infected cells incubated at 39.5°C following mock transfection (A) or transfection with 10 ug RNA transcribed from pT7-NBgl/EcoRI DNA (B).

The first suggestion that transfected RNA may be able to complement the defect in tsNC11 was from a simple observation of the amount of virus induced fusion detected in tsNC11-infected/pT7-NBgl RNA transfected cultures compared to mock transfected virus- infected cells (Fig. 3). The number and extent of fusion seen in mock transfected, tsNC11-infected cells that had been shifted to the nonpermissive temperature at 4 hours after infection, was very low (Fig. 3A). In contrast, cells which had been infected with tsNC11 and transfected with 10 ug of capped RNA synthesized from pT7-NBgl/Eco RI DNA showed multiple, large syncytia characteristic a productive MHV infection (Fig. 3B).

To determine if the amount of syncytia formation correlated with the amount of virus produced from these cultures, plaque assays were performed on the supernatants collected from the cultures. As shown in Figure 4, there was greater than 3-fold increase in the amount of virus produced following transfection with 10 ug of pT7-NBgl/Eco RI RNA. This data indicates that the RNA or the gene product translated from the transfected RNA may act in trans to complement the defect in tsNC11 and thereby increase the production of infectious viral particles at the nonpermissive temperature. Interestingly, although there was a corresponding 3-fold increase in the amount of virus that produced plaques at the nonpermissive temperature, we did not detect a significantly higher reversion frequency (the number of infectious virus particles detected at 39.5°C divided by the number detected at 33.0°C). This data indicates that the increase in virus titer is probably due to complementation rather than a recombination event which would repair the temperature sensitive defect in the original tsNC11 genome., and therefore produce large amounts of virus which are no longer temperature sensitive.

	33.0°C	39.5°C	39.5°C/33.0°C
lipofectin alone	8.3×10^3	4.8×10^1	5.8×10^{-3}
1.0 ug RNA	7.0×10^3	9.5×10^1	14.0×10^{-3}
5.0 ug RNA	15.5×10^3	13.8×10^1	8.9×10^{-3}
10.0 ug RNA	28.8×10^3	12.0×10^1	4.2×10^{-3}

Figure 4. Determination of infectious particles released into the supernatant from tsNC11-infected/RNA transfected DBT cells. Virus titer is expressed in PFU/ml.

In summary, we have developed a transfection assay system which allows us to compliment the defect in a MHV RNA(-) temperature sensitive mutant, tsNC11. The data indicate that a gene product from the first 5.3 kb of gene 1 is sufficient to compliment the defect in RNA synthesis. We plan to use this assay to map the location of the ts lesion. In addition, we plan to exploit the high rate of recombination of coronaviruses to recover "repaired" recombinant viruses (11, 12), and thereby definitively show that the putative ts lesion is responsible for the particular ts phenotype of the virus. These experiments will provide information on the role of individual protein subunits in coronavirus RNA synthesis and for the mechanism of recombination of coronaviruses.

ACKNOWLEDGMENTS

We thank Jeanette Paintsil for assistance with the transfection experiments. This work was supported by Public Health Service research grant AI32065 from the National Institutes of Health (to S.C.B.) and JFRA-371 from the American Cancer Society (to S.C.B.).

REFERENCES

1. Lai, M.M.C. 1990. Annu. Rev. Microbiol. 44: 303-333.
2. Lee, H.J., C.-K. Shieh, A.E. Gorbalenya, E.V. Koonon, N. La Monica, J. Tuler, A. Bagdzhadzhyan, and M.M.C. Lai. 1991. Virology 180: 567-582.
3. Pachuk, C.J., P.J. Bredenbeck, P.W. Zoltick, W.J.M. Spaan, and S.R. Weiss. 1989. Virology 171: 141-148.
4. Soe, L.H., C.-H. Shieh, S.C. Baker, M.-F. Chang, and M.M.C. Lai. 1987. J. Virol. 61: 3968-3976.
5. Baker, S.C., C.-K. Shieh, L.H. Soe, M.-F. Chang, D. M. Vannier, and M.M.C. Lai. J. Virol. 63: 3693-3699.
6. Denison, M.R., and S. Perlman. 1986. J. Virol. 60: 12-18.
7. Denison, M.R., P.W. Zoltick, S.A. Hughes, B. Giangreco, A.L. Olson, S. Perlman, J.L. Leibowitz, and S.R. Weiss. 1992. Virology 189: 274-284.
8. Baric, R.S., K. Fu, M.C. Schaad, and S.A. Stohlman. 1990. Virology 177: 646-656.
9. Schaad, M.C., S.A. Stohlman, J. Egbert, K. Lum, K. Fu, T. Wei, and R.S. Baric. 1990. Virology 177: 634-645.
10. Makino, S. M. Joo and J.K. Makino. 1991. J. Virol. 65: 6031-6041.
11. van der Most, R. G., L. Heijnen, W. J. M. Spaan and R. J. de Groot. 1992. Nuc. Acids Res. 20: 3375-3381.
12. Liao, C.-L. and M. M. C. Lai. 1992. J. Virol. 66: 6117-6124.

A NEWLY IDENTIFIED MHV-A59 ORF1a POLYPEPTIDE p65 IS TEMPERATURE SENSITIVE IN TWO RNA NEGATIVE MUTANTS

Scott A. Hughes[1], Mark R. Denison[2], Pedro Bonilla[1], Julian L. Leibowitz[3], Ralph S. Baric[4], and Susan R. Weiss[1]

[1]University of Pennsylvania School of Medicine, Phila., PA., [2]Vanderbilt University Medical Center, Nashville, TN., [3]University of Texas, Houston, TX., and [4]UNC School of Public Health, Chapel Hill, NC

ABSTRACT

Polypeptide products of MHV-A59 gene 1 have been identified in infected DBT cells and in the products of *in vitro* translations of genome RNA (1,2). In this paper we report the identification in infected cell lysates of a 65-kDa polypeptide (p65) encoded in ORF 1a. Studies on the kinetics of appearance and processing of p65 show that p65 is detectable after p28 but before the appearance of p290, p240 and p50. No homologue of the p65 polypeptide identified in infected cell lysates was immunoprecipitated from *in vitro* translations of genomic RNA, providing further evidence that *in vitro* processing of polypeptides encoded in ORF 1a of gene 1 differs from that which occurs late in infection of DBT cells. Although the function of p65 is unknown, two MHV-A59 ts mutants isolated and characterized by Baric et al. (3,4) do not produce detectable levels of p65 at the non-permissive temperature indicating that p65 may play an important role in the virus life cycle.

INTRODUCTION

The coronavirus mouse hepatitis virus, strain A59 (MHV-A59) contains a single stranded positive-sense RNA of approximately 32-kb (5). The genome of MHV contains 7 genes, each of which encode one or more viral proteins (6,7,8). Because MHV virions do not contain RNA polymerase (9), this enzyme must be synthesized from the incoming viral genome RNA. Based on sequence analysis (10) and RNA recombination studies using temperature-sensitive (ts) mutants (3), it is believed that gene 1 may encode proteins which are either directly or indirectly involved in viral RNA synthesis. Gene 1 contains two open reading frames, ORF 1a and ORF 1b which overlap by 75 nucleotides. ORF 1b is translated in the -1 frame with respect to ORF 1a and is translated via a ribosomal frameshifting mechanism (11). Together, ORF 1a and ORF 1b have the capacity to encode an as yet undetected polyprotein of greater than 750kDa. Amino acid sequence analysis of gene 1 of MHV-JHM and MHV-A59 suggests that ORF 1a contains a picornaviral 3C-like protease domain and two papain like protease domains, and ORF 1b contains polymerase, helicase and zinc finger motifs (6,7,8).

Identification of gene 1 products and characterization of subsequent processing events should provide a better understanding of viral protease and polymerase activities as well as coronavirus replication in general.

Coronaviruses, Edited by H. Laude and J.F. Vautherot
Plenum Press, New York, 1994

MATERIALS AND METHODS

Virus and Cells

MHV-A59 (12) was propagated in the murine astrocytoma cell line, DBT. Temperature sensitive (ts) mutants of MHV-A59 obtained from Ralph Baric were plaque purified and maintained as previously described (4).

Preparation and analysis of infected cell lysates

Confluent monolayers of DBT cells were infected with wild type or ts mutants of MHV-A59 and radiolabeled with [^{35}S] methionine as previously described (2,4). At the times indicated in the individual experiments the infected radiolabeled cells were placed on ice and processed, after which the cell lysate was immunoprecipitated as previously described (2). Immunoprecipitated sample was resuspended in 50 ul of 2x Laemmli buffer, boiled for 5 minutes and electrophoresed on a 5-18% gradient SDS-polyacrylamide gel (gSDS-PAGE).

In vitro Translation of MHV-A59 Genome RNA

Purification of MHV-A59 genome RNA, translation in rabbit reticulocyte lysate, immunoprecipitation, and electrophoresis were all performed as previously described (1).

Preparation of Antibodies

Polyclonal antiserum (UP102) directed against polypeptides encoded in the first 1.8 kb of ORF1a was raised in rabbits using as an immunogen virally encoded protein expressed in *E. coli*. To generate the immunogen the first 1.8 kb of ORF1a was placed under the control of a T7 promoter in a pET 3B vector and expressed in *E. coli* using T7 RNA polymerase(13). The resulting insoluble fusion protein was recovered from the pellet following bacterial cell lysis and used directly and after denaturation with 2% SDS and 5% 2-mercaptoethanol as an immunogen. Antisera anti-p28 and 81043 were generated as previously described (2,14). The locations of viral sequences encoding the polypeptides used to raise these antisera are shown in table 1.

TABLE 1. Antisera used for Immunoprecipitation of MHV-A59 ORF1A Products

Antiserum	Peptide type	Peptide size (aa)	Nucleotides
p28	Oligopeptide	14	287-329
UP102	Fusion protein	~600	182-1984
81043	Fusion protein	~400	2879-3968

RESULTS

Immunoprecipitation of MHV-A59 ORF1a Encoded Polypeptides

Antisera (Table 1) directed against ORF1a encoded polypeptides were used to immunoprecipitate MHV nonstructural polypeptides from [^{35}S] methionine labeled infected cell lysates. As previously reported (2) anti-p28 serum precipitated a 28 kDa (p28) polypeptide and 81043 antisera precipitated a 290 kDa (p290) precursor polypeptide. The 240-kDa (p240) and 50-kDa (p50) cleavage products of p290 were also detected by 81043 (Fig. 1). Based on potential antibody recognition domains it had been hypothesized (2) that a polypeptide of greater than 30 kDa is encoded between p28 and p290. To identify polypeptides encoded between p28 and p290, antiserum (UP102) was generated against polypeptides encoded between nucleotides 182 and 1984. It was predicted that UP102 could potentially precipitate p28, a polypeptide of greater than 30kDa and large precursor proteins. As predicted, UP102 precipitated p28, which was also precipitated by anti-p28. UP102 also precipitated a 65kDa polypeptide (p65) that is not precipitated by either anti-p28 or 81043 (Fig. 1), indicating that p65 is encoded between p28 and p290.

Pulse chase labeling of infected DBT cells was performed to determine specific precursor-product relationships of ORF1a products (data not shown). As previously reported p28 was detected immediately after a 20 minute pulse label and the amount of p28 detected remained constant throughout the 120 minute chase period. The 65-kDa polypeptide was detected at 15 minutes of chase and increased in amount detected throughout the 120 minute chase period. A heterogeneous band of greater than 400-kDa that was precipitated by UP102 throughout the first 15 minutes of the chase decreased in amount detected as p65 increased in amount detected. The disappearance of the heterogeneous band as p65 begins to appear indicates that p65 is processed from an as yet unidentified precursor protein.

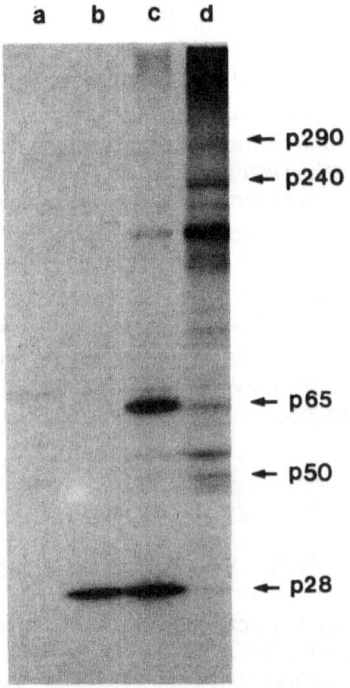

Fig. 1 Immunoprecipitation of MHV-A59 ORF1a products. Immunoprecipitates of [^{35}S] methionine labeled, (lane A) mock infected and (lanes B,C,D) MHV-A59 infected DBT cells were analyzed by 5-18% gradient polyacrylamide gel electrophoresis. Lane A and C, antiserum UP102; lane B, anti-p28 antiserum; lane D, 81043 antiserum. Antisera are as described in Table 1. Specific precipitation products are to the right of the gel, with arrows indicating these products.

Immunoprecipitation of *In Vitro* Translation Products

We have previously hypothesized (2) that intracellular translation and processing of ORF1a products differs from that which occurs during *in vitro* translation of viral genome. The major ORF1a *in vitro* translation products previously reported are p28, p220 and their precursor p250 (15). *In vitro* translation of viral genome and subsequent immunoprecipitation of translation products using UP102 (data not shown) results in the immunoprecipitation of p28 and p220. Antiserum UP102 also precipitates several other bands which were also precipitated by either anti-p28 or 81043. No *in vitro* homologue of p65 was precipitated by UP102, providing further data that differences exist between *in vivo* and *in vitro* translation and processing of ORF1a products.

Polypeptide Analysis of MHV-A59 Temperature Sensitive Mutants

Temperature sensitive mutants have been utilized in several different animal virus systems to characterize the processing and function of viral proteins (16,17). MHV-A59 ts mutants have been partially characterized and genetic recombination analysis has revealed 5 RNA minus (A thru E) and 1 RNA plus (group F) complementation groups (3,4) a selection of MHV-A59 ts mutants obtained from Ralph Baric were analyzed for defects in ORF1a polypeptide translation and processing. DBT cells were infected with ts MHV-A59 for 5.5 hours at the permissive temperature of 32 °C and then the infected cells were shifted to the non-permissive temperature of 39.5 °C and labeled with [35S] methionine at various times after the temperature shift. Ts mutants NC11 (group B) and LA16 (group A/B) were found to be defective for translation and/or processing of p28 and p65 at the non-permissive temperature of 39.5 °C (data not shown). Both NC11 and LA16 showed greatly reduced levels of p28 and undetectable levels of p65 at the non-permissive temperature as compared to high levels of p28 and p65 detected at the permissive temperature. A revertant of NC11 also obtained from Ralph Baric produced wild type levels of both p28 and p65 at 32 °C and 39.5 °C, indicating that the translation and/or processing defect has been corrected in this revertant. The protein defect(s) seen in NC11 and LA16 appear to be unique to these two mutants because other members of group A and B produce detectable levels of p28 and p65 at both 32 °C and 39.5 °C. To determine if p28 and p65 were thermolabile at the non-permissive temperature for NC11 and LA16, infected cells were radiolabeled at 32 °C and then chased with cold methionine at the non-permissive temperature. The amounts of p28 and p65 produced at 32 °C were stable (data not shown) throughout the chase period at 39.5 °C, indicating that decreased levels of p28 and p65 at 39.5 °C in NC11 and LA16 were not due to the thermolability of either polypeptide. The exact cause of the defect seen for NC11 and LA16 has yet to be determined.

DISCUSSION

Detection of MHV-A59 ORF1a Encoded Polypeptides

In this study we report the identification of p65, an ORF1a encoded polypeptide. UP102 precipitated p28, a polypeptide (p65) which was hypothesized to be encoded directly downstream of p28 and a heterogeneous band of greater than 400kDa which is believed to contain a putative precursor of p65.

Immunoprecipitation of ORF1a *in vitro* translation products using UP102 failed to detect an *in vitro* homologue of p65 identified in infected cell lysates, providing further evidence that differences exist between *in vivo* and *in vitro* translation and processing of ORF1a encoded polypeptides. Possible cause(s) of differences that exist in the pattern of ORF1a polypeptide synthesis *in vivo* and *in vitro* may be in part due to: cellular proteases present *in vivo* and not *in vitro* that are responsible for processing of virally encoded polypeptides; temporal regulation of polypeptide synthesis and processing *in vivo*; or incomplete translation of ORF1a *in vitro*.

Processing of ORF1a Encoded Polypeptides

To elucidate the synthesis and processing pattern of ORF1a encoded polypeptides pulse chase experiments were performed. As previously reported p28 was detected first and levels of p28 remained constant throughout the chase period. After 15 minutes of chase, p65 was detected and increased in amount throughout the chase period. After p65 was detected, p290 and its cleavage products p240 and p50 were also detected (Fig.2). The specific proteases and target sites responsible for the intracellular cleavage of p65 are not yet known. If p65 is encoded directly adjacent to p28, N terminal cleavage of p65 maybe mediated in *cis* or *trans* by putative papain-like proteases encoded within ORF1a. The Sindbis virus papain protease nsP2 has been shown to cleave at the dipeptides GA or GG (18). A GG dipeptide at residue 196 in ORF1a maybe a possible papain protease cleavage site. Another predicted cleavage site is at one of the YG dipeptides at either residues 258 or 273 (19). A vaccinia recombinant containing the p28 coding sequence and terminating at the second YG made a protein of approximately 32 kDa. This suggests that the true cleavage site is upstream of the second YG dipeptide at residue 273. (20)We are currently attempting to immuno-purify p65 and sequence its N-terminus to clearly ascertain the cleavage site.

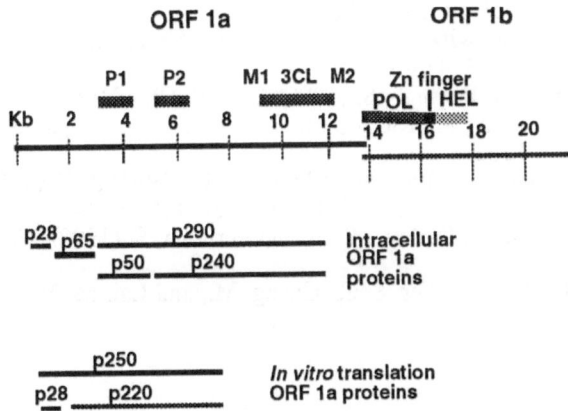

Fig. 2 **Model of ORF1a translation and processing in MHV-A59 infected cells and in *in vitro* translation of MHV-A59 genome RNA.** The size of ORF1a, ORF1b, location of putative functional domains, the predicted sizes of the polypeptides and the possible cleavage sites are all shown to scale. General alignment of polypeptides is based on antisera specificity and has not been confirmed by protein sequencing.

Polypeptide Processing in MHV-A59 ts Mutants

Two ts mutants obtained from Ralph Baric, NC11 and LA16 are defective for translation and/or processing of p28 and p65. It is not yet clear if the defect is in the translation or processing of ORF1a encoded polypeptides. Currently, we are attempting to immunuprecipitate large precursor proteins produced by NC11 and LA16 at the non-permissive temperature. If precursor polypeptides are made at the non-permissive temperature it would support the hypothesis that the defect(s) present in NC11 and LA16 are affecting polypeptide processing. We believe that the ts mutants will be valuable tool in studies which attempt to elucidate processing and function of ORF1a encoded polypeptides.

REFERENCES

1. Denison, M. R., Zoltick, P. W., Leibowitz, J. L., Pachuk, C. J., and Weiss, S. R. (1991) *J. Virol.* **65,** 3076-3082
2. Denison, M. R., Zoltick, P. W., Hughes, S. A., Giangreco, B., Olson, A. L., Perlman, S., Leibowitz, J. L., and Weiss, S. R. (1992) *J. Virol.* **189,** 274-284
3. Baric, R. S., Fu, K., Schaad, M. C., and Stohlman, S. A. (1990) *Virology* **177,** 646-656
4. Schaad, M. C., Stohlman, S. A., Egbert, J., Lum, K., Fu, K., Wei, T. Jr., and Baric R. S. (1990) *Virology* **177,** 634-645

5. Pachuk, C. J., Zoltick, P. J., Spaan, W. J. M., and Weiss, S. R. (1989) *Virology* **171**, 141-148

6. Lee, H.-J., Shieh, C.-K., Gorbalenya, A. E., Koonin, E, V., LaMonica, N., Tuler, J., Bagdzhadzhyan, A., and Lai, M. M. C. (1991) *Virology* **180**, 567-582.

7. Breedenbeek, P. J., Pachuk, C. J., Noten, A. F. H., Charite, J., Luytjes, W., Weiss, S. R., and Spaan, W. J. M. (1990) *Nucleic Acids Res.* **18**, 1825-1832

8. Bonilla, P., and Weiss, S. R. Unpublished.

9. Brayton, P., R., Lai, M. M. C., Patton, C. D., and Stohlman, S., A. (1982) *J. Virol.* **42**, 847-853

10. Gorbalenya, A. E., Koonin, E. V., Donchenko, A. P., and Blinov, V. M. (1989) *Nucleic Acids Res.* **17**, 4847-4861

11. Brierley, I., Boursnell, M. E. G., Binns, M. M., Billimoria, B., Blok, V. C., Brown, T. D. K., and Inglis, S. C. (1987) *EMBO J.* **6**, 3779-3785

12. Sturman, L., S., and Takemoto, K. K. (1972) *Infect. Immun.* **6**, 501-507

13. Studier, W. F., Rosenberg, A. H., Dunn, J. J., and Dubendorff, J. W. (1990) *Methods in Enzymology* **185**, 60-89

14. Zoltick, P. W., Leibowitz, J. L., De Vries, J. R., LWeinstock, G. M., and Weiss, S. R. (1989) *Gene* **85**, 413-420

15. Denison, M, R., and Perlman, S. (1986) *J. Virol.* **60**, 12-18

16. Hahn, Y. S., Grakoui, A., Rice, C.M., Strauss, E. G., and Strauss, J. H. (1989) *J. Virol.* **63**, 1194-1202

17. Burns, C. C., Richards, O. C., and Ehrenfeld, E. (1992) *J. Virol.* **189**, 568-582

18. Shirako, Y., and Strauss, J. H. (1990) *Virology* **177**, 54-64

19. Mobley, J., Evans, G., Dailey, M. O., and Perlman, S. (1992) *Virology* **187**, 443-452

20. Soe, L. H., Shieh, C., Baker, S. C., Chang, M., and Lai, M. M. C. (1987) *J. Virol.* **61**, 3968-3976

PROTEOLYTIC PROCESSING OF THE N-TERMINAL REGION OF THE EQUINE ARTERITIS VIRUS REPLICASE

Eric J. Snijder, Alfred L.M. Wassenaar, and Willy J.M. Spaan

Department of Virology, Institute of Medical Microbiology
Faculty of Medicine, Leiden University
Postbus 320, 2300 AH Leiden, The Netherlands

ABSTRACT

A papainlike cysteine protease (PCP) domain in the N-terminal region of the equine arteritis virus (EAV) replicase was identified by *in vitro* translation and mutagenesis studies. The EAV protease was found to direct an autoproteolytic cleavage at its C-terminus which leads to the production of an approximately 30K N-terminal replicase product (nsp1) containing the PCP domain. Amino acid residues Cys^{164} and His^{230} of the EAV replicase polyprotein were identified as the most likely candidates for the role of PCP catalytic residues. It was shown that cleavage occurs *in cis* between Gly^{260} and Gly^{261}.

INTRODUCTION

Equine arteritis virus (EAV) is an enveloped positive-stranded RNA virus. Its isometric nucleocapsid core contains a nonsegmented 12.7 kb genome (for a recent review: see reference 1). The morphological characteristics and genome size of EAV are most comparable to those of togaviruses. However, we have recently described[2,3] that the EAV replication strategy is similar to that of coronaviruses (for a review: see reference 4) and toroviruses[5,6], which possess 25-31 kb positive-stranded RNA genomes. Among their common features are a polycistronic genome organization, the same basic gene order (5'-replicase gene-envelope protein genes-nucleocapsid protein gene-3'), and the production of a 3'-coterminal nested set of 4 to 7 subgenomic mRNAs. The 5' part of the genomes of these viruses is occupied by two large open reading frames (ORF1a and ORF1b) which encode the viral replicase[2,6,7,8,9]. Both ORF1a and ORF1b are expressed from the genomic RNA, the latter by means of a ribosomal frameshifting mechanism[2,6,10]. The predicted

ORF1b products of corona-, toro-, and arteriviruses contain a number of homologous protein domains[2,6,8] which, in addition to the other similarities described above, indicate that these viruses are evolutionarily related. We have therefore proposed these viruses to be members of a coronaviruslike superfamily of positive-stranded RNA viruses[2,6]. The large EAV replicase gene product (345K) is presumed to be a polyprotein precursor which is posttranslationally cleaved into smaller functional units. In the 187K ORF1a amino acid sequence putative papainlike cysteine protease and trypsinlike serine protease domains were identified[2] (Fig. 1). We have now initiated a study of the coronaviruslike replicase using the relatively small EAV replicase gene as a model.

RESULTS

Reconstruction and *in vitro* translation of EAV ORF1a

The analysis of the posttranslational processing of the EAV replicase polyprotein was started by reconstructing ORF1a from seven overlapping cDNA clones. In transcription vector pEAV1a a full-length cDNA copy of ORF1a is located downstream of the T7 RNA polymerase promoter. In addition, a set of pEAV1a deletion mutants (p1aΔ1 through p1aΔ8) was produced which contained termination codons for translation at various positions in ORF1a. To create termination codons for translation, *Nhe*I linkers (5' CTAGCTAGCTAG 3') were inserted into the following pEAV1a restriction sites: *Sac*I (nt 858), *Hin*dIII (nt 1501), *Kpn*I (nt 1802), *Sal*I (nt 2608), *Nhe*I (nt 2878), *Apa*I (nt 3688), *Eco*RV (nt 4263), and *Bam*HI (nt 5115). Nucleotide (nt) and amino acid (aa) sequence numbers refer to the EAV genomic and protein sequences which we have published previously[2].

Vector pEAV1a and the p1aΔn constructs were used to transcribe a set of 3'-truncated RNAs from which an increasing part of ORF1a could be translated. *In vitro* translation was carried out in a rabbit reticulocyte lysate (Promega) in the presence of [35S]methionine (1 h at 30°C). A direct SDS-PAGE analysis of the translation products of the p1aΔn series and pEAV1a is shown in Fig. 1. Only p1aΔ1, which contained a termination codon in the center of the putative PCP domain, produced a protein of the predicted size. All other constructs gave rise to a prominent product of about 30K and to accompanying bands which were smaller than predicted from the (partial) ORF1a amino acid sequence. These data indicated that an approximately 30K protein was cleaved from the N-terminus of the EAV ORF1a product. The putative PCP domain, which resided in this cleavage product, could be involved in this proteolytic event.

Analysis and mutagenesis of the EAV PCP domain

Typical papainlike cysteine proteases show a requirement for at least one cysteine and one histidine residue[11]. On the basis of amino acid sequence comparison of the ORF1a sequence and cellular and viral papainlike thiol proteases, Cys[164] had been proposed as active site residues of an EAV papainlike cysteine protease[2]. His[219] and His[230] were considered the most likely candidates for the role of catalytic His. To prove that the EAV PCP domain was responsible for the observed proteolytic processing of the ORF1a protein, amino acid substitutions were introduced into the ORF1a sequence. Transcription vector pCP0 (Fig. 2), encoding a 46K product which is cleaved into 30K and 16K polypeptides, was used as a basis for these experiments. Derivatives of this construct were used to test

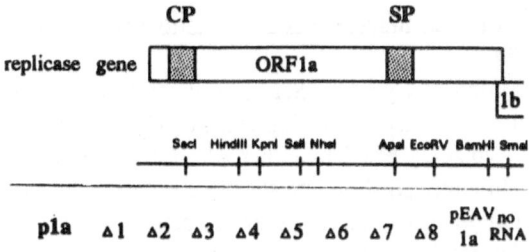

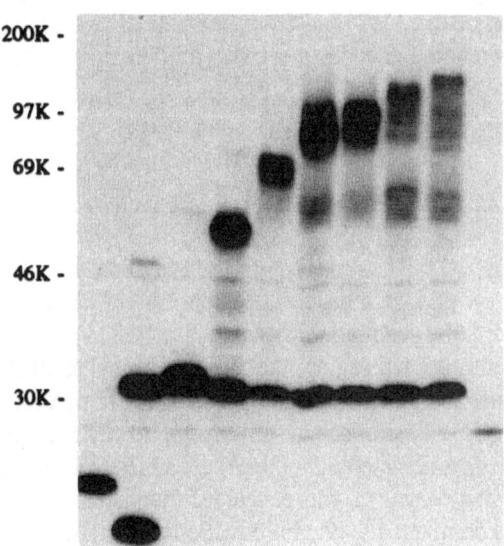

Fig. 1. *In vitro* translation results from EAV ORF1a expression constructs.

A schematic representation of the ORF1a region of the EAV replicase gene is shown. The positions of the predicted papainlike cysteine protease (CP) and trypsinlike serine protease (SP) domains are indicated. The restriction sites were used in the construction of the expression plasmids p1aΔ1 through p1aΔ8 and pEAV1a. Plasmids p1aΔ1 through p1aΔ8 and pEAV1a were used for *in vitro* transcription of a 3'-truncated set of RNAs from which an increasing part of ORF1a could be translated. The RNAs were translated in a rabbit reticulocyte lysate in the presence of [^{35}S]methionine and direct analysis of translation products was performed by SDS-PAGE.

Fig. 2. Identification of possible active site residues of the EAV papainlike cysteine protease.

Amino acid substitutions were introduced into construct pCP0 (p1aΔ2 from Fig. 1) and tested by *in vitro* translation and SDS-PAGE. Mutations are indicated at the top of each lane; the name of the corresponding expression plasmid is shown at the bottom of the lane. The 46K full-length translation product and its 30K and 16K cleavage products are indicated.

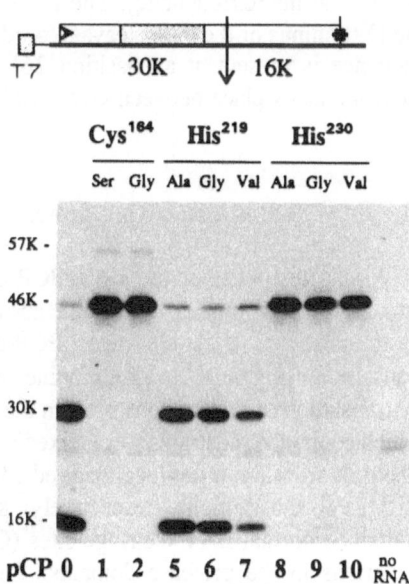

the influence of substitutions at the positions of Cys[164], His[219], and His[230]. Nucleotide changes were introduced by oligonucleotide-directed mutagenesis and mutations were tested by *in vitro* transcription and translation (Fig. 2).

Both the rather conservative substitution of Cys[164] by Ser as well as the Cys[164] to Gly mutation completely abolished proteolytic activity, indicating that this Cys residue is indeed essential for the protease function. Replacing His[219] by either Ala, Gly, or Val did not affect cleavage to a significant extent. In contrast, the same set of substitutions at the position of His[230] exhibited an effect similar to that observed after replacing Cys[164]: the His[230]-Val substitution completely inhibited proteolytic activity, and only traces of cleavage products could be detected after translation of the Ala[230] and Gly[230] mutants. These results confirmed that the 5' region of the EAV genome encodes a proteolytic domain which is responsible for the observed cleavage event.

Activity of the EAV PCP domain in *E. coli* and identification of the PCP cleavage site

The bacterial expression vector pGEX-2T (Pharmacia) was used to express the N-terminus of the EAV ORF1a polypeptide as part of a bacterial fusion protein. The 86K fusion protein contained a pGEX-derived 26K glutathione *S*-transferase (GST) moiety, followed by 9 aa encoded by the EAV 5' untranslated region, the 528 N-terminal aa of the ORF1a product (including the PCP domain), and 33 aa encoded by a short in-frame vector sequence. Remarkably, the construct did not only produce an 86K fusion protein: it also yielded 55K and 32K bands. In view of the data obtained with the pla∆n series (see above) and the 26K size of the GST part of the fusion protein, it was concluded that the EAV PCP domain was functional in *E. coli*. This was confirmed by the introduction of mutations into the PCP domain of the fusion protein.

The 55K bacterial expression product represented the N-terminal cleavage product consisting of the GST part and the previously observed 30K EAV ORF1a protein. The amount of the 32K C-terminal cleavage product was sufficient to allow purification for N-terminal microsequencing. The following amino acid residues were found to constitute the N-terminus of the 32K cleavage product: Gly-Tyr-Asn-Pro-Pro-Gly-Asp-Gly-Ala. This sequence is present at aa position 261-269 in the EAV ORF1a protein, indicating that cleavage takes place between Gly[260] and Gly[261] in the EAV sequence.

The EAV PCP is a *cis*-acting protease

To test whether the EAV PCP can function *in trans*, the protease domain and a number of the mutants described above were used in *trans*-cleavage assays. To test posttranslational *trans*-cleavage, PCP-containing proteins were prepared by *in vitro* translation using unlabeled methionine, while labeled (uncleaved) substrates were prepared by translation of mutant constructs in the presence of [^{35}S]methionine. Translation reactions which contained substrates were mixed with equal volumes of protease-containing reactions. The 46K translation products derived from pCP1 and pCP10 (carrying the Cys[164]-Ser and His[230]-Val substitutions, respectively; see also Fig. 2) were used as substrates. They contained normal EAV cleavage sites (Gly[260]-Gly[261]), which remained unprocessed due to mutations in the protease domains. Though mixtures of substrates and proteases were incubated at 30°C for up to 7 h, both nsp1 and its precursor were unable to produce detectable amounts of 30K and 16K cleavage products. The possibility of cotranslational *trans*-cleavage was excluded, by mixing transcripts encoding substrates and proteases and translating them in the same reaction.

DISCUSSION

Although the available coronaviruslike replicase sequences have been searched for possible protease domains extensively, - even their cleavage sites have already been predicted[9,12,13] - there is little experimental evidence which supports these theoretical analyses. The presence of a number of ORF1a- and ORF1b-encoded proteins in MHV-infected cells has been reported[12,14]. However, coronavirus proteases have not yet been studied in detail, and replicase cleavage sites and processing pathways remain to be elucidated.

In this paper the first detailed analysis of a protease domain which is located in a coronaviruslike replicase is reported. The identification of a Cys and a downstream His as putative active site residues indicates that this EAV protease is related to the cellular papainlike enzymes, and can therefore be added to the short but growing list of viral papainlike cysteine proteases. In a recent review, Gorbalenya et al.[15] discriminated between a group of viral PCPs which mediate the production of a single N-terminal cleavage product (the so-called 'leader proteases') and PCPs which are thought to be 'main' proteases involved in multiple processing steps. Of the latter group only the PCP domain residing in the Sindbis virus (SIN) nsp2 has actually been shown to possess proteolytic activity[16]; it is responsible for the production of the non-structural SIN proteins from a polyprotein precursor.

The EAV PCP clearly belongs to the group of leader proteases, which (among others) includes the potyvirus helper component protease[17,18] (HC-Pro) and the p29 protease of the hypovirulence-associated virus (HAV) of the chestnut blight fungus[19,20]. Although the overall sequence similarity between the EAV domain and these two well-studied leader PCPs is very limited, some striking similarities can be observed. The spacing between the putative catalytic Cys and His residues is similar and clearly different from the spacing in cellular papainlike proteases. In addition, all three proteases efficiently cleave a Gly-Gly dipeptide which is located 30-41 aa downstream of the putative active site His residue.

We have failed to detect other cleavage events after *in vitro* translation of ORF1a-specific RNAs. This is surprising since, in addition to the PCP domain, the larger constructs from the p1a△n series (p1a△7 and p1a△8) and the full-length ORF1a construct contain the coding information for the putative EAV serine protease[2] (aa position 1090-1210). Although several additional bands were observed (Fig. 1), none of them can be explained as the obvious result of proteolytic processing.

We are currently preparing antibodies specific for ORF1a-encoded proteins, which will enable us to study the *in vivo* processing of the ORF1a product. Using an anti-peptide rabbit antiserum directed against the N-terminal part of the ORF1a product, we have recently identified the 30K EAV nsp1 in infected cells. This may enable us to study the function of nsp1 which contains a cysteine-rich N-terminal region, in addition to the C-terminal PCP domain. Also EAV nsp2 (starting with Gly[261]) will be the subject of future study.

ACKNOWLEDGMENTS

We thank Johan den Boon, Twan de Vries, Alexander Gorbalenya, and R. Amons for technical assistance and helpful discussions.

REFERENCES

1. P.G.W. Plagemann and V. Moennig, Adv. Virus Res. 41:99.(1991).

2. J.A. den Boon, E.J. Snijder, E.D. Chirnside, A.A.F. de Vries, M.C. Horzinek and W.J.M. Spaan, J. Virol. 65:2910 (1991).

3. A.A.F. de Vries, E.D. Chirnside, P.J. Bredenbeek, L.A. Gravestein, M.C. Horzinek and W.J.M. Spaan, Nucleic Acids Res. 18:3241 (1990).

4. W.J.M. Spaan, D. Cavanagh and M.C. Horzinek, J. Gen. Virol. 69:2939 (1988).

5. E.J. Snijder, M.C. Horzinek and W.J.M. Spaan, J. Virol. 64:331 (1990).

6. E.J. Snijder, J.A. den Boon, P.J. Bredenbeek, M.C. Horzinek, R. Rijnbrand and W.J.M. Spaan, Nucleic Acids Res. 18:4535 (1990).

7. M.E.G. Boursnell, T.D.K. Brown, I.J. Foulds, P.F. Green, F.M. Tomley and M.M. Binns, J. Gen. Virol. 68:57 (1987).

8. P.J. Bredenbeek, C.J. Pachuk, J.F.H. Noten, J. Charité, W. Luytjes, S.R. Weiss and W.J.M. Spaan, Nucleic Acids Res. 18:1825 (1990).

9. H.J. Lee, C.K. Shieh, A.E. Gorbalenya, E.V. Koonin, N. la Monica, J. Tuler, A. Bagdzhadzhyan and M.M.C. Lai, Virology 180:567 (1991).

10. I. Brierley, P. Diggard and S.C. Inglis, Cell 57:537 (1989).

11. L. Polgar and P. Halasz, Biochem. J. 207:1 (1982).

12. M.R. Denison, P.W. Zoltick, J.L. Leibowitz, C.J. Pachuk and S.R. Weiss, J. Virol. 65:3067 (1991).

13. A.E. Gorbalenya, E.V. Koonin, A.P. Donchenko and V.M. Blinov, Nucleic Acids Res. 17:4847 (1989).

14. M.R. Denison, P.W. Zoltick, S.A. Hughes, B. Giangreco, A L. Olson, S. Perlman, J.L. Leibowitz and S.R. Weiss, Virology 189:274 (1992).

15. A.E. Gorbalenya, E.V. Koonin and M.M.C. Lai, FEBS Lett. 288:201 (1991).

16. J.H. Strauss and E.G. Strauss, Semin. Virol. 1:347 (1991).

17. J.C. Carrington and K.L. Herndon, Virology 187:308 (1992).

18. C.-S. Oh and J.C. Carrington, Virology 173: 692 (1989).

19. G.H. Choi, D.M. Pawlyk and D.L. Nuss, Virology 183:747 (1991).

20. E.V. Koonin, G.H. Choi, D.L. Nuss, R. Shapira and J.C. Carrington, Proc. Natl. Acad. Sci. USA 88:10647 (1991).

The data presented in this paper are derived from a publication by the same authors in the Journal of Virology, volume 66, pages 7040-7048 (published by the American Society for Microbiology in december 1992).

Chapter 4

Coronaviruses, Toroviruses and Arteriviruses:Common and Distinctive Features

THE CORONAVIRUSLIKE SUPERFAMILY

Eric J. Snijder[1], Marian C. Horzinek[2], and Willy J.M. Spaan[1]

[1]Department of Virology, Institute of Medical Microbiology, Faculty of Medicine, Leiden University, Postbus 320, 2300 AH Leiden, The Netherlands, and [2]Department of Virology, Veterinary Faculty, University of Utrecht, Yalelaan 1, 3584 CL Utrecht, The Netherlands

> *If superior creatures from space ever visit earth,*
> *the first question they will ask, in order to*
> *assess the level of our civilization, is:*
> *'Have they discovered evolution yet?'*
>
> Richard Dawkins
> *The Selfish Gene*

INTRODUCTION

In the past three years, the increasing knowledge of viral genome organizations, replication strategies, and nucleotide sequences has had its impact on coronavirus taxonomy. The 'superfamily' concept[1,2], which is based on evolution and phylogeny, and which had already closed the gaps between other virus groups (e.g. the alphaviruslike and picornaviruslike superfamilies), has now been found to apply to a group of coronaviruslike viruses as well.

The sequence analysis of the genomes of the 'classic' coronaviruses infectious bronchitis virus (IBV) and mouse hepatitis virus (MHV), the Berne torovirus (BEV), and the 'unclassified togavirus' equine arteritis virus (EAV) revealed unexpected evolutionary links. The common features of these viruses are centered around the 'coronaviruslike' replicase gene and the replication and expression strategy which is associated with it. In this short comparative review, toroviruses and arteriviruses will be introduced to the coronavirologist, and similarities and differences with coronaviruses will be discussed, using MHV as the standard coronavirus for comparison throughout this paper.

Coronaviruses, Edited by H. Laude and J.F. Vautherot
Plenum Press, New York, 1994

TOROVIRUSES

In 1972, a virus was isolated from a diarrheic horse during routine diagnostic work at the University of Berne, Switzerland. The isolate, designated P138/72, displayed an unusual morphology but was not studied in more detail until a similar virus was isolated from diarrheic calves in Breda, Iowa, U.S.A., in 1979[3,4]. Berne virus (BEV) and Breda virus (BRV) were found to be antigenically related[3]. In later years, a second BRV serotype was identified[4] and similar pleiomorphic viruses were found in the stools of children and adults with gastroenteritis in Birmingham, U.K., and Bordeaux, France[5]. Immune EM experiments indicated that these human viruses were serologically related to BRV and BEV[6].

The unique morphology of BEV and BRV (see below) and the physicochemical properties of BEV[7] initially led to the proposal of a new family of enveloped animal viruses, the *Toroviridae*[8]. BEV, the only torovirus so far which can replicate in cultured cells, was chosen as the torovirus prototype.

ARTERIVIRUSES

Equine arteritis virus (EAV) was first isolated from a fetus aborted during an endemic disease outbreak in pregnant mares[9]. Serological evidence suggests that the virus is widespread in the horse population and only rarely causes disease. However, in pregnant mares abortion is common[10,11]. A carrier state exists in seropositive stallions in which EAV is produced in semen[12]. These 'shedding stallions' may consequently infect broodmares by a venereal route. Field isolates are rare, may be difficult to propagate in cell culture, and consequently the biology of EAV is poorly understood. The biological and clinical properties of EAV have been reviewed recently[13].

The molecular characterization of lactate dehydrogenase-elevating virus (LDV) of mice[13,14] and swine infertility and respiratory syndrome virus[15] (SIRSV or 'Lelystad virus'), which has been reported recently, has revealed that these two viruses are closely related to EAV. Though it has not yet been characterized at the molecular level, simian hemorrhagic fever virus (SHFV) may be the fourth member of this virus group[13].

The morphological characteristics and genome size of EAV (12.7 kb) are most comparable to those of togaviruses. However, as will be described below, the arterivirus replication strategy[16] is similar to that of coronaviruses and toroviruses, which possess 25-31 kb genomes.

VIRION ARCHITECTURE

Torovirions are pleiomorphic particles which measure 120-140 nm in their largest diameter[3]. Spherical, oval, elongated, and kidney-shaped virions have been observed. Their two most striking morphological features are the spikes on the viral envelope, which resemble the peplomers of coronaviruses, and the tubular nucleocapsid of helical symmetry, which seems to determine the shape of the virion[3]. The presence of nucleocapsids in the form of a doughnut, a shape described in Latin by the word 'torus', led to the proposal of the name toroviruses.

The morphogenesis of BEV has been studied by EM methods[17]. Preformed, tubular nucleocapsids were found to bud at intracellular membranes, predominantly those of the Golgi system. A morphological change seems to take place during virus maturation: prior to budding the nucleocapsids are straight, but in extracellular virus the characteristic torus-shape is prevalent.

Three structural BEV proteins have been identified and characterized (see also below): an 18K nucleocapsid (N) protein, a 26K transmembrane protein (initially named E), and a 180K spike glycoprotein (initially named P). In view of the recent classification of the toroviruses as a *genus* in the coronavirus family, the E and P protein will from now on be referred to as M and S protein, respectively. The current structural model of the BEV particle is shown in Fig. 1.

Equine arteritis virus was initially classified as a member of the togavirus family[18].

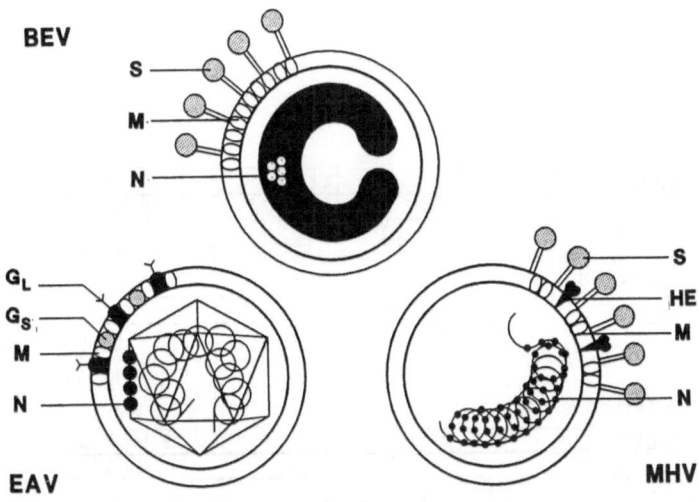

Fig. 1. Schematic representation of the structure of BEV (torovirus), EAV (arterivirus), and MHV (coronavirus). The structural proteins of each virus group are included.

The spherical enveloped equine arteritis virus particle has a diameter of 50-70 nm[19]. It consists of an icosahedral core structure of 35 nm surrounded by an envelope carrying ring-like structures with a diameter of 12-15 nm[20]. The identification and characterization of four virion proteins have been reported recently[21]: a 12K nucleocapsid (N) protein, an unglycosylated 18K transmembrane protein (M), a 25K glycoprotein G_S, and a second glycoprotein, G_L, which, due to heterogeneous glycosylation, gives rise to proteins with sizes between 30K and 42K. The current model of the EAV particle is shown in Fig. 1.

In summary, nucleocapsid architecture - a classic trait for viral taxonomy, with the same ranking as nucleic acid type or the presence of an envelope - is icosahedral in EAV, helical in coronaviruses, and tubular in toroviruses. An important additional difference at the structural level is the fact that the EAV envelope does not bear the elongated spikes which are so characteristic for both coronaviruses and toroviruses. However, as will be discussed in the next paragraph, the genome organization and expression of the three virus groups are strikingly similar and evidence for common ancestry was obtained by comparison of replicase amino acid sequences.

GENOME ORGANIZATION AND EXPRESSION

The BEV genome (probably) contains six open reading frames (ORFs). As in coronaviruses, the two most 5' reading frames (ORF1a and 1b) are expressed from the genomic RNA and constitute the replicase gene. Assuming that the 5' part of the BEV replicase gene, which has yet to be sequenced, contains a single open reading frame (ORF1a) of about coronaviral size, the toroviral genome measures approximately 25 kb.

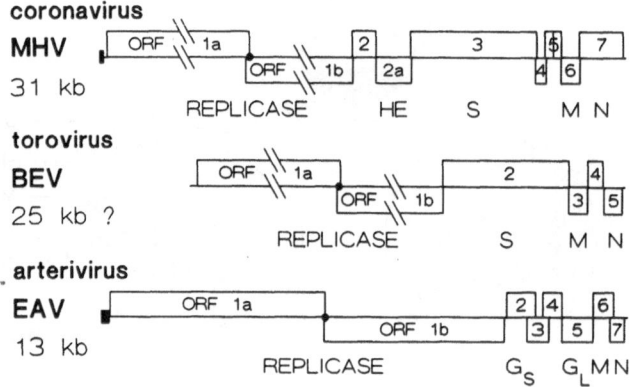

Fig. 2. Genome organization of MHV (coronavirus), BEV (torovirus), and EAV (arterivirus). The genes encoding the viral replicases and structural proteins are indicated.

The other reading frames are expressed by the generation of a 3'-coterminal nested set of four mRNAs[22]. ORFs 2, 3, and 5 encode the viral structural proteins S[23], M[24], and N[25], respectively. In view of the observed sequence similarities with other viral surface proteins, the apparently truncated ORF4[26] should probably be considered a (structural) pseudogene. The BEV genome organization is summarized in Fig 2.

Also during EAV replication a 3'-coterminal nested set of viral mRNAs is produced, ranging in size from genome length (12.7 kb) to 0.8 kb[16,27]. Viral subgenomic (sg) RNAs are composed of leader and body sequences which are not contiguous on the

EAV genome, the 207 nt leader sequence being derived from the extreme 5' end[16,27]. Again two replicase ORFs (1a and 1b) are expressed from the genomic RNA. Of the reading frames from the 3' end of the genome, ORFs 2, 5, 6, and 7 have now been shown to encode the G_S, G_L, M and, N proteins, respectively. The characteristics of the products of ORFs 3 and 4 are typical of membrane proteins, but no information on the function of these proteins has been obtained until this moment. The EAV genome organization is shown in Fig. 2.

Unlike most other ORFs of coronavirus-like genomes, ORF1b is not expressed from a separate sgRNA but by ribosomal frameshifting during the translation of genomic RNA[16,28,29]. The ORF1b product contains a number of domains which are likely to be indispensable during the early stages of viral replication, e.g. synthesis of negative-stranded RNA and the onset of sgRNA synthesis. The ORF1a/ORF1b frameshift mechanism and the RNA structures involved in this process are remarkably conserved in corona-, toro-, and arteriviruses. This indicates that translational frameshifting is an ancient and probably essential regulating step in replicase gene expression. Apparently the respective levels of ORF1a- and ORF1b-derived proteins in infected cells, and possibly also the level of ORF1a/ORF1b fusion product(s), have to be regulated. Nevertheless, the frameshifting efficiency in a reporter gene varies: figures of 25-40%, 20-30%, and 15-20% have been reported for corona-[28,30], toro-[29], and arteriviruses[16], respectively.

RNA SYNTHESIS

The generation of an extensive 3'-coterminal nested set of mRNAs from an unsegmented genome distinghuishes the members of the coronaviruslike superfamily from the viruses from most other groups of positive-stranded RNA viruses. The ancestral relationship between corona-, toro-, and arteriviral replicases[16,29] (see below) suggests that this replication strategy is dictated by the properties of the coronaviruslike replicase.

Since both corona- and arteriviral mRNAs contain a common leader sequence at their 5' end, the absence of such a leader in BEV RNAs[22,31] would form a conspicuous difference. For all three virus groups conserved sequences which are assumed to be involved in sgRNA transcription have been described. For the coronavirus MHV this is the intergenic 5' AAUCuAuAC 3' motif which has been identified as the site of leader to body fusion[32]. A similar though shorter junction sequence (5' UCAAC 3') has been reported for the arterivirus EAV[16,27]. Though also the genome of the torovirus BEV contains conserved intergenic sequences (5' uaUcUUUAGa 3'), no evidence for the presence of a common leader sequence has been obtained[22,31]. BEV mRNAs appear to terminate at or just upstream of the conserved intergenic sequence. In terms of replication however, the consequences of this dissimilarity could be limited to the initiation of positive-stranded RNA transcription only: direct binding of the polymerase to the various BEV 'core promoters' on a negative-stranded template may simply replace the leader-primed initiation of this transcription process which is used by coronaviruses and EAV.

Recent findings suggest that coronaviruses may utilize the fact that their sgRNAs contain 5'- and 3'-terminal sequences which are identical to those of the genomic RNA: transcriptionally active negative-stranded sgRNAs have been detected in infected cells[33,34]. This implies that coronaviral sgRNAs may function as replicons. Until now, attempts to demonstrate negative-stranded viral RNA in torovirus-infected cells have remained unsuccessful, but from EAV-infected cells subgenomic RF RNAs can be isolated[35]. In view of the similar replication strategies, analogous transcription mechanisms, possibly with minor variations, may very well be used by corona-, toro-, and arteriviruses.

STRUCTURAL PROTEINS

The nucleocapsid (N) protein

In view of the very different nucleocapsid structures which have been described for corona-, toro-, and arteriviruses, it is not surprising that their N proteins have little in common. No sequence similarities were detected except for the fact that all three N proteins are rich in basic amino acid residues. The N protein sizes are very different: the coronavirus N protein has a characteristic molecular weight of 45K-55K[32], BEV contains an 18K N protein[23], and the EAV N protein is only 12K in size[16,21].

The membrane (M) protein

Structurally similar transmembrane proteins are found in corona-, toro-, and arterivirus particles. These membrane (M) proteins all lack an N-terminal signal sequence. Instead they contain three membrane-spanning domains in their N-terminal half which are attached to a rather amphipathic C-terminal part. The topological model which has been proposed for the coronavirus M protein[36], has been found to apply to the orientation of the BEV M protein in the membrane as well[24]. In addition to these structural similarities, the MHV and BEV M proteins are about the same size (26K) and contain a small box of sequence similarity[24]. In our opinion, these similarities and the linkage of the coronaviral and toroviral M proteins to homologous replicase genes (see below) indicate that these proteins are homologous and not analogous.

For the EAV triple-spanning M protein the evidence for common ancestry is less convincing: the protein is smaller (18K) and no obvious sequence similarities have been detected. However, whether homologous or not, the role of the EAV M protein may be similar to that of its coronaviral and toroviral equivalents. An interesting common feature of these three virusgroups is their intracellular maturation. The M protein of coronaviruses has been implicated to play a crucial role in the budding process[37] and its localization. It was shown to accumulate in intracellular membranes[38], which is probably also true for the BEV M protein[24]. Therefore, the M proteins of corona-, toro-, and possibly arteriviruses may contain properties which are essential for virus assembly and which have been conserved during the evolution of these intracellularly budding RNA viruses.

The surface glycoproteins

The surface glycoproteins of arteriviruses are clearly different from those of corona- and toroviruses. The characterization of the small (G_S) and large (G_L) EAV glycoproteins was reported recently[21]. The 25K G_S protein is a minor protein in virus particles (1-2 %). The G_L protein, which is observed as a 30K-42K smear due to heterogeneous N-acetyllactosamine addition, is about equally abundant in virions as the M and N proteins.

The envelopes of both corona- and torovirus particles are studded with drumstick-shaped projections of similar size. Heterogeneous 75K-100K protein material from BEV particles was recognized by both neutralizing and hemagglutination-inhibiting monoclonal antibodies and was therefore assumed to represent the spike (S) protein[39]. This N-glycosylated protein is derived from processing of a 200K precursor which is found in infected cells, but not in virions. However, size, post-translational cleavage, and extensive N-glycosylation are not the only similarities between toro- and coronaviral S proteins. Both glycoproteins also form oligomers and contain hydrophobic domains and heptad repeat sequences at corresponding positions in their sequence[25]. Therefore, the coiled-coil structural model, which has been put forward to explain the elongated shape of the coronavirus spike[40], probably also applies to the torovirus surface projections.

In conclusion, the BEV envelope proteins (formerly named E and P) have been shown to be structurally similar to the coronaviral M and S proteins. This phenomenon could possibly be explained by invoking convergent evolution. However, in view of the evolutionary relationship between toro- and coronaviruses (see below), we postulate that these similarities reflect common ancestry. The absence of antigenic relationships and amino acid sequence homologies is indicative for the large evolutionary distance between both virus groups. The arteriviruses clearly contain a completely different set of envelope proteins, with the possible exception of the M protein.

THE CORONAVIRUSLIKE REPLICASE

Genome replication and assembly of new virions are generally considered to be the two fundamental processes in the viral life cycle. The replicase gene is the best candidate for the title 'core of the virus' since replicase proteins (and the replication strategy which they impose upon a viral genome) are conserved among seemingly disparate groups of plant and animal viruses[1,2]. This is (again) exemplified by the replicases of the members of the coronaviruslike superfamily, which have been discussed extensively elsewhere[16,29].

The evolutionary relationship between corona- and toroviruses was first recognized during the sequence analysis of the BEV ORF1b region. Conserved domains (up to 50% amino acid sequence identity) were identified[29] which are present in the same order in the ORF1b sequences of the coronaviruses MHV and IBV. The importance of these homologous domains was underlined by their subsequent detection in the replicase sequence of EAV[16], which is, however, much shorter and more distantly related (up to 30% amino acid sequence identity in the most conserved regions). The organization and conserved regions of the various coronaviruslike replicases are shown in Fig. 3. Of course, the conservation of two domains (polymerase and helicase) which are common to all positive-stranded RNA viruses is not very surprising; their presence indicates that all these viruses may have descended from the same RNA virus prototype. It is remarkable, however, that only in coronaviruslike replicases the helicase domain is located downstream of the

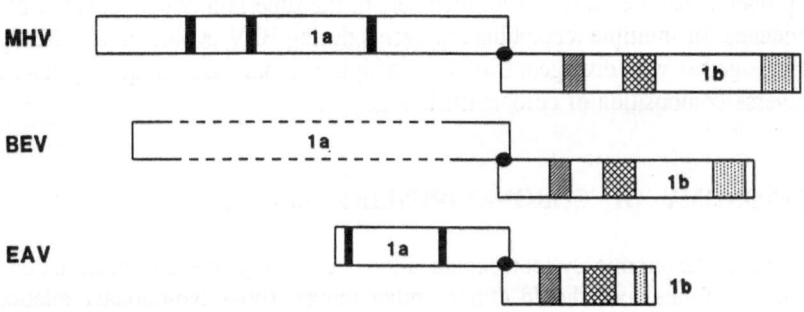

Fig. 3. Schematic representation of coronaviruslike replicases. Filled, hatched, cross-hatched, and dotted boxes represent protease, polymerase, helicase and C-terminal ORF1b domains, respectively.

polymerase motif. Also the conservation of additional replicase domains, for which no homologue can be found in other viral replicases, clearly indicates that the coronaviruslike replicases are more related to each other than to any other group of positive-stranded RNA viruses.

Ribosomal frameshifting during coronaviruslike replicase gene expression produces a large ORF1a/ORF1b fusion product (345K for EAV, 741k for IBV, 810K for MHV). As described for other viral replicases, these large replicase proteins are proteolytically cleaved into smaller active units. The proteases responsible for this posttranslational processing are thought to be located in the ORF1a protein (it should be noted that the torovirus ORF1a sequence is not yet available). A number of replicase cleavage products has been detected in coronavirus-[41,42] and arterivirus-infected[43,44] cells. However, the characterization of the viral proteases involved, which are thought to belong to the papainlike, trypsinlike, and picornavirus 3C-like classes of proteolytic enzymes, has only just begun.

THE EVOLUTION OF THE CORONAVIRUSLIKE SUPERFAMILY

Features which are shared by the members of the coronaviruslike superfamily are: the basic genome organization replicase-envelope proteins-nucleocapsid protein, the production of 3'-coterminal nested sets of mRNAs, and the conserved organization of the replicase gene (homologous replicase domains at comparable positions in the protein; two reading frames connected by a frameshift site). Noticeable differences are the dissimilar N proteins and nucleocapsid structures, the (probable) absence of a common 5' leader sequence in the BEV mRNAs, the much smaller genome size of arteriviruses, and the absence of a large spike glycoprotein in this same virus group.

The common ancestry of the coronaviruslike replicase proteins (and probably also the replication strategy connected with it) is evident. Since viral structural proteins are known to evolve at a higher rate than nonstructural proteins, it is likely that the structural similarities of coronavirus and torovirus S and M proteins also reflect common ancestry, despite the absence of convincing sequence similarities. Similar observations can be made for the EAV M protein. This leaves the different glycoproteins of arteriviruses and the diverse N proteins of coronaviruslike viruses to account for.

The coupling of different sets of structural genes to the same replicase gene has been explained by recombination of complete genes or gene sets (modules)[1,2,45]. The RNA recombination frequency during coronavirus replication has been shown to be remarkably high[32,46], which is thought to be determined by the replication strategy and replicase properties. The different sets of structural genes (and the varying number of 'additional' genes) which are now known to be linked to the coronaviruslike replicase gene indicate that also this property may be shared by all members of the superfamily ('modular' evolution). Direct evidence for multiple recombination events during BEV evolution has already been obtained[26]. Together with divergent evolution, a high recombination frequency can account for the diverse composition of coronaviruslike genomes.

THE TAXONOMY OF CORONAVIRUSLIKE VIRUSES

A useful taxonomic system should allow us to organize our information about viruses. Such a framework should (among other things) show evolutionary relationships between *species*. Until recently, phylogeny has not been an important criterion in animal virus taxonomy[47]. The taxonomic system discriminated only three hierarchical levels:

families (sometimes subfamilies), *genera*, and *species*[48]. This explains why we have so gratefully utilized the unofficial 'superfamily' or 'supergroup' category which was introduced by Strauss & Strauss[2] and Goldbach & Wellink[1].

Our increasing knowledge of viral genomic sequences and expression strategies will necessitate the introduction of additional higher taxonomic categories to permit a true phylogenetic classification of all viruses. At the VIIIth International Congress of Virology (ICV) in Berlin (august 1990) the International Committee on the Taxonomy of Viruses (ICTV) has recognized the common ancestry of *Rhabdoviridae*, *Paramyxoviridae*, and *Filoviridae* (which incidentally also display diverse nucleocapsid morphologies) by bringing them together in a new (higher) taxonomic category: the order of the *Mononegavirales*.

Although a number of *species* may have to be reclassified, the introduction of phylogenetic taxonomy does not have to cause a revolution in virus systematics: many of the existing virus families and groups can probably be maintained. However, their correct organization into higher categories will require some serious consideration.

A meaningful classification of the members of the coronaviruslike superfamily clearly requires four hierarchical levels: the coronavirus and torovirus *species* have now been classified into two *genera* which belong to the *family Coronaviridae*. The obvious evolutionary link of this family to the arteriviruses would be reflected most accurately by promoting the present *genus* arterivirus to the family status and by establishing an *'order'* (to replace the 'superfamily') comprising the *Coronaviridae* and *Arteriviridae* families. Classification of the arteriviruses as a third genus of the coronavirus family is a less attractive alternative, because this would not recognize the more distant position of the arteriviruses (which do not carry a 'corona', anyway). However, this problem could be circumvented by establishing two subfamilies (*Coronavirinae* and *Arterivirinae*) and changing the family name *Coronaviridae* into something more appropriate.

The recently proposed polythetic definition of virus *species*[49] could also be applied to higher order taxa and would be flexible enough to accomodate the observed differences and similarities between the various members of the coronaviruslike superfamily. A polythetic class is defined by a large number of properties, each of which might also be absent in a member of the class or present in a member of another class[49]. The description of a virus *species* as 'a polythetic class of viruses constituting a replicating lineage and occupying a particular ecological niche' incorporates aspects from previous *species* definitions, but it can accommodate biological variability and genetic recombination more easily. From this *species* definition it is clear that, like in other areas of biology, the taxonomy of viruses should have a genetic basis[50]: as exemplified by the members of the coronaviruslike superfamily, virus evolution is governed by heredity (the passing of genetic information from parent to progeny) and the processes of mutation, recombination, and selection.

REFERENCES

1. R. Goldbach and J. Wellink, Intervirology 29:260 (1988).
2. J.H. Strauss and E.G. Strauss, Annu. Rev. Microbiol. 42:657 (1988).
3. M. Weiss, F. Steck and M.C. Horzinek, J. Gen. Virol. 64:1849 (1983).
4. G.N. Woode, L.J. Saif, M. Quesada, H.J. Winand, J.F.L. Pohlenz and N. Gourley, Am. J. Vet. Res. 46:1003 (1985).
5. G.M. Beards, C. Hall, J. Green, T.H. Flewett, F. Lamouliatte and P. du Pasquier, The Lancet ii:1050 (1984).

6. G.M. Beards, D.W.G. Brown, J. Green and T.H. Flewett, J. Med. Virol. 20:67 (1986).

7. M. Weiss and M.C. Horzinek, Vet. Microbiol. 11:41 (1986).

8. M.C. Horzinek, T.H. Flewett, L.J. Saif, W.J.M. Spaan, M. Weiss and G.N. Woode, Intervirology 27:17 (1987).

9. E.R. Doll, J.T. Bryans, W.H.M. McCollum and M.E. Wallace, Cornell. Vet. 47:3 (1957).

10. E.R. Doll, R.E. Knappenberger and J.T. Bryans, Cornell. Vet. 47:69 (1957).

11. W. Golnik, A. Moraillon and J. Golnik, J. Vet. Med. B 33:413 (1986).

12. P.J. Timoney, W.H. McCollum, A.W. Roberts and T.W. Murphy, Res. Vet. Sci. 41:279 (1986).

13. P.G.W. Plagemann and V. Moennig, Adv. Virus Res. 41:99 (1991).

14. E.K. Godeny, L. Chen, S. Kumar, S.L. Methven and M.A. Brinton, Virology, submitted.

15. J.J.M. Meulenberg M.M. Hulst, E.J. de Meijer, P.L.J.M. Moonen, A. den Besten, E.P. de Kluyver, G. Wensvoort and R.J.M. Moormann, Virology, in press (1992).

16. J.A. den Boon, E.J. Snijder, E.D. Chirnside, A.A.F. de Vries, M.C. Horzinek and W.J.M. Spaan, J. Virol. 65:2910 (1991).

17. M. Weiss and M.C. Horzinek, J. Gen. Virol. 67:1305 (1986).

18. E.G. Westaway, M.A. Brinton, S.Y. Gaidamovich, M.C. Horzinek, A. Igarashi, L. Kaariainen, D.K. Lvov, J.S. Porterfield, P.K. Russel and D.W. Trent, Intervirology 24:125 (1985).

19. B. Hyllseth, Arch. Ges. Virusforsch. 40:177 (1973).

20. M.C. Horzinek, J. Maess and R. Laufs, Arch. Ges. Virusforsch. 33:306 (1971).

21. A.A.F. de Vries, E.D. Chirnside, M.C. Horzinek and P.J.M. Rottier, J. Virol. 66: in press (1992).

22. E.J. Snijder, M.C. Horzinek and W.J.M. Spaan, J. Virol. 64:331 (1990).

23. E.J. Snijder, J.A. den Boon, W.J.M. Spaan, G.M.G.M. Verjans and M.C. Horzinek, J. Gen. Virol. 70:3363 (1989).

24. J.A. den Boon, E.J. Snijder, J. Krijnse Locker, M.C. Horzinek and P.J.M. Rottier, Virology 182:655 (1991).

25. E.J. Snijder, J.A. den Boon, W.J.M. Spaan, M. Weiss and M.C. Horzinek, Virology 178:355 (1990).

26. E.J. Snijder, J.A. den Boon, M.C. Horzinek and W.J.M. Spaan, Virology 180:448 (1991).

27. A.A.F. de Vries, E.D. Chirnside, P.J. Bredenbeek, L.A. Gravestein, M.C. Horzinek and W.J.M. Spaan, Nucleic Acids Res. 18:3241 (1990).

28. I. Brierley, P. Diggard and S.C. Inglis, Cell 57:537 (1989).

29. E.J. Snijder, J.A. den Boon, P.J. Bredenbeek, M.C. Horzinek, R. Rijnbrand and W.J.M. Spaan, Nucleic Acids Res. 18:4535 (1990).

30. P.J. Bredenbeek, C.J. Pachuk, J.F.H. Noten, J. Charité, W. Luytjes, S.R. Weiss and W.J.M. Spaan, Nucleic Acids Res. 18:1825 (1990).

31. E.J. Snijder, J.A. den Boon, M.C. Horzinek and W.J.M. Spaan, J. Gen. Virol. 72:1635 (1991).

32. W.J.M. Spaan, D. Cavanagh and M.C. Horzinek, J. Gen. Virol. 69:2939 (1988).

33. S.G. Sawicki and D.L. Sawicki, J. Virol. 64:1050 (1990).

34. P.B. Sethna, S.L. Hung and D.A. Brian, Proc. Natl. Acad. Sci USA 86:5626 (1989).

35. E.J. Snijder, unpublished results.

36. P.J.M. Rottier, G.W. Welling, S. Welling-Wester, H.G.M. Niesters, J. Lenstra and B.A.M. van der Zeijst, Biochemistry 25:1335 (1986).

37. M.E. Dubois-Dalcq, E.W. Doller, M.V. Haspel and K.V. Holmes, Virology 119:317 (1982).

38. P.J.M. Rottier and J.K. Rose, J. Virol. 61:2042 (1987).

39. M.C. Horzinek, J. Ederveen, B. Kaeffer, D. de Boer and M. Weiss, J. Gen Virol. 67:2475 (1986).

40. R.J. de Groot, W. Luytjes, M.C. Horzinek, B.A.M. van der Zeijst, W.J.M. Spaan and J.A. Lenstra, J. Mol. Biol. 196:963 (1987).

41. M.R. Denison, P.W. Zoltick, J.L. Leibowitz, C.J. Pachuk and S.R. Weiss, J. Virol. 65:3067 (1991).

42. M.R. Denison, P.W. Zoltick, S.A. Hughes, B. Giangreco, A.L. Olson, S. Perlman, J.L. Leibowitz and S.R. Weiss, Virology 189:274 (1992).

43. E.J. Snijder, A.L.M. Wassenaar and W.J.M. Spaan, J. Virol. 66: in press (1992).

44. E.J. Snijder, A.L.M. Wassenaar and W.J.M. Spaan, unpublished results.

45. D. Zimmern, in: RNA Genetics, vol. 2:211, J.J. Holland, E. Domingo, and P. Ahlquist, Eds., CRC Press, Boca Raton, USA (1987).

46. M.M.C Lai, Microbiol. Rev. 56:61 (1992).

47. D.W. Kingsbury, Intervirology 29:242 (1988).

48. R.E.F. Matthews, Intervirology 17:1 (1982).

49. M.H.V. van Regenmortel, Intervirology 31:241 (1990).

50. D.H.L. Bishop, Intervirology 24:79 (1985).

EQUINE ARTERITIS VIRUS CONTAINS A UNIQUE
SET OF FOUR STRUCTURAL PROTEINS

Antoine A.F. de Vries[1], Ewan D. Chirnside[2], Marian C. Horzinek[1], and Peter J.M. Rottier[1]

[1]Institute of Virology, Department of Infectious Diseases and Immunology, Veterinary Faculty, University of Utrecht, Yalelaan 1, 3584 CL Utrecht, The Netherlands, and [2]Equine Virology Unit, The Animal Health Trust, Lanwades Park, Kennett, Suffolk, England CB8 7PN

INTRODUCTION

Equine arteritis virus (EAV) belongs to a group of small enveloped positive-stranded RNA viruses, provisionally designated arteriviruses, which further includes lactate dehydrogenase-elevating virus (LDV), simian hemorrhagic fever virus (SHFV), and swine infertility and respiratory syndrome virus (SIRSV) or Lelystad virus (LV). These viruses resemble togaviruses in their physicochemical properties, the polarity and size (13-15 kb) of their genomes, and the size (50-70 nm) and morphology (isometric) of the virions[1,2] but virus assembly occurs by intracellular budding[1,3]. In contrast, the genome organization and gene expression strategy are similar to those of corona- and toroviruses[2,4,5,6,7,8].

The 5' three-fourths of the arterivirus genome consists of two overlapping open reading frames (ORFs 1a and 1b) which are translated from the genomic RNA (RNA 1) and encode the viral replicase. ORF 1b is expressed by ribosomal frameshifting and possesses four amino acid domains characteristic for the polymerases of corona- and toroviruses[2,4,6]. The 3' part of the arterivirus genome contains six overlapping ORFs expressed from a 3' coterminal nested set of subgenomic mRNAs (RNAs 2 to 7). The subgenomic mRNAs possess a common leader sequence derived from the 5' end of the viral genome. The leader sequence is joined to the coding parts of the mRNAs at conserved sequence motifs located upstream of ORFs 2 to 7[2,5,7]. The identity of the expression products of ORFs 2 to 7 has not yet been established except for LDV ORF 7 which encodes the nucleo-capsid protein[9].

To further elucidate the evolutionary relationship of arteriviruses with corona- and toroviruses we have studied the structural proteins of EAV and identified the corresponding genes.

Coronaviruses, Edited by H. Laude and J.F. Vautherot
Plenum Press, New York, 1994

MATERIALS AND METHODS

Cells and viruses

Cells were maintained in Dulbecco's modified Eagle's medium (DMEM) containing 10% fetal calf serum (FCS), 100 IU/ml penicillin, and 100 µg/ml streptomycin. A stock of the Bucyrus strain of EAV[10] was prepared in BHK-21 cells. A recombinant vaccinia virus expressing the bacteriophage T7 RNA polymerase gene (VTF7-3) was obtained from Dr. B. Moss.

Plasmid construction

ORF 6 was cloned behind a T7 promoter by insertion of a 556 bp ClaI-FspI fragment from cDNA clone 106[5] into ClaI- and SmaI-digested pBluescript SK(-). The resulting plasmid was designated pAVI16. ORF 5 was placed downstream of a T7 promoter by ligating a 1243 bp DraI-HindIII fragment from cDNA clone 008[5] into SmaI- and HindIII-digested pBS(+) to yield plasmid pAVI05. Subsequently, plasmid pAVI15 was constructed by cloning the 0.9kb SacI-XbaI fragment from pAVI05 into the polylinker region of pBluescript KS(+).

Preparation of antisera

Antisera directed against the carboxy termini of the ORF 7, 6, and 2 translation products were produced by subcutaneous injection of rabbits at monthly intervals with synthetic peptides SP07 (NH$_2$-Tyr-Trp-Val-Pro-Thr-Lys-Gln-Ile-Gln-Arg-Lys-Val-Ala-Pro-Pro-Ala-Gly-Pro-COOH), SP06 (NH$_2$-Tyr-Ala-Gly-Arg-Leu-Phe-Ser-Lys-Arg-Thr-Ala-Ala-Thr-Ala-Tyr-Lys-Leu-Gln-COOH), and SP02 (NH$_2$-Cys-Pro-Ser-Arg-Arg-Thr-Ser-Ser-Gly-Thr-Leu-Pro-Arg-Arg-Lys-Ile-Leu-COOH), respectively. The peptides were coupled to keyhole limpet hemocyanin prior to immunization. A serum directed against the structural proteins of EAV was prepared by multiple inoculations of a rabbit with sucrose gradient purified virus. For the primary immunizations the antigens were emulsified in Freund's complete adjuvant, for the booster injections Freund's incomplete adjuvant was used.

Expression of ORF 5 and 6

BHK-21 cells were transfected with plasmid pAVI16 or pAVI15 using cationic liposomes[11]. After incubation for 5 h at 37°C, the cells were infected at a MOI of 10 with the vaccinia virus recombinant vTF7-3 expressing the bacteriophage T7 RNA polymerase[12]. Metabolic labeling was started at 10 h after transfection. The labeling regimen and cell lysis procedure were as described for EAV-infected BHK-21 cells.

Radioactive labeling of intracellular proteins

Subconfluent monolayers of BHK-21 cells were infected with EAV at a MOI ≥ 20 as described previously[13]. The cells were starved in methionine free medium for 30 min at 7½ h p.i. and subsequently labelled for 20 min at 37°C by addition of 100 µCi/ml Tran[^{35}S]-label (> 1000 Ci/mmol; ICN). RK-13, LLC-MK2, and VERO C1008 cells were similarly labelled at 6, 8, and 12 h p.i., respectively. In some experiments a 2 h chase was performed with DMEM-10% FCS-2 mM methionine. The cells were lysed in 20 mM Tris-HCl (pH 7.6), 150 mM NaCl, 1% Nonidet P-40 (NP-40), 0.5% sodium deoxycholate, 0.1% sodium dodecyl sulfate (SDS) containing

1 µg/ml of aprotinin, leupeptin, and pepstatin A. The lysates were cleared by centrifugation and supplemented with EDTA to a 5 mM final concentration.

Preparation of radiolabeled virions

Subconfluent monolayers of BHK-21 cells were infected with EAV at a high MOI. After incubation for 7 h at 39.5°C the medium was replaced by methionine-free or cysteine-free medium containing 2% FCS and 100 µCi/ml L-[^{35}S]methionine or L-[^{35}S]cysteine (> 1000 Ci/mmol; ICN), respectively. At 10½ h p.i. one fifth a volume of DMEM-10% FCS was added and the cells were further incubated for 3½ h at 39.5°C. Finally, an excess methionine or cysteine was added and the incubation was continued for 30 min. The medium was then collected and the virus was purified by PEG precipitation followed by sucrose gradient fractionation[13].

Immunoprecipitation and gel electrophoresis

Protein samples were diluted in immunoprecipitation buffer (20 mM Tris-HCl [pH 7.6], 150 mM NaCl, 5 mM EDTA, 0.5% NP-40, 0.1% sodium deoxycholate, 1 µg/ml protease inhibitors) containing either 0.1% SDS (anti-virion and anti-SP07 serum) or 0.25% SDS (anti-SP06 and anti-SP02 serum); 3 µl rabbit serum was added and the samples were incubated overnight at 4°C. The immune complexes were adsorbed to Pansorbin^R (Calbiochem) and collected by centrifugation. Pellets were washed three times in 20 mM Tris-HCl (pH 7.6), 150 mM NaCl, 5 mM EDTA, 0.1% NP-40 and once in 20 mM Tris-HCl (pH 7.6), 0.1% NP-40. The immune complexes were dissolved in 25 µl Laemmli sample buffer and analyzed in SDS-15% polyacrylamide gels[14]. Alternatively, the immunoprecipitates were subjected to endoglycosidase treatment before gel electrophoresis.

Endoglycosidase treatment

Endoglycosidase F/N-glycosidase F (glyco F) (Boehringer Mannheim) digestions were carried out in 50 mM sodium phosphate (pH 6.8), 20 mM EDTA, 1% NP-40, 0.15% SDS, 1% 2-mercaptoethanol containing 1 µg/ml protease inhibitors. The incubations were done overnight at 30°C with 100 mU enzyme per reaction. Endo-ß-galactosidase (endo ß) (from *Bacteroides fragilis*; Boehringer Mannheim) digestions were performed overnight at 37°C using 2 mU enzyme per reaction in 50 mM sodium acetate (pH 5.75), 200 µg/ml acetylated BSA (Biolabs), 1 µg/ml protease inhibitors.

RESULTS

Identification of the structural proteins

To define the structural proteins of EAV, [^{35}S]methionine labeled virus was isolated from the culture medium of EAV-infected BHK-21 cells and analyzed by SDS-polyacrylamide gel electrophoresis (PAGE). A set of four virion proteins with molecular weights of 14, 16, 25, and 30 to 42 K was consistently observed (Fig. 1). The 14-, 16-, and 30- to 42-kDa proteins were recognized by an antiserum directed against sucrose gradient purified virus (Fig. 1). Glyco F treatment of radiolabeled virions demonstrated that the two largest proteins are N-glycosylated and confirmed the presence of four structural proteins (data not shown).

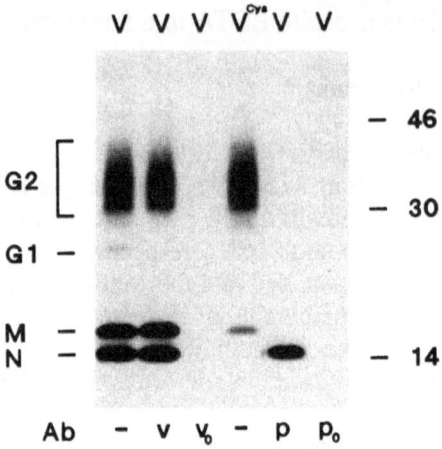

Figure 1. Protein composition of EAV and identification of the ORF 7 product. [^{35}S]methionine (V) and [^{35}S]cysteine (V^{Cys}) labeled virus were analyzed by SDS-PAGE directly or after immunoprecipitation with antibodies (Ab) directed against purified virus (v) or a peptide specific for the ORF 7 protein (p). The corresponding presera (v$_0$, p$_0$) did not recognize any of the virion proteins. The positions of the 14-kDa protein (N), the 16-kDa protein (M) and the 25-kDa protein (G$_S$) are indicated; the bracket depicts the size limits of the heterogeneously glycosylated 30- to 42-kDa protein (G$_L$). The position and size (in kDa) of marker proteins analyzed in the same gel are indicated at the right.

ORF 7 encodes the 14 kDa nucleocapsid (N) protein

The nucleocapsid of EAV is composed of a single-stranded, polyadenylated genomic RNA and a 12- to 14-kDa core protein[4,15]. Since ORF 7 encodes a basic protein with a predicted molecular weight of 12.3 K[4] and since the in vitro translation product of RNA 7 comigrated in gel with the core protein[16], it probably is the nucleocapsid gene. To test this hypothesis, we exploited the fact that only ORF 7 specifies a viral protein lacking cysteine[4]. Virions were labelled with [^{35}S]methionine or with [^{35}S]cysteine, purified in sucrose gradients and the protein patterns were compared. The nucleocapsid protein of 14 kDa prominent in [^{35}S]methionine labeled virus was indeed absent in the [^{35}S]cysteine labeled preparation (Fig. 1). Moreover, the [^{35}S]methionine labeled 14-kDa protein was also specifically precipitated using an antiserum directed against the 18 carboxy terminal amino acids of the deduced ORF 7 product (Fig. 1).

ORF 6 encodes the 16 kDa membrane (M) protein

The predicted ORF 6 product has a molecular weight close to 16 K, lacks N-glycosylation sites and contains five methionines and one cysteine. In vitro translation of ORF 6 yielded a protein of the anticipated size that was recognized by the anti-virion serum (data not shown). In combination with its relatively low incorporation of [^{35}S]cysteine (Fig. 1) and its insensitivity to glyco F treatment, these data suggested that the 16-kDa virion protein is encoded by ORF 6.

To prove this assumption, BHK-21 cells were transfected with plasmid pAVI16, which has ORF 6 cloned behind a T7 promoter, and infected with recombinant vaccinia virus vTF7-3 synthesizing the bacteriophage T7 RNA polymerase. The cells were metabolically labeled and the ORF 6 product was precipitated using the anti-virion serum or a specific anti-peptide serum. A 16-kDa protein indistinguishable from that in the [³⁵S]methionine labeled virus preparation was detected with both antisera but not with the corresponding presera (Fig. 2). The anti-virion serum also recognized a protein of 28 kDa. This protein is probably a dimer of the ORF 6 product formed during the analytical procedures. A strong tendency to aggregate has also been reported for the M protein of coronaviruses which shares many characteristics with the EAV ORF 6 product (see discussion). The inability of the anti-peptide serum to immunoprecipitate the dimer may be attributed to the stringent immunoprecipitation conditions (0.25% SDS) that would prevent complex formation. Alternatively, the anti-peptide serum may fail to precipitate the dimer because aggregation renders the carboxy terminus of the ORF 6 protein inaccessible for antibodies.

To complete the evidence that ORF 6 encodes the 16-kDa virion protein, [³⁵S]methionine labeled virus was subjected to immmunoprecipitation using the anti-peptide serum. The 16-kDa virion protein was clearly recognized by the anti-peptide serum (Fig. 2), but a substantial amount of the 30- to 42-kDa protein was precipitated too. The corresponding preserum did not recognize this 30- to 42-kDa smear. The co-precipitation of the 30- to 42-kDa protein presumably results from a specific association with the 16-kDa protein as it is unlikely that both proteins share a common epitope recognized by the anti-peptide serum. This may also explain why the anti-virion serum only weakly binds the ORF 6 product synthesized in pAVI16-transfected cells.

ORF 5 encodes the heterogeneously glycosylated 30-42 kDa (G$_L$) protein

The predicted ORF 5 product has a molecular weight of 28.7 K and contains one potential N-glycosylation site[4]. In vitro translation of ORF 5 in the presence of microsomes yielded a 30-kDa protein which was strongly recognized by the anti-virion serum. Its apparent molecular weight was reduced to 27 K after glyco F treat-

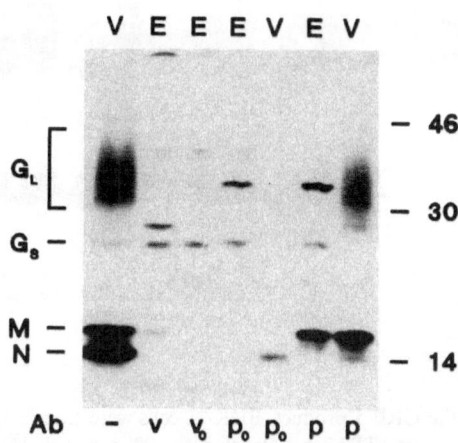

Figure 2. Characterization of the ORF 6 product. BHK-21 cells were transfected with pAVI16, infected with recombinant vaccinia virus vTF7.3 and labeled with [³⁵S]methionine. The ORF 6 product was immunoprecipitated from the cell lysate using the anti-virion serum (v) or a specific anti-peptide serum (p) and the immunoprecipitates were analyzed by SDS-PAGE (lanes E). The protein patterns were compared with those of [³⁵S]methionine labeled virus applied to the gel directly or after immunoprecipitation with the anti-peptide serum (lanes V). The corresponding presera (v$_0$, p$_0$) failed to immunoprecipitate the ORF 6 protein from both the transfected cells and the virus preparation.

ment (data not shown). Since the same molecular weight was found for the glyco F digested 30- to 42-kDa virion protein (data not shown) we speculated that ORF 5 encodes the 30- to 42-kDa smear.

To study the ORF 5 expression product, its coding sequence was cloned behind a T7 promoter. The resulting plasmid, designated pAVI15, was expressed in BHK-21 cells which were then labeled for 20 min with Tran[^{35}S]-label. Immunoprecipitates were prepared using the anti-virion serum and treated or mock-treated with glyco F. A discrete 30-kDa product was detected whose molecular weight was reduced to 27 K by the glycosidase (Fig. 3). A 30-kDa protein was also immunoprecipitated from a lysate of EAV-infected cells labeled for the same period. Glyco F digestion converted this protein into a 27-kDa species which comigrated with the deglycosylated form of both the ORF 5 expression product and the 30- to 42-kDa virion protein. To determine whether the 30- to 42-kDa protein derives from the 30-kDa protein by maturation of the N-glycan, we performed a pulse-chase experiment with EAV-infected BHK-21 cells. The 30-kDa protein observed after pulse labeling was processed during a 2 h chase to a 28- to 42-kDa smear (Fig. 3). After treatment with glyco F, the smear was changed into the 27-kDa protein identified before.

To investigate whether the broad size distribution of the 30- to 42-kDa protein is caused by polylactosaminoglycan modification, [^{35}S]methionine labeled virions were treated with endo ß. This enzyme cleaves the internal ß1→4 galactosidic linkages of oligosaccharides with the general sequence R_1-GlcNAcß1→3Galß1→4 GlcNAc/Glc-R_2[17]. Whereas the 25-kDa glycoprotein was resistant to endo ß treatment, the 30- to -42 kDa smear was converted to a discrete band of 30 kDa (Fig. 3). The slight size difference between the glyco F and endo ß treated ORF 5 protein reflects the fact that the latter leaves part of the N-linked oligosaccharide side chain

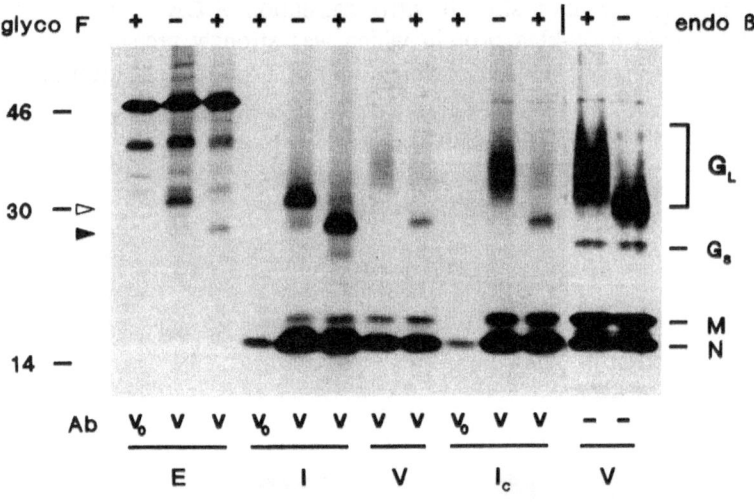

Figure 3. Characterization of the ORF 5 product. BHK-21 cells were transfected with pAVI15, infected with recombinant vaccinia virus vTF7.3 and labeled for 20 min with [^{35}S]methionine (lanes E). EAV-infected BHK-21 cells were labeled with [^{35}S]methionine for the same period (lanes I). Alternatively, the 20 min pulse labeling of EAV-infected cells was followed by a 2 h chase (lanes I_c). The ORF 5 protein was immunoprecipitated from the cell lysates and from a [^{35}S]methionine labeled virus preparation (lanes V) using the anti-virion serum (v). Immunoprecipitations with the corresponding preserum (v_0) were included as a control. The immunoprecipitates were treated (+) or mock-treated (-) with glyco F and analyzed by SDS-PAGE. The position of the glycosylated ORF 5 product obtained after a 20 min pulse is indicated by an open arrowhead. The closed arrowhead points to the glyco F treated ORF 5 protein. The two lanes on the right display [^{35}S]methionine labeled virus treated (+) or mock-treated (-) with endo ß.

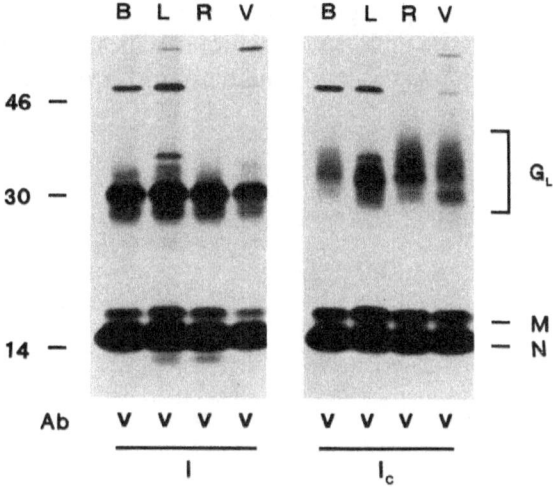

Figure 4. Maturation of the ORF 5 protein in different cell lines. EAV-infected BHK-21 (B), LLC-MK2 (L), RK-13 (R), and VERO C1008 (V) cells were labelled with [³⁵S]methionine for 20 min (I) and chased for 2 h (I_c). The ORF 5 protein was imunoprecipitated from the cell lysates with the anti-virion serum (v) and analyzed by SDS-PAGE.

intact. Since the nature and extent of the polylactosaminoglycan modification of proteins is determined by the particular cell line in which it is synthesized[18,19], we infected four different cell types with EAV and analyzed the ORF 5 products. After pulse labeling for 20 min a single protein of 30 kDa was detected with each cell line. During a 2 h chase the 30-kDa protein was converted into a smear whose appearance in gel was different for distinct cell types (Fig. 4). The heterogeneity in N-glycosylation of the ORF 5 protein was also found in the virions produced by these cell lines (data not shown).

Significantly more of the ORF 6 protein was immunoprecipitated after a 2 h chase than after a pulse (Fig. 3 and 4). Efficient precipitation of the ORF 6 protein by the anti-virion serum may rely upon its interaction with the ORF 5 protein; complex formation between these two may require (partial) maturation of the ORF 5 and/or ORF 6 protein. The failure of the ORF 6 specific anti-peptide serum to recognize the expression product of ORF 5 (data not shown) together with its ability to co-precipitate a substantial portion of the ORF 5 protein from a virus preparation (Fig. 2) provides further evidence for an interaction between the ORF 5 and 6 proteins.

ORF 2 encodes the glycosylated 25 kDa (G_S) protein

The deduced ORF 2 product possesses one possible N-glycosylation site and has a calculated molecular weight of 22.7 K after removal of the putative amino terminal signal sequence[4]. In vitro translation of ORF 2 in the presence of microsomes yielded a product which comigrated with the 25-kDa virion protein. Since the molecular weight of both the 25-kDa virion protein and the translation product of ORF 2 was decreased to 22 K by glyco F treatment (data not shown), ORF 2 was thought to encode the 25-kDa virion protein. For this reason, an anti-peptide serum was raised against the predicted carboxy terminus of the ORF 2 product. The antiserum was applied to a preparation of [³⁵S]methionine labeled virions. The resulting immunoprecipitate was divided in two portions one of which was treated with glyco F. The anti-peptide serum specifically recognized the 25-kDa virion protein which was converted to a protein of 22 kDa by glyco F (Fig. 4).

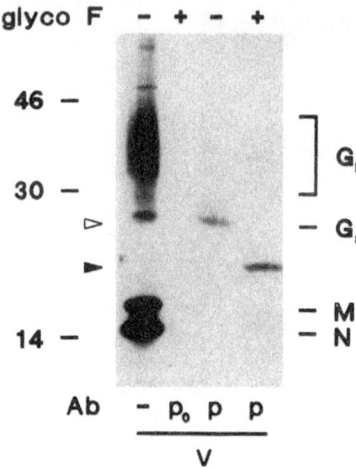

Figure 5. Characterization of the ORF 2 product. The ORF 2 protein was immunoprecipitated from a [^{35}S]methionine labeled virus preparation (lanes V) with a specific anti-peptide serum (p). An immunoprecipitation with the corresponding preserum (p_0) was carried out as a control. The immunoprecipitates were treated ($+$) or mock-treated ($-$) with glyco F and analyzed by SDS-PAGE together with [^{35}S]methionine labeled virus. The open arrowhead marks the position of the glycosylated ORF 2 product, the closed arrowhead points to the ORF 2 product obtained after digestion with glyco F.

DISCUSSION

In this paper we showed that the protein composition of EAV is entirely different from that of corona- and toroviruses. From the combination of our results with previous data[15] it follows that virions contain a 14-kDa nucleocapsid protein (N) and three envelope proteins designated M, G_S, and G_L. M is an unglycosylated protein of 16 kDa, G_S and G_L are N-glycosylated proteins of 25 kDa and 30 to 42 kDa, respectively. N, M, and G_L are major structural proteins, G_S is only a minor virion constituent. Although both glycoproteins contain a single N-glycan and pass the same intracellular compartments as part of a virion, maturation of their N-linked oligosaccharide side chains is different. The mature G_L protein is susceptible to digestion with endo-ß-galactosidase which indicates that it becomes glycosylated by the addition of variable numbers of lactosamine (Galß1→4GlcNAcß1→3) repeats to the N-linked core oligosaccharide. G_L is modified in a distinct way in different cell lines as is the case for other proteins that acquire polylactosaminoglycans[18,19]. The extracellular G_S protein is resistant to treatment with endo-ß-galactosidase and endoglycosidase H (data not shown) and therefore probably acquires a regular complex type N-glycan. To date the only other viral proteins known to acquire a polylactosaminoglycan are the NB protein of influenza B virus[18] and the SH proteins of human and bovine respiratory syncytial viruses[19]. The function of the poly-N-acetyllactosamine modification attached to these proteins is unknown.

We also identified the genes for the structural proteins of EAV and compared them with those of corona- and toroviruses. ORF 7 codes for N, ORF 6 for M, ORF 5 for G_L, and ORF 2 for G_S. No amino acid sequence similarities between the structural protein genes of EAV and those of corona- and toroviruses were observed. However, the hydrophobicity plot of the EAV ORF 6 protein appears to be very similar to that of the membrane protein (M) of coronaviruses[20] and of the envelope protein (E) of toroviruses[21], which suggests that it is also a type III integral membrane proteins containing three successive transmembrane helices. Since the EAV ORF 6 protein accumulates in intracellular membranes too (data not shown), we anticipate that the M (E) proteins of arteri-, corona- and toroviruses are functional homologs. Moreover, the positions of the structural protein genes in the EAV genome correspond with a corona- and toroviruslike gene order: 5'-replicase-envelope glycoprotein(s)-membrane protein (M)-nucleocapsid protein (N)-3'.

The collective data illustrate that arteriviruses resemble corona- and toroviruses in their replication and transcription mechanism although the structural proteins are largely different from those of corona- and toroviruses. The universal presence of a triple-spanning membrane protein among the nested set viruses may relate to their specific mode of intracellular budding. A detailed analysis of the virion structure of arteriviruses and the properties of their structural proteins may uncover yet unsuspected similarities with corona- and toroviruses.

REFERENCES

1. P.W.G. Plagemann, and V. Moennig, *Adv. Vir. Res.* 41:99 (1992).
2. J.J.M. Meulenberg, M.M. Hulst, E.J. de Meijer, P.L.J.M. Moonen, A. den Besten, E.P. de Kluyer, G.Wensvoort, and R.J.M. Moormann, *Virology*, in press.
3. J.M.A. Pol, unpublished results.
4. J.A. den Boon, E.J. Snijder, E.D. Chirnside, A.A.F. de Vries, M.C. Horzinek, and W.J.M. Spaan, *J. Virol.* 65:2910 (1991).
5. A.A.F. de Vries, E.D. Chirnside, P.J. Bredenbeek, L.A. Gravestein, M.C. Horzinek, and W.J.M. Spaan, *Nucleic Acids Res.* 18:3241 (1990).
6. E.K. Godeny, L.Chen, S. Kumar, S.L. Methven, and M.A. Brinton, *Virology*, submitted.
7. L. Kuo, Z. Chen, R.R.R. Rowland, K.S. Faaberg, and P.G.W. Plagemann, *Virus Res.* 23:55 (1992).
8. E.K. Godeny, unpublished results.
9. E.K. Godeny, D.W. Speicher, and M.A. Brinton, *Virology* 177:768 (1990).
10. E.R. Doll, J.T. Bryans, W.H. McCollum, and M.E.W. Crowe, *Cornell Vet.* 47:3 (1957).
11. P.L. Felgner, T.R. Gadek, M. Holm, R. Roman, H.W. Chan, M. Wenz, J.P. Northrop, G.M. Ringold, and M. Danielsen, *Proc. Natl. Acad. Sci. USA* 84:7413 (1987).
12. T.R. Fuerst, E.G. Niles, F.W. Studier, and B. Moss, *Proc. Natl. Acad. Sci. USA* 83:8122 (1986).
13. M.F. van Berlo, M.C. Horzinek, and B.A.M. van der Zeijst, *Virology* 118:345 (1982).
14. U.K. Laemmli, *Nature (London)* 227:680 (1970).
15. B. Hyllseth, *Arch. Ges. Virusforsch.* 40:177 (1973).
16. M.F. van Berlo, P.J.M. Rottier, W.J.M. Spaan, and M.C. Horzinek, *J. Gen. Virol.* 67:1543 (1986).
17. P. Scudder, K.-i. Uemura, J. Dolby, M.N. Fukuda, and T. Feizi, *Biochem. J.* 213:485 (1983).
18. M.A. Williams, and R.A. Lamb, *Mol. Cell. Biol.* 8:1186 (1988).
19. K. Anderson, A.M.Q. King, R.A. Lerch, and G.W. Wertz, *Virology* 191:417 (1992).
20. P.J.M. Rottier, G.W. Welling, S. Welling-Wester, H.G.M. Niesters, J.A. Lenstra, and B.A.M. van der Zeijst, *Biochemistry* 25:1335 (1986).
21. J.A. den Boon, E.J. Snijder, J. Krijnse Locker, M.C. Horzinek, and P.J.M. Rottier, *Virology* 182:655 (1991).

THE *CORONAVIRIDAE* NOW COMPRISES TWO GENERA, *CORONAVIRUS* AND *TOROVIRUS*: REPORT OF THE *CORONAVIRIDAE* STUDY GROUP

D. Cavanagh[1], D.A. Brian, M.A. Brinton, L. Enjuanes, K.V. Holmes, M.C. Horzinek, M.M.C. Lai, H. Laude, P.G.W. Plagemann, S.G. Siddell, W.J.M. Spaan, F. Taguchi and P.J. Talbot

[1]Division of Molecular Biology
AFRC Institute for Animal Health
Compton Laboratory
Compton, Newbury RG16 0NN, UK

ABSTRACT

At the April 1992, mid-term meeting of the International Committee on Taxonomy of Viruses (ICTV) a proposal from the *Coronaviridae* Study Group (CSG) to include the *torovirus* genus in the *Coronaviridae* was accepted. Following another proposal, the *arterivirus* genus was removed from the *Togaviridae* but not assigned to another family. The arteriviruses have some features in common with the *Coronaviridae* but also have major differences. After much debate, culminating in September 1992, it was decided that the CSG would not recommend inclusion of *arterivirus* in the *Coronaviridae*. It was agreed that (a) the nomenclature used for coronavirus genes, mRNAs and polypeptides (Cavanagh et al., 1990) should be used for toroviruses, (b) that the small (about 100 amino acids) membrane-associated protein, which is distinct from the integral membrane glycoprotein M, associated with virions of infectious bronchitis (Liu & Inglis, 1991) and transmissible gastroenteritis (Godet et al., 1992) coronaviruses would be referred to by the acronym sM (lower case 's') and (c) that 'pol' (polymerase) should be used as a working term for gene 1, which comprises open reading frames (ORFs) 1a and 1b in both genera of the *Coronaviridae*.

COMPARISON OF CORONAVIRUSES AND TOROVIRUSES

The main features of the toroviruses, of which there are two known species (Berne virus and Breda virus), and coronaviruses are described in detail in the chapter by Snijder

Coronaviruses, Edited by H. Laude and J.F. Vautherot
Plenum Press, New York, 1994

et al. in this volume. The decision to include the toroviruses in the *Coronaviridae* was in recognition of the many characteristics shared by these two genera, most of the similarities having come to light only recently with the sequencing of a large part of the genome of Berne virus, the type species and most thoroughly studied torovirus. Some of the features of the two genera are summarized in Table 1. Both genera comprise viruses which are enveloped and have prominent spike (S) glycoproteins of ~200 kDa which exhibit a coiled-coil structure in the carboxy-terminal, membrane-anchoring half. Each virus possesses an integral membrane protein (M)(~25 kDa) with three membrane-spanning sequences in the amino-terminal half. The genomes are single-stranded, non-segmented, positive-sense RNAs of ~30 kb, the first two-thirds being the gene (number 1) which encodes the presumptive RNA-dependent RNA polymerase. Gene 1 codes for two overlapping ORFs, 1a and 1b, ribosomal frameshifting being involved in the translation of the second ORF. Five or more subgenomic mRNAs are generated forming a 3'co-terminal nested set. Only the 5' sequence not possessed by the next smaller mRNA is translated. Overall genome organisation is similar, the gene order being 5'-pol-S-M-N-3', where N is the nucleocapsid protein. There are additional genes, some of which are not common between the two genera and, indeed, not possessed by all members of one genus e.g. the haemagglutinin-esterase glycoprotein of some coronaviruses.

A number of features require that the coronaviruses and toroviruses should be in separate genera. There is virtually no sequence similarity between the two groups. The N proteins differ greatly in size (Table 1) and form differently shaped nucleocapsids. The viruses are of similar size, about 130 nm in diameter, the coronaviruses being pleomorphic but roughly spherical in shape. In negatively-stained preparations toroviruses can look very similar to coronaviruses but in ultra-thin sections toroviruses can have disc-, kidney- or rod shapes. Leaders are present on the 5'termini of coronavirus mRNAs but these have not been found on Berne virus mRNAs.

ARTERIVIRUSES

The genus *arterivirus* includes equine arteritis and lactate dehydrogenase-elevating virus. Like members of the *Coronaviridae*, the arteriviruses are enveloped, with a single-stranded, non-segmented, positive-sense RNA genome and with similar genome organisation (Table 1). A 3' co-terminal nested set of five or more mRNAs are produced, ribosomal frameshifting is involved in translation of the pol gene and the integral membrane protein has a triple membrane-spanning domain. However, there are several major differences from the other two genera. The arterivirus nucleocapsid is icosahedral, the virions being only 50-70 nm in diameter. The surface glycoprotein is neither prominent nor does it have a coiled-coil structure and comprises a much smaller polypeptide, as do M and N, than the corona- and torovirus counterparts. The genome is only ~13 kb. The arteriviruses are described in more detail in the chapters by Snijder *et al.* and de Vries *et al.* in this volume.

The CSG would have liked to have assigned *arterivirus* to a taxon such that it drew attention to the similarities with the corona- and toroviruses while at the same time emphasizing the important differences. No proposal met with general approval. Consequently it was agreed that no formal proposal regarding the taxonomy of the arteriviruses would be made by the CSG and that the matter would be referred back to the ICTV Executive Committee.

Table 1. Comparison of some features of coronaviruses, toroviruses and arteriviruses.

Feature	*coronavirus*	*torovirus*	*arterivirus*
Enveloped	+	+	+
Nucleocapsid	helical	tubular	isometric
Positive ssRNA	+	+	+
Genome size (kb)	⁻30	⁻30	⁻13
3'co-terminal nested set of mRNAs	+	+	+
Leader on mRNAs	+	not detected	+
Ribosomal frameshifting during pol translation	+	+	+
Prominent spikes	+	+	−
Coiled-coil structure in spike protein	+	+	−
Size of some virion proteins (kDa):			
large surface glycoprotein	200	200	42
integral membrane protein	25	26	18
nucleocapsid protein	60	18	12
Integral membrane protein with triple membrane-spanning sequences	+	+	+

REFERENCES

1. D. Cavanagh, D.A. Brian, L. Enjuanes, K.V. Holmes, M.M.C. Lai, H. Laude, S.G. Siddell,W.J.M. Spaan, F. Taguchi and P.J. Talbot. *Virology* 176:306 (1990).
2. D.X. Liu and S.C. Inglis. *Virology* 185:911 (1991).
3. M. Godet, R. L'Haridon, J-F. Vautherot and H. Laude. *Virology* 188:666 (1992).

Chapter 5

Cellular Receptors for Coronaviruses

CORONAVIRUS RECEPTOR SPECIFICITY

Kathryn V. Holmes,[1] Gabriela Dveksler,[1] Sara Gagneten,[1] Curtis Yeager,[2] Sue-Hwa Lin,[3] Nicole Beauchemin,[4] A. Thomas Look,[5] Richard Ashmun[5,6] and Carl Dieffenbach[7]

[1]Department of Pathology and
[2]Department of Microbiology
 Uniformed Services University of the Health Sciences
 Bethesda, MD 20814-4799
[3]Department of Biochemistry
 Baylor University School of Medicine
 Houston, TX 77030
[4]McGill Cancer Center, Montreal, Canada, H3G 1Y6
[5]Departments of Experimental Oncology and Tumor Cell Biology,
 St Jude Children's Research Hospital
 Memphis, TN 38105
[6]Department of Pediatrics, University of Tennessee College of Medicine
 Memphis, TN 38163
[7]Division of AIDS, NIAID, NIH, Bethesda, MD 20852

INTRODUCTION

Bang and his colleagues first showed the importance of the host in determining susceptibility to coronavirus infection, using mouse hepatitis virus MHV-2 as a model system which could cause death of Pri mice but not of C3H mice or macrophages isolated from them (1). Subsequently, as additional coronaviruses were recognized, it became clear that these viruses generally have narrow host ranges and show marked tissue specificity for replication (2). We have studied the importance of virus-receptor interactions in the host range and tissue tropism of coronavirus infections.

IDENTIFICATION OF A RECEPTOR FOR MOUSE HEPATITIS VIRUS MHV-A59

Our first approach to identifying a receptor for MHV-A59 utilized a solid phase virus binding assay in which virus bound to undenatured intestinal brush border membranes (BBM) from MHV-A59-susceptible BALB/c mice (3). In contrast, the virus did not bind to BBM from adult SJL/J mice, which are much more resistant to MHV-A59 infection than BALB/c mice (3). This suggests that differences in virus binding to membranes of these mouse strains might play a role in the observed difference in susceptibility to MHV infection. Incubation of viruses with BBM proteins separated by SDS-PAGE in a virus-overlay protein blot assay (VOPBA) showed that MHV-A59 bound to a 110-120 kDa glycoprotein and, less stongly, to a 58 kDa glycoprotein in BALB/c BBM, but not to BBM proteins from SJL/J mice. This 110 kDa glycoprotein was a candidate for a virus receptor.

Antibody that recognized the 110 kDa glycoprotein was raised by immunizing SJL/J mice with BBM from BALB/c mice (4). Both the polyclonal mouse antibody and a monoclonal antibody, MAb-CC1, recognized the 110 kDa and 58 kDa BALB/c glycoproteins in immunoblots of BBM proteins and blocked virus infection of several mouse fibroblast and macrophage cell lines (5). Anti-receptor MAb-CC1 also partially protected infant mice from infection with MHV-A59 (6). Williams et al. isolated the glycoprotein from Swiss Webster mouse liver by immunoaffinity chromatography with MAb-CC1, and showed by N-terminal amino acid sequencing that the 110 kDa glycoprotein was a murine member of the carcinoembryonic antigen (CEA) family of glycoproteins in the immunoglobulin superfamily (5).

CEA-related glycoproteins are expressed on many epithelial cell membranes where they are believed to play a role in intercellular adhesion (7). A radioimmunoassay (RIA) that measured binding of radiolabeled MAb-CC1 to crude membrane preparations from many tissues showed that the MHV receptor glycoproteins were present in highest amounts on BALB/c colon, small intestine and liver which are the principal targets for virus replication in vivo (5). This RIA was not sensitive enough to detect the receptor on membranes of other tissues or of cultured mouse cells that are susceptible to MHV-A59 virus infection. The presence of this receptor antigen on membranes of mouse cell lines was demonstrated by the finding that MAb-CC1 could block MHV-A59 infection of the cells. These experiments suggest a correlation between the tissue tropism of MHV-A59 in vivo with the degree of expression of a receptor moiety.

The cDNA encoding the 110 kDa BALB/c protein (MHVR) that binds MHV-A59 was cloned and expressed in BHK hamster kidney cells and human RD cells (8). These cell lines are normally resistant to MHV-A59 infection, but they became susceptible to infection following transfection with MHVR cDNA. This indicates that MHV-A59 cannot infect hamster cells because they do not express a suitable receptor for the virus. The importance of receptor specificity in determining the host range of mouse coronavirus MHV was further supported by the observation that MHV-A59 did not bind detectably to intestinal BBM of any species except mice (9).

A series of recombinant MHVR proteins with deletions of specific domains was made and transiently expressed in BHK cells which were then tested for susceptibility to MHV-A59 infection (10). These experiments showed that both MHV-A59 virus and MAb-CC1 bound to the first 133 amino acids of MHVR corresponding to the N-terminal immunoglobulin-like domain.

As shown for many other CEA-related glycoproteins, we found that many different isoforms of MHVR glycoproteins can be co-expressed in murine cells in various combinations and ratios. As described in the accompanying paper (11), we examined the ability of each of 5 of these murine glycoproteins to serve as functional receptors for MHV-A59 when expressed in hamster cells. We found that MHVR and several splice variants were functional receptors as was a glycoprotein from SJL/J mice that is homologous to the 58 kDa glycoprotein of MHVR (10). This is the first report of multiple alternative receptors for an enveloped virus. It raises the important questions of whether all of these glycoprotein isoforms are equally effective as receptors for MHV, or whether some strains of virus may preferentially utilize different receptor isoforms. If there were differences in receptor utilization among various MHV strains, then differences in tissue tropism and virulence of these virus strains might be related to the specificity of virus-receptor interactions. Quantitative evaluation of the affinity of MHV S glycoproteins with the alternative MHV receptors will be required to address this novel and important aspect of the role of virus receptors in virus strain differences.

STUDIES ON THE HOST RANGE AND TISSUE TROPISM OF RAT CORONAVIRUS

Rat coronaviruses are closely related to MHV serologically, but they cause quite different diseases. Sialodacryadenitis virus (RCV-SDAV) and Parker's rat coronavirus (RCV-P) infect the respiratory epithelium, salivary and lacrimal glands of rats, but not intestine or liver (12,13). RCV-SDAV was found to bind to membrane proteins of rat respiratory epithelium, salivary and lacrimal glands but not to rat or mouse intestine or liver. Thus, the specificity of binding of the rat coronavirus to membranes reflects both the host range and tissue tropism of the virus in vivo (14).

Percy and his colleagues showed that rat coronaviruses could replicate in mouse L2 cells which also support the replication of MHV-A59 (15). To determine whether RCV-SDAV could utilize the MHV receptor to enter L2 cells, we pre-treated the cells with MAb-CC1 and then challenged them with the virus (14). We found that MAb-CC1

Table 1. Susceptibility of Mouse Strains to Rodent Coronavirus Infection and Protection by Anti-receptor Antibody MAb-CC1

Cell Line	Virus	Susceptibility to Infection	Protection by MAb-CC1
L2(Percy)	MHV-A59	Yes	Yes
	RCV-SDAV	Yes	No

prevented MHV-A59 infection of these cells, but failed to inhibit replication of RCV-SDAV (Table 1). These observations suggest that rat coronavirus may utilize not MHVR or its splice variants but a different receptor on L2(Percy) cells.

Unlike MHV-A59, rat coronaviruses did not bind to only one major membrane protein from its target tissues in VOPBAs. Rat tissues do, however, express CEA-related glycoproteins homologous to MHVR (16). To determine whether a rat CEA-related glycoprotein could serve as a receptor for RCV-SDAV, we used this virus to challenge COS cells transfected with the rat glycoprotein, called Ecto-ATPase (17), that is a homolog of mouse MHVR and human biliary glycoprotein. No infection was detected by immunofluorescence of cells with anti-viral antibody at 8 hours after virus challenge.

Additional studies on the virus receptor activities of other CEA-related glycoproteins of the rat are in progress. These studies raise the interesting possibility that closely related coronaviruses of the mouse and rat utilize two different types of receptors on the membranes of their host cells, and that availability of these receptors is an important determinant of virus susceptibility.

IDENTIFICATION OF A RECEPTOR FOR HUMAN CORONAVIRUS HCV-229E

Human coronaviruses (HCVs) are found in two different antigenic groups of coronaviruses. Although HCV-OC43 is serologically related to MHV and RCV, its receptor has not yet been identified. HCV-229E is in a serogroup with feline infectious peritonitis virus, canine coronavirus, and porcine enteric coronavirus TGEV (18,19). We prepared monoclonal antibodies directed against human cell lines that are susceptible to infection with HCV-229E, and tested them for their ability to protect these cells from HCV-229E infection. MAb-RBS was found to block infection, and this antibody immunoprecipitated a 150 kDa glycoprotein from membranes of the human cell lines (20). As found for TGEV (21), a receptor for HCV-229E is aminopeptidase N (APN), a zinc-

Table 2. Correlation of Expression of Coronavirus Receptors Tissues in their Natural Hosts with Sites of Virus Infection.

Virus	Receptor	Properties	Sites of	
			Receptor Expression	Infection
MHV	MHVR MHVR(2d), mmCGM2	110 kDa 58 kDa	Liver Intestinal epithelium Respiratory epithelium Brain Spleen	Liver Intestinal epithelium Respiratory epithelium Brain Spleen
RCV	Unknown	?	Salivary glands Lacrimal glands Respiratory epithelium	Salivary glands Lacrimal glands Respiratory epithelium
HCV-229E	Human amino-peptidase N	150 kDa	Respiratory epithelium Enteric epithelium Monocytes Granulocytes Synapses	Respiratory epithelium

binding metalloprotease found on monocytes, granulocytes, intestinal brush border and nerve synapses (22,23) and respiratory epithelium (24). Hamster or mouse cells are resistant to infection with HCV-229E unless they are tranfected with the cDNA encoding hAPN. Both the activity of the enzyme and virus infection can be blocked by treatment of the cells with MAb-RBS or by chelating the zinc ions from the medium (20,22). These experiments show that the receptor for HCV-229E is hAPN, a cell membrane glycoprotein unrelated to MHVR.

CONCLUSIONS

The recognition of different types of cellular receptors by the spike glycoproteins of MHV-A59 and HCV-229E appears to be an important determinant of tissue tropism and host range of these coronaviruses. Accordingly, mutations in the gene encoding the S glycoprotein might be expected to result in alterations in receptor specificity and altered host range and/or tissue tropism of the virus. Naturally occurring strains of MHV and IBV and mutants selected with monoclonal antibodies have been shown to have major differences in the size, amino acid sequence and antigenicity of their S glycoproteins. In some cases these mutations have been correlated with alterations in biological properties (25-27). Development of methods to introduce recombinant mutant glycoproteins into the coronavirus genome will permit more detailed studies on the biological significance of receptor specificity as a determinant of tissue tropism and host range.

The levels of receptor expression in different tissues and selective use of alternative receptors by different virus strains are probably important factors of virus-receptor interactions that affect the biological activity of receptors. Although the principal target tissues for MHV-A59 are the tissues in which the receptors are expressed in the highest amounts (5), other tissues in which MHVR-related glycoproteins are also expressed are not common sites of virus infection or symptomatology. Table 2 shows the murine and human coronavirus receptors and some of the cell types in which they are known to be expressed (28-30). It is likely that in some of these tissues, host factors that affect steps in virus replication subsequent to virus attachment to its receptor will be found to determine susceptibility to coronavirus infection.

ACKNOWLEDGEMENTS

The authors are grateful to P. Elia, H. Nguyen, A. Basile and C. Cardellichio for excellent technical assistance. This work was supported in part by NIH grants AI25231 and AI26075. The opinions and assertions expressed herein are those of the authors and do not reflect the views of the Uniformed Services University or the Department of Defense.

REFERENCES

1. F.B. Bang and A. Warwick, (1960). *Proc. Natl. Acad. Sci. U.S.A.* 46: 1065-1075 .
2. H. Wege, S. Siddell and V. Ter Meulen, (1982) *Curr.Top. Microbiol. Immunol.* 99: 165-200.
3. J.F. Boyle, D.G. Weismiller and K.V. Holmes, (1987). *J.Virol.* 61: 185-189.
4. R.K. Williams, G.S. Jiang, S.W. Snyder, M.F. Frana, K.V. and Holmes, (1990). *J. Virol.* 64 :3817-3823 .
5. R.K. Williams, G.S. Jiang and K.V. Holmes, (1991). *Proc. Natl. Acad. Sci. U.S.A.* 88: 5533-5536 .
6. A.L. Smith, C.B. Cardellichio, D.F. Winograd, M.S. Desouza, S.W. Barthold and K.V. Holmes, (1991). *J. Infect. Dis.* 163: 879-882 .
7. A. Yachi and J.E. Shively, "The Carcinoembryonic Gene Family", A. Yachi and J.E. Shively (eds), Elsevier Science Publishers BV, New York, NY (1989).
8. G.S. Dveksler, M.N. Pensiero, C.B. Cardellichio, R.K. Williams, G.S. Jiang, K.V. Holmes and C.W. Dieffenbach, (1991). *J. Virol.* 65: 6881-6891.
9. S.R. Compton, C.B. Stephensen, S.W. Snyder, D.G. Weismiller and K.V. Holmes, (1992). *J. Virol.* in press .
10. G.S. Dveksler, M.N. Pensiero, C.W. Dieffenbach, C.B. Cardellichio, A.A. Basile, P.E. Elia and K.V. Holmes, (1992). *Proc. Natl. Acad. Sci. U.S.A.* in press .

11.G.S. Dveksler, A.A. Basile, C.B. Cardellichio, N. Beauchemin, C.W. Dieffenbach and K.V. Holmes. in :"Coronaviruses: Molecular Biology and Pathogenesis", H. Laude and J.F. Vautherot (eds), Plenum Publishing Corp.NY in press (1993).

12. D.H. Percy, P.E. Hanna, F.X. Paturzo, (1984). *Lab. Anim. Sci.* 34: 255-260 .

13. D.H. Percy and K.L. Williams, (1990). *Lab.Anim. Sci.* 40(6): 603-607 .

14. S. Gagnetan. Ph.D. Dissertation, Uniformed Services University, Bethesda, MD (1992).

15. D. Percy, S. Bond and J. Macinnes, (1989). *Arch. Virol.* 104: 323-333.

16. B. Obrink, (1991). *Bioessays* 13: 227-234 .

17. S.H. Lin and G. Guidotti, (1989). *J. Biol. Chem.* 264: 14408-14414.

18. N.C. Pedersen, I. Ward and W.L. Mengeling, (1978). *Arch. Virol.* 58: 45-53.

19. D.A. Tyrrell, J.D. Almeida, C.H. Cunningham, W.R. Dowdle, M.S. Hofstad, K. McIntosh, M. Tajima, L.Y.A. Zakstelskaya, B.C. Easterday, A.Z. Kapikian and R.W. Bingham, (1975). *Intervirology* 5: 76-82.

20. C.L. Yeager, R. Ashmun, R.K. Williams, C.B. Cardellichio, L.H. Shapiro, A.T. Look and K.V. Holmes, (1992). *Nature* 357: 420-422.

21. B. Delmas, J. Gelfi, R. L'Haridon, L.K. Vogel, H. Sjöström, O. Noren and H. Laude, (1992). *Nature* 357: 417-420.

22. R.A. Ashmun and A.T. Look, (1990). *Blood* 75: 462-469.

23. R.A. Ashmun, A.T. Look, W.M. Roberts, M.F. Roussel, S. Seremetis, M. Ohtsuka and C.J. Sherr, (1989). *Blood* 73: 827-837.

24. C. Yeager, Ph.D. Dissertation, Uniformed Services University, Bethesda, MD (1992).

25. J.K. Fazakerley, S.E. Parker, F. Bloom and M.J. Buchmeier, (1992). *Virology* 187: 178-188.

26. T.M. Gallagher, C. Escarmis and M.J. Buchmeier, (1991). *J. Virol.* 65: 1916-1928.

27. F. Taguchi, P.T. Massa and V. ter Meulen, (1986). *Virology* 155: 267-270.

28. J. Thompson, S. Barnert, B. Berling, S. von Kleist, V. Kodelja, K. Lucas, E.M. Mauch, F. Rudert, H. Schrewe, M. Weiss and W. Zimmermann, The Carcinoembryonic Antigen Gene Family, A. Yachi and J.E. Shively (eds), 65-74, Elsevier Science Publishers BV (Biomedical Division) (1989).

29. J.Q. Huang, C. Turbide, E. Daniels, S. Jothy and N. Beauchemin, (1990). *Development* 110: 573-588.

30. L.H. Shapiro, R.A. Ashmun, W.M. Roberts and A.T. Look, (1991). *J. Biol. Chem.* 266: 11999-12007.

EXPRESSION OF MHV-A59 RECEPTOR GLYCOPROTEINS IN SUSCEPTIBLE AND RESISTANT STRAINS OF MICE

G.S. Dveksler[1], A.A. Basile[1], C.B. Cardellichio[1], N. Beauchemin[2] , C.W. Dieffenbach[3] and K.V. Holmes[1]

Department of Pathology, Uniformed Services University of the Health Sciences, Bethesda, MD   20814-4799[1], McGill Cancer Centre, McGill University, Montreal, Canada H3G IY6[2], and Division of AIDS, NIAID, National Institutes of Health, Bethesda, MD 20892[3]

Introduction

Band and Warwick showed that inbred mouse strains differed in their susceptibility to virulent strain of mouse hepatitis virus, MHV-2, and that peritoneal macrophages cultured from these mouse strains reflected their differences in susceptibility to the virus (1). Adult SJL/J mice are highly resistant to MHV-JHM and MHV-A59 (2,12,13). Our laboratory showed that membranes from the liver and intestinal epithelial cells of BALB/c mice could bind MHV-A59 virions in a solid phase assay using undenatured membrane proteins, whereas membranes from SJL/J mice did not bind virus, suggesting that differences in binding of virus to cells from different mouse strains could account for their differences in susceptibility to MHV (3). In virus-overlay protein blot assays of intestinal brush border and liver membrane proteins, MHV-A59 virions bound strongly to a 58 kDa glycoprotein from BALB/c mice, but the virus did not recognize any proteins from SJL/J mouse tissues (3,4). Anti-receptor monoclonal antibody MAb-CC1 that blocked infection of murine cell lines with MHV-A59 recognized the 110-120 kDa and 58 kDa membrane glycoproteins of BALB/c mice, but no proteins from SJL/J membranes (6,7).

The cDNA coding for the 110 kDa glycoprotein detected by VOPBA was isolated from a BALB/c liver cDNA library and named MHVR (5). Amino acid sequence analysis and cross-reactivity with anti-human carcinoembryonic antigen (CEA) antibodies indicated that MHVR is a member of the murine CEA family of glycoproteins, in the immunoglobulin superfamily (5,6,7). MHVR is a 424-amino acid glycoprotein with four immunoglobulin-like domains, a transmembrane domain and a short intracytoplasmic tail (Figure 1). When MHVR was transfected into MHV-resistant human or hamster cells, they became susceptible to MHV-A59 as well as other MHV strains (5). Although MHV-A59-resistant adult SJL mice do not express a protein that binds MAb-CC1 or MHV-A59 virions, their intestinal brush border and liver membranes express two related glycoproteins that are recognized by antibody directed against the N-terminal 15 amino acids of MHVR and by antibodies to human CEA-related glycoproteins (6,7). This paper describes the cloning of cDNAs encoding for other CEA gene family members (CGMs) isolated from BALB/c, SJL/J and CD-1 mice and shows that they can also act as functional receptors for MHV-A59.

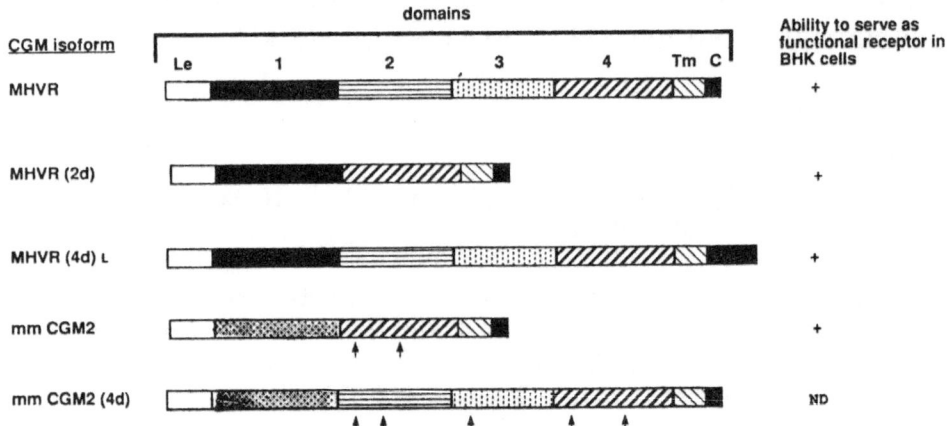

Figure 1. Schematic representation of the CEA gene family members (CGM) isolated from BALB/c and SJL/J mice. Le = leader peptide, Tm = transmembrane domain and C = cytoplasmic domain. The arrows indicate the presence of an amino acid difference between the mmCGM$_2$ isoforms and MHVR isoforms.

Results and Discussion

Characterization of splice variants of MHVR in BALB/c mice. Northern blot analysis of transcripts from adult BALB/c and SJL/J colon revealed two major MHVR-related transcripts of 3.4 and 1.7 kb and multiple additional faint bands. In these tissues, the 3.4 kb mRNA species was more abundant than the 1.7 kb species which was detected after prolonged exposure of the blots.

To isolate the cDNA encoding the 58 kDa glycoprotein detected by immunoblot analysis and VOPBA in intestine and liver membranes of MHV-A59-susceptible strains, we designed primers based on the highly conserved leader peptide and the last amino acids of MHVR, including the stop codon. A new cDNA clone, named MHVR (two domain) or MHVR(2d) was obtained by RNAPCR using BALB/c colon RNA as starting material. MHVR(2d) contains the leader, domains 1 and 4, the transmembrane sequence and the same cytoplasmic domain as MHVR. This clone is probably a splice variant of MHVR since differential splicing is a common phenomenon for members of the human CEA family (8). Besides the absence of the second and third domains in MHVR(2d), the only sequence difference between the open reading frames of the two cDNAs is the result of the splicing that created a new codon for glutamine absent in MHVR. A third member of the family that shares the same N-terminal domain of MHVR was isolated from a colon cDNA library of CD-1 mice (Figure 1). This clone, named MHVR(4d)$_L$, has the same amino acid sequence as MHVR up to the intracytoplasmic domain. In the intracytoplasmic domain of MHVR(4d)$_L$, the last three amino acids of MHVR are substituted with D,Q, and R, and 62 additional amino acids extend the length of this domain.

Cloning and expression studies of the MHVR-related cDNAs in the SJL/J mouse strain. Immunoblots with polyclonal anti-MHVR showed that the two SJL/J glycoproteins homologous to MHVR had slightly lower molecular weights (3 to 5 kDa) in SDS-PAGE than the BALB/c proteins MHVR and MHVR(2d) (7). Although the glycoproteins have different molecular weights, Northern blot analysis of RNA isolated from colon and small intestine of the two mouse strains showed no difference in the sizes of the mRNAs that were detected upon probing with MHVR (9). On the other hand, when MHVR was used to probe Southern blots of DNA isolated from these strains, restriction fragment length polymorphisms were observed upon digestion of the DNA with BamHI, EcoRI, and SstI (unpublished). To obtain cDNA clones encoding the CGM isoforms expressed in SJL/J mice, we PCR amplified cDNAs obtained from colon and small intestine with oligonucleotide primers derived from the MHVR sequence. The open reading frames of the two SJL/J clones obtained showed numerous amino acid differences when compared to the sequence of MHVR and MHVR(2d). Like BALB/c mice, the CGMs of SJL/J mice included both a four domain type and a two domain splice variant that results from the deletion of the second and third constant domains. The sequence of the open reading frame of the SJL/J two domain protein is identical to that of mmCGM$_2$, a cDNA clone isolated from colon of CD-1 mice (10). Most of the differences between MHVR(2d) and mmCGM$_2$ mapped to the N-terminal domain (Figure 2) and two amino acid changes were found in the fourth domain. In the four domain isoforms, MHVR and mmCGM$_2$(4d), in addition to the differences in the first and fourth domains, two additional amino acids in the second domain and one amino acid in the third domain distinguished the amino acids sequences (Figure 1). This comparison between the BALB/c and SJL/J clones indicated that the lower molecular weights of the SJL/J glycoproteins probably result from the absence of the first potential N-linked glycosylation site in the N-terminal domain of MHVR (Figure 2).

CD-1 mice are outbred, and cloning showed that they express CGMs with both types of N-terminal sequences, the one present in MHVR and that in mmCGM$_2$. We investigated whether inbred mouse strains expressed only one N-terminal domain or the other, which would be expected if mmCGM$_2$ and MHVR were alleles of the same gene. For that purpose we designed oligonucleotide probes that, under special washing conditions, were specific for the MHVR or the SJL/J

N-DOMAIN

```
MHVR         EVTIEAVPPQVAEDNNVLLLVHNLPLALGAFAWYKGNTTAIDKEI
                                                  ::::::
mmCGM2       EVTIEAVPPQVAEDNNVLLLVHNLPLALGAFAWYKGNPVSTNAEI


ARFVPNSNMNFTGQAYSGREIIYSNGSLLFQMITMKDMGVTTLDMTDENYRRTQATVRFHVH
::  :::  :::  : :     ::      : ::  :  :  : ::     :   :   :
VHFVTGTNKTTTGPAHSGRETVYSNGSLLIQRVTVKDTGVYTIEMTDENFRRTEATVQFHVH
```

Figure 2. Amino Acid sequence comparison of the N-terminal domains of MHVR and mmCGM$_2$ (5,10). Potential N-linked glycosylation sites are underlined.

sequences, and we took advantage of the different restriction enzyme site recognition sequences present in domain one of these clones. The study was extended to other inbred strains, C57/BL6 and C3H mice. The results were further confirmed by RNAse protection with RNA isolated from liver, colon and small intestine of the four mouse strains. For RNAse protection experiments, we used riboprobes specific for the N-terminal domains and a riboprobe from the fourth domain that hybridizes to all of the known isoforms. The SJL/J mice expressed only the N-terminal domain of the $mmCGM_2$ isoform, while the other strains of mice expressed only that of the MHVR isoform (Table 1) (9). Thus, the $mmCGM_2$ and MHVR isoforms and their splice variants appear to be encoded by different alleles of the murine CEA-related glycoprotein.

Table 1. Expression of CGM isoforms in different strains of mice.

N-terminal domains of MHVR and $mmCGM_2$ expressed in liver, colon and small intestine of different mouse strains

Probe specificity	BALB/c	C57BL/6	SJL/J	C3H
MHVR	+	+	-	+
$mmCGM_2$	-	-	+	-
MHVR and $mmCGM_2$	+	+	+	+

The presence (+) or absence (-) of expression of transcripts encoding the N-terminal domains of MHVR and $mmCGM_2$ was determined by RNase protection assays and RNAPCR.

Ability of splice variants of MHVR and $mmCGM_2$ to act as functional receptors for MHV-A59. Deletion mutagenesis studies showed that the N-terminal domain of MHVR is responsible for its interactions with the virus and with the MAb-CC1 (11). However, the first domain alone is not enough to act as a functional receptor when transfected into hamster cells, and the presence of either domain 2 or 4 is required (11). Since both MHVR(2d) and MHVR(4d)$_L$ share the same N-terminal domain as MHVR and have at least one extra constant domain, we expected that they would also serve as functional MHV-A59 receptors and be recognized by the MAb-CC1. The ability of the two new isoforms to act as virus receptors was studies in transient transfection assays. When MHVR(2d) and MHVR(4d)$_L$ where expressed in receptor-negative BHK-cells, they became susceptible to MHV-A59 as determined by the presence of viral antigens in the cytoplasm of the transfected cells by immunofluorescence. Our studies indicate that several MHVR isoforms from BALB/c or CD-1 mice are functional receptors for MHV-A59. The mRNAs for the 2 and 4 domain isoforms were detected by RNAPCR in all target tissues for MHV-A59 infection that were examined including brain, liver, colon, and small intestine of BALB/c mice.

We tested the two domain CGM, $mmCGM_2$ that is expressed in SJL/J mice for its ability to function as a receptor for MHV-A59. When BHK cells were transiently or stably transfected with mmCGM$_2$, they became susceptible to MHV-A59 infection. This result was surprising since the animals are resistant to MHV infection virus and MAb-CC1 did not bind in VOPBAs to SJL/J liver and intestine membranes which express both the 55 kDa glycoprotein encoded by mmCGM$_2$ and the four domain variant mmCGM$_2$(4d) (3,6). Anti-MHVR MAb-CC1 did not protect hamster cells transfected with mmCGM$_2$ from infection with MHV-A59, indicating that this antibody reacts with an MHVR-specific epitope and suggesting that the virus and MAb-CC1 binding domains may not be identical. Studies to determine whether the mmCGM$_2$(4d) glycoprotein of SJL/J mice is also a functional receptor for murine coronaviruses are in progress.

A continuous cell line of SJL/J embryo fibroblasts was prepared, the mRNA encoding its MHVR-related glycoproteins was analyzed, and its ability to support MHV-A59 and MHV-JHM infection was explored. RNAPCR analysis showed that the cell line expressed both the two and four domain isoforms of mmCGM$_2$. At several passage levels, the cells were completely resistant to infection with MHV-A59, as shown by immunofluorescence with antiviral antibody at 7.5 hours after virus challenge. To determine whether resistance of the SJL/J cell line was due to an SJL/J-specific post-translational modification in processing of these glycoproteins, we transfected cDNAs encoding the MHVR, MHVR(2d), MHVR(4d)$_L$, or mmCGM$_2$(2d) isoform into these cells and tested whether expression or overexpression of these proteins rendered the cells susceptible to infection with MHV-A59. Surprisingly, expression of each of these isoforms rendered the SJL/J cell line susceptible to infection, indicating that the SJL/J fibroblasts can process the glycoproteins in an appropriate manner to permit their use as receptors for MHV-A59. Thus, overexpression of one of the isoforms already expressed in the SJL/J tissues made the cells susceptible to infection. Quantitative studies on the binding affinities of the receptor isoforms and their relative efficiency in facilitating virus penetration in cell lines will be undertaken in order to elucidate the mechanism for resistance of the SJL/J cell line and adult animals to MHV-A59 and MHV-JHM infection.

Acknowledgements

This research was supported in part by grants from NIH (AI25231), USUHS (C074EF), Medical Research Council of Canada (PG11410), and the Cancer Research Society, Inc. The opinions or assertations contained herein are the private views of the authors and should not be construed as official or necessarily reflecting the views of the Uniformed Services University of the Health Sciences or the Department of Defense.

References

1. Bang, F.B. & Warwick, A. (1960) *Proc. Natl. Acad. Sci. U.S.A.* **46**, 1065-1075.

2. Stohlman, S.A., Frelinger, J.A. & Weiner, L.P. (1980) *J. Immunol.* **124**, 1733-1739.

3. Boyle, J.F., Weismiller, D.G. & Holmes, K.V. (1987) *J. Virol.* **61**, 185-189.

4. Compton, S.R.,Stephensen, C.B.,Snyder, S.W.,Weismiller, D.G. & Holmes, K.V. (1992) *J. Virol.* **in press**.

5. Dveksler, G.S., Pensiero, M.N., Cardellichio, C.B., Williams, R.K., Jiang, G.S., Holmes, K.V. & Dieffenbach, C.W. (1991) *J. Virol.* **65**, 6881-6891.

6. Williams, R.K.,Jiang, G.S. & Holmes, K.V. (1991) *Proc. Natl. Acad. Sci. U.S.A.* **88**, 5533-5536.

7. Williams, R.K., Jiang, G.S., Snyder, S.W.,Frana, M.F. & Holmes, K.V. (1990) *J. Virol.* **64**, 3817-3823.

8. Barnett, T.R., Kretschmer, A., Austen, D.A., Goebel, S.J., Hart, J.T., Elting, J.J. & Kamarch, M.E. (1989) *J. Cell Biol.* **108**, 267-276.

9. Dveksler, G.S., Dieffenbach, C.W., Cardellichio, C.B., McCuaig, K., Pensiero, M.N.,Jiang, G.S., Beauchemin, N., & Holmes, K.V. (1993), *J. Virol.* in press.

10. Turbide, C., Rojas, M., Stanners, C.P. & Beauchemin, N. (1991) *J. Biol. Chem.* **266**, 309-315.

11. Dveksler, G.S.,Pensiero, M.N.,Cardellichio, C.B.,Dieffenbach, C.W. & Holmes, K.V. (1992) *Proc. Natl. Acad. Sci. U.S.A.*, **in press**.

12. Barthold, S.W.,Beck, D.S. & Smith, A.L. (1986) *Arch. Virol.* **91**, 247-256.

13. Stohlman, S.A. & Frelinger, J.A. (1978) *Immunogenetics* **6**, 277-281.

MOUSE HEPATITIS VIRUS INFECTION UTILIZES MORE THAN ONE RECEPTOR AND REQUIRES AN ADDITIONAL CELLULAR FACTOR

Kyoko Yokomori and Michael M. C. Lai

Howard Hughes Medical Institute and Department of Microbiology
University of Southern California, School of Medicine, Los Angeles
California 90033-1054

INTRODUCTION

Mouse hepatitis virus (MHV) causes hepatitis, encephalomyelitis, respiratory and gastrointestinal ailments. Our laboratory has been particularly interested in the mechanism of neuropathogenesis of MHV. Different MHV strains have different capacities to cause central nervous system (CNS) infection. Among the neurotropic strains, there is also a large degree of variation in their ability to cause encephalitis or demyelination. Various virus variants have been obtained from the neurotropic MHV strains which show altered neuropathogenicity; for example, many of the variants which escape neutralization by the spike (S) protein-specific monoclonal antibodies differ from the parental viruses in not causing encephalitis, but still retain the ability to cause demyelination (Fleming et al., 1986; Dalziel et al., 1986). These variants have either point mutations or deletions in the viral spike protein (Gallagher et al., 1990; Wang et al., 1992). Since the S protein presumably interacts with the viral receptor on the surface of target cells, it is reasonable to assume that the difference in the viral neurotropism is caused by the variations in their ability to interact with the receptors. Therefore, it is conceivable that the MHV receptors in different cell types, e.g. neurons, astrocytes or oligodendrocytes, in CNS might be different.

The MHV receptors have been identified as a member of murine homologue of carcinoembryonic antigen (CEA), mmCGM1 (Williams et al., 1991). It has been shown that this receptor molecule binds MHV in an in vitro virus binding assay (Boyle et al., 1987), and that SJL mouse, which is resistant to infection by A59 and JHM strains of MHV, lacks a functional receptor protein capable of binding virus in vitro (Williams et al., 1990). Thus, mmCGM1 may be the molecule controlling viral tissue tropism. However, mmCGM1 was not detected in the mouse brain; thus, the possibility arises that mmCGM1 may not be the universal receptor for MHV, and different MHV strains may utilize different receptors in different tissues.

MHV UTILIZES A DIFFERENT MEMBER OF CEA AS THE RECEPTOR IN THE BRAIN

We reasoned that if MHV utilizes an alternative receptor in the brain, this receptor would be related to mmCGM1 to a certain extent. To identify this possible receptor, we performed reverse transcriptase-polymerase chain reaction (RT-PCR) analysis of C57BL/6 mouse brain and liver RNA by using two primers specific for the conserved sequences at the 5'- and 3'-ends of the mmCGM1 molecule. The results showed that liver RNA yielded

Coronaviruses, Edited by H. Laude and J.F. Vautherot
Plenum Press, New York, 1994

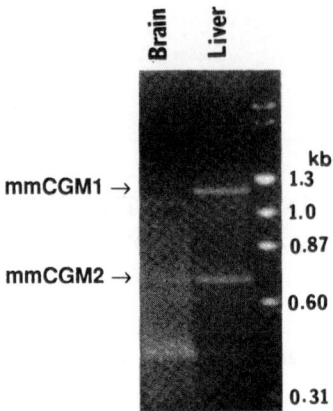

Figure 1. RT-PCR analysis of mmCGM-like molecules in C57BL/6 mouse liver and brain.

at least two PCR products, while brain RNA yielded only one of these two plus a few additional bands (Fig. 1) (Yokomori and Lai, 1992a). Cloning and sequencing of these PCR products showed that the largest PCR product of liver RNA represents the mmCGM1-derived PCR product, whereas the small one represents a similar molecule lacking 564 nucleotides in the middle. This molecule is similar in structure to another member of the CEA family, mmCGM2. Sequence analysis suggested that these two molecules probably represent the RNA species derived from alternative splicing of the same CEA gene. The identity of other molecules has not been determined.

The capacity of these CEA molecules to serve as MHV receptors was determined by cloning these PCR products into an expression vector with SV40 T antigen promoter (pECE) (Ellis et al., 1986) and transfecting them into COS cells, which were then tested for their infectability by MHV-A59 and JHM. The results indicated that both of them rendered the COS cells susceptible to MHV infection (Table 1). Thus, we concluded that both of them can be used as MHV receptors at roughly the same efficiency. Since the mouse brain expresses only mmCGM2 but not mmCGM1, the former is most likely the receptor used by MHV in CNS infection, in contrast to liver infection. This study thus shows that MHV could utilize different receptors in the brain and in the liver.

Table 1. Receptor functions of CEA molecules from B6 and SJL mice in Cos 7 cells[a].

		Virus titer (PFU/ml)	
Transfectant		JHM	A59
mmCGM1:	B6	1.1×10^3	4.4×10^3
	SJL	2.1×10^2	2.0×10^3
mmCGM2:	B6	3.1×10^2	3.1×10^2
	SJL	3.8×10^2	2.1×10^2
Vector (pECE)		0	2.0×10^1
None		0	3.3×10^1

[a]Cos 7 cells were transfected with various DNAs and infected with either JHM or A59 at 40 h posttransfection. Viruses were harvested at 24 h after infection, and plaque assayed on DBT cells.

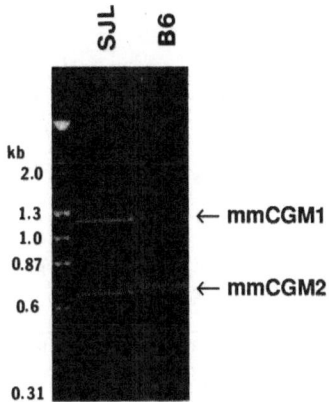

Figure 2. RT-PCR analysis of mmCGM-like molecules in SJL and C57BL/6 mouse liver and brain.

THE RESISTANCE OF SJL MICE TO MHV INFECTION IS NOT DUE TO THE DEFECTIVENESS OF THE MHV RECEPTOR

SJL mice have been known to be resistant to infection by MHV-A59 and JHM but susceptible to MHV-3 (Wilson and Dales, 1988). The viral resistance appears to be an intrinsic property of the cells, since the primary macrophages of SJL mice in culture also showed similar resistance (Stohlman et al., 1980). The mechanism of resistance has been controversial: one study indicated that the defect lies in the virus dissemination (Wilson and Dales, 1988), whereas another study revealed that the viral receptor is defective (Williams et al., 1990). The availability of cDNA probes for MHV receptors (mmCGM1 and 2) allowed us to re-examine this issue directly. We isolated the CEA-related molecules from SJL mouse liver by using RT-PCR techniques with the same set of primers as described above. Two PCR products indistinguishable from those from the susceptible mouse C57BL/6 were obtained (Fig. 2) (Yokomori and Lai, 1992b). Furthermore, these two molecules served as the receptors for MHV-A59 and JHM as efficiently as those from C57BL/6 mice, when they were transfected into COS cells (Table 1). Thus, the MHV receptor molecules from SJL mice are functional. Furthermore, Northern blot analysis of SJL mouse liver RNA indicated that these molecules are expressed in comparable amounts to that in C57BL/6 mice (data not shown). These results thus suggest that a cellular factor other than the viral receptor is defective in SJL mice, and is responsible for the viral resistance of this mouse strain.

THE EVIDENCE FOR THE REQUIREMENT OF A SECOND CELLULAR FACTOR FOR VIRUS ENTRY

To determine the mechanism of viral resistance of SJL mice, we examined several murine cell lines with regard to their virus infectability. The use of cultured cell lines simplified the parameters involved in the establishment of viral infection. We examined the established cell lines from not only the resistant SJL mice but also the susceptible C57BL and BALB/C mice. These cells were infected with either A59 or JHM viruses, and the kinetics of virus growth were then examined. Three representative cell lines are shown in Fig. 3. DBT cells, an astrocytoma cell line derived from BALB/C mice (Hirano et al., 1974), are fully susceptible to both A59 and JHM viruses. The other two cell lines are MC7, which was derived from C57BL/6, and BXS, which is derived from B10x SJL F1 mice. These two cell lines were susceptible to A59 virus infection, but, surprisingly, resistant to JHM infection. These results are different from the viral susceptibility of the parental mice. Thus, there is no strict correlation between the viral susceptibility in animal and that in cell culture, and that there is a selective resistance of most of the cell lines to

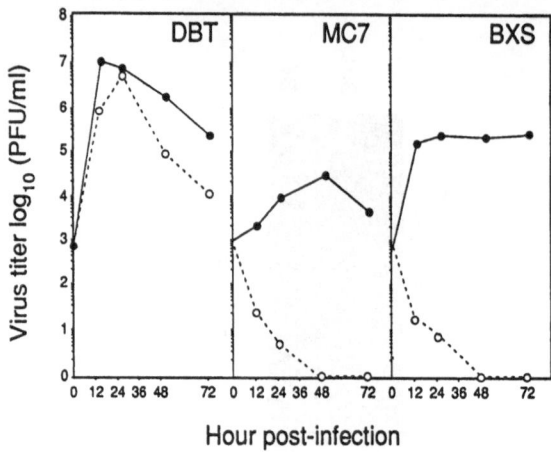

Figure 3. Viral growth kinetics on different cell lines. Virus titers released from each cell line at different time points postinfection are indicated. Solid line: A59. Dashed line: JHM.

JHM strain, but not A59. Since the viral receptors work as equally efficient receptors for both virus strains, these results suggest that there is another cellular factor which can regulate the susceptibility to viral infections.

We next examined whether these cell lines express functional MHV receptors. By RT-PCR analysis using primers specific for mmCGM2, we showed that all of these cell lines expressed at least the same or higher level of mmCGM2 molecules than that in the DBT cell line, which is highly susceptible to MHV infection (data not shown). We have also molecularly cloned the full-length mmCGM2 from one of the cultured cell lines and expressed this cDNA in COS cells. It was found that the expression of this cDNA rendered the COS cells susceptible to both A59 and JHM infection. We conclude that the MHV receptors are functional and properly expressed in these cell lines.

We then studied the step of viral replication which is blocked in these cell lines. We found that no viral RNA or protein synthesis could be detected in the JHM-infected cell lines except in DBT cells. Furthermore, the transfection of purified viral genomic RNA into these cell lines led to the production of fully infectious virus particles, indicating that viral RNA replication can proceed in these cell lines. Thus, we conclude that viral replication is blocked early in viral replication in these cell lines, most likely in the viral entry step.

PERSPECTIVES

These studies indicated that MHV can utilize more than a single CEA molecule as the viral receptor. The flexibility of MHV to utilize different molecules as the receptor in different tissues should allow the virus to infect more tissues than is otherwise possible. The CEA molecules are expressed as the alternatively spliced molecules in different tissues and possibly in different developmental periods. This spatial and temporal regulation of CEA molecules could conceivably determine the target cell specificity of MHV. Thus, further studies on the pattern of expression of CEA should shed light on the tissue tropism of MHV.

The second conclusion of this study is the demonstration that MHV entry requires an additional cellular factor other than the cellular receptor. This factor appears to be more sensitively regulated than the receptor, and is able to discriminate different virus strains. Thus, this factor exerts a very fine control on the tissue tropism of MHV strains. Conceivably, the neurotropism or hepatotropism of different MHV strains may be determined by this second factor, inasmuch as the MHV receptors expressed in the brain and liver function equally efficiently for both A59 and JHM. The nature of this factor is not yet clear. Preliminary data using recombinant viruses between A59 and JHM strains suggested that the viral gene responsible for the ability of MHV to grow in these cell lines is the spike protein gene. Thus, the second factor may be interacting with the viral spike protein directly or indirectly, and may trigger the cellular signal to initiate the penetration or uncoating of virus particles. Future studies on this factor should shed further light on the mechanism of virus entry and viral resistance.

ACKNOWLEDGMENT

We thank Daphne Shimoda for the preparation of the manuscript. This work was supported in part by a research grant from the U.S. Public Health Service. K.Y. is a Research Associate and M.M.C.L. is an Investigator of Howard Hughes Medical Institute.

REFERENCES

Boyle, J.F., Weismiller, D.G., and Holmes, K.V., (1987). *J. Virol.* 61:185-189.

Dalziel, R.G., Lampert, P.W., Talbot, P.J., and Buchmeier, M.J., (1986). *J. Virol.* 59:463-471.

Ellis, L. Clauser, E., Morgan, D.O., Edery, M., Roth, R.A., and Rutter, W.J., (1986).*Cell* 45: 721-732.

Fleming, J.O., Trousdale, M.D., El-Zaatari, F., Stohlman, S.A., and Weiner, L.P., (1986).*J. Virol.* 58:869-875.

Gallagher, T.M., Parker, S.E., and Buchmeier, M.J., (1990).*J. Virol.* 64:731-741.

Hirano, N., Fujiwara, K., Hino, S., and Matsumoto, M., (1974). *Arch. Gesamte Virusforshung* 44:298-302.

Stohlman, S.A., Frelinger, J.A., and Weiner, L.P., (1980). *J. Immunol.* 124:1733-1739.

Wang, F.I., Fleming, J.O., and Lai, M.M.C., (1992)..*Virology* 186:742-749.

Williams, R.K., Jiang, G.-S., and Holmes, K. V., (1991). *Proc. Natl. Acad. Sci. U.S A* 88: 5533-5536.

Williams, R.K., Jiang, G.-S., Snyder, S.W., Frana, M.F., and Holmes, K.V., (1990).*J. Virol.* 64: 3817-3823.

Wilson, G.A.R., and Dales, S., (1988). *J. Virol.* 62: 3371-3377.

Yokomori, K., and Lai, M.M.C., (1992a). *J. Virol.* 66: 6194-6199.

Yokomori, K., and Lai, M.M.C., (1992b). *J. Virol.* 66: 6931-6938.

DISSEMINATION OF MHV4 (STRAIN JHM) INFECTION DOES NOT REQUIRE SPECIFIC CORONAVIRUS RECEPTORS

T. M. Gallagher(1), M. J. Buchmeier(1) and
S. Perlman(2)

(1)Division of Virology
Department of Neuropharmacology
The Scripps Research Institute
La Jolla, CA 92037
(2)Departments of Pediatrics and Microbiology
University of Iowa
Iowa City, IA 52242

ABSTRACT

In this report, we demonstrate the syncytial spread of MHV4 (strain JHM) infection through non-murine cell cultures which lack a specific MHV4 receptor and are therefore resistant to infection by free virions. This was achieved by allowing infected murine cells to settle onto confluent monolayers of non-murine cells in a straightforward infectious center assay. Receptor-independent syncytium formation induced by cells expressing the MHV4 spike (S) from recombinant vaccinia viruses (VV) indicated that spread was mediated by this coronavirus glycoprotein. We conclude that the S protein of MHV4 is so potently fusogenic that it does not require prior binding to a virus-specific surface receptor to induce fusion of closely-opposed plasma membranes.

INTRODUCTION

It is well known that the murine coronaviruses exhibit tropism for murine hosts; these viruses are transmissible to rats or monkeys (1) only under highly artificial conditions of intracerebral inoculation. In addition, considerable evidence indicates variation in the susceptibility of murine cell types to murine coronavirus infection (2). Tropism for murine cells appears to be due to the absence of a suitable plasma membrane receptor for murine coronavirus on non-murine cells (3). The additional restriction to selected mouse cell lines

Coronaviruses, Edited by H. Laude and J.F. Vautherot
Plenum Press, New York, 1994

may be due to requirements subsequent to receptor binding, such as those necessary for the appropriate exposure of infectious RNA to the cytosol (4, 5). Such species and tissue preference contrasts with the extended host range of viruses from other families. For example, the myxoviruses bind and productively infect a wide range of cell types, presumably because many configurations of ubiquitous sialic acid are acceptable receptor ligands (6), and because subsequent genome delivery is relatively unrestricted.

However, two observations suggested to us that murine coronavirus infections may actually involve more cell types than is commonly believed. First, for years now we have observed clear evidence of syncytia formation in non-murine cells infected with VV-S recombinants, suggesting that specific spike:receptor binding is not a prerequisite for MHV-induced fusion activity. This finding has in fact been recently documented for insect and rabbit cells (7, 8). Second, Murray et al. (9) recently demonstrated MHV infection of primates, with resultant clinical symptoms, upon intracerebral inoculation with lysates of infected murine cells. This result, along with a similar finding in rats (10) suggested an unexpectedly large in vivo host range for MHV. These discoveries prompted our investigation into whether an authentic MHV4 infection could disseminate and amplify in cultured cell monolayers of non-murine origin.

METHODS AND RESULTS

Expression of MHV4 spike from recombinant VV-infected cells results in syncytial spread on non-murine cells

Previous studies (11) involved synthesis of the MHV4 S glycoprotein from cDNA inserted in a plasmid containing a bacteriophage T7 RNA polymerase promoter. The plasmid, designated pGEM4Z-S, successfully expressed S molecules when transfected into cells infected with vTF7.3 (12) a recombinant VV that provided requisite T7 RNA polymerase activity. S protein expressed in this manner clearly induced syncytium formation on DBT (mouse) cell monolayers, but failed to fuse Hela (human), BHK (hamster) or RK13 (rabbit) cell monolayers, suggesting a requirement for MHV receptor in fusion. These findings were in agreement with previous reports involving expression of murine coronavirus S glycoproteins from weak VV promoters (13), and most likely reflected poor translational activity of the RNA transcribed from the pGEM T7 promoter (14).

In an attempt to increase the level of glycoprotein expression from cDNA, we excised S cDNA from pGEM4Z-S and inserted it into two different vaccinia virus (VV) insertion-expression vectors, pTM1 (14) and p1200 (15). pTM1 carries both a T7 RNA polymerase promoter and an efficient picornaviral ribosome binding site, thereby providing the cis elements necessary for high efficiency, early transcription and translation of insert cDNA. p1200 harbors a late cowpox promoter whose activation in VV-infected cells gives rise to abundant late transcripts.

For insertion into pTM1, PCR mutagenesis (16) was performed to create an NcoI site at the S initiation codon, and then the S cDNA (NcoI-BamHI) was ligated into the polylinker region. Insertion into p1200 involved modification of S cDNA (HindIII-BamHI) using linkers to generate ends compatible with ClaI overhangs followed by ligation into the unique ClaI site of the vector. Standard methods involving homologous recombination and bromodeoxyuridine selection (17) were used to prepare VV-spike (vTM1-S and v1200-S) recombinants.

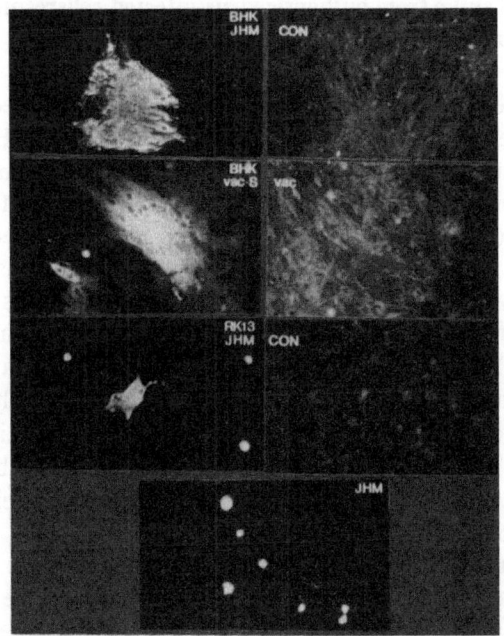

Figure-1. MHV can spread to resistant BHK and RK13 cells. After trypsinization, DBT cells were re-suspended to 10^6 cells/ml and either mock-infected ("con") or infected with either MHV (0.5 PFU/cell) ("JHM") or vTF7.3 (which expresses T7 RNA polymerase) and vTM1-S ("vac-S") or vTF7.3 alone ("vac"). After 1 hr at 37°C, cells were extensively washed to remove unbound virus. 1.0×10^5 cells aliquots were then seeded onto confluent lawns of either BHK rows (1, 2), RK13 (row 3) or onto empty wells (row 4). After 11 hrs (cells dually infected with VV recombinants) or 22 hrs (cells infected with MHV), cells were fixed with 10% formalin, permeabilized with 2% NP40 and incubated with a mixture of murine monoclonal antibodies specific for the nucleocapsid, matrix and spike proteins. Antiviral antibody was detected with fluoroscein-conjugated goat anti-mouse antibody (Fab')^2 fragment.

Coomassie-stained electropherograms representing the protein from cells dually infected with vTF7.3 and vTM1-S did not reveal S glycoprotein (data not shown), indicating that S expression was substantially lower than previously reported for this system (14). Despite this finding, we did observe fusion of Hela, BHK and RK13 cells by 12 hours postinfection (hpi). Figure 1 depicts syncytium formation on a BHK cell monolayer. Similar results (not shown) were obtained from v1200-S infected cells at 48 hpi. Thus, it appeared that the requirement for a specific plasma membrane ligand for S in the fusion process could be overcome by increasing the surface density of the S fusogen.

Expression of MHV4 spike from authentic virus-infected cellsresults in syncytial spread on non-murine cells

Discovery of vS-mediated syncytium formation across a broad spectrum of cell types led to the obvious question of whether S expressed from authentic MHV4 could function similarly. Human, hamster and rabbit cells are known to be resistant to infection with MHV virions, and using hamster cells, this resistance has been shown to be due to lack of cell surface receptor (3). To test the fusion function of S expressed from MHV4 we performed infectious center assays. Susceptible murine DBT cells were infected in suspension with MHV4 and were seeded onto confluent monolayers of DBT, BHK or RK13 cells. Foci of fused cells containing viral antigens developed in all three different cell monolayers. The results with BHK and RK13 cells are shown in Figure 1. By quantitating immunofluorescence signals, we found that essentially all productively infected DBT cells were able to form foci on all cell lines. Syncytium size, however, varied with cell type, with DBT being largest (50-100 nuclei/focus by 11 hr pi), BHK being intermediate (20-30 nuclei/focus by 22 hr pi) and RK13 being smallest (1-10 nuclei/focus by 22 hr pi).

Progeny virus production in these spreading foci was assayed at 22 hpi (Table 1). On average, the respective yields from DBT, BHK and RK13 cells were 7800, 250 and 21 times higher than from control cultures of infected cells plated in the absence of an underlying cell monolayer.

Table 1. Progeny virus titers after recruitment of cell monolayers into MHV4 infection.

Underlying cell monolayer	Titer(a)
None (Control)	2.47 +/- 0.31
RK13	3.79 +/- 0.28
BHK	4.87 +/- 0.56
DBT	6.36 +/- 0.27

10^5 DBT cells infected in suspension with MHV4 (0.5 FU/cell) were diluted at 1 hr pi into 3 ml of complete growth medium and overlaid on to 10 cm^2 plates containing the indicated confluent underlying monolayers. After 22 hrs, virus was released from cells by three freeze-thaw cycles and the resulting extracts were titered on DBT indicator cells.

(a) Titers are expressed as geometric means +/- one standard deviation per ml.

DISCUSSION

Receptor-independent dissemination of infection may have in vivo relevance

Our in vitro results suggest that MHV4 may spread within infected animals without requirement for receptor. Infection of this type may operate in the rat, which lacks MHV receptor (18), yet can by infected if cell lysates or brain tissue homogenates are used as inocula (10). Our results also suggest an important role for multinucleated giant cells in the infected animal. Limited data concerning this possibility is available (19), and careful studies employing mutants which vary in fusogenicity, as well as in situ hybridization and immunohistochemistry techniques, will be required to identify in vivo spread of infection.

Virions and infected cells may establish infection in Fundamentally different ways

Considerable concentration of virus particles from supernatant fluids to cell surface is a requisite initial step in infection by virions. In contrast infected cells display exocytosed S glycoprotein at the plasma membrane at high density. Only low affinity interactions permitting juxtaposition of opposing cells would be required to bring S into range for fusion activity. Adhesion molecules normally present on cell surfaces may promote the obligate interactions.

Receptor-independent spread only requires the presence of a fusogenic surface molecule and could potentially be important in the spread of paramyoxoviruses, herpes viruses and lentiviruses. It has been shown recently that the fusogenic gp41 transmembrane protein of HIV is capable, in the absence of all other HIV proteins, of initiating syncytium formation in normally resistant cells (20). This suggests that receptor-independent spread may be functional in other systems under the appropriate circumstances.

The mechanism of murine cell resistance to infection by MHV4 virions remains unknown, although previous results have pointed toward the importance of events after receptor binding but before RNA synthesis (2, 4, 5). Such events include virion:cell membrane fusion, internalization of virus and nucleocapsid delivery to the cytoplasm. Thus it remains to be determined whether the delivery of infected cytosol to resistant cells can be achieved by cell:cell fusion.

Functional domains of the spike can be defined

With the infectious center assay described in this report it will be possible to study the membrane fusion function of individual spike glycoproteins in the absence of their receptor binding function. Previous crude assays comparing the syncytium-forming capacity of variant S proteins (11) did not account for variations between the proteins for their cognate receptor. Variant spikes on the surface of vS-infected cells can now be tested in the absence of receptor. These assays will permit unambiguous identification of those residues participating in the membrane fusion process.

Acknowledgements: The authors thank Chris Baugh and Michelle Zandonatti for technical assistance. We also thank Dr. B. Moss for providing vTF7.3 and pTM1 and Dr. D.J. Pickup for providing p1200. This work was performed while S.P. was on sabbatical leave at The Scripps Research Institute. Supported by grants from the NIH--NS 12428, NS 22347 (MJB) and NS 24401 (SP), and from the Multiple Sclerosis Society (SP). SP was supported by a Research Career Development Award from the NIH. This is publication No. 8583-NP from the Scripps Research Institute.

REFERENCES

1. H. Wege, S. Siddell, and V. ter Meulen. Current Topics in Microbiology and Immunology 99:165-200 (1982).
2. L. Sturman, and K. Holmes. Adv. in Virus Res. 28: 35-112 (1985).
3. G. Dveksler, C. Pensiero, C. Cardellichio, R. Williams, G. Jiang, K. Holmes, and C. Dieffenbach. J. Virol. 65:6881-6891 (1991).
4. I. Shif, and F. Bang. J. Exp. Med. 131: 843-850 (1970).
5. C. Kooi, L. Mizzen, C. Alderson, M. Daya, and R. Anderson. J. Gen. Virol. 69: 1125-1135 (1988).
6. J. Paulson, J. Sadler, and R. Hill. J. Biol. Chem. 254:2120-2124 (1979).
7. F. Taguchi, T. Ikeda, and H. Shida. J. Gen. Virol. 73:1065-1072 (1992).
8. D. Yoo, M. Parker, L.A. Babiuk. Virology 180:395-399 (1991).
9. R. Murray, G. Cai, K. Hoel, J. Zhang, K. Soike, G. Cabirac. Virology. 188: 274-284 (1992).
10. O. Sorensen, D. Perry, and S. Dales. Arch. Neurol. 37:478-484 (1980).
11. T. Gallagher, C. Escarmis, and M. Buchmeier. J. Virol. 65:1916-1928 (1991).
12. T. Fuerst, E. Niles, F. Studier, and B. Moss. Proc. Natl. Acad. Sci. USA 83: 8122-8126 (1986).
13. H. Vennema, L. Heijnen, A. Zijderveld, M. Horzinek, and W. Spaan. J. Virol. 64: 339-346 (1990).
14. O. Elroy-Stein, and B. Moss. Proc. Natl. Acad. Sci. USA 87: 6743-6747 (1990).
15. D. Patel, C. Ray, R. Drucker, and D. Pickup. Proc. Natl. Acad. Sci. USA 85: 9431-9435 (1988).
16. H. Kadowaki, T. Kadowaki, F. Wondisford, and S. Taylor. Gene 76: 161-166 (1989).
17. M. Mackett, G. Smith, and B. Moss. J. Virol. 49:857-864 (1984).
18. K. Holmes, R. WIlliams, C. Cardellichio, S. Comptorn, C. Stephensen, S. Shyder, M. Frana, G-S. Jiang, A. Smith, R. Knobler. Adv. Exp. Med. Biol. 276: 37-44 (1990).
19. N. Goto, K. Takahashi, K. Huang, K. Katami, and K. Fujiara. Japan. J. Exp. Med. 49: 169-177 (1979).
20. L.G. Perez, M.A. O'Donnell, E.B. Stephens. J. Virol. 66:4134-4143 (1992).

VIRUS-LIGAND INTERACTIONS OF OC43 CORONAVIRUS WITH CELL MEMBRANES

Arlene R. Collins

Department of Microbiology
State Univ. of NY at Buffalo
Buffalo, NY 14214

ABSTRACT

The binding of human coronavirus OC43 to human rhabdomyosarcoma cells which are highly susceptible to infection was studied by a solid phase virus binding assay and a receptor blockade assay. It was observed that whole virions and S(spike) bound to a 90 kD glycoprotein of RD cells even after treatment of the substrate with neuraminidase or 0.1 M NaOH. A second receptor of 45 kD also bound virus and was identified as HLA class I antigen. Antibody to both receptors reduced the virus yield in a receptor blockade assay. Sera from four patients with multiple sclerosis contained receptor blocking activity which correlated with antibodies to HLA. No receptor blocking antibodies to the 90 kD RD cell protein were found in human sera.

INTRODUCTION

Human coronaviruses are the cause of acute respiratory illness and have been associated with enteric disease in man. As such, their target cells are likely to be the epithelial cells lining the respiratory and enteric tract. Attachment of virus to specific receptors on the cell surface is a major determinant of virus tropism and pathogenesis [1]. Recently, aminopeptidase-N was identified as a cell receptor for the human coronavirus 229E [2]. A cell receptor for OC43 coronavirus has not been identified. The pathogenesis of human coronaviruses is also of concern in association with the autoimmune disease, multiple sclerosis (MS). Recently, interest in this association has been rekindled by the report of Murray et al. [3] in which probing MS and control brain with probes specific for human, murine, porcine and bovine coronaviruses by in situ hybridization resulted in the detection of coronavirus RNA in 12 or 22 MS samples, five of which reacted with the OC43 probe. Also, the coronavirus strains isolated from MS brain have been shown to cause focal demyelinating lesions in monkeys [4]. A study of virus-ligand interactions of OC43 with human rhabdomyosarcoma (RD) cells, which are highly susceptible to infection, was undertaken to identify possible cell receptors and to assess interactions of the virions with host cell molecules.

Coronaviruses, Edited by H. Laude and J.F. Vautherot
Plenum Press, New York, 1994

METHODS

Viruses and Cells

Human coronaviruses OC43 and 229E were plaque purified in MRC-5, human diploid lung cells (Viro Med, Minnetonka, MN 55343). Stock virus was maintained by passage in MRD-5 cells. Human rhabdomyosarcoma (RD) cells were obtained from R. Crowell, Hanneman Medical College, Philadelphia, PA. Monkey kidney, CV-1 cells were obtained from R. Hughes, RPCI, Buffalo, NY. For plaque assays virus suspensions were diluted in EMEM and duplicate volumes of 0.2 ml were allowed to adsorb to MRC-5 cells in 24 well trays (Costar, Cambridge, MA) for 1 h at 37°C. The monolayers were then overlayed with 0.5 ml of 0.5% agarose with EMEM plus 0.2% serum. Cultures were stained with neutral red, and plaques were counted after 4 days at 33°C. The titer was determined as the average number of plaques per dilution times the dilution.

Antiserum Production

The procedure of Knudsen [5] was followed. Briefly, the 90 K RD protein was separated by PAGE electrophoresis and eluted from the gel by transfer to nitrocellulose paper. The protein-bearing nitrocellulose was solubilized with dimethyl sulfoxide and used as an immunogen in rabbits. An equal volume of Freund's adjuvant, either complete (first injection) or incomplete (subsequent injections) was added to the mixture which was then injected subcutaneously in four sites in the rabbit. A total of four injections was given at two week intervals. Antisera were tested by a Western blot assay and in the receptor blocking assay.

Monoclonal Antibodies

Monoclonal anti-human HLA class I-ABC, clone w6/321, mouse IgG2a was obtained as ascites fluid (10 mg/ml) from C-six Diagnostics, Inc., Mequon, WI 53092. Monoclonal anti-bovine coronavirus antibody to 110/120 peplomer, clone Bio-09, was purchased from Biosoft, AMAC, Inc., Westbrook, ME 04092.

Human Sera

Sera from multiple sclerosis patients (ages 16-53) were collected at the W.C. Baird Multiple Sclerosis Centre, Dr. Lawrence Jacobs, Director, Buffalo, NY. Sera from normal subjects who tested CMV positive (cytomegalovirus antibody positive) were obtained from the Regional Red Cross Blood Centre.

Virus-Immunoblot Assay

Growth and purification of virus: HCV-OC43 was propagated in RD cells. Supernatant virus was harvested after 4-5 days incubation at 33°C, clarified by low speed centrifugation (1000 x g, 10 min.) and sedimented by ultra-centrifugation at 100,000 x g for 4 h in a Beckman T35 rotor. The pellet was resuspended at 100 x concentration in TMEN buffer (50 mM Tris-maleate, 1 mM EDTA, 100 mM NaCl pH 6.5). For virus purification, concentrated virus was layered onto a sucrose gradient (20-60% w/w in TMEN) and centrifuged in an SW41 rotor at 60,000 x g for 4 h. The virus band (about 3 cm from the bottom) was collected, diluted in phosphate buffered saline (PBS, 0.1M, pH 7.2) and sedimented under the same conditions. The pellet was resuspended in PBS and stored at -70°C. Spike and HE peplomers of the virions were obtained by disrupting the virions in 1% n-octylglucopyranoside and separation on a 10-30% w/w sucrose gradient centrifuged at 105,000 x g for 18 h [6].

Substrate. Cells grown in T75 flasks (Costar, Cambridge MA 02139) were washed twice with rinse buffer (20 mM Tris-HCl, pH 9.0, 137 mM NaCl, 1 mM CaCl2 and 0.5mM MgCl2) and solubilized on ice for 20 min. in 1 ml ice cold lysis buffer (rinse buffer with 1% v/v NP40, 10% v/v glycerol, 1% v/v aprotinin). Insoluble cell debris was pelleted at 1000 x g for 10 min. and the clarified cell lysates were stored at -70°C.

Preparation of Solid Phase Substrate. Purified virus in 2 x sample buffer, L+H molecular weight markers (BioRad, Richmond CA 94804) in 2 x sample buffer were boiled 5 min and loaded along with cell lysates in lysis buffer (30-45 μg protein per lane) onto 10% SDS gels, 0.75 mm thickness. Proteins were separated at 10 mA per slab for 1-2 h and immediately transferred onto Immobilon-P membranes (Millipore, Bedford, MA 01730) for 25-30 min in transfer buffer (0.1 M Tris, 0.129M glycine pH 8.4, 20% methanol, 0.1% SDS) at 100 mA constant current. Membranes containing transferred proteins were soaked in 5% bovine serum albumin in TBS buffer (0.01 M Tris-HCl pH 7.5, 0.15M NaCl) at 37°C for 30 min and washed three times in TBS with 0.1% bovine serum albumin (TBS-BSA).

Solid Phase Binding. The interaction of OC43 with cell membrane proteins was performed as follows: OC43 virus (512 HAU) in 5 ml TBS-BSA was incubated with solid phase substrate at 25°C for 1 h with gentle shaking. To detect bound virus the membranes were washed three times with TBS-BSA and then incubated with a 1:200 dilution of rabbit anti-OC43 polyclonal antibody (prepared by immunization with gradient purified virus from infected suckling mouse brain) for 1 h at 25°C. Virus bound antibody was visualized by incubation with horseradish peroxidase-conjugated goat anti-rabbit IgG 1:1000 dilution (BioRad) and developed in TBS buffer containing 0.5 mg/ml 4-chloro-1-naphthol and 0.03% H_2O_2.

Receptor Blocking Assay

RD cells grown in 24 well culture trays (1.4 x 10^5 cells per well), were incubated for 1 h at 37°C with 0.5 ml of EMEM containing heat inactivated antibody diluted twofold beginning at 1:100. The antibody was removed and the cells were infected with virus at a multiplicity of 1 in 0.2 ml inoculum for 1 h at 38°C. After washing the monolayer twice to remove unbound virus, the cells were incubated for 24 h at 37°C in EMEM containing the antibody. Cells were then scraped into the medium and frozen at -70°C for titration by plaque assay. Receptor blockade was expressed as percent inhibition of virus yield and was determined by calculating the difference in virus yield between the virus titer in the presence of each antibody dilution and the virus titer of the untreated control. When pre-immune rabbit serum was tested in the receptor blocking assay, 50% inhibition was given by the 1:50 dilution but <10% inhibition at the twofold greater dilution.

RESULTS

Multiple RD cell proteins bound OC43 virus in a virus immunoblot assay comparing RD cell membranes and CV-1 cell membranes (Figure 1). In the RD membrane preparation, the virus-binding bands were observed at approximately 90 kD, 45 kD and 36 kD. Some binding activity is visible with the CV-1 membrane preparations. This was also seen in the control substrate when incubated with rabbit antibody alone. It was further observed that treatment of the substrate with neuraminidase (from C. perfringens, 10 units in 0.1 M acetate, pH 6.0) or sodium hydroxide (0.1 M) for 30 min did not remove the virus binding activity of the 90 kD RD protein.

In order to characterize the virus binding proteins of RD cells, antibodies that would specifically block each virus-protein interaction were sought. For the 90 kD RD cell protein, a rabbit antibody was raised by immunization with the protein band excised from nitrocellulose. For the 45 kD RD cell protein, a monoclonal antibody, w6/32, to HLA class I antigen was found to be reactive. The 36 kD RD cell protein is still uncharacterized. Binding of virus to specific RD cell proteins was repeatable with gradient purified virus.

Identification of the virus binding-RD cell proteins by specific antibodies

RD cell proteins were separated on SDS-PAGE gels and eluted from the gel by transfer to Immobilon membranes. After blocking non-specific sites on the membrane by incubation with 5% bovine serum albumin, the membranes were reacted with test antibodies at 1:100, 1:200 and 1:400 dilution along with pre-immune rabbit serum. One lane of RD cell protein substrate was reacted with the second antibody (goat anti-rabbit or goat anti-mouse conjugated to horseradish peroxidase, 1:1000 dilution) alone. The rabbit antibody raised to the 90 K RD protein bound to a single protein as shown by the band in the 90,000 molecular weight region of the gel (Figure 2a). Pre-immune rabbit serum at 1:100 dilution was negative. The w6/32, anti human HLA class I antibody bound to a single RD cell protein shown by the band at 45,000, the approximate molecular weight of the HLA class I heavy chain. No bands were visible in the CV-1 cell control lane (Figure 2b).

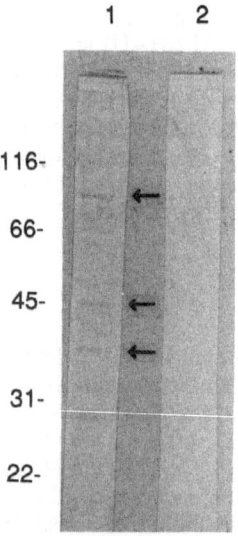

Figure 1. Binding of OC43 to RD cell proteins. Solubilized RD cell proteins after SDS-PAGE were transferred to Immobilon (Millipore) and reacted with OC43 virions (lane 1) or without virions (lane 2). Virus binding was detected by rabbit anti-OC43 antibody, goat anti-rabbit HRP conjugate and the reaction was developed with 4-chloro-1-naphthol. The virus binding bands are at 90 kD, 45 kD and 36 kD as indicated by the arrows.

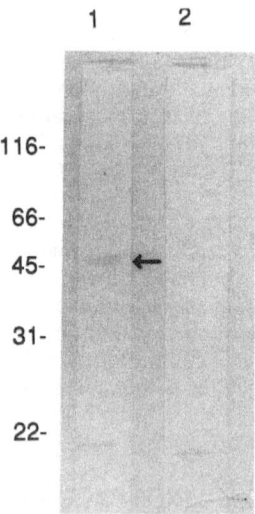

Figure 2. Binding of antibodies to RD cell receptor proteins. Solubilized cell proteins (lane 1 RD cells, lane 2 CV-1 cells), after SDS-PAGE, were transferred to Immobilon (Millipore) and reacted with (A) 90 kD RD cell antibody or (B) monoclonal anti-HLA class I antibody (10μ/ml). Binding of primary antibody was detected with goat anti-mouse or rabbit-HRP conjugate and the reaction was developed with 4-chloro-1-naphthol. The 90 kD and 45 kD bands are indicated by arrows.

Identification of multiple virus receptors by receptor blockade

It was then important to identify those virus-binding reactions that could lead to infection of RD cells. Using RD cell monolayers as target cells, the ability of antibodies to 90 kD RD, HLA class I (w6/32) and BCV 110/120 kD peplomer to block OC43 virus infection when present in the supernatant medium was compared. Serial dilutions of antibodies were incubated with the RD cells for 1 h prior to infection and for 24 h after infection and the effect on virus yield was determined by comparing virus plaque titers in samples with antibody to the virus titer of an untreated control (Figure 3). Rabbit antibody to the 90 k RD protein blocked the production of infectious virus in these experiments. The highest concentration (1:100) gave incomplete protection from virus infection showing a 96% reduction in the total yield. The pre-immune rabbit serum gave less than 10% inhibition at this concentration. w6/32 monoclonal antibody was very effective in blocking infectious virus production. At 1:400 dilution of antibody, 25 µg/ml, less than 10^2 plaque forming units of virus were synthesized. Monoclonal antibody to the BCV 110/120 kD peplomer gave a 40% reduction in OC43 virus yield at the 1:100 dilution.

Using the 229E virus, a human coronavirus belonging to another group, w6/32 antibody reduced the virus yield by less than 10% at the 1:100 dilution, indicating that the HLA class I antigen was not a receptor for the 229E virus.

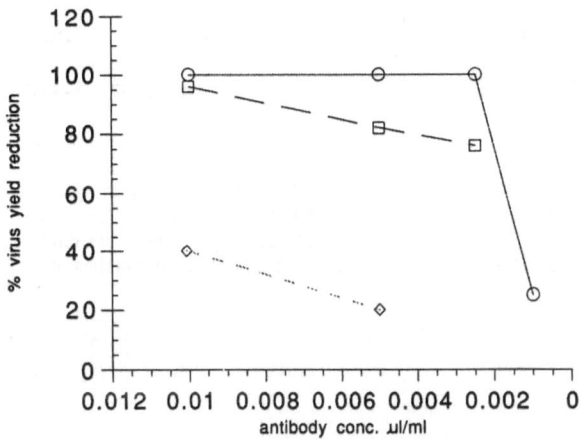

Figure 3. Inhibition of OC43 virus replication by anti-receptor antibody. RD cell monolayers were treated for 1 h with various dilutions of antibody, infected with OC43 virus at a multiplicity of 1 for 1 h at 37°C, washed twice and reincubated for 24 h at 33°C with medium containing antibody. The total virus yields were then determined by plaque assay and the % yield reduction due to antibody was calculated by comparison to the virus controls (without antibody). Open circle--anti HLA, open square--anti 90 kD RD, ◊—anti BCV 110/120 kD protein.

Correlation of cell receptor antibody with blocking activity in Human sera

In order to study the possibility that human sera contain antibodies to 90 kD RD cell protein, twenty human sera (10 MS, 10 CMV) were screened for antibodies which bound to the protein in the solid phase assay. Six out of 20 adult human sera contained antibodies which recognized 90 kD RD. When these sera were tested in the receptor blockade assay, four of the six sera were able to reduce virus yield by 30-100% when present during virus replication in vitro. Another five of the 20 sera (indicated by the open circles in Figure 4a) inhibited virus replication but did not contain antibodies to 90 kD RD. Similarly, the human sera were also tested for antibodies to the 45 kD HLA class I antigen and correlation of the presence of these antibodies with inhibition of virus replication was examined (Figure 4b). Out of eight human sera with antibodies which bound to 45 kD HLA class I antigen, four inhibited virus replication by 50-120%. All four sera came from multiple sclerosis patients. The other four sera with HLA class I antibodies which did not inhibit virus replication, came from CMV positive blood donors.

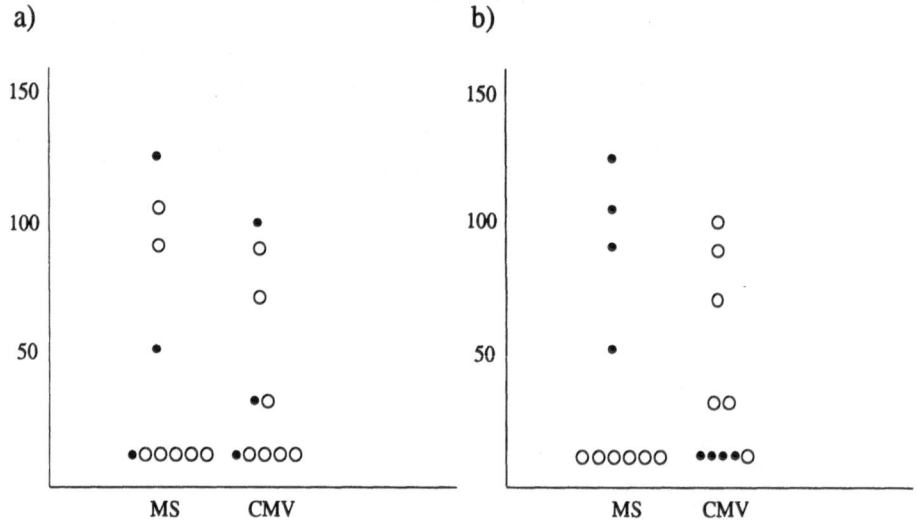

Figure 4. Association of antibodies to virus binding proteins with receptor blocking activity in human sera. MS, sera from multiple sclerosis patients; CMV sera from normal blood donors who are CMV antibody positive. Inhibition of virus yield (percent) by sera (diluted 1:200) with (solid circles) and without (empty circles) antibodies to the 90 kD RD cell protein are shown in (A) and antibodies to HLA class I antigen are shown in (B).

DISCUSSION

The preliminary results presented here indicate that multiple receptors for OC43 coronavirus are found on RD cells. At least three bands (at 90 kD, 45 kD and 36 kD molecular weight) were seen by virus-immunoblotting of RD cell proteins. While membrane proteins were thought to be the ligand in this assay, the possibility that cytoplasmic proteins bound to the virus was not excluded. Virus binding to intracellular proteins would also be of interest since such interactions could lead to alterations in cell function. Another method was employed to identify those proteins likely to be on the cell surface.

The receptor blockade test employed antibodies of known specificity to identify the cell proteins with receptor activity. The ability of receptor antibody to block infection is indirect evidence of receptor activity. Introduction of the receptor molecule into previously negative cells would be direct proof. However, blocking infectivity by antibody gives preliminary evidence of receptor activity. Previous studies with human Coxsackie B3, rhinovirus 14 and murine coronavirus A59 have shown that antibodies to a specific receptor reduced viral replication and modulated disease [1,6,7].

The experiments reported here indicate that two receptor specificities, 90 kD RD and HLA class I are involved in OC43 infection of RD cells. Rabbit antibody raised to the 90 kD RD cell protein incompletely blocked virus infection. The reason may be that the antigen used for immunization lacked conformational domains present in the native molecule or that more than one receptor is involved. HLA class I antigen appears to be an essential component of the virus attachment complex since monoclonal antibody w6/32 completely blocked infectivity. Previously, monoclonal antibody w6/32 was used to show that measles virus-specific T cell clones were HLA class II restricted and not HLA class I [8]. HLA class I is a member of the immunoglobulin gene superfamily which includes carcinoembryonic antigen, the receptor for mouse hepatitis virus (MHV). The MHV receptor plays an important role in the pathogenesis of MHV infections. SJL/J mice which lack a functional virus receptor, are resistant to MHV infection and develop chronic demyelination [9]. MHV and OC43 belong to the same serogroup. It would be reasonable to expect that the two viruses could use receptors which are structurally similar.

In the experiments to screen human sera for anti-receptor antibody, a correlation between the presence of antibodies to HLA class I and inhibition of virus replication was observed for four sera, all from MS patients. Another four sera with HLA class I antibodies did not inhibit virus replication. These were from CMV positive blood donors. CMV is known to express a glycoprotein similar to HLA with 13 potential glycosylation sites in contrast to only one-three in HLA molecules [10]. This glycoprotein may stimulate the antibody response which was observed. HLA A, B and C specific antibodies are normally found in multiparous women and in recipients of multiple blood transfusions. This pilot study suggests that an environmental exposure in some MS cases may produce HLA specific antibodies that block coronavirus receptors.

The involvement of more than one cell protein as receptor is of considerable interest since the process of virus attachment and penetration for this virus may be quite complex. The envelope of OC43 virus contains two types of surface peplomers: a spike protein which recognizes 9-0-acetylated sialic acid as a receptor determinant and a hemagglutinin-esterase which is a less efficient hemagglutinin and an esterase capable of cleaving the 9-0-acetyl bond which comprises the receptor determinant [11]. Virus binding the 90 kD RD protein is esterase resistant suggesting that this interaction is likely to be irreversible. Utilization of HLA antigen as a virus receptor would involve a mechanism different from antigen presentation which is the normal function of HLA in infected cells. The relatively large mass of individual virions and the mobility of the HLA antigens in the membrane suggests that multiple HLA receptor sites may be occupied by the virus and may undergo conformational changes making available other receptors embedded further in the plasma membrane. Further studies are underway to identify these complexes.

REFERENCES

1. F.G. Hayden, J.M. Gwaltney and R.J. Colonno, *Antiviral Res.* 9:233-247 (1988).
2. C.L. Yeager, R.A. Ashmun, R.K. Williams, C.B. Cardellichio, L.H.Shapiro, A.T. Look and K.V. Holmes, *Nature* 357:420-422 (1992).
3. S.R. Weiss, *Virology* 126:669-677 (1983)
4. R.S. Murray, B. Brown, D. Brian and G.F. Cabirac, *Ann. Neurol.* 31:525-533 (1992).
5. K. Kundsen, *Anal. Biochem.* 147:285-288 (1985).
6. A.H. Weller, K. Simpson, M. Herzum, N. van Houten and S.A. Huber, J.*Immunol.* 143:1843-1850 (1989).
7. A.L. Smith, C.B. Cardellichio, D.F. Winograd, M.S. deSouza, S.W.Barthold and K.V. Holmes, J. *Infect. Dis.* 163:879-882 (1991).
8. S. Jacobson, T. Nepom, J.R. Richert and H.F. McFarland, J. *Exp. Med.* 161:263-268 (1985).
9. R. Williams, G-S. Jiang and K.V. Holmes, *Proc. Natl. Acad. Sci. USA* 88:5533-5536 (1991).
10. S. Beck and B.G. Barrell, *Nature* 331:269-272 (1988).
11. B. Schultze, H-J. Gross, R. Bossmer and G. Herrler, J. *Virol.* 65:6232-6237 (1991).

FURTHER CHARACTERIZATION OF AMINOPEPTIDASE-N AS A RECEPTOR FOR CORONAVIRUSES

Bernard Delmas,[1] Jacqueline Gelfi,[1] Hans Sjöström,[2]
Ove Noren,[2] and Hubert Laude[1]

[1]Unité de Virologie et Immunologie Moléculaires
Institut National de la Recherche Agronomique
78352 Jouy-en-Josas, France
[2]Department of Biochemistry C, Panum Institute
University of Copenhagen, DK2200 Denmark

ABSTRACT

We recently reported that porcine aminopeptidase-N (pAPN) acts as a receptor for transmissible gastroenteritis virus (TGEV). In the present work, we addressed the question of whether TGEV tropism is determined only by the virus-receptor interaction. To this end, different non-permissive cell lines were transfected with the porcine APN cDNA and tested for their susceptibility to TGEV infection. The four transfected cell lines shown to express pAPN at their membrane became sensitive to infection. Two of these cell lines were found to be defective for the production of viral particles. This suggests that other factor(s) than pAPN expression may be involved in the production of infectious virions. The pAPN-transfected cells were also tested for their susceptibility to several viruses which have a close antigenic relationship to TGEV. So far, we failed to evidence permissivity to the feline infectious peritonitis coronavirus FIPV and canine coronavirus CCV. In contrast, we found clear evidence that porcine respiratory coronavirus PRCV, a variant of TGEV which replicates efficiently in the respiratory tract but to a very low extent in the gut, may also utilise APN to gain entry into the host cells. This suggests that the switch between TGEV and PRCV tropisms in vivo may involve other determinant(s) than receptor recognition.

INTRODUCTION

Coronaviruses are characterized by a restricted host range and tissue tropism[1]. Thus, TGEV replicates selectively in the enterocytes covering the villi of the small intestine[2], whereas human coronavirus 229E (HCV-229E) multiplies in the upper respiratory tract[3]. It was recently reported that APN/CD13 acts as a receptor for these two viruses[4,5]. APN/CD13 is expressed in a large variety of tissues and cells -notably epithelial and myeloid cells- (6,7 and references therein). The highest APN activity is associated with the brush border membrane of the enterocytes and of the renal proximal tubule cells. The distribution of APN and the site of multiplication of TGEV in the intestine are strongly correlated. In other organs like liver, lung, kidney and in cells of the myeloid lineage where APN is expressed, TGEV may replicate, but without causing the histological damages observed in the intestine[8]. HCV-

229E multiplies in the epithelium of the trachea where APN is assumed to be expressed, but in contrast to TGEV, HCV-229E enteric infections have not been clearly identified[3]. Taken together, these observations strongly suggest that APN expression is a prerequisite to allow virus multiplication, but is not the sole determinant of the tissue tropism in vivo. In the first part of this paper, we report data from in vitro experiments which suggest that other factor(s) than APN expression might modulate the susceptibility to TGEV.

The second part deals with experiments aiming to establish whether APN expression would confer susceptibility to three viruses antigenically related to TGEV: the feline infectious peritonitis virus FIPV, the canine coronavirus CCV and the porcine respiratory coronavirus PRCV, the respiratory variant of TGEV. Both FIPV and CCV have been reported to replicate in the small intestine of experimentally infected piglets[9], thus suggesting that they might use porcine APN as a receptor. In contrast, PRCV replicates selectively in the lung alveolar epithelium and at a very low level in the small intestine enterocytes. The fact that PRCV spike gene encodes a truncated protein has led to the hypothesis that the altered tropism of PRCV could relate to an impaired interaction with the APN molecule[10].

MATERIAL AND METHODS

Viruses. The isolate RM4 of PRCV, the CCV strain K378/20 (both supplied by Rhône-Mérieux, Lyon) and the FIPV strain 79.1146 (supplied by M. Horzinek, Utrecht) were used as a source of virus.

Cell transfections. The cDNA encoding the pAPN was subcloned downstream of the ubiquitin promotor in the BamHI site of the pTEJ4 expression vector[4]. MDCK, HRT-18, BHK-21 and Vero cells were cotransfected with this construct and pSV2neo by CaPO4 precipitation (MDCK cell line) or by lipofection (other cell lines). Cell clones resistant to the neomycin analogue G418 were selected and assayed for APN expression by measuring a APN activity (MDCK cell clones) or by TGEV susceptibility acquisition (other cell clones).

PRCV-induced cytopathic effect. The assays were performed 16h after infection at a multiplicity of 0.1 PRCV plaque-forming units (p.f.u.) per cell. Monolayers were fixed and stained with a crystal violet solution (10% alcohol). The dye associated to intact cells was quantified by optical absorbance after solubilization in acetic acid[11].

RESULTS AND DISCUSSION

ANOTHER FACTOR THAN AMINOPEPTIDASE-N EXPRESSION INVOLVED IN TGEV PERMISSIVITY ?

Clones derived from four cell lines (MDCK, HRT-18, BHK-21 and Vero cells) and expressing recombinant pAPN were selected. They were assayed for their susceptibility to TGEV infection by several approaches: detection of viral antigens synthesized in infected cells by immunofluorescence or immunoprecipitation assays, measurement of the cytopathic effect (c.p.e.) induced by the infection, titration of infectious viral progeny and quantification of viral particles produced. All the pAPN-expressing clones derived from the four cell lines studied became TGEV-sensitive as determined by the synthesis of viral antigens and by the c.p.e. observed 12 to 18h post infection (p.i.), after infection at a multiplicity of 10-20 p.f.u. per cell (not shown). BHK- and Vero-derived clones produced infectious virions in the same range than the swine testis cell line ST, which is highly permissive to TGEV (5.8×10^6 to 1.4×10^7 p.f.u./ml). In contrast, infectivity titers recovered from MDCK- and HRT-derived clones 20h p.i. did not differ significantly from the titer of the virus measured after viral adsorption (Fig. 1). Thus, MDCK and HRT cell clones failed to produce infectious particles. To confirm this observation, MDCK cell monolayers were infected at different multiplicities of infection

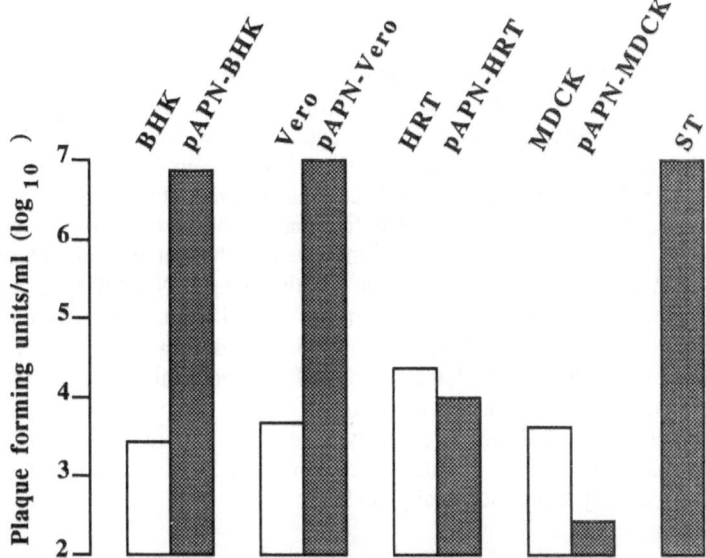

Figure 1. Infectious virus production in different cell lines expressing constitutively porcine aminopeptidase-N. Cell cultures were infected with TGEV, inoculum was rinsed 1h after adsorption and the total infectivity titer was determined 20h post infection by plaque assay on ST cells.

Table 1. TGEV cytopathic effect[a] as a function of the multiplicity of infection in two pAPN-expressing cell lines.

Cell clone	Multiplicity of infection				
	10	1	0.1	0.01	0.001
pAPN-BHK	++[b]	++	++	++	+
pAPN-MDCK	++	+	-	-	-

[a] Observed 48h post infection.

[b] Degree of lysis of the cell sheets: >90% (++); between 90 and 10% (+); <10% (-).

(m.o.i.) and the viability of the cells was measured 48 h p.i. so as to allow several cycles of virus multiplication. Table 1 shows that MDCK monolayers became destroyed only at a high m.o.i. (>1), in contrast to BHK monolayers, which were lysed at a markedly lower m.o.i. In the same way, we tried to identify formation of viral particles by metabolic ^{35}S-labeling of infected MDCK cells followed by PAGE analysis of the labeled material obtained through high-speed sedimentation of cell culture supernatants. We failed to detect labeled virus between 5 to 11h p.i. whereas viral particles were detected 6h p.i. in a control infection of ST cells (not shown).

Altogether, these results suggest that the presence of APN on the cell surface is not sufficient by itself to allow a complete cycle of replication. The defectiveness in viral production of MDCK clones seems to involve a late event in TGEV replication cycle since accumulation of structural viral proteins was clearly evidenced in infected cells. There are several hypotheses to explain such results. MDCK cell restriction could be associated to the presence (or the absence) of a cellular factor. The fact that all the porcine kidney cell lines checked in our laboratory are permissive to TGEV indicate that this putative factor would be species-specific. Another hypothesis is based on the fact that the pAPN-MDCK clones were selected on their capacity to express large amounts of pAPN: 50 to 100 fold the basal expression of APN in the MDCK cell line. It is possible that during the process of virus maturation, the newly synthetized particles are captured, via the viral spikes, by the APN molecules present in large amount in the endoplasmic reticulum, thus resulting in a block of virion maturation and/or transport. To explore this hypothesis, we plan to obtain clones of ST cells, naturally permissive to TGEV replication and expressing large quantities of APN in order to examine their capacity to produce viral particles.

PORCINE RESPIRATORY CORONAVIRUS UTILIZES PIG AMINOPEPTIDASE-N AS A RECEPTOR

To show if FIPV and CCV can use pig APN as a receptor, we performed infections of BHK cells expressing recombinant pig APN. No c.p.e. could be observed in these cells after infection with FIPV or CCV (not shown). These results would indicate that neither FIPV nor CCV are able to infect pAPN-expressing BHK clones. They are consistent with the fact that our attempts to infect any of the pig cell lines available in the laboratory with FIPV and CCV have met with no success. Nevertheless, the possibility remains that FIPV and CCV still may bind porcine APN, thus implying other factor(s) than receptor recognition could modulate the viral infection. In this respect, we observed that MDCK-derived clones expressing human APN (hAPN) were not susceptible to the human coronavirus 229E, whereas hAPN-3T3 clones became sensitive to HCV-229E (5 and data not shown). Alternatively, it is tempting to speculate that FIPV and CCV recognize feline and canine APN in a species-specific manner, as it seems to be the case for TGEV and HCV-229E.

To establish if PRCV can use pAPN as a receptor, PRCV infections of MDCK and BHK cells expressing pAPN were carried out. As shown in Fig. 2, a specific c.p.e. was observed in a pAPN-BHK cell clone whereas BHK cells behave as unsensitive cells. A similar result was obtained with MDCK-derived cell clones (not shown). Preincubation of the pAPN-BHK cell clone with the anti-APN antibody G43 resulted in a complete block of TGEV infection. In addition, we analyzed the synthesis of the PRCV nucleoprotein N in infected cells by immunofluorescence assay. A cytoplasmic fluorescence was observed in pAPN-BHK cells only (not shown). To confirm these results, ST cells were incubated with dilutions of the six anti APN antibodies before infection. Table 2 shows that these antibodies are able to block PRCV infection in the same range than they do with TGEV.

Altogether, these results provide strong evidence that the respiratory virus PRCV, like the enteropathogenic virus TGEV, recognize pAPN on target cells. We previously demonstrated, using two different binding assays, that TGEV binds directly to pig APN[4]. To complete these studies, we performed a binding assay between TGEV and pure pAPN preincubated with the anti-S neutralizing antibody 48.1. This antibody was able to block binding between virions and APN, at the same extent as an anti-receptor antibody did

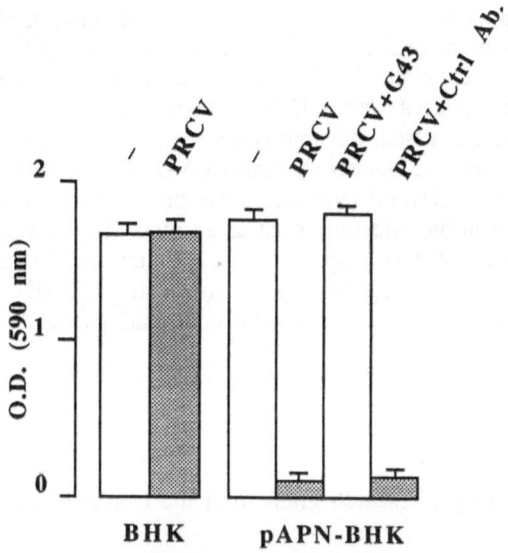

Figure 2. Colorimetric quantification of pAPN-BHK and BHK cells survival after infection or mock-infection (-) by PRCV. The dye incorporated in surviving cell monolayers was measured by optical absorbance. In two assays, G43 or a control antibody were added before infection. The data are given as mean values + s.d.m. (n=4).

Table 2. Neutralizing activity of anti-APN antibodies towards TGEV and PRCV in ST cells.

Antibody	Neutralization titer [a]	
	TGEV	PRCV
G 3	160	<20
G 18	5,000	7,500
G 43	20,000	40,000
I 31	320	20
T 35	160	<20
W 26	20,000	20,000

[a] Expressed as the reciprocal of the last dilution protecting against the viral cytopathic effect.

297

binding between virions and APN, at the same extent as an anti-receptor antibody did (unpublished results). Thus, the viral S protein binds specifically to APN. The S gene of different PRCV strains is deleted of 672 (or 681) nucleotides, resulting in a shortening of 224 (or 227) amino acid stretch in the N terminal part of the protein[10]. Since PRCV is able to infect BHK cells expressing pAPN, we conclude that this amino acid stretch missing in the S PRCV protein is not directly involved in the binding of the S protein on APN.

In conclusion, it appears that the difference of tropism between PRCV and TGEV cannot be simply explained by the unability of PRCV to interact with APN. PRCV replicates selectively in the respiratory tract : in the alveolar epithelial cells, which are assumed to express APN and in the alveolar macrophages where CD13, a marker of the myeloid lineage shown to be identical to APN[7], is also expressed. It is thus conceivable than PRCV uses APN to gain entry in these cells. The fact that the respiratory virus HCV-229E also uses human APN as a receptor also favors this view. Several hypotheses can be made to explain the defect of productive PRCV infection in the intestine, such as an instability of PRCV spike in the digestive tract or a low capacity of PRCV spike protein to induce membrane fusion between viral particles and enterocyte membranes. Finally, the conversion of ORF3a to a pseudogene in PRCV genome has also to be considered as possibly contributing to its altered tropism.

Acknowledgments

B.D. was supported in part by a research grant from the Commission of the European Communities.

REFERENCES

1. S. Siddell, H. Wege and V. Ter Meulen, The biology of coronaviruses, *J. gen. Virol.* 64:761 (1983).
2. M. Pensaert, E.O. Haelterman and T. Burstein, Transmissible gastroenteritis of swine: virus-intestinal cell-interactions, *Arch. Gesamte. Virusforch.* 31:321 (1970).
3. K. McIntosh, in *Virology* 2nd edn (eds Fields , D.N. *et al.*) pp857, Raven, New York (1990).
4. B. Delmas, J. Gelfi, R. L'Haridon, L. K. Vogel, H. Sjöström, O. Noren and H. Laude, Aminopeptidase N is a major receptor for the enteropathogenic coronavirus TGEV, *Nature* 357:417 (1992).
5. C.T. Yeager, R.A. Ashmun, R.K. Williams, C.B. Cardellichio, L.H. Shapiro, A.T. Look and K.V. Holmes, Human aminopeptidase N is a receptor for human coronavirus 229E, *Nature* 357:420 (1992).
6. A.J. Kenny and S. Maroux, Topology of microvillar membrane hydrolases of kidney and intestine, *Physiol. Rev.* 62:91 (1982).
7. A.T. Look, R.A. Ashmun, L.H. Shapiro and S.C. Peiper, Human myeloid plasma membrane glycoprotein CD13 (gp150) is identical to aminopeptidase N, *J. clin. Invest.* 83:1299 (1989).
8. E.O. Haelterman, On the pathogenesis of transmissible gastroenteritis of swine, *J. Am. vet. med. Assoc.* 160:534 (1972).
9. R.D. Woods, N.F. Cheville and J.E. Gallagher, Lesions in the small intestine of newborn pigs inoculated with porcine, feline and canine coronaviruses, *Am. J. vet. Res.* 42:1163 (1981).
10. H. Laude, K. Van Reeth and M. Pensaert, Porcine respiratory coronavirus: molecular features and virus-host interactions, *Vet. Res.* 24:125 (1993).
11. W. Kueng, E. Silber and U. Eppenberger, Quantification of cells cultured on 96-well plates, *Analyt. Biochem.* 182:16 (1989).

RECOGNITION OF N-ACETYL-9-O-ACETYLNEURAMINIC ACID BY BOVINE CORONAVIRUS AND HEMAGGLUTINATING ENCEPHALOMYELITIS VIRUS

Beate Schultze and Georg Herrler

Institut für Virologie, Philipps-Universität Marburg
Robert-Koch-Str.17, 3550 Marburg, Germany

ABSTRACT

The S protein of hemagglutinating encephalomyelitis virus is shown to be a hemagglutinin requiring N-acetyl-9-O-acetylneuraminic acid as a receptor determinant on the surface of erythrocytes. The ability of bovine coronavirus to recognize 9-O-acetylated sialic acid was used to establish a binding assay for the detection of glycoproteins containing this type of sugar. The assay is very fast, because it uses the acetylesterase of the viral HE protein to localize bound virus.

INTRODUCTION

Bovine coronavirus (BCV) has been known for some years to require 9-O-acetylated sialic acid on the cell surface for agglutination of erythrocytes (1,2). Recently, it has been shown that BCV uses N-acetyl-9-O-acetylneuraminic acid as a receptor determinant also on cultured cells to initiate infection (3). The binding of BCV to receptors containing 9-O-acetylated sialic acid can be mediated by two viral surface proteins. The HE protein, which also has acetylesterase activity, is able to agglutinate mouse or rat erythrocytes (4,5); these cells contain a high proportion of N-acetyl-9-O-acetylneuraminic acid (50%

and more) among their surface-bound sialic acids. The S protein has recently been shown to recognize 9-O-acetylated sialic acid, too (5). In fact, it requires fewer receptors on the surface of erythrocytes for agglutination than does the HE protein. Therefore, the S protein is the major hemagglutinin of BCV.

Here we report that hemagglutinating activity can be assigned not only to the S protein of BCV, but also to the S protein of hemagglutinating encephalomyelitis virus (HEV). In addition we show that the activities of the surface proteins S (binding to N-acetyl-9-O-acetylneuraminic acid) and HE (acetylesterase) can be used in a sensitive assay for the detection of glycoproteins containing 9-O-acetylated sialic acid.

MATERIALS AND METHODS

Viruses and Cells. Strain L-9 of BCV and strain NT-9 of HEV were grown in MDCK I cells as described previously (2).

Isolation of Viral Glycoproteins. The S and HE protein of HEV were isolated as reported recently (6).

Hemagglutination Assay. The hemagglutinating activity of virus and isolated glycoproteins was determined according to published procedures (2,5,6).

Binding Assay. The immobilization of glycoproteins on nitrocellulose and the binding of BCV have been described recently (5). Bound virus was detected according to the method described for influenza C virus by Zimmer et al. (7).

RESULTS AND DISCUSSION

The S Protein of BCV and HEV has hemagglutinating Activity

The isolation of the viral surface proteins S and HE has been described for both BCV and HEV (6). Following removal of the detergent octylglucoside by dialysis, the S protein was analyzed for hemagglutinating acitivity. As shown in Table 1, the same result was obtained with HEV as has been reported previously for BCV. Erythrocytes from adult chickens were agglutinated by virus and by isolated S, but not by HE Protein. No hemagglutination was observed with erythrocytes from one-day old chicken. The latter cells lack 9-O-acetylated sialic acid - in contrast to erythrocytes from adult chicken (8). The

Table 1. Hemagglutinating activity of BCV, HEV, and the isolated proteins S and HE.

| | hemagglutinating activity (HA-units/ml) with erythrocytes from | | | |
| | adult chicken | | one-day old chicken | |
	BCV	HEV	BCV	HEV
virus	256	128	< 2	< 2
S protein	1024	256	< 2	< 2
HE protein	< 2	< 2	< 2	< 2

result from Table 1 indicates that isolated S protein of HEV uses the same type of receptors as does the S protein of BCV, i.e. it is dependent on the presence of N-acetyl-9-O-acetylneuraminic acid.

Use of BCV for the Detection of Glycoproteins containing
N-acetyl-9-O-acetylneuraminic acid

We have previously reported that BCV can be used as a lectin to identify glycoproteins which contain N-acetyl-9-O-acetylneuraminic acid (5). For this purpose, glycoproteins were immobilized by transfer from a SDS-polyacrylamide gel to nitrocellulose. After incubation with BCV, virus which had bound to individual proteins, was identified by a enzyme-linked immunoreaction. This detection system is rather time-consuming, because it requires several consecutive incubations (first antibody, second antibody, enzyme, substrate) with intermittent washings. This assay was modified such that the acetylesterase of BCV was used to detect bound virus as has been described recently for influenza C virus (7). The principle is shown in Fig.1. After BCV has bound to the immobilized glycoproteins, the substrate α-naphthylacetate is added. The acetylesterase activity of HE cleaves the substrate releasing α-naphthol. The latter compound reacts with Fast Red, which is included in the substrate solution, resulting in a colored insoluble complex. The colored precipitate indicates the presence of bound virus

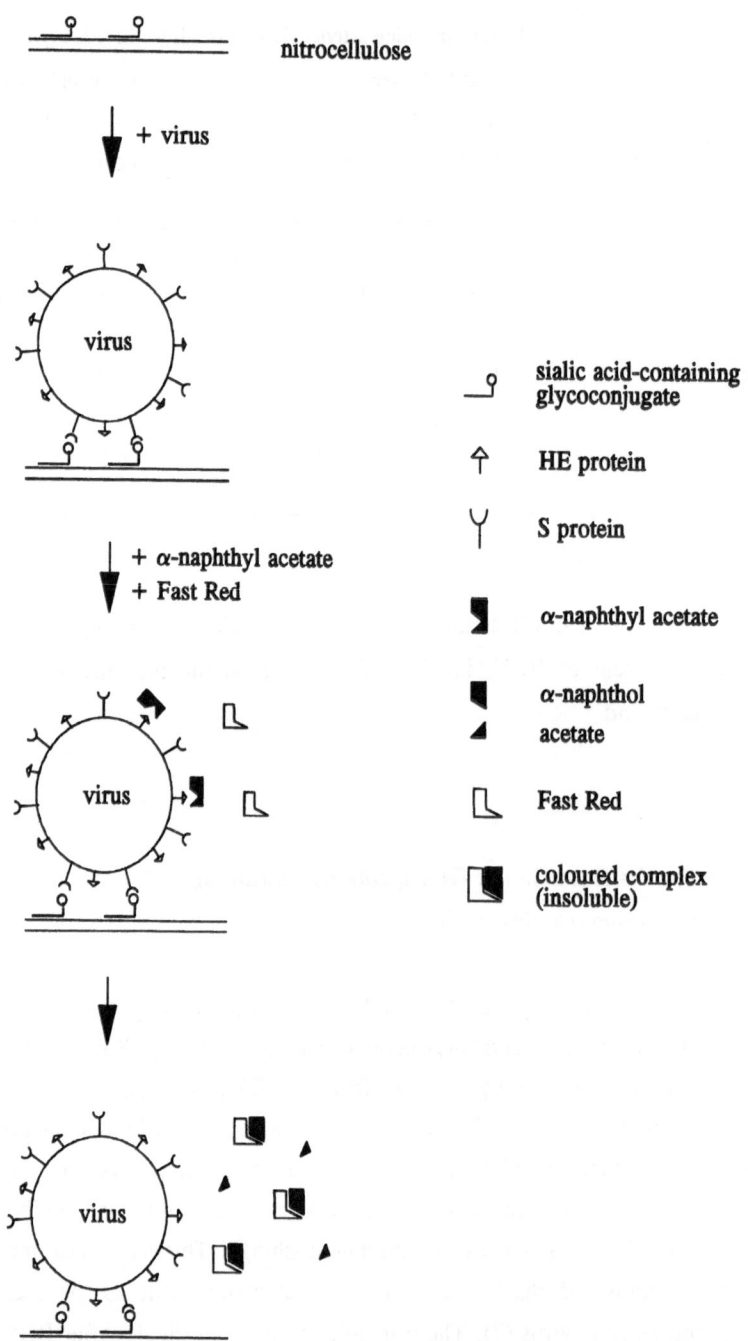

nitrocellulose

+ virus

virus

+ α-naphthyl acetate
+ Fast Red

virus

virus

⌐9   sialic acid-containing
       glycoconjugate

⇑   HE protein

Y   S protein

▰   α-naphthyl acetate

▰   α-naphthol
◢   acetate

⌐L   Fast Red

▰▰   coloured complex
       (insoluble)

Figure 1. Schematic illustration of the binding assay with BCV for the detection of glycoproteins containing N-acetyl-9-O-acetylneuraminic acid.

and therefore, the presence of glycoproteins containing 9-O-acetylated sialic acid.

This assay was applied to the glycoproteins of rat serum which are known to contain N-acetyl-9-O-acetylneuraminic acid. As shown in Fig. 2, the staining pattern indicates that BCV bound to many bands. In order to show that the binding of BCV to these proteins is due to the presence of N-acetyl-9-O-acetylneuraminic acid, a control sample was pretreated with sodium hydroxide. Under these alkaline conditions O-acetyl groups

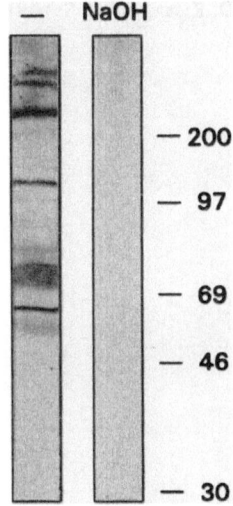

Figure 2. Detection of glycoproteins containing N-acetyl-9-O-acetylneuraminic acid in rat serum after immobilization of the proteins on nitrocellulose.

of sialic acid are released while the remainder of the oligosacharides is unaffected. The lack of staining of rat serum proteins after incubation with sodium hydroxide indicates that bands revealed by the binding assay with BCV are actually due to the presence of N-acetyl-9-O-acetylneuraminic acid.

By applying this assay to cell surface proteins it should be possible in the future to identify potential receptors for BCV and related coronaviruses among cell surface proteins.

ACKNOWLEDGMENTS

We thank Gert Zimmer for help with the binding assay. This work was supported by Deutsche Forschungsgemeinschaft (He 1168/2-2).

REFERENCES

1. R. Vlasak, W. Luytjes, W. Spaan, and P. Palese, P., Proc.Natl.Acad.Sci.USA 85:4526 (1988).

2. B. Schultze, H.-J. Gross, R. Brossmer, H.-D. Klenk, and G. Herrler, Virus Res. 16:185 (1990).

3. B. Schultze, and G. Herrler, 1992, J.Gen.Virol. 73:901 (1992).

4. B. King, B.J. Potts, and D.A. Brian, Virus Res. 2:53 (1985).

5. B. Schultze, H.-J. Gross, R. Brossmer, and G. Herrler, J.Virol. 65:6232 (1991).

6. B. Schultze, K. Wahn, H.-D. Klenk, and G. Herrler, 1991a, Virology 180:221 (1991).

7. G. Zimmer, G. Reuter, and R. Schauer, Eur.J.Biochem. 204:209 (1992).

8. G. Herrler, G. Reuter, R. Rott, H.-D. Klenk, and R. Schauer, Biol.Chem.Hoppe-Seyler 368:451 (1987).

N-ACETYLNEURAMINIC ACID PLAYS A CRITICAL ROLE FOR THE HAEMAGGLUTINATING ACTIVITY OF AVIAN INFECTIOUS BRONCHITIS VIRUS AND PORCINE TRANSMISSIBLE GASTROENTERITIS VIRUS

Beate Schultze,[1] Luis Enjuanes[2], Dave Cavanagh[3], and Georg Herrler[1]

[1]Institut für Virologie, Philipps-Universität Marburg
Robert-Koch-Str.17, 3550 Marburg, Germany
[2]Centro de Biologia Molecular, CSIC Universidad Autonoma
Canto Blanco, 28049 Madrid, Spain
[3]A.F.R.C. Institute for Animal Health, Compton, nr Newbury
Berkshire, RG16 ONN, United Kingdom

ABSTRACT

Porcine transmissible gastroenteritis virus (TGEV) was found to resemble avian infectious bronchitis virus (IBV) in its interaction with erythrocytes. Inactivation of the receptors on erythrocytes by neuraminidase treatment and restoration of receptors by reattaching N-acetylneuraminic acid (Neu5Ac) to cell surface components indicated that $\alpha2,3$-linked Neu5Ac serves as a receptor determinant for TGEV as has been reported recently for IBV (1). Similar to IBV, the haemagglutinating activity of TGEV is evident only after pretreatment of virus with neuraminidase indicating that inhibitors on the virion surface have to be inactivated in order to induce the HA-activity of these viruses. A model is presented to explain why the HA-activity of untreated virus is masked and how neuraminidase treatment results in the unmasking of this activity.

INTRODUCTION

Only a few members of the family Coronaviridae are efficient in agglutinating red blood cells: human coronavirus OC43, bovine coronavirus, porcine haemagglutinating encephalomyelitis virus, and some strains of murine coronaviruses. Each of these viruses contains an HE protein, which is not found in other coronaviruses (2). As the HE protein has haemagglutinating (HA) activity (3), the presence or absence of this protein seemed to account for the difference in the agglutinating ability of coronaviruses. However, HE protein is only able to agglutinate mouse and rat erythrocytes, which contain a large amount of N-acetyl-9-O-acetylneuramininc acid, the receptor determinant recognized by BCV. Chicken erythrocytes, which are less rich in 9-O-acetylated sialic acid, are not agglutinated by HE protein, though they are agglutinated by BCV (4). Recently it has been shown that the S protein is able to agglutinate chicken erythrocytes using the same receptor determinant as BCV and HE protein (4). As the HA-activity of S parallels that of the intact virus, it is the actual haemagglutinin of BCV.

An S protein is present on all coronaviruses. Therefore, the question remains, why viruses lacking an HE protein are very poor haemagglutinins or even devoid of agglutinating activity. An answer may come from findings obtained with infectious bronchitis virus, which acquires HA-activity after enzymatic pretreatment of the virus (5). This virus uses $\alpha 2,3$-linked N-acetylneuraminic acid (Neu5Ac) as a receptor determinant for attachment to erythrocytes (1). The HA-activity is evident only after treatment of the virus with neuraminidase. Here we show that TGEV resembles IBV in this respect : (i) the haemagglutinating activity is induced by neuraminidase treatment of the virus; (ii) $\alpha 2,3$-linked Neu5Ac serves as a receptor determinant.

MATERIALS AND METHODS

Viruses and Cells. Strain M41 of IBV was grown in embryonated chicken eggs (6). The Purdue strain of TGEV was grown in LLC-PK1 cells.

Haemagglutination Assay. The haemagglutinating activity was determined according to published procedures (7).

Neuraminidase treatment. Viruses or cells were treated with neuraminidase as described recently (1).

Resialylation of erythrocytes. Erythrocytes were resialylated to contain $\alpha 2,3$-linked Neu5Ac as described (1).

RESULTS AND DISCUSSION

TGEV has been reported to be a poor haemagglutinating agent (8). Significant HA-titres have been obtained only with virus preparations which had been concentrated by ultracentrifugation. Based on the results obtained with IBV, we analyzed whether the HA-activity of TGEV can be enhanced by neuraminidase treatment. As shown in Table 1, TGEV behaved in the same way as IBV. Purified virus was unable to agglutinate chicken erythrocytes. High haemagglutination titres were observed, however, when the virus was pretreated with neuraminidase. Both the enzyme from *Vibrio cholerae* and Newcastle disease virus were effective. Among the two common linkage types of sialic acid, Neu5Acα2,3Gal and Neu5Acα2,6Gal, the viral neuraminidase has a preference for the cleavage of the former linkage type. This result indicates that α2,3-linked sialic acid has to be removed from the viral surface in order to induce the haemagglutinating activity of both coronaviruses.

Table 1. Induction of the haemagglutinating activity of IBV and TGEV by neuraminidase treatment.

pretreatment of virus	haemagglutinating activity (HA-units/ml)	
	IBV	TGEV
none	< 2	< 2
VC-neuraminidase	256	512
NDV-neuraminidase	256	512

Purified virus preparations were treated with neuraminidase from *Vibrio cholerae* (VC) or Newastle disease virus (NDV) and analyzed for their ability to agglutinate chicken erythrocytes.

The similarity between IBV and TGEV in the induction of the HA-activity suggested that TGEV may use the same type of receptors for attachment to erythrocytes. In fact, asialo cells obtained by treatment with neuraminidase from *Vibrio cholerae* were resistant to agglutination by both IBV and TGEV (Table 2) indicating a crucial role of sialic acid in the interaction of both viruses with erythrocytes. This conclusion was confirmed by the finding that receptors for TGEV can be restored by resialylation of asialo cells. Following attachment of Neu5Ac in an α2,3-linkage to the surface of erythrocytes, the

Table 2. Inactivation of receptors for TGEV on chicken erythrocytes by neuraminidase treatment and restoration of receptors by resialylation of the cells.

| erythrocytes | haemagglutinating activity (HA-units/ml) | |
	IBV-NA	TGEV-NA
control	512	256
asialo	< 2	< 2
resialylated Neu5Acα2,3Galß1,3GalNAc	256	64

IBV and TGEV had been pretreated with neurminidase to induce the HA-activity of bothe viruses. Asialo cells were obtained by treatment of chicken erythrocytes with neuraminidase from *Vibrio cholerae*. Resialylated cells were obtained by incubation of asialo cells with sialyltransferase and CMP-Neu5Ac.

cells became susceptible to agglutination by TGEV (Table 2) as has been reported recently for IBV (1).

The results show that not only coronaviruses of the BCV serogroup are potent haemagglutinating agents (BCV, HCV-OC43, HEV and some strains of murine coronaviruses), but also viruses of other serogroups (TGEV and IBV). Future work has to show whether haemagglutinating activity can be induced with human coronavirus 229E, feline infectious coronavirus and canine coronavirus. There are some similarities in the HA-activity of the BCV serogroup on one side and of IBV/TGEV on the other side. For both groups of viruses, the S protein is the haemagglutinin and sialic acid serves as receptor determinant. The difference is in the type of sialic acid recognized - N-acetyl-9-O-acetylneuraminic acid in the case of BCV and Neu5Ac in the case of IBV/TGEV - and in the presence or absence, respectively, of a receptor-destroying enzyme on the virus particle. A virus using sialic acid as a receptor determinant faces the problem that such a common sugar is present not only on the cellular receptors but also on a variety of other glycoproteins and glycoplipids. These glycoconjugates may act as inhibitors, because they prevent the virus from finding sialic acid on the target cell. Viruses like BCV contain a receptor-destroying enzyme and are, therefore, able to inactivate such inhibitors. In the case of IBV and TGEV, which lack a comparable enzyme, the inhibitors have to be inactivated by exogenous enzyme.

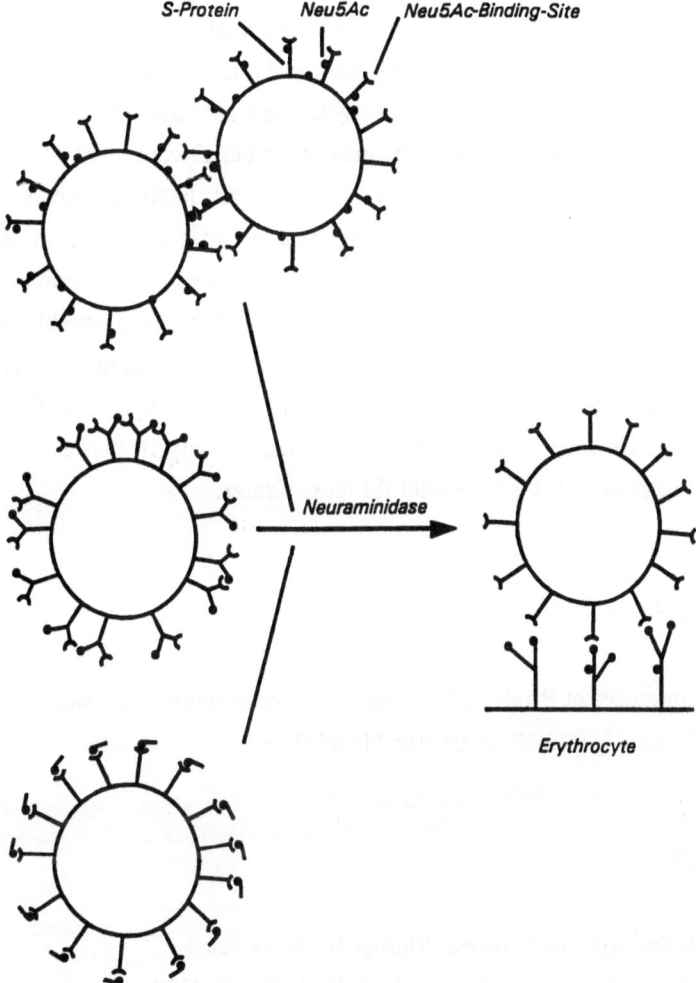

Figure 1. Schematic illustration of different possibilities to explain the lack of HA-activity of untreated IBV and TGEV and the induction of the activity by neuraminidase treatment.

There are several ways to explain how the inhibition of the HA-activity may occur (Fig. 1). The inhibitory sialic acid may be part of a viral component, e.g. of the S protein, and this may result in aggregate formation, because the virions can attach to each other. This possibility appears unlikely, because there is no evidence that IBV or TGEV have a greater tendency to form aggregates than BCV. Another possibility is that the inhibition is due to a viral component of the same virus particle, i.e. the sialic acid binding-site of the S protein binds to a sialic acid residue of a neighbouring glycoprotein. However, the glycoproteins M and S of egg-grown IBV have been shown to contain only oligosaccharides of the mannose-rich type which are sensitive to endo H treatment. Therefore, no sialic acid is expected to be present on the glycoproteins of IBV (9). For this reason, a third possibility appears to be the most likely explanation. Cellular compounds containing $\alpha2,3$-linked sialic acid may be attached to the S protein and prevent it from binding to erythrocyte receptors. Future work has to show whether this explanation is correct. It remains also to be shown whether the sialic acid binding activity plays a role in the infectious cycle. Aminopeptidase has been shown to serve as a receptor for TGEV (10). Whether the binding to sialic acid may have a supporting function is not known at present. However, the fact that this binding activity is conserved in different coronaviruses suggests that it is important for these viruses.

ACKNOWLEDGMENTS

The technical assistance of Birgit Döll is gratefully acknowledged. This work was supported by Deutsche Forschungsgemeinschaft (He 1168/2-2).

REFERENCES

1. B. Schultze, D. Cavanagh, and G. Herrler, Virology 189:792 (1992).

2. W. Spaan, D. Cavanagh, and M.C. Horzinek, J.Gen.Virol. 69:2939 (1988).

3. B. King, B.J. Potts, and D.A. Brian, Virus Res. 2:53 (1985).

4. B. Schultze, H.-J. Gross, R. Brossmer, and G. Herrler, J.Virol. 65:6232 (1991).

5. R.W. Bingham, M.H. Madge, and D.A.J. Tyrrell, J.Gen.Virol. 28:381 (1975).

6. D. Cavanagh, and P.J. Davis, J.Gen.Virol. 67:1443 (1986).

7. B. Schultze, H.-J. Gross, R. Brossmer, H.-D. Klenk, and G. Herrler, Virus Res. 16:185 (1990).

8. M. Noda, F. Koide, M. Asagi, and Y. Inaba, Arch.Virol. 99:163 (1988).

9. D. Cavanagh, J.Gen.Virol. 64:1187 (1983)y

10. B. Delmas, J. Gelfi, R. L'Haridon, L.K. Vogel, H. Sjöström, O. Noren, and H. Laude, Nature 357:417 (1992).

Chapter 6

Pathogenesis: Mechanisms and Diversity

MOUSE HEPATITIS VIRUS AND HERPES SIMPLEX VIRUS MOVE ALONG DIFFERENT CNS PATHWAYS

Stanley Perlman[1,2], Edward Barnett[2] and
Gary Jacobsen[3]

Departments of Pediatrics and Microbiology[1]
Neuroscience Program[2], College of Dentistry
University of Iowa
Iowa City, IA 52242

ABSTRACT

The spread of mouse hepatitis virus, strain JHM and herpes simplex virus type 1 in the central nervous system after inoculation into the nares and main olfactory bulb has been examined. The results show that each virus infects a subset of the possible connections of the olfactory bulb and that the subset infected by each virus is different. Thus, both viruses will be useful for studying the neuroanatomic connections of the olfactory bulb, and possibly for functional analyses as well.

INTRODUCTION

After intranasal inoculation into the murine nares, mouse hepatitis virus (MHV) enters the central nervous system (CNS) via the olfactory nerve and spreads to many of the neuroanatomic connections of this nerve (1-5). Virus spreads trans-synaptically, since surgical removal of the olfactory bulb prevents virus entry into the CNS (6). Analysis of the CNS structures infected by MHV suggests that virus spread occurs primarily by retrograde transport, since structures such as the diagonal band of Broca which project to the main olfactory bulb (MOB) but do not receive input from this structure are heavily labelled in the MHV infection. In addition, the olfactory tubercle, which receives input from the olfactory bulb but does not project to it is only minimally labelled after intranasal inoculation (1). Noticeably absent from the list of olfactory connections infected by MHV is the locus coeruleus (LC), a prominent source of afferent noradrenergic fibers to the MOB.

Coronaviruses, Edited by H. Laude and J.F. Vautherot
Plenum Press, New York, 1994

Herpes simplex virus, type 1 (HSV-1) is a neurotropic virus that also spreads trans-synaptically after intranasal inoculation (7-10). Previous reports have shown that HSV-1 infects the LC, suggesting that MHV and HSV-1 may infect different structures in the CNS after intranasal inoculation. In this study, the spread of these two viruses after intranasal inoculation and after direct inoculation into the MOB was determined. Our results indicate that the two viruses infect a partially non-overlapping set of CNS structures connected to the MOB. Our results also indicate that the entry into the CNS via the trigeminal nerve occurs commonly after intranasal inoculation of HSV-1, but does not occur after inoculation with MHV.

MATERIALS AND METHODS

Mice: Pathogen-free C57BL/6 and BALB/c mice were purchased from Jackson Laboratories and Sasco Laboratories.

Viruses: MHV, strain JHM, was propagated and purified as previously described. HSV-1, strain 17, obtained from Dr. Moira Brown, was propagated in RK13 cells and titered in Vero cells. For intranasal infection, 0.5-1 x 10^5 PFU MHV-JHM or 8 x 10^5 PFU HSV-1 were inoculated in a volume of 10 uL. For intrabulbar inoculation, 0.2 uL containing either 2 x 10^3 PFU MHV or 1.6 x 10^4 PFU HSV-1 was administered using a glass micropipette attached to a 1 uL Hamilton syringe.

For inoculation into the dental pulp, the free marginal gingiva was removed from the left mandibular central incisor with the beveled edge of a 18G needle. In early experiments, a high speed handpiece with a 1/4 round bur was used to expose the pulp. In later experiments, a slow speed hand piece with 1/4 round bur was used instead, to minimize heat generation and obstruction of the pulp chamber. 0.1 to 1 uL MHV-JHM (10^4-5 x 10^5 PFU) was deposited into the chamber with a 30G needle.In situ hybridization: Murine brains and skulls were harvested at various times after infection, and frozen in O.C.T. medium (Miles Laboratory) prior to analysis. MHV-JHM was detected by in situ hybridization as previously described (11). HSV-1 was also detected by in situ hybridization, using an antisense RNA probe from the VP5 gene (a DNA clone encoding VP5 was kindly provided by Dr. E. Wagner). The same method was used as for MHV except that samples were treated at 95°C prior to annealing in order to denature HSV DNA. Hybridization was performed at 55°C.

In some experiments, cells were double labeled for tyrosine hydroxylase and viral nucleic acid. In these experiments, mice were perfusion-fixed with PLP (paraformaldehyde-lysine-periodate) and brains were post-fixed with PLP. Cryostat sections were immunolabeled for tyrosine hydroxylase and then assayed by in situ hybridization for viral nucleic acid as described (Barnett and Perlman, manuscript in preparation).

RESULTS

MHV Spread Into The CNS After Intranasal And Intrabulbar Inoculation

As reported previously, MHV infects the MOB and many of its primary and secondary connections, including the anterior olfactory nucleus, piriform cortex, medial septal nucleus, medial habenula, ventral pallidum, ventral tegmental area, diagonal band of Broca, lateral hypothalamus, parabrachial nucleus, dorsal raphe nucleus and pontine reticular nuclei after intranasal inoculation (1-5, Barnett, Cassell and Perlman, in preparation). After direct inoculation into the MOB, the same structures are labeled, supporting the notion that MHV spreads to primary and secondary connections of the olfactory bulb by a trans-synaptic route.

Unexpectedly, no additional areas of the brain were labeled after intranasal as compared to intrabulbar inoculation. Previous analyses had suggested that MHV also enters the CNS via the trigeminal nerve, based on apparent labeling of the mesencephalic nucleus of the trigeminal nerve as well as labeling of the nerve itself after intranasal inoculation (4) The mesencephalic nucleus should be labeled after intranasal but not after intrabulbar inoculation if MHV really enters the brain via the trigeminal nerve.

These results suggested that the trigeminal nerve was not an important site of viral entry. Several additional observations supported this conclusion. First, more detailed analysis of coronal sections after intranasal inoculation showed that the structure which had been identified as the mesencephalic nucleus was in fact the parabrachial nucleus. Second, previous results showed that surgical ablation of one olfactory bulb prevented CNS entry on the side of the surgery after intranasal inoculation (6). Virus could be detected bilaterally at later times, consistent with crossover via the anterior commissure. More recently, we have surgically ablated both olfactory bulbs in 5 mice. In three of these mice, no virus could be detected at all in the CNS, and MHV was only present in the anterior part of the brain in the other two mice, consistent with incomplete ablation. No labeling that could be attributed to the trigeminal nerve was present in any animal. Third, treatment of the olfactory epithelium with Triton X- 100, which causes reversible destruction of the olfactory receptor neurons (12) prevented all entry of MHV into the CNS for 5 to 10 d after treatment, the exact duration of protection depended upon the age of the animal. After 5 to 10 d, the same structures were infected as in untreated mice. These results showed that either the olfactory and trigeminal nerves showed exactly the same time course of destruction and recovery with regard to susceptibility to MHV or more likely, that only the olfactory nerve was an important site of entry.

MHV Enters The Inferior Alveolar And Trigeminal Nerves After Inoculation Into The Dental Pulp

These results all suggested that while MHV infects the trigeminal nerve, it is unable to enter the CNS via this nerve. As a direct assay of MHV entry via the trigeminal nerve, virus was inoculated into the dental pulp of central mandibular incisors of susceptible 4 week old C57BL/6 mice. Since the dental pulp is innervated only by the inferior alveolar nerve, a branch of the trigeminal nerve, this approach should produce selective labeling of only the trigeminal system. In initial experiments, large amounts of virus (2×10^4-2×10^5 PFU) were administered after a cavity was created using a high speed drill. Under these conditions virus was detected in the inferior alveolar nerve (Figure-1) or trigeminal nerve in some mice (Table-1) but could not be detected in any CNS nucleus of the trigeminal nerve.

Two additional modifications was made in subsequent experiments. First, smaller amounts of virus (5-10×10^3 PFU) were administered to decrease the likelihood of olfactory spread. Second, a slow speed drill was substituted for the high speed drill, since excessive heat production and creation of a smear layer obstruction of the pulp cavity would potentially compromise viral entry into the dental pulp. As in the first set of experiments, MHV could only be detected in the inferior alveolar nerve or trigeminal nerve but not in the brain after these modifications were made. As a control, we have determined that HSV-1 enters the brain after inoculation into the tooth pulp (Jacobsen, Barnett, Cassell and Perlman, manuscript in preparation).

We are presently in the process of evaluating additional methods of virus delivery into the dental pulp using pulpotomies with the objective of increasing viral infection in the inferior alveolar nerve and possibly the brain. We are pursuing this approach since many of the positive inferior alveolar nerves were lightly labeled in the first experiments.

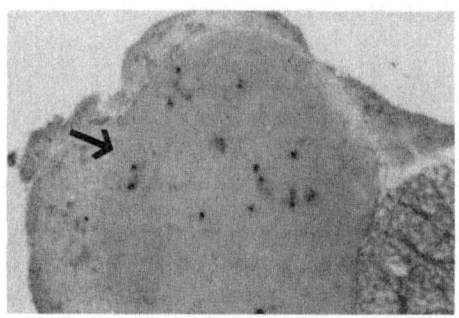

Figure 1-The head of an MHV-infected 4 week old C57BL/6 mouse was frozen and analyzed by in situ hybridization. Viral RNA was detected in the inferior alveolar nerve at the level of the parotid gland (arrow).

Table 1. MHV entry after inoculation into exposed pulp

Treatment	# days p.i.	# mice	# positive-AN/TG1
Dental pulp	3	8	4
inoculation	4	13	5
	5	22	9
	6	<u>11</u>	<u>6</u>
Total		54	24 (44%)
Oral inoculation	5	3	0 (0%)

1AN/TG-alveolar nerve or trigeminal nerve

Spread Of HSV-1 Into The CNS After Intranasal And Intrabulbar Inoculation.

Previous reports showed that HSV-1 could enter the murine CNS after intranasal inoculation (7-10). Entry occurred via both the olfactory and trigeminal nerves. Of particular interest, HSV-1 infected the LC, a structure not infected by MHV. With the goal of comparing directly the spread of MHV and HSV-1 within the murine CNS, we adapted our in situ hybridization techniques to the study of the HSV-1 infected brain, as described in Materials and Methods.

After both types of inoculation, several connections of the MOB were labeled by both viruses, including the anterior olfactory nucleus, the horizontal band of Broca, the piriform cortex, insular cortex, and dorsal raphe. Several structures were infected predominantly by one of the two viruses. Of particular interest, the locus coeruleus which was not labeled after MHV infection was infected by HSV-1. Other structures, such as the piriform cortex, were labeled differentially. Layer 2 of this structure was heavily labeled in HSV-1 infected brains whereas layers 1 and 3 were preferentially labeled after infection with MHV.

The LC consists of noradrenergic neurons and this specificity was used to confirm the infection of the LC by HSV-1 and not MHV. Infected brains were assayed initially for tyrosine hydroxylase using immunohistochemical staining as described in Materials and Methods and then processed by in situ hybridization for either HSV-1 or MHV nucleic acid. The LC is strongly immunolabeled by the antibody under these conditions. Analysis of the sections revealed that many double labeled cells were present in the HSV-1 infected brains in the region of the LC, but few or no double labeled cells were noted when brains were simultaneously assayed for tyrosine hydroxylase and MHV. MHV instead infected the adjacent parabrachial complex. These results confirm the preferential infection of the LC by HSV-1.

DISCUSSION

The experiments described above show that both HSV-1 and MHV enter the CNS via the olfactory nerve, but that only HSV-1 enters via the trigeminal nerve. MHV can clearly infect the trigeminal nerve after either intranasal inoculation or direct inoculation into dental pulp, but is unable to enter the CNS.

The results also show that HSV-1 and MHV infect a different subset of the neuroanatomic connections of the MOB. The difference is precise and reproducible and may reflect:

a) lack of MHV or HSV-1 receptor on the neurons that are not infected by the particular virus in question. The receptor for MHV is one or more members of the carcinoembryonic antigen family (13), but MHV may infect cells lacking this receptor if virus is presented in a cell-associated form (14).

b) lack of access. As an example, the noradrenergic input from the LC terminates predominantly on granule cells in the MOB. If MHV is unable to cross dendrodendritic synapses such as occurs between mitral and granule cells, this would limit access of virus to terminals of the LC. Similarly, if MHV does not infect granule cells, then it would not have access to the efferent fibers from the LC. Arguing against this hypothesis is the fact that the LC has connections with multiple other structures that are infected by MHV and thus it is unlikely that the virus does not have access to LC terminals.

c) different sites of egress from infected cells. Preferential exit of MHV or HSV-1 from one part of a neuron, such as a dendrite or axon, would limit exposure of virus to particular synaptic inputs on the cell.

d) anterograde versus retrograde transport. Some strains of HSV-1 preferentially travel in neurons in an anterograde direction whereas other strains travel in a retrograde fashion (15). Differential transport is most likely not important in our experiments, but must be considered in analyzing the subset of neuroanatomic connections of the MOB infected by any virus.

Acknowledgements: This research was supported by grants from the N.I.H. (NS24401) and National Multiple Sclerosis Society (RG2117). S.P. was the recipient of an Research Career Development Award from the N.I.H. G.J. was supported by a stipend from the Dows Research Institute. E.B. was supported by an NRSA from the N.I.H.

REFERENCES

1. E. Barnett, M. Cassell, and S. Perlman. Abstracts Society for Neuroscience 17:1511 (1991).
2. N. Goto, N. Hirano, M. Aiuchi, T. Hayashi, and K. Fujiwara. Jpn. J. Exp. Med. 47:59-70 (1977).
3. E. Lavi, P.S. Fishman, M.K. Highkin, and S.R. Weiss. Lab. Invest. 58:31-36 (1988).
4. S. Perlman, G. Jacobsen, and A. Afifi. Virol. 170:556-560 (1989).
5. S. Perlman, G. Jacobsen, A.L. Olson, and A. Afifi. Virol. 175:418-426 (1990).
6. S. Perlman, G. Evans, and A. Afifi. J. Exp. Med. 172:1127-1132 (1990).
7. K. Kristensson. Acta Neuropathol. 16:54-63 (1970).
8. A.H. Tomlinson, and M.M. Esiri. J. Neurol. Sci. 60:473-484 (1983).
9. W.G. Stroop, D.L. Rock, and N.W. Fraser. Lab. Invest. 51:27-38 (1984).
10. J.H. McLean, M.T. Shipley, and D.I. Bernstein. Brain Res. Bulletin. 22:867-881 (1989).
11. S. Perlman, G. Jacobsen, and S. Moore. Virol. 166:328-338 (1988).
12. S. Rochel, and F.L. Margolis. J. Neurochem. 35:850-860 (1980).
13. G.S. Dveksler, M.N. Pensiero, C.B. Cardellichio, R.K. Williams, G.S. Jiang, K.V. Holmes, and C.W. Dieffenbach. J. Virol. 65:6881-6891 (1991).
14. T.M. Gallagher, M.J. Buchmeier, and S. Perlman. (Virology, in press) (1992).
15. M.C. Zemanick, P.L. Strick, and R.D. Dix. Proc. Natl. Acad. Sci. 88:8048-8051 (1991).

MHV-JHM INFECTIONS OF RODENT NEURONAL CELLS: REPLICATION AND TRAFFICKING OF STRUCTURAL PROTEINS AND PROGENY VIRIONS

John Pasick and Samuel Dales

Cytobiology Group
Department of Microbiology and Immunology
Health Sciences Center
University of Western Ontario
London, Ontario, Canada
N6A 5C1

INTRODUCTION

A number of studies have signified the importance of neurons in the pathogenesis of the acute as well as chronic forms of MHV-induced neurologic disease in rodents. Following intranasal inoculation, MHV strains A59 and JHM invade the murine CNS by way of the olfactory and trigeminal nerves.[1,2,3] Accordingly, interneuronal spread was hypothesized to be the most likely mechanism to explain viral CNS penetration by this route. Subsequent spread to other CNS regions was shown to also involve specific neuronal populations and tracts.[2,4] Immunohistochemistry and in situ hybridization have demonstrated tropism of MHV-A59 for neurons within the olfactory nuclei, nuclei of the amygdala, central tegmental nucleus, entorhinal cortex, subiculum, claustrum, lateral habenular nucleus, subthalamic nucleus, basal ganglia, substantia nigra and septal nuclei among others in mice[2,5] and of JHMV for hippocampal and cerebellar Purkinje neurons in rats.[6,7] Although JHMV inoculated rats can develop a delayed onset, demyelinating encephalomyelitis with no prior evidence of acute encephalitis, histopathological and in situ hybridization studies have implied that clinically silent neuronal involvement likely precedes the development of the more chronic, demyelinating form of disease.[6,8] In addition to their potential importance as vehicles for MHV penetration and spread and as targets for cytopathic processes, neurons may also serve as reservoirs for viral persistence. The suggestion has been made that neurons may in fact be particularly well suited for this, due to insufficient expression of MHC class I molecules on their surfaces thus enabling them to avoid recognition by virus specific cytotoxic T cells.[9] For these reasons, studies focusing on the interactions between neurotropic strains of MHV and rodent neurons should enhance our understanding of neuropathogenic mechanisms in a fundamental way.

Assembly of JHMV and MHV-A59 virions in fibroblastic cells occurs within the

Coronaviruses, Edited by H. Laude and J.F. Vautherot
Plenum Press, New York, 1994

perinuclear region by budding into a smooth membrane compartment transitional between the rough endoplasmic reticulum and Golgi apparatus.[10,11] Furthermore, budding from this site was concluded to be controlled both temporally and spatially by the accumulation of the viral M structural protein.[11] Similar results were found with infection of murine pituitary tumor AtT20 cells where progeny virions were observed to exit cells via the constitutive rather than the regulated exocytic pathway.[12] Progeny JHM virions have likewise been shown to mature in close association with the Golgi apparatus in cultured mouse spinal cord neurons.[13]

The mechanism by which progeny virions exit infected cells in these examples likely involves a microtubule-dependent process. In this regard neurons are strikingly specialized. Each neuron is characteristically asymmetric, possessing two different types of neurites each performing specific functions. The single axon is responsible for conducting electrical signals away from the cell soma, while several dendrites serve as the main signal reception apparatus. Associated with this polarity in neurite function, are underlying differences in their membrane proteins and cytoskeletal frameworks.[14] Furthermore, since the biosynthetic functions of the neuron are restricted to the somato-dendritic domain, movement of materials to and from the apical or axonal domain, which may extend more than a meter in length, must involve an efficient process. This is accomplished by two microtubule-dependent processes: fast transport of membranous organelles and slow transport of cytosolic and particularly cytoskeletal proteins.[14] The present study deals with the growth and trafficking of JHMV structural proteins and virions in two rodent neuronal systems: cultured rat hippocampal neurons and murine OBL21 neuronal cells.[15]

RESULTS AND DISCUSSION

Embryonic day 18 to 20 Wistar Furth rats were used for the preparation of dissociated hippocampal neuron cultures following previously described procedures[16] with minor modifications. Cells were plated on either poly-L-lysine coated glass coverslips or plastic petri dishes at densities ranging between 3×10^4 to 6×10^4 viable cells per cm^2 and grown in a serum-free medium (DMEM and Ham's F12 1:1 with N2 supplements[17] plus Na pyruvate). Type-1 astrocytes were grown in co-culture with hippocampal neurons to provide trophic support and to buffer against the excitotoxic effects of the amino acid neurotransmitter glutamate. The astrocyte feeders, which were physically separated from neuronal cultures, were grown as monolayers in 60 mm petri dishes for neurons situated on 12 mm diameter glass coverslips, or grown on 24.5 mm diameter Costar transwell inserts (catalogue # 3425) for neuronal cultures situated in Costar 6 well cluster plates.

Hippocampal neurons were inoculated with JHMV at between 9 and 11 days in vitro (D.I.V.) with moi's that ranged between 5 and 30 pfu/cell. At this stage in culture most axons and dendrites have matured in terms of their functional and molecular properties and synaptogenesis is well under way.[16] In addition to wt JHMV, two variants, ATllf cord and V5A13 (88), which possess overlapping deletions of 441 and 447 nucleotides respectively within the S1 portion of the S gene[18,19] were also used. These variants were of special interest to us due to their reported attenuation in neurovirulence as judged by the shift in the pattern of disease they induced from one of fatal encephalomyelitis to chronic demyelination.[20,21] Moreover, this shift was suggested to result from the loss of tropism for neurons but not glia.[20] This system consequently provided an ideal opportunity to assess their tropism directly.

Cell density appeared to be a more important determinant than moi with respect to establishing and maintaining viral growth in hippocampal neuronal cultures. In addition, although infection could be established with all three viruses, the number of neurons initially infected and subsequent growth tended to be greater with the two variants (Table 1).

Table 1. Growth of wt JHMV and S Deletion Variants in Dissociated Hippocampal Neuron Cultures.

		PFU/ml of Culture Supernatant[*] Days Post-Inoculation				
		1	2	3	4	5
Experiment # 1[a]	wt JHMV	1.6 ±0.5	4.4 ±2.0	12.6 ± 8.0	47.0 ±12.7	55.5 ±17.7
	AT11f cord	3.1 ±0.1	33.0 ±12.1	116.0 ± 26.5	485.0 ± 57.3	239.7 ± 37.9
	wt JHMV	0.2 ±0.2	<0.1	0	0.6 ±1.0	1.4 ±2.4
Experiment # 2[b]	AT11f cord	1.6 ±1.0	3.3 ±1.2	2.1 ±1.0	15.8 ± 9.7	42.3 ± 2.3
	V5A13 (88)	4.1 ±1.1	12.0 ± 6.2	16.6 ± 6.1	48.7 ±22.6	79.0 ± 9.0

[*]Expressed as x 10^2

[a]In Experiment # 1 cells dispersed from E20 Wistar Furth rat hippocampi were seeded at a density of 60,000 viable cells/cm^2. At 11 D.I.V. cultures were inoculated with virus at a moi of 5 pfu/cell.

[b]In Experiment # 2 cells dispersed from E20 Wistar Furth rat hippocampi were seeded at a density of 30,000 viable cells/cm^2. At 10 D.I.V. cultures were inoculated with virus at a moi of 30 pfu/cell.

Thus, if in vitro virus-neuron interactions accurately reflect the situation occurring in vivo, one can make the provisional conclusion that loss of neuronal tropism is not associated with these truncations involving the S glycoprotein. Furthermore, although our wt JHMV expresses the HE glycoprotein,[19] the variants do not,[19] suggesting that this glycoprotein is likely not necessary for neuronal infections. Hence, the neuroattenuation associated with these variants in vivo, may not be easily explained by a shift in tropism. In some culture preparations endogenous astrocytes proliferated to represent a significant proportion of the total cell population. However, they did not usually become infected to any significant extent until after 2 dpi, a finding in agreement with our previous work.[22] Furthermore, astroglial infections were usually associated with the formation of syncytia while neuronal infections were not.

Viral proteins appeared to localize to both tapering and branching dendritic-like as well as fine axonal-like processes of infected neurons. Dual immunolabelling with antibodies specific for either N or S and tau was carried out to further assess the distribution of viral proteins to the two neuronal domains (Fig.1).

Tau along with MAP2 are microtubule-associated proteins which appear to form cross-bridges between adjacent microtubules.[23] Immunocytochemical studies on cultured rodent neurons have demonstrated that tau and MAP2 topographically segregate to predominantly axonal and somato-dendritic domains respectively[24] in agreement with the situation existing in vivo.[25] This is slightly complicated by the fact that axons tend to arise most commonly as branches from the proximal portions of dendrites rather than directly from the cell soma in hippocampal neurons grown in culture,[26] and that during the early stages of culture, MAP2 and tau immunoreactivity co-distribute within

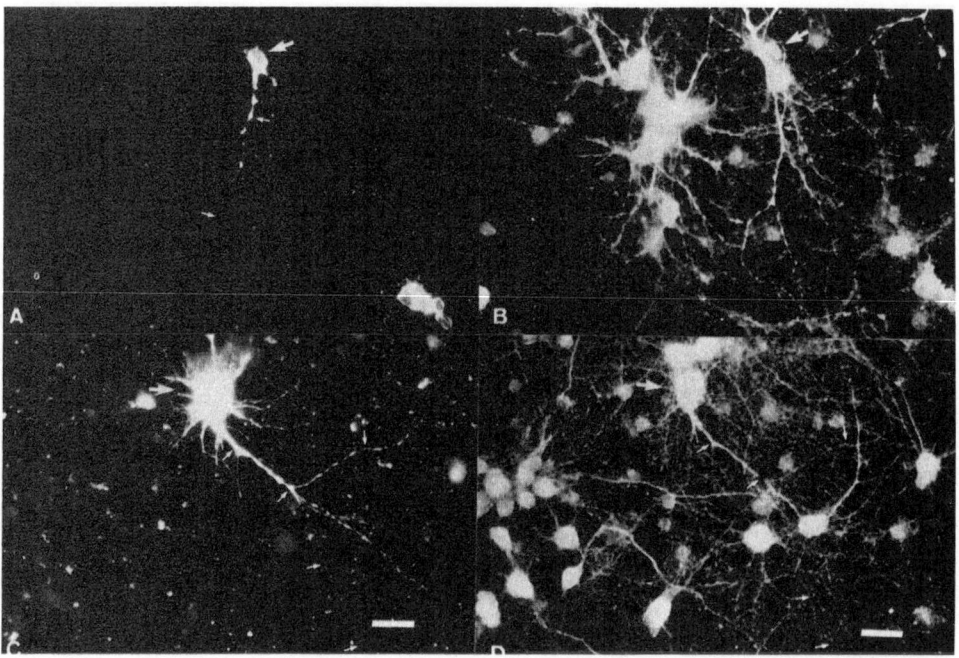

Figure 1. E20 hippocampal neuronal cultures infected with V5A13 (88) variant at 10 D.I.V. and immunocytochemically labelled 1dpi. A, an infected neuron is labelled by using MAb 4B6.2 specific for N and FITC conjugated secondary antibody. B, cells in the same field as A are labelled with a rabbit polyclonal anti-tau antibody (Sigma) and Texas Red conjugated secondary antibody. C, an infected neuron is labelled with MAb 5B17.0 specific for S and FITC conjugated secondary antibody. D, identical field to that in C is labelled as described for B. Arrows point to complementary neurites in A and B and C and D. Bar represents 19 μm.

both axon- and dendrite-like neurites.[24,27] However, in cultures in which prominent fasciculation of axons occurred, viral N and S immunoreactivity could be localized within axon bundles, eliminating some of this ambiguity. Electron microscopic examination of cultures revealed the association of nucleocapsid core components with microtubule arrays as reported previously.[22] Occasionally virions were observed within neurites (Fig. 2). Studies of JHMV infected polarized ependymal cells indicate that progeny exit the cell baso-laterally judging by the pattern of spread to adjacent ependyma as well as subependymal tissues.[28] However, arguments were also put forward suggesting the apical exit of progeny JHMV from infected ependymal cells.

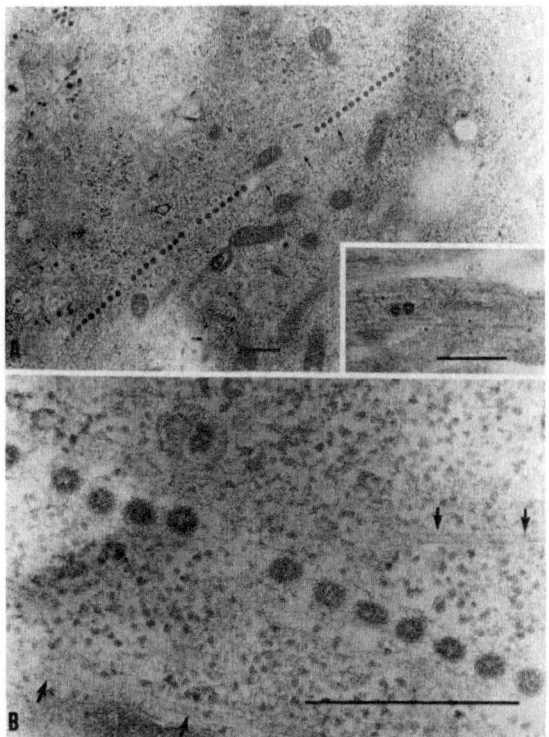

Figure 2. A demonstrates progeny JHM virions forming linear arrays within cisternae of infected OBL21 cells. The solid arrows point to an immediately adjacent microtubule. The inset shows two virions within a neurite from an infected hippocampal neuronal culture. B is a higher resolution image of the area bordered by the open arrows in A. The arrows in B point to microtubules. The bars in all panels represent 0.5 μm.

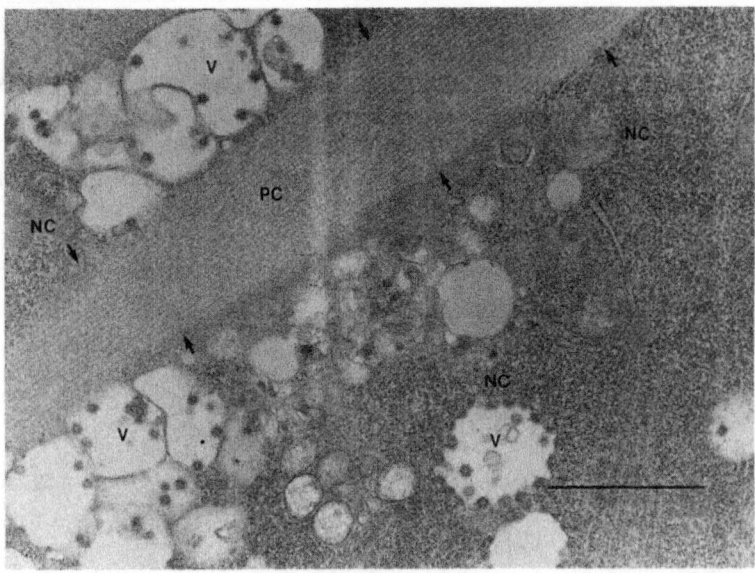

Figure 3. OBL 21 neuronal cell cultures inoculated with wtJHMV at a moi of 10 pfu/cell were treated with 10 μg/ml vinblastine sulphate at 36 hours post–inoculation then processed for electron microscopy 12 hours later. Arrows outline a tubulin paracrystal, PC, around which nucleocapsid core components, NC, and virions within vesicles, V, have closely associated. Bar represents 1 μm.

The pattern of distribution of N and S in hippocampal neurons reported here indicates that assembly and/or trafficking takes place somato-dendritically as observed previously[13,29] and in concordance with the reported presence of Golgi elements within dendrites.[30] The observed presence of both viral proteins in axonal-like neurites indicates one of two possibilities: assembly and/or trafficking may be nonpreferentially targeted toward both axons and dendrites or, viral proteins present within axons may result from transsynaptic transmission with subsequent retrograde spread to the cell soma. JHMV penetration by olfactory and trigeminal nerves implying transport in the anterograde direction, lends support to the former assumption,[3] although the latter while less likely, is also plausible. Of possible relevance to the above observations is the intriguing protein sequence homology between N and tau in which a respective 42% and 20% amino acid sequence similarity and identity exists. This was found by Dr. Michael Clarke of the Department of Microbiology and Immunology at UWO using the nucleotide sequence of N to search the Gen Bank data base. Optimal alignment of a putative tau microtubule binding motif was found between amino acids 328 and 340 inclusive within the carboxy-terminal portion of N.

To further assess the importance of microtubules in transporting newly assembled virions to the cell exterior, studies were carried out using the neuronal cell line OBL21. This cell line was clonally derived from CD.1 mouse olfactory bulb cultures immortilized with a replication-defective retrovirus vector carrying the avian myc gene.[15] The OBL21 line is a stable, homogeneous line expressing neurofilament but not glial fibrillary acidic protein and has maintained this phenotype while in our hands (data not shown). As with JHMV infected hippocampal neurons, assembly and/or trafficking of viral products N and S occurred in the neurite-like processes of OBL21 cells. In some infected cells mature virions were organized in precise linear arrays within cisternae adjacent to parallel microtubules (Fig. 2). To demonstrate virus-microtubule association more rigorously, infected OBL21 cultures were treated with vinblastine, a mitotic inhibitor which also inhibits axonal transport by promoting the depolymerization of microtubules and concomitant formation of tubulin paracrystals.[31] The resulting collapse of the microtubule framework was associated with the formation of massive aggregates of viral core components and membrane bound progeny around the tubulin paracrystals (Fig. 3).

In summary, neurons are targets for JHMV in vivo and in vitro. Variants characterized as having neuroattenuated phenotypes in vivo don't appear to possess altered tropism for primary hippocampal neurons in culture suggesting that the basis for in vivo neuroattenuation may be more complicated than initially thought. Trafficking of viral structural proteins in primary hippocampal neurons, as well as the involvement of microtubules in the movement of progeny virions in OBL21 cells, strengthens the supposition that rodent neurons may be capable of spreading the infection transneuronally.

ACKNOWLEDGEMENTS

We wish to thank Michael Buchmeier of the Scripps Clinic and Research Foundation for generously providing MAbs 4B6.2 and 5B17.0, Constance Cepko of Harvard Medical School for kindly providing the OBL21 cell line, Andrea Hanington for maintaining the rodent colony, and Jennifer Fraser and Sharon Wilton for technical assistance.

This work was supported by grants from the Medical Research Council of Canada and the Multiple Sclerosis Society of Canada to S.D. J.P was the recipient of a postdoctoral fellowship from the M.S. Society of Canada.

REFERENCES

1. S.W. Barthold, Acta Neuropath. 76:502 (1988).
2. E. Lavi, P.S. Fishman, M.K. Highkin, and S.R. Weiss, Lab Invest. 58:31 (1988).
3. S. Perlman, G. Jacobsen, and A. Afifi, Virology 170:556 (1989).
4. S. Perlman, G. Jacobsen, A.L. Olson, and A. Afifi, Virology 175:418 (1990).
5. P.S. Fishman, J.S. Gass, P.T. Swoveland, E. Lavi, M.K. Highkin, and S.R. Weiss, Science 229:877 (1985).
6. O. Soresen and S. Dales, J. Virol. 56:434 (1985).
7. D. Parham, A. Tereba, P.J. Talbot, D.P. Jackson, and V.L. Morris, Arch. Neurol. 43:702 (1986).
8. M. Koga, H. Wege, and V. ter Meulen, Neuropath. Appl. Neurobiol. 10:173 (1984).
9. E. Joly, L. Mucke, and M.B.A. Oldstone, Science 253:1283 (1991).
10. A. Massalski, M. Coulter-Mackie, R.L. Knobler, M.J. Buchmeier, and S. Dales, Intervirol. 18:135 (1982).
11. J. Tooze, S. Tooze, and G. Warren, Eur. J. Cell Biol. 33:281 (1984).
12. J. Tooze, S.A. Tooze, and S.D. Fuller, J. Cell Biol. 105:1215 (1987).
13. M.E. Dubois-Dalcq, EW. Doller, M.V. Haspel, and K.V. Holmes, Virology 119:317 (1982).
14. R.B. Vallee and G.S. Bloom, Ann. Rev. Neurosci. 14:59 (1991).
15. E.F. Ryder, E.Y. Snyder, and C.L. Cepko, J. Neurobiol. 21:356 (1990).
16. G. Banker and K. Goslin, in: "Culturing Nerve Cells," G. Banker and K. Goslin, eds., MIT Press, Cambridge, Mass. (1991).
17. J.E. Bottenstein and G.H. Sato, Proc. Natl. Acad. Sci. U.S.A. 76:514 (1979).
18. S.E. Parker, T.M. Gallagher, and M.J. Buchmeier, Virology 173:664 (1989).
19. N. La Monica, L.R. Banner, V.L. Morris, and M.M.C. Lai, Virology 182:883 (1991).
20. R.G. Dalziel, P.W. Lampert, P.J. Talbot, and M.J. Buchmeier, J. Virol. 59:463 (1986).
21. V.L. Morris, C. Tieszer, J. MacKinnon, and D. Percy, Virology 169:127 (1989).
22. J.M.M. Pasick and S. Dales, J. Virol. 65:5013 (1991).
23. M. Goedert, R.A. Crowther, and C.C. Garner, Trends Neurosci. 14:193 (1991).
24. K.S. Kosik and E.A. Finch, J. Neurosci. 7:3142 (1987).
25. L.I. Binder, A. Frankfurter, and L.I. Rebhun, J. Cell Biol. 101:1371 (1985).
26. W.P. Bartlett and G.A. Banker, J. Neurosci. 4:1944 (1984).
27. A. Caceres, G.A. Banker, and L. Binder, J. Neurosci. 6:714 (1986).
28. F-I. Wang, D.R. Hinton, W. Gilmore, M.D. Tousdale, and J.O. Fleming, Lab. Invest. 66:744 (1992).
29. R.L. Knobler, M. Dubois-Dalcq, M.V. Haspel, A.P. Claysmith, P.W. Lampert, and M.B.A. Oldstone, J. Neuroimmunol. 1:81 (1981).
30. C.G. Dotti and K. Simons, Cell 62:63 (1990).
31. S. Dales, K.C. Hsu, and A. Nagayama, J. Cell Biol. 59:643 (1973).

PERSISTENCE OF VIRAL RNA IN THE CENTRAL NERVOUS SYSTEM OF MICE INOCULATED WITH MHV-4

John O. Fleming,[1,2] Jacqueline J. Houtman,[1] Hulya Alaca,[1] Harry C. Hinze,[1] Debbie McKenzie,[1] Judd Aiken,[1] Thomas Bleasdale,[3] and Susan Baker[3]

[1]University of Wisconsin, Madison WI 53792
[2]Wm S. Middleton Memorial Veterans Hospital, Madison, WI 53705
[3]Loyola University Medical Center, Chicago IL 60153

ABSTRACT

In order to study the role that viral persistence may play in chronic central nervous system (CNS) disease induced by murine coronaviruses, we have used the reverse transcriptase-polymerase chain reaction (RT-PCR) to study viral RNA in the brains of mice after intracerebral inoculation of JHM virus (JHMV or MHV-4). Quantitative RT-PCR showed that JHMV RNA decreased from approximately 2 ng/ug total brain RNA at day 6 post-inoculation (PI) to 0.1 pg/ug total brain RNA at 360 days PI. Double-stranded viral RNA could be detected up to day 20 PI. By the selective use of upstream or downstream primers during the RT step, it was possible to measure negative sense and positive sense JHMV RNA respectively, and we found that there was a marked rise in the ratio of positive to negative sense JHMV RNA after day 13 PI. Analysis of amplified products by dideoxy DNA sequencing showed that the characteristic mutation of our input virus (at position 3340 of gene 3) is maintained to at least day 42 PI. Taken together, these results favor a model of JHMV persistence in vivo in which viral RNA is present as double stranded forms initially and predominantly as single stranded, positive sense forms at late timepoints. Further analysis of this model in quantitative terms may contribute to our understanding of the biological significance of coronavirus persistence in the CNS.

INTRODUCTION

An interesting feature of coronaviruses has been their ability to readily establish persistent infections. Many elegant studies have demonstrated this phenomenon in vitro [1,2]. Also, other investigations have indicated that coronaviruses may persist in vivo as well[3-7], and recent reports have indicated that coronavirus persistence, particularly in the

Coronaviruses, Edited by H. Laude and J.F. Vautherot
Plenum Press, New York, 1994

CNS, may contribute to human diseases, such as multiple sclerosis[8,9]. Because of the important biological questions raised by these findings, as well as their possible relevance to clinical disorders, there is a need to develop additional systems in which coronavirus persistence in vivo may be studied experimentally.

In this regard, Lavi and Weiss[7] have comprehensively reviewed available investigations of coronavirus persistence in vivo. Most - but not all - of these reports indicate that infectious virus can only be isolated transiently after inoculation, while, by contrast, viral RNA itself appears to be sequestered in tissues for prolonged periods. Despite these studies, many important questions remain, such as the state of viral RNA during persistence, specific mutations that may be associated with or permit persistence, and the biological consequences of persistence. In an initial attempt to answer these questions in a murine model of coronavirus persistence in the CNS, we have "recovered" and analyzed viral RNA by the RT-PCR method.

METHODS AND RESULTS

Mice and Viruses

Unless stated otherwise, six week old male C57BL/6J mice seronegative for MHV were inoculated with 10^3 plaque forming units (PFU) of JHMV, variant 2.2-V-1, intracerebrally (i.c.). This variant was selected for resistance to a neutralizing monoclonal antibody directed against the S protein; we have previously shown that gene 3, coding for the spike protein of this variant, is identical to that of parental or "wild type" JHMV, except for a point mutation at nucleotide 3340[10]. This variant is attenuated with regard to lethality, but causes marked demyelination in almost all mice inoculated [11]; in this respect it resembles the original (1949) JHMV isolates.

RNA Extraction and RT-PCR

Mice were sacrificed and total brain RNA was isolated using phenol/chloroform extraction, followed by ethanol precipitation in the presence of 3 M ammonium acetate. Total brain RNA was then subjected to RT and PCR using JHMV-specific oligonucleotide primers as indicated. Extensive precautions were taken to avoid contamination, including separate rooms and pipettes for RNA extraction, RT-PCR preparation, PCR cycling, and product analysis. The quality of RNA was assured by RT-PCR amplification of β actin sequences (data not shown).

Initial RT-PCR assays utilized primers for JHMV gene 7 (primers AM8 1-2, nucleotides 672 - 1237). (These primers were used in all assays below, unless stated otherwise.) Plasmid AM8, containing most of the JHMV nucleocapsid gene sequences, was generously provided by Dr. Michael Lai and served as a positive control for these assays. Because of the co-terminal, 3' nested-set structure of coronavirus RNAs, primers AM8 1-2 will recognize all JHMV RNA species. As shown in figure 1, JHMV-specific RNA could be detected in murine brain as late as 360 days PI. By contrast, infectious virus could not be detected beyond days 10-12 PI. Assay of non-CNS tissues showed low levels of JHMV RNA at day 42 PI in liver but not in spleen.

In order to test the generality of these results, MHV RNA was detected at day 42 in the brains of mice in two further experiments: 1) after i.c. inoculation of C57BL/6J mice with 10^2 PFU of MHV A59 or 2) after i.c. inoculation of BALB/cJ mice with 10^3 PFU of

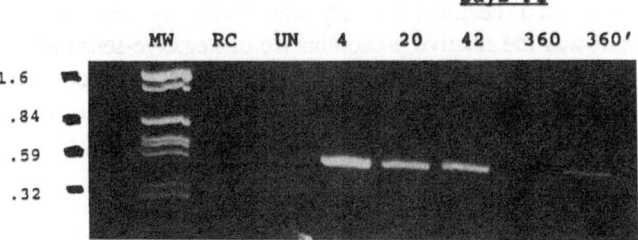

Figure 1. JHMV RNA persists in mouse brain for 360 days PI. RC = reagent control and UN = uninfected mouse. Two mice were studied at one year PI (animals 360 and 360').

JHMV 2.2-V-1. In both cases, infectious virus could not be recovered beyond the period of acute disease. These results indicate that the phenomenon of CNS persistence of murine coronavirus RNA does not depend upon a particular strain of mouse or MHV.

Form of JHMV RNA Persisting in Mouse Brain

Having shown that MHV RNA persists in the murine CNS for prolonged periods of time, we next sought to determine in what form the nucleic acid persisted. Initially, DNA was extracted from the brains of mice 13 days PI and subjected to PCR assay without prior reverse transcription. No product was obtained in these reactions, indicating that JHMV is unlikely to persist in DNA form in vivo.

In a further attempt to characterize persisting JHMV nucleic acids, brain RNA samples from 0 to 20 days PI were treated with RNase A in order to digest single-stranded RNA prior to RT-PCR. As shown in figure 2, reverse transcription of RNase-treated RNA was primed with either random hexamer primers, upstream primer (AM8 1, recognizing negative sense JHMV RNA), or downstream primer (AM8 2, recognizing positive sense JHMV RNA). The PCRs were then performed with both upstream and downstream primers (AM8 1-2).

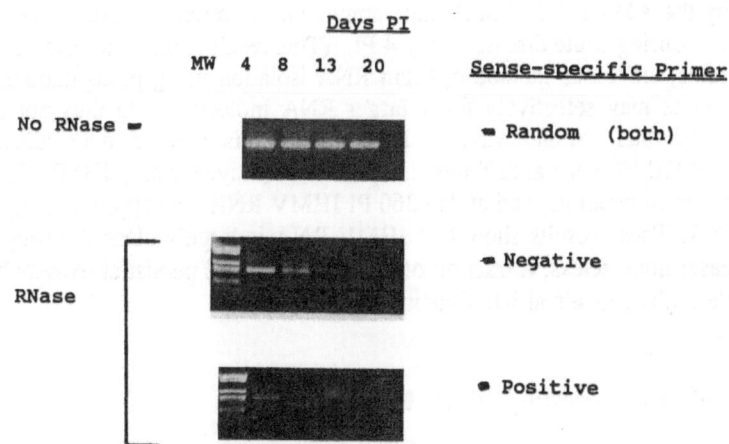

Figure 2. Assay for double-stranded JHMV RNA, days 4 to 20 PI. MW = molecular weight standards, and numerals indicate the number of days PI.

The data shown in Figure 2 indicate that RNase-resistant, double-stranded RNA was present at day 4 PI and declined to very low levels by day 20 PI. A surprising, asymmetrical result was the relative predominance of negative-sense RNase-resistant RNA in this period. Although these results are preliminary, they are consistent with a model proposed by Sethna and colleagues[12] to explain similar data obtained during in vitro study of the replication of transmissible gastroenteritis virus (TGEV) replication (please see discussion).

Using RT primed by either AM8 1 or AM8 2 primer (no RNase treatment), followed by PCR using both primers, it was possible to estimate the relative abundances of total positive and negative sense JHMV species in brain. Preliminary results of these assays indicate a relative rise in the amount of negative sense RNA present near day 12 PI, the time of transition from infectious virus to persisting RNA forms. After this period, total JHMV RNA is predominantly of positive sense.

Quantitative RT-PCR for Persisting JHMV RNA

In order to determine the amount of JHMV-specific RNA that persists in mouse brain after viral inoculation, a quantitative RT-PCR based on the method of Gilliland et al.[13] was developed. In this assay, a series of reactions is performed in which varying amounts of competitor or "false template" RNA is added to aliquots of RNA from an infected brain from the timepoint to be analyzed. The competitor RNA was transcribed in vitro from a pGEM (Promega) plasmid containing gene 7 sequences (AM8) which are identical to genuine JHMV sequences, except for the introduction of a Bgl II restriction site in the target region for amplification. Thus, the amplified products obtained in each RT-PCR depend on the relative concentration of the two competing template RNAs. The two RT-PCR products may be distinguished after RT-PCR by digestion with Bgl II.

An important point to emphasize is that each tube in the quantitative RT-PCR serves as its own control in the sense that both RNA templates are subjected to identical reaction conditions. Since the concentration of added synthetic RNA in each tube in the series of RT-PCR reactions is known, the tube in which both products are equally abundant indicates the concentration of viral RNA in a particular infected brain sample.

Our results with quantitative RT-PCR to date indicate that JHMV RNA, as represented by the AM8 1-2 565 nucleotide amplicon, is present at approximately 2 ng/ug brain total RNA during acute disease at day 4 PI. (This result is likely to be an overestimate in absolute terms, since our method of brain RNA isolation using precipitations with high salt concentrations may selectively favor larger RNA molecules and thus not be entirely representative of total cellular RNA. Nonetheless, this method does allow accurate quantification of JHMV RNA at different timepoints in relative terms.) JHMV RNA in brain fell rapidly at later timepoints, and at day 360 PI JHMV RNA was approximately 0.1 pg/ug total brain RNA. These results show that JHMV RNA is largely cleared during acute and subacute disease; nonetheless, a fraction of JHMV RNA does persist at extremely low, but nonetheless detectible, levels at late timepoints.

Sequence Stability of Persisting JHMV RNA

As indicated above, the JHMV strain used in our studies has a mutation in gene 3 (spike protein) at nucleotide 3340. In order to determine if this mutation, present in our "input virus", could be found in RNA persisting in vivo, total brain RNA samples from mice at days 4, 20, and 42 PI were subjected to RT-PCR using gene 3 oligonucleotides which prime the amplification of a product corresponding to nucleotides 3246 to 3468. This product was cloned into pTZ18u plasmid and sequenced by the dideoxy method (Sequenase). We found that the mutation present in JHMV 2.2-V-1 (C to T at nt 3340) is present in brain RNA extracted from murine brain at day 4 PI; identical results were obtained at days 20 and 42 PI (data not shown). At present, this analysis of RNA sequence stability is being extended to later timepoints and other regions of the JHMV genome.

DISCUSSION

In the studies above, we have demonstrated a transition from 1) an acute, productive CNS infection with JHMV to 2) a state in which viral RNA may be detected in the apparent absence of infectious virus. This experimental paradigm may mimic putative coronavirus persistence in human diseases, such as multiple sclerosis[8,9].

We were able to detect double stranded JHMV RNA forms in the CNS until day 20 PI (Figure 2). This result may imply active replication or transcription of the viral genome. Analysis of the relative abundances of negative and positive sense species that are detected in these assays for double stranded forms suggests a replicative species which is similar to that proposed by Sethna and colleagues[11] to explain data obtained with TGEV in vitro. In this model, a replicative intermediate consisting of negative sense template is associated with multiple nascent positive sense progeny RNAs. Upon digestion with RNase, the entire negative strand would remain and be amplified by RT-PCR; however, the regions of protected positive sense RNA would be too small for amplification.

The period from day 12 to 20 PI is likely to be critical in several ways. First, infectious virus may only be detected until approximately day 12 PI, while double stranded JHMV RNA is seen until day 20 PI. This suggests a short period of time in which viral RNA is being actively replicated, but infectious virus is absent. Second, there appears to be a relative rise in the abundance of total negative sense JHMV RNA near day 12 PI; thereafter, the predominant species appears to be single stranded positive sense JHMV RNA. The finding of positive sense RNA predominance at late timepoints is consistent with the pattern of RNA forms observed in vitro [2] in studies of persistent infection by bovine coronavirus. Whether these changes in the relative abundances of different JHMV RNA forms are due to intrinsic viral factors, host cell adaptations, or the influence of the immune system is at present unknown. Certainly, viruses which have become attenuated with regard to transcription or translation would be expected to have a selective survival advantage subacutely, during the period of maximal anti-JHMV immune response.

At late timepoints, viral RNA may be detected in low amounts in the CNS and trace amounts in the liver. We do not know whether or not there is an interplay of CNS and systemic compartments during chronic disease; possibly there could be trafficking of very low levels of infectious virus or of activated immune cells between the CNS and the systemic compartments in this period.

Also, we do not currently know whether persisting JHMV RNA is of genomic and/or subgenomic length, since our RT-PCR primers (AM8 1-2) will recognize both. Experiments are underway to characterize the persisting RNA in this regard. Hofmann and colleagues[2] have postulated that subgenomic RNA forms may participate in the establishment of persistent coronavirus infections in vitro by in effect acting as defective interfering RNAs. If subgenomic forms predominate in our in vivo system, our findings would be consistent with this hypothesis.

Persistence of JHMV RNA in vivo does not appear to be limited to the viral isolate and mouse strain used in the majority of our experiments; that is, similar findings occurred when MHV A59 or BALB/cJ mice were used. Possibly, the patterns of persistence of viral RNA in CNS we have demonstrated may characterize the majority of chronic CNS coronavirus infections . Further, in studies of the S2 region of gene 3 to date, RNAs isolated from persistently infected mice appear to have retained their original sequences. Additional sequencing studies are underway in order to determine if critical mutations in this or other genes play a role in the establishment or maintenance of JHMV persistence.

The preliminary experiments above demonstrate the feasibility of RT-PCR rescue of JHMV RNA during chronic CNS disease caused by this virus. We hope that this experimental system will yield insights into the pathogenesis of JHMV and suggest research strategies which may be fruitfully applied to human disease.

ACKNOWLEDGEMENTS

We thank Joan Suter and Amy Herbst for excellent technical assistance. This work was supported by Multiple Sclerosis Society Research grant 2283-A-2 and the Wm. S. Middleton Memorial Veterans Hospital.

REFERENCES

1. Siddel ST, Wege H, ter Meulen V. In Cooper M, et al. (eds), Current Topics in Microbiology and Immunology, Volume 99. Springer-Verlag, Berlin, 1982, pp. 131-163.
2. Hofmann MA, Sethna PB, Brian DA. J Virol 64:4108-4114, 1990.
3. Herndon RM, Griffin DE, McCormick U, Weiner LP. Arch Neurol 32:32-35,1975.
4. Knobler RL, Lampert PW, Oldstone MB. Nature 298:279-280, 1982.
5. Perlman S, Reis D. Microbial Pathogenesis 3:309-314, 1987.
6. Perlman S, Jacobsen G, Olson AL, Afifi A. Virology 175:418-426, 1990.
7. Lavi E, Weiss SR. In Gilden DH, Lipton HL (eds.), Clinical and Molecular Aspects of Neurotropic Virus Infection. Kluwer Academic Publishers, Norwell, MA, 1989, pp. 101-139.
8. Burks JS, DeVald BL, Jankovsky LD, Gerdes JC. Science 209:933-934,1980.
9. Murray RS, Brown B, Brian D, Cabirac GF. Ann Neurol 31:525-533, 1992.
10. Wang FI, Fleming JO, Lai MMC. Virology 186:742-749, 1992.
11. Fleming JO, Trousdale MD, El-Zaatari FAK, Stohlman SA, Weiner LP. J Virol 58:869-875, 1986.
12. Sethna PB, Hung SL, Brian DA. Proc Natl Acad Sci USA 86:5626-5630, 1989.
13. Gilliland G, Perrin S, Bunn HF. In Innis MA et al. (eds.), PCR Protocols. Academic Press, San Diego, 1990, pp. 60-69.

THE ROUTE OF TRANSMISSION OF HEMAGGLUTINATING ENCEPHALOMYELITIS VIRUS (HEV) 67N STRAIN IN 4-WEEK-OLD RATS

Norio Hirano[1], Satoshi Haga[1] and Kosaku Fujiwara[2]

[1]Department of Veterinary Microbiology, Iwate University
Morioka 020, Japan
[2]Department of Veterinary Pathology II, Nihon University
Fujisawa 252, Japan

ABSTRACT

Four-week-old Wistar rats were inoculated with HEV by different routes. Animals died of encephalitis after intraperitoneal (i.p.), subcutaneous (s.c.) and intravenous (i.v.) as well as intracerebral (i.c.) and intranasal (i.n.) inoculation. However when inoculated subcutaneously, rats died a few days earlier than those inoculated i.p. and i.v., suggesting that the virus might be transmitted to the central nervous system (CNS) by the neuronal route rather than by blood stream.

Rats which were inoculated subcutaneously at the site of the neck (group A) began to die on day 4 p.i., a few days earlier than animals inoculated in the foot pad of the right leg (group B). On day 2 and 3 after inoculation, the virus titer in the brain was higher in group A, but group B animals showed higher virus titers in the lumber region of spinal cord than group A animals.

In order to follow the virus spread from the peripheral nerve to the brain, the virus was inoculated into the sciatic nerve of rats. The inoculated rats developed clinical signs on day 4 and began to die on day 6. On day 2, virus was detected in the posterior half of the spinal cord and migrated toward the anterior half and in the brain where it was present on day 3. The highest virus titers in the brain were recorded on day 4 to 6, meanwhile the virus titers in the spinal cord tend to decrease. By immunohistochemical study, antigen positive neurons were found in the spinal cord and brain on day 4. Viral specific antigen was detected in the neurons in the cerebral cortex and mid brain of some animals which had survived viral infection more than 24 days.

INTRODUCTION

The HEV 67N strain causes encephalomyelitis or vomiting and wasting syndrome in piglets [1,2]. In experimental infection of piglets, the virus spreads along the nerve pathways to the CNS, and virus replication is restricted to the neurons [3]. In our experimental studies of HEV 67N strain [4], the virus produced encephalomyelitis in mice when inoculated by several routes and propagated mainly in the nerve cells of the CNS. However, 20-day-old or older mice were resistant to the virus injected i.n., i.p. or s.c.

Coronaviruses, Edited by H. Laude and J.F. Vautherot
Plenum Press, New York, 1994

Recently, the authors reported the successful propagation and plaque assay of HEV 67N strain in established cell line, SK-K cell culture [5].

The SK-K-passaged virus caused encephalitis in mice and in adult rats. Four-week-old rats inoculated i.p., i.v., s.c., i.c. died of encephalitis, but s.c. inoculated rats died a few days earlier than i.p. and i.v. ones. This suggests that the virus might spread to the CNS by the neuronal route rather than by blood stream. Rats were shown to be more sensitive than mice and thus might be useful for studying the pathogenesis of HEV infection.

In order to follow the virus spread from the peripheral nerve to the brain, the virus was inoculated into the sciatic nerve of the rats. The virus growth and its distribution in the spinal cord and the brain was studied by virus titration and immunohistochemistry in the different organs.

MATERIALS AND METHODS

Virus. Plaque-purified HEV 67N strain was propagated in SK-K cells and was assayed for infectivity in a plaque assay, as described [5].

Animal inoculation. Four-week-old male rats were obtained from a breeder colony, which was serologically negative for murine coronaviruses. Rats were inoculated i.c. and i.n with 0.02 ml of viral suspension and with 0.2 ml for i.p., s.c. and i.v inoculations. For inoculation into the sciatic nerve, animals were deeply anesthetized with pentobarbital and the right thigh was surgically operated to inoculate with 0.02 ml of the virus.

Infectivity of the brain and spinal cord. Ten percent tissue homogenates (w/v in Eagle's MEM) were prepared from the brain and spinal cord, and assayed for infectivity in SK-K cell system as described [5]. The spinal cord was divided into 2 pieces, anterior half and posterior one for infectivity assay.

Immunohistochemical study. Immunohistochemistry was performed on paraffin sections by ABC method using anti-67N mouse antibody (1:1000). For histopathological study, serial sections were stained with hematoxylin and eosin (HE).

RESULTS

Five rats were tested for susceptibility to the virus by each route of inoculation, and mortality and clinical signs were recorded over 14 days. Most rats died of encephalitis (Table 1) ; rats inoculated i.c. were most susceptible to the virus, whereas i.p. inoculated ones were most resistant. Among rats inoculated by other routes, s.c. inoculated rats began to die a few days earlier than i.n., i.p., and i.v. inoculated ones. This suggested the possibility that the virus might be transmitted to the CNS by neuronal route.

To test this hypothesis, 2 groups of rats (A and B) were inoculated s.c., at 2 different sites, with 10 PFU of virus. Rats in group A which had been inoculated at the site of neck, began to die on day 4, a few days earlier than animals in group B which had received the virus in the foot pad of right leg. The virus was also detected earlier in the brain and anterior half of the spinal cord of rats in group A, and the infectivity titers of these organs were higher. In contrast to this, the virus was isolated one day earlier from the posterior half of the spinal cord of rats in group B.

The virus was never detected from the liver and spleen. Taken together, these results suggest that the virus might spread to the brain via peripheral nerves.

To follow the virus spread from the peripheral nerve to the brain, 1000 PFU of HEV were inoculated into the sciatic nerve of right leg. On day 4 p.i., the inoculated animals developed clinical signs consisting of ataxia, hypersensitivity and flapping ears, and they began to die on day 6. Some of severely diseased rats survived for more than 24 days.

The virus was first isolated only from the posterior half of the spinal cord on day 2 (Fig. 1). On day 3, the virus was also detectable in the anterior half of the spinal cord and the brain. The virus titers in the spinal cord increased to reach a maximum on day 4 and decreased thereafter. On day 5, the highest titers in the brain ranged from $10^{6.5}$ to 10^7 PFU/0.2g. The virus could not be detected in the liver, spleen or blood.

Table 1. Susceptibility of 4-week-old rats to HEV 67N

Route of inoculation	Inoculum (PFU)	Dead/Tested	Mean time to death
i.c.	1×10^5	5/5*	4.8 (4-5)**
	1×10^3	5/5	5.0 (4-6)
	1×10	1/5	6.3 (5-7)
i n.	1×10^5	5/5	7 8 (6-12)
	1×10^3	5/5	9.0 (9-11)
	1×10	1/5	10.0 (10)
i.p.	1×10^6	4/5	6.8 (6-9)
	1×10^4	1/5	10.0 (10)
	1×10	1/5	10.0 (10)
s.c.	1×10^6	5/5	5.2 (4-6)
	1×10^4	5/5	5.4 (4-6)
	1×10	1/5	6.0 (6)
i.v.	1×10^6	5/5	6.0 (6)
	1×10^4	4/5	6.2 (6-7)
	1×10	1/5	7.0 (7)

* No. of animals dead/tested, 14 days postinoculation.
** Mean time to death in days (range).

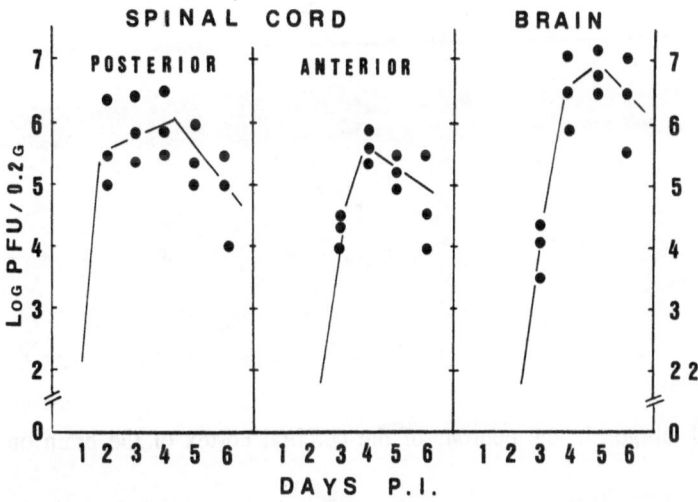

Fig. 1. Virus growth in the spinal cord (posterior half and anterior half) and in the brain of 4-week-old rats after inoculation with 1000 PFU of HEV 67N strain into the sciatic nerve.

Using the ABC method, viral specific antigens were detected on day 3 in the spinal cord of the infected rats (Fig. 2). On day 4, viral antigen positive cells were also found in the cerebral cortex of the brain (Fig. 3). Although the brain and spinal cord showed high infectivity titers, the virus antigen positive cells were not widely distributed in the CNS. Some neurons became necrotic in the spinal cord and brain, and a few mononuclear cells were seen in the perivascular and subarachnoidal space of the rats sacrificed on day 4 (HE staining). Antigen positive cells were identified as neurons by their distribution and shape.

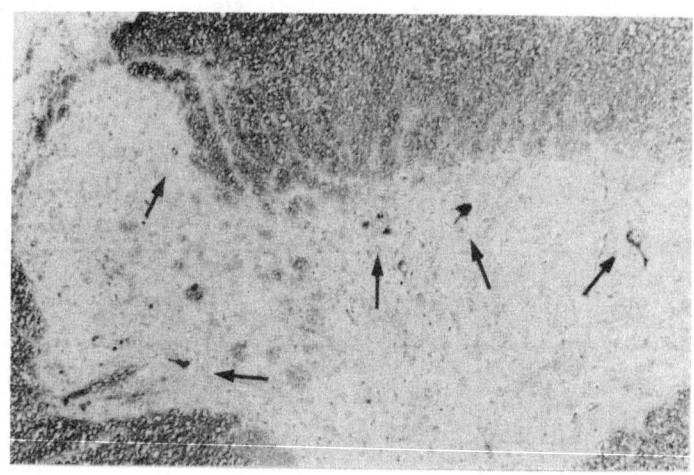

Fig. 2. Viral antigen in the neurons (arrow) in the spinal cord of rat sacrificed on day 4.

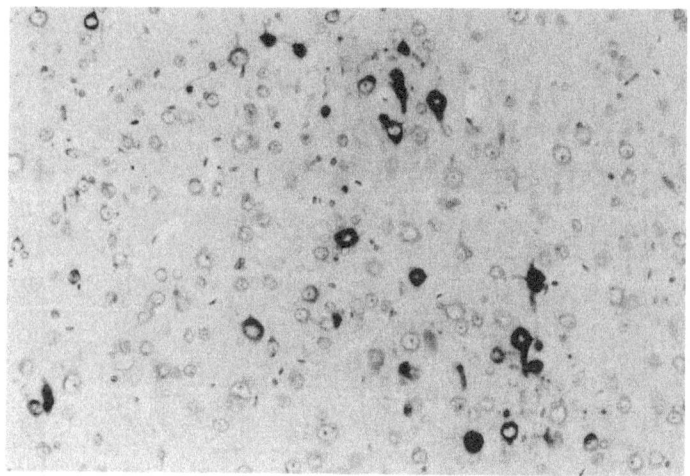

Fig. 3. Viral antigen in the neurons of the cerebral cortex of the brain on day 4 after inoculation.

As mentioned above, some diseased rats survived for more than 24 days after developed severe CNS symptoms. Viral antigens were detected in neurons of the cerebral cortex in the brain of rats sacrificed on day 24 (Fig. 4), and also in the neurons of the midbrain and area around the 3rd ventricle. In the 3rd ventricle, ependymal cells were antigen negative, suggesting that the virus spread was not established via the central canal. In the cerebral cortex, antigen positive neurons apparently increased in number and were more widely distributed than in the early stage of infection. These findings suggest strongly that a persistent infection of HEV was established, and that the virus spread widely in the CNS with time in the survived rats. The direct inoculation with a small dose of the virus into the nerve might lead to a persistent infection.

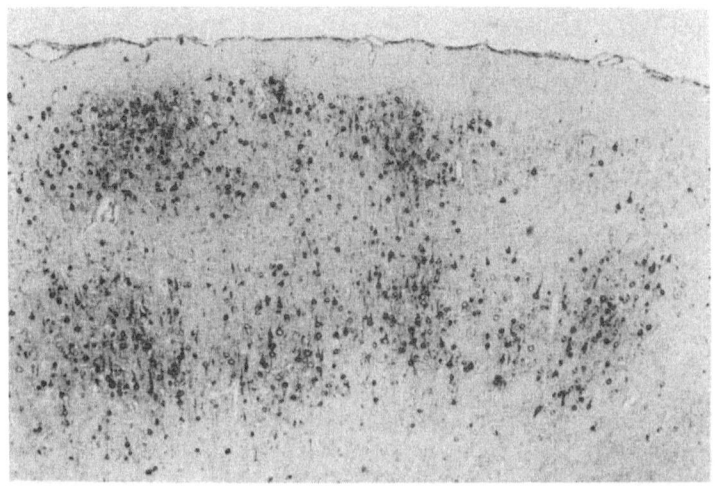

Fig. 4. Virus antigen in the numerous neurons in the cerebral cortex of the survived rats after developed severe clinical signs on day 24

DISCUSSION

HEV is a neurotropic coronaviruses causing a vomiting and wasting disease in piglets. In the diseased animals, the virus reaches the CNS through the nerve pathways and then spreads within brainstem and spinal cord. Virus replication was shown to be restricted to neurons [3].

In our previous studies on HEV infection in mice [4], all mice inoculated i.c. died of encephalitis regardless of age but 20-day-old or older mice did not die following i.n., i.p. or s.c. inoculation. In the diseased mice, the virus propagated mainly in the CNS and neuronal cells were the main targets of virus replication.

In the present study, we demonstrate that rats inoculated s.c. with the virus began to die a few days earlier than those inoculated i.n., i.p. and i.v. and that the virus spreads from peripheral nerve to the CNS.

Our experiments showed that the virus spread from the foot to the brain in 3 days. The results of virus isolation from the liver and spleen and the identification of virus growth in the brain strongly suggest that the virus might spread by neuronal route.

After inoculation with a small dose into the sciatic nerve, animals developed clinical signs on day 4, when viral titers were at their peak value in the spinal cord, and began to die on day 6 when the brain infectivity reached a maximum of 10^7 PFU/0.2g. The virus growth in the brain was concomitant with the development of clinical signs and this virus growth was restricted to the neurons which served as a main target of the virus replication.

The present study demonstrates that HEV spread via neuronal route from peripheral nerve to the brain, and that persistent infection of HEV was established in rats after direct inoculation in the sciatic nerve. The virus spread in the CNS and persistent infection in the rat would be a useful model for studying the pathogenesis of chronic and persistent infection with human and animal coronaviruses.

ACKNOWLEDGMENT

The authors thank Professor Y. Iwasaki, Department of Neurological Science, School of Medicine, Tohoku University for his valuable advice.

REFERENCES

1. C.K. Roe, and T.J.L. Alexander. Canad. Comp. Med. 22, 305-307 (1958)
2. K. Andries, and M.B. Pensaert. Am. J. Vet. Res. 41, 1372-1385 (1980)
3. K. Andries, and M.B. Pensaert. Adv. Exp. Med. Biol. 142, 399-408 (1981)
4. K. Yagami, K. Hirai, and N. Hirano, J. Comp. Pathol. 96, 645-657 (1986)
5. N. Hirano, K. Ono, H. Takasawa, T. Murakami, and S. Haga, J. Virol. Methods 27, 91-100 (1990)

NEUROTROPISM OF HUMAN CORONAVIRUS 229E

Pierre J. Talbot, Sophie Ékandé, Neil R. Cashman[*],
Samir Mounir and Janet N. Stewart

Centre de Recherche en Virologie
Institut Armand-Frappier, Université du Québec
Laval, Québec, Canada H7N 4Z3

and

[*]Montreal Neurological Institute
McGill University
Montreal, Québec, Canada H3A 2B4

ABSTRACT

The 299E prototype strain of human coronavirus (HCV-229E) has so far been mainly associated with infections of the respiratory tract. In the present study, we show evidence for infection of the central nervous system (CNS) by HCV-229E, both *in vitro* and *in vivo*.

Various human cell lines of CNS origin were tested for their susceptibility to infection by HCV-229E. Production of viral antigens was monitored by indirect immunofluorescence with monoclonal antibodies and infectious progeny virions by plaque assay on the L132 human embryonic lung cell line. The SK-N-SH neuroblastoma and H4 neuroglioma cell lines were highly susceptible to infection. The U-87 MG and U-373 MG astrocytoma cell lines were also infectable by HCV-229E. We could also demonstrate infection of the MO3.13 cell line, which was established by fusion of human oligodendrocytes with a thioguanine-resistant mutant of the TE671 (RD) human rhabdomyosarcoma cell line. An apparently more extensive infection of the MO3.13 cells, when compared to the parental cells, supports the notion that human oligodendrocytes are differentially susceptible to infection by this virus.

We also tested for HCV-229E gene expression in pathological brain specimens. For that purpose, we developed a reverse transcription-polymerase chain reaction (RT-PCR) assay to amplify a portion of the mRNA encoding the viral nucleocapsid protein. Using stringent laboratory conditions, viral RNA was detectable in brain tissue of 4 of 11 multiple sclerosis patients and none of 6 neurological and 5 normal controls.

These results strongly suggest neurotropism on the part of HCV-229E and emphasize the importance of further studies on the possible involvement of human coronaviruses in neurological diseases such as multiple sclerosis.

Coronaviruses, Edited by H. Laude and J.F. Vautherot
Plenum Press, New York, 1994

INTRODUCTION

Human coronaviruses (HCV) are recognized as respiratory pathogens that are responsible for 15 to 35 % of common colds[1]. Two prototype strains, named 229E and OC43, represent the two known serotypes. Besides infections of the respiratory tract, other pathologies have sporadically been associated with human coronaviruses. A seroepidemiological study has linked them with some pneumoniae, perimyocardites, meningites and radiculites[2]. Also, their involvement in enteric infections was suggested from various reports of their presence in stool specimens of infants and children with severe diarrhea[3,4,5].

Another disease association that remains to be confirmed is the possible involvement of HCVs in neurological disorders, specifically multiple sclerosis (MS). The original report of coronavirus-like particles in the brain of an MS patient[6] was rapidly followed by the description of two coronaviruses isolated from two MS patients[7]. Despite the closer relationship of these isolates to murine coronaviruses[8], it was recently shown that their replication could be detected in brain tissue of some MS patients[9] and that they could cause a demyelinating disease in primates[10]. Moreover, antibodies to both serotypes of human coronaviruses were detected in the cerebrospinal fluids of MS patients, which could result from replication of HCVs within the central nervous system[11]. Finally, the neurotropism of some murine strains of coronavirus is well recognized and actually provides a very useful animal model of virus-induced demyelinating disorders, given the close analogy of the clinical manifestations of the disease in rodents to multiple sclerosis[12]. Together, these indications constitute enough stimulus for renewed interest in the study of the neurotropism of human coronaviruses and their possible involvement in multiple sclerosis.

Very few studies have attempted to verify the replication of human coronaviruses in neural and glial cells of the nervous system. Pearson and Mims[13] reported a productive infection of murine neurons in primary cultures by HCV-OC43, as well as the presence of viral antigen in astrocytes. However, infection of myelin-producing oligodendrocytes was not detected. In the same report, the authors also reported that human embryo brain cells, including astrocytes were susceptible to HCV-OC43 infection but did not produce infectious virus[13]. Collins and Sorensen[14] showed that the U87-MG human glioblastoma cell line could be persistently infected with HCV-OC43. No studies have been reported so far on the neurotropism of HCV-229E, a virus isolated from the human respiratory tract in 1966[15].

In the present study, we report on the replication of HCV-229E in neural and glial cell lines and the presence of its genome in parts of the brains of some MS patients.

MATERIALS AND METHODS

Virus and Cell Lines

The 229E strain of HCV was originally obtained from the American Type Culture Collection (ATCC; Rockville, MD), plaque-purified twice and grown on the human embryonic lung cell line L132, as described previously[16,17]. The following human neural and glial cell lines were obtained from the ATCC: H4 (neuroglioma, brain), SK-N-SH (neuroblastoma, metastasis to bone marrow), U-373 MG and U-87 MG (glioblastoma, astrocytoma). The MO3.13 human-human hybrid cell line was derived by lectin-enhanced polyethylene glycol-mediated somatic cell fusion between the thioguanine-resistant rhabdomyosarcoma mutant RD-TG.6, derived from the TE671 (RD) cell line (ATCC), and primary human oligodendrocytes obtained from cultures of human adult temporal lobectomies[18]. Similar to oligodendrocytes but unlike the parent tumor rhabdomyosarcoma line, MO3.13 cells have been demonstrated to express myelin basic protein and proteolipid protein by immunohistochemistry, Western immunoblotting and Northern blotting, and show surface immunoreactivity for galactocerebroside and myelin-associated glycoprotein.

These cell lines were grown as monolayers at 37°C, in a humidified atmosphere containing 5 % (v/v) CO_2, using Dulbecco's modified Eagle's medium containing high glucose, L-glutamine and sodium pyruvate, and supplemented with 10 % (v/v) heat-inactivated fetal calf serum, without antibiotics (Gibco BRL Life Technologies, Inc., Burlington, Ontario, Canada). Cells were passaged every four days at a concentration of 100,000 cells per ml.

Immunofluorescence Assay

For immunofluorescence, approximately 1.5×10^6 cells (obtained by trypsinization of cell monolayers grown on plastic 25-cm^2 flasks, followed by pelleting) were mixed with an equal volume of HCV-229E virus stock diluted to give an MOI of 1. Twenty-five microliters of this suspension was deposited into each well of a 12-well glass slide (Flow, ICN Biomedical Canada Ltd., Mississauga, Ontario, Canada) and infection allowed to continue for 20 or 40 h at 33°C (the optimal HCV-229E growth temperature[17]) and 5 % (v/v) CO_2. Slides were washed twice in Dulbecco's phosphate buffered saline (PBS) and fixed with cold acetone at -20°C for 20 min. Viral antigen was detected by adding a 1/10 dilution of ascites fluids containing HCV-229E-specific monoclonal antibodies, designated 3-10H.5 or 4-9H.5 (ELISA titers: 1/144,000 and 1/20,000, respectively), or control ascites fluids prepared with the parental myeloma cells. After incubation at room temperature in a humidified chamber for 2 h, the slides were washed twice with PBS and a 1/100 dilution of fluorescein isothiocyanate-conjugated F(ab')$_2$ goat antibodies to mouse immunoglobulins (Cappel, Organon Tecknika Inc., Scarborough, Ontario, Canada) were added for another 30 min. Fluorescence was observed with a Leitz fluorescence microscope after mounting the slides with glass coverslips, using glycerol:PBS (9:1).

Preparation of RNA and Reverse-Transcription Polymerase Chain Reaction

Brain tissues were collected from a total of 11 patients diagnosed with multiple sclerosis and 11 control patients, five with normal autopsy reports and six with indications of other neurological diseases (Montreal Brain Bank, Montreal, Quebec and University Hospital, London, Ontario). Total RNA was extracted from coded central nervous system (CNS) tissues by the method of Chomczynski and Sacchi[19]. Briefly, 50 to 300 mg of tissue were thawed and homogenized in 0.5 ml of 4 M guanidinium thiocyanate, extracted with phenol and chloroform and precipitated twice with ethanol. After washing with 70% (v/v) ethanol, the pellets were air-dried and resuspended in water. The extracted RNA was first tested for the presence of human myelin basic protein or actin mRNAs by reverse-transcription - polymerase chain reaction (RT-PCR) to insure that it was undegraded[20].

For RT-PCR, HCV-229E primers were designed to amplify either of two regions of about 300 bases in the gene coding for the nucleocapsid protein of this virus, at positions 498 to 806 or 964 to 1265[21]. To confirm the identity of the amplified products, hybridization was performed with oligonucleotide probes derived from sequences located between each pair of primers (bases 693-716 and 1080-1103, respectively[21]). Approximately 1 µg of RNA was reverse transcribed at 37°C for 35 min with 20 U of Moloney murine leukemia virus reverse transcriptase (Pharmacia Canada Inc., Baie d'Urfé, Québec, Canada) and 50 pmol of both up- and downstream primers (to target both positive- and negative-stranded RNA) in 10 mM Tris-HCl (pH 8.8), 50 mM KCl, 0.1 % (v/v) Triton X-100 (1X *Taq* polymerase buffer; BIO/CAN, Mississauga, Ontario, Canada), 1.0 mM (each) dATP, dCTP, dGTP and dTTP (Pharmacia), 40 U of RNAguard (Pharmacia) and 4.0 mM $MgCl_2$. Twenty microliters of the reverse transcription was added to 80 µl of a PCR mix overlaid with mineral oil. This mixture contained 1X *Taq* polymerase buffer (BIO/CAN), 2.5 U of *Taq* polymerase (BIO/CAN), 50 pmol of both up- and downstream primers, 0.25 mM (each) dATP, dCTP, dGTP and dTTP (Pharmacia) and 2.4 mM $MgCl_2$. PCR was performed using a modification of the original method[22]. An amplification cycle of 1 min at 94°C, 2 min at 60°C and 2 min at 72°C was repeated 30 times and was followed by an extension period of 7 min at 72°C. Twenty µl of reaction products were loaded onto 1.5 % (wt/vol) agarose gels, allowed to migrate and transferred to nitrocellulose filters according to the method of Southern[23]. Blots were hybridized with

^{32}P-end-labeled oligonucleotide probes at 50°C for 16 h in a buffer containing 6x SSC, 1x Denhardt's solution, 0.05% (wt/vol) pyrophosphate and 100 µg/mL sonicated salmon sperm DNA. The blots were washed 3 x 15 min at room temperature and for 20 min at 60°C in 6x SSC, 0.05% (wt/vol) pyrophosphate and exposed to X-ray film (Kodak, Rochester, NY) at -70°C for 48 or 96 h.

The detectability level of the RT-PCR assay was evaluated by cloning the amplification product of RNA prepared from HCV-229E-infected L132 cells into the *Sma*I site of the pGEM 3Z vector (Promega, Fisher, Montréal, Québec, Canada). Five µg of *Hind*III-linearized plasmid were transcribed *in vitro* in the presence of 20 mM Tris-HCl (pH 8.0), 4 mM MgCl$_2$, 1 mM spermidine, 25 mM NaCl (transcription buffer; Stratagene, La Jolla, California), 0.4 mM (each) ATP, GTP and UTP, 20 U T7 DNA polymerase and 2 U RNase Block II (Stratagene) and 3 µM [α-^{32}P]CTP. The amount of transcript was evaluated by ^{32}P incorporation and the number of molecules detectable was estimated by performing RT-PCR on ten-fold serial dilutions of the transcript.

RESULTS AND DISCUSSION

Infectability of neural and glial cell lines

We found that all neural and glial cell lines tested could be infected with HCV-229E, as revealed by immunofluorescence with virus-specific monoclonal antibodies (MAbs). Representative results are shown in Fig. 1.

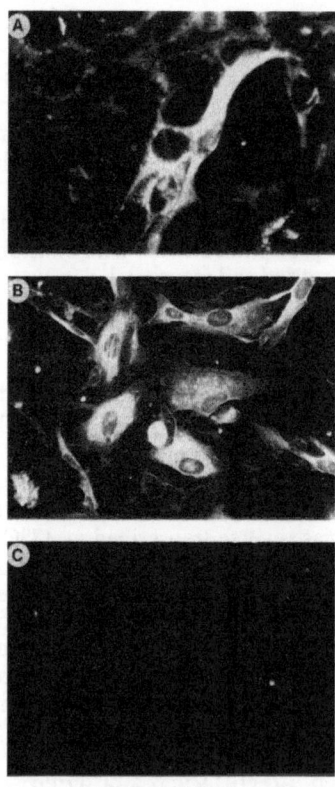

Figure 1. Immunodetection of HCV-229E antigens in neural and glial cell lines. Panel A: MO3.13 cells at 40 h post-infection, revealed with the 3-10H.5 MAb. Panel B: SK-N-SH cells at 20 h post-infection, revealed with the 4-9H.5 MAb. Panel C: SK-N-SH cells at 20 h post-infection, revealed with control antibody (magnification = 100X).

Varying degrees of antigen expression were observed in the different human neural and glial cells lines, as summarized in Table 1.

It is noteworthy that cells related to neurons, astrocytes and oligodendrocytes were susceptible to infection by the human coronavirus 229E, since the same situation is observed in the mouse with the neurotropic strains of murine hepatitis virus[24]. Indeed, this is the first report on the infection of such cells by HCV-229E. Interestingly, it appears that immortalized human oligodendrocytes were infected with this virus. The more extensive infection observed in these cells when compared to the parental rhabdomyosarcoma cell line, at least at 40 h post-infection (Table 1), suggests that the oligodendrocyte phenotype

Table 1. Percentage of infected cells after infection with an MOI of 1 for 20 or 40 h.

Cells		Percent infected cells at each time post-infection (%)	
Line designation	Type	20 h	40 h
SK-N-SH	Neural	90	90
H4	"	18	20
U-373 MG	Astrocyte	3	60
U-87 MG	"	2	60
MO3.13	Oligodendrocyte	6-10	20
TE671 (RD)	MO3.13 parental	6	3
L132	Lung	70-80	80

was involved in susceptibility. Even though final proof on the susceptibility of human oligodendrocytes to coronavirus infection will require the use of primary oligodendrocyte cultures and immunohistochemistry on human CNS tissue sections, the infection of these myelin protein-producing cells is indeed relevant to the possibility of the induction of demyelinating disease in humans by coronaviruses.

Detection of viral RNA in brain samples

Having demonstrated infection of human and glial cells *in vitro* by HCV-229E, we wanted to test for the presence of this virus in CNS tissue of multiple sclerosis patients. A previous study had used classical hybridization to search for the RNA of HCV-OC43 in four MS brain autopsy samples, with negative results[25]. Thus, we wanted to develop a more sensitive assay, capable of detecting very low amounts of viral nucleic acid, characteristic of persistent infections. For that purpose, we developed a polymerase chain reaction (PCR) assay, modified to include a reverse transcription step from extracted RNA (RT-PCR). As shown in Fig. 2, the detectability level of our assay was less than 60,000 molecules. Moreover, this number is most likely an overestimation since the *in vitro* transcription may have produced a proportion of incomplete transcripts which could not be amplified by RT-PCR. Similar detectability levels of RT-PCR were reported in other systems[26]. Since the target sequence is located on the most abundant viral mRNA as well as on the genome, and since we designed the assay to amplify both positive- and negative-sense RNAs, we estimate that a single infected cell should be detectable in our procedure.

We applied our HCV-229 RT-PCR assay to RNA extracted from frozen brain autopsy samples from both MS and control patients. Positive signals were obtained from four of eleven MS patients and none of eleven controls, which included five histopathologically normal patients, and six patients whose autopsy report indicated Alzheimer's disease (four patients), subacute meningoencephalitis (one patient) and ischemic vascular disease (one patient). Representative results are shown in Fig.3.

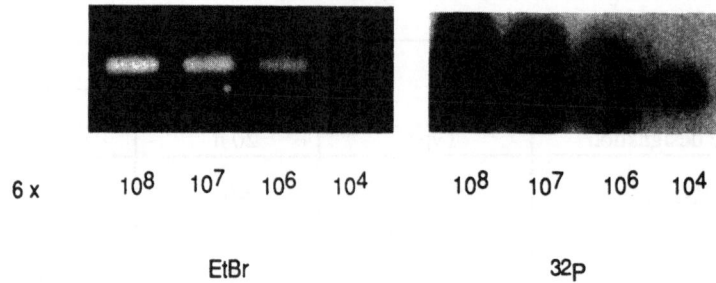

6 x 10^8 10^7 10^6 10^4 10^8 10^7 10^6 10^4

EtBr   32p

Figure 2. Determination of the detectability of the HCV-229E RT-PCR. The estimated amounts of transcripts indicated on the figure were amplified by RT-PCR and analyzed on agarose gels, which were stained with ethidium bromide (EtBr), followed by Southern hybridization with an internal radiolabeled oligonucleotide probe (^{32}P).

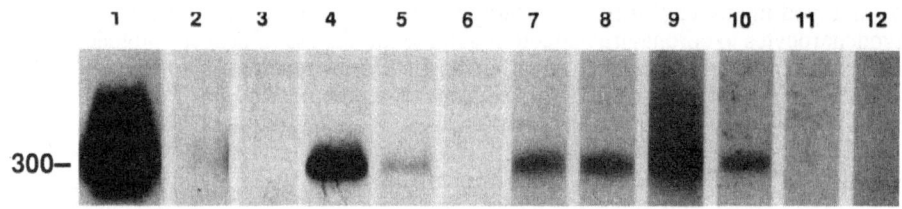

Figure 3. Detection of HCV-229E in brain samples by RT-PCR. The templates in each lane are RNA extracted from: lane 1, L132 cells infected with HCV-229E; lane 2, mock-infected L132 cells; lane 3, no RNA; lane 4, mouse brain mixed with 170 pg RNA extracted from HCV-229E-infected L132 cells; lane 5, mouse brain mixed with 1.7 pg RNA extracted from HCV-229E-infected L132 cells; lane 6, mouse brain; lanes 7-12, specimens from MS patients (positive samples were from: lanes 7 and 9, white matter; lane 8, gray matter; lane 10, plaque tissue). Blots were exposed to X-ray film for 48 (lanes 1-6) or 96 h (lanes 7-12). The migration position of the 300 bp HCV-229E amplified product is indicated on the left.

The results of our study indicate for the first time that the 229E strain of human coronavirus, which has so far only been associated formally with respiratory infections, has the capacity to infect cells of the central nervous system, as monitored by *in vitro* infection of transformed cell lines. Moreover, its genome is expressed in some human brains. Interestingly, it was shown in the animal model of coronavirus-induced demyelinating disease that neurotropic coronaviruses can gain access to the central nervous system from the respiratory tract, through the olfactory and trigeminal nerves[27]. Thus, it is conceivable that the same type of spread from the upper respiratory tract to the CNS could occur in humans and be relevant to the development of a neurologic disease in genetically predisposed individuals[28,29].

An association of HCV-229E with multiple sclerosis cannot be inferred from our pilot RT-PCR study, given the small number of clinical samples analyzed. Moreover, the establishment of an etiologic link with the disease will require large scale studies involving an epidemiological component. Nevertheless, we believe that research efforts are needed to establish the pathological relevance of the HCV-229E neurotropism suggested by our study. Interestingly, we recently found evidence for the sharing of amino acid sequences between HCV-229E and myelin basic protein[16]. Since this protein appears to be a target for the immune system of MS patients[30], the possibility that molecular mimicry of myelin antigens by a virus, be it a coronavirus or another virus sharing similar mimetic properties, could be involved in the establishment of this degenerative autoimmune neurologic disease also warrants attention.

ACKNOWLEDGMENTS

This work was supported by grant MT-9203 from the Medical Research Council of Canada to P.J.T., who also gratefully acknowledges salary support in the form of a University Research Scholarship from the National Sciences and Engineering Research Council of Canada (NSERC). Funding for the developement of oligodendrocyte cell lines was provided by the Network for Neural Regeneration and Recovery, one of fifteen Networks of Centres of Excellence supported by the Government of Canada. J.N.S. is grateful to the Institut Armand-Frappier for studentship support. We thank Francine Lambert for excellent technical assistance. We are grateful to Dr. Jack P. Antel (Montreal Neurological Hospital) for his help in setting-up the RT-PCR study, to George P.A. Rice, Vince L. Morris and George Ebers (University of Western Ontario, London, Ontario), and to Karen Hellauer (The Montreal Brain Bank) for providing frozen CNS tissue. We also thank Dr. Gordon S. Francis (Montreal Neurological Hospital) for help in selecting some samples and Dr. Mark S. Freedman (Montreal Neurological Institute) for helpful advice.

REFERENCES

1. K. McIntosh, *In*: "Virology, 2nd edn.", B.N. Fields, D.M. Knipe *et al.*, eds., p. 857, Raven Press, New York (1990).
2. N. Riski, and T. Hovi. *J. Med. Virol.* 6: 259 (1980).
3. S. Resta, J.P. Luby, C.R. Rosenfeld, and J.D. Siegel. *Science* 229: 978 (1985).
4. M. Battaglia, N. Passarini, A. DiMatteo, and G. Gerna. *J. Inf. Dis.* 155: 140 (1987).
5. M.L. Mortensen, C.G. Ray, C.M. Payne, A.D. Friedman, L.L. Minnich, and C. Rousseau. *Amer. J. Dis. Child.* 139: 928 (1985).
6. R. Tanaka, Y. Iwasaki, and H.J. Koprowski. *J. Neurol. Sci.* 28: 121 (1976).
7. J.S. Burks, B.L. DeVald, L.D. Jankovsky, and J.C. Gerdes. *Science* 209: 933 (1980).
8. S. Weiss. *Virology* 126: 669 (1983).
9. R.S. Murray, B. Brown, D. Brian, and G.F. Cabirac. *Ann. Neurol.* 31: 525 (1992).
10. R.S. Murray, G.-Y. Cai, K. Hoel, J.Y. Zhang, K.F. Soike, and G.F. Cabirac. *Virology* 188: 274 (1992).
11. A. Salmi, B. Ziola, T. Hovi, and M. Reunanen. *Neurology* 32: 292 (1982).

12. V. ter Meulen, P.T. Massa, and R. Dörries. *In*: "Handbook of Clinical Neurology: Viral Disease, Revised Series", P.J. Vinken, G.W. Bruyn, and H.L. Klawans, eds., Vol. 12 (56), p. 439, Elsevier, New York (1989).

13. J. Pearson, and C.A. Mims. *J. Virol.* 53: 1016 (1985).

14. A.R. Collins, and O. Sorensen. *Microbial Path.* 1: 573 (1986).

15. D. Hamre, and J.J. Procknow. *Proc. Soc. Exp. Biol. Med.* 121: 190 (1966).

16. P. Jouvenne, S. Mounir, J.N. Stewart, C.D. Richardson, and P.J. Talbot. *Virus Res.* 22: 125 (1992).

17. N. Arpin, and P.J. Talbot. *In*: "Coronaviruses and Their Diseases", D. Cavanagh and T.D.K. Brown, eds., p. 73, Plenum Press, New York (1990).

18. G. Trudel, J. Antel, and N.R. Cashman. *Soc. Neurosci. Abst.* 17: 31(1991).

19. P. Chomczynski, and N. Sacchi. *Anal. Biochem.* 162: 156 (1987).

20. J.N. Stewart, S. Mounir, and P.J. Talbot. *In*: "Diagnosis of Human Viruses by the Polymerase Chain Reaction, Frontiers in Virology, Vol. 1", Y. Becker and G. Darai, eds., p. 316, Springer Verlag, Berlin (1992).

21. S.S. Schreiber, T. Kamahora, and M.M.C. Lai. *Virology* 169: 142 (1989).

22. R.K. Saiki, D.H. Gelfand, S. Stoffel, S.J. Scharf, R. Higuchi, G.T. Horn, K.B. Mullis, and H.A. Erlich. *Science* 239: 487 (1988).

23. E. Southern. *Methods Enzymol.* 68: 152 (1979).

24. H. Wege, S. Siddell, and V. ter Meulen. *Curr. Top. Microbiol. Immunol.* 99: 165 (1982).

25. O. Sorensen, A. Collins, W. Flintoff, G. Ebers, and S. Dales. *Neurology* 36: 1604 (1986).

26. R.E. Gama, P.J. Hughes, C.B. Bruce, and G. Stanway. *Nucl. Acids Res.* 16: 9346 (1988).

27. S. Perlman, G. Jacobsen, A.L. Olson, and A. Afifi. *Virology* 175: 418 (1990).

28. T.H. Maugh II. *Science* 195: 768 (1977).

29. P. Talbot, and P. Jouvenne. *Médecine/ Sciences* 8: 119 (1992).

30. K. Ota, M. Matsui, E.L. Milford, Mackin, G.A., H.L. Weiner, and D.A. Hafler. *Nature* 346: 183 (1990).

CORONAVIRUS JHM OMP1 PATHOGENESIS IN OWL MONKEY CNS AND CORONAVIRUS INFECTION OF OWL MONKEY CNS VIA PERIPHERAL ROUTES

Gary F. Cabirac[1,2], Kenneth F. Soike[3], Catalin Butunoi[1], Kristen Hoel[1], Steven Johnson[1], Guang-Yun Cai[1], and Ronald S. Murray[1]

[1]Rocky Mountain Multiple Sclerosis Center, Colorado Neurological Institute and Swedish Medical Center, Englewood, CO 80150
[2]Department of Biochemistry, Biophysics and Genetics, University of Colorado Health Sciences Center, Denver, CO 80262
[3]Department of Microbiology, Tulane Regional Primate Research Center Covington, LA 70433

ABSTRACT

Two separate studies are described in this report. First, 5 Owl monkeys were inoculated intracerebrally (IC) with coronavirus JHM OMP1; this virus isolate was cultured from the brain of an animal inoculated with uncloned MHV JHM. Two of the animals became neurological impaired and were sacrificed; these animals had developed severe encephalomyelitis as previously described[1]. Two of the remaining 3 healthy animals were inoculated IC again at 90 days post-inoculation (DPI) and all 3 were sacrificed approximately 5 months after the first virus inoculation. Despite the lack of detectable infectious virus, viral RNA and antigen, all 3 animals had significant white matter inflammation and areas of demyelination in the spinal cord. In the second study 4 Owl monkeys were inoculated intranasally (IN) and ocularly and 4 inoculated intravenously (IV) with JHM OMP1. The animals were sacrificed between 16 and 215 DPI with 2 IN and 2 IV animals receiving a second IV inoculum at 152 DPI. Viral RNA and/or antigen was detected in the brains of all animals and the distribution corresponded to areas of inflammation and edema. One of the animals that received the second inoculum developed neurological impairment and subsequent analysis of tissues showed viral antigen in both brain and spinal cord. Viral products were predominantly found in blood vessels suggesting hematogenous spread with entry into the central nervous system (CNS) through endothelium.

Coronaviruses, Edited by H. Laude and J.F. Vautherot
Plenum Press, New York, 1994

INTRODUCTION

In humans, coronaviruses are a leading upper respiratory tract pathogen[2] and are associated with gastroenteritis[3] but have never been proven to cause serious human illness. However, we have reported that coronavirus RNA and antigen are detectable in demyelinating lesions of human multiple sclerosis brain[4] and that, following IC inoculation, coronaviruses can productively infect and disseminate in the brains of primates causing encephalomyelitis and demyelination[1]. In addition, two recent reports demonstrate that human coronavirus (HCV) 229E RNA can be detected in multiple sclerosis brain by the polymerase chain reaction (PCR) assay[5] and that the cell surface glycoprotein aminopeptidase N, a polypeptide expressed on nerve synapses, functions as a receptor for 229E[6]. Collectively these data suggest that coronaviruses are capable of natural infection of the human central nervous system (CNS) and hence may be involved in neurologic disease.

In an effort to further define the interaction of coronaviruses with human CNS tissue, we have utilized primates to investigate the disease potential of these viruses. The first part of this report will describe results of IC inoculation of JHM OMP1 into Owl monkeys. Data on two of the animals that developed neurological impairment have already been published[1]; presented here are results from the remaining animals that were sacrificed at later times. Secondly, we demonstrate that JHM OMP1 can gain access to primate CNS following IN or IV inoculation.

METHODS AND RESULTS

IC Inoculated Animals

Five outbred young adult Owl monkeys were inoculated IC as previously described[1] with JHM OMP1. Assays for infectious virus and neutralizing antibody, *in situ* hybridization, immunohistochemical and histochemical staining were performed as described[1]. Two of the five animals, K191 and K063, were observed with apparent neurological dysfunction on 10 and 12 DPI, respectively, and sacrificed. The results of analysis of tissue from these two animals has previously been described[1]. Briefly, infectious virus was cultured from the brains of both animals and from the blood of K063, abundant amounts of viral RNA and antigen were detected in both grey and white matter cells and pathologic examination of tissues revealed that both animals had developed severe encephalomyelitis. The remaining 3 animals, K189, K171 and K072, appeared normal and by 90 DPI had no detectable neutralizing antibody. Two of these animals, K189 and K171, were inoculated IC with approximately 5×10^4 TCID50 of JHM OMP1 at this time. Both animals seroconverted after this second inoculation developing low titers of 1:20 and 1:10 for K189 and K171, respectively. These two animals and K072 were sacrificed 150 days after the initial inoculation; all three animals appeared healthy at the time of sacrifice.

Infectious virus could not be cultured from brain, CSF or blood obtained from these three animals. Histochemical staining of tissues revealed that all three animals had white matter inflammation, demyelination in the spinal cord and meningitis. However, no virus products could be detected in these affected areas by *in situ* hybridization or immunohistochemical staining methods. PCR amplification of RNA extracted from tissue sections failed to produce coronavirus specific products. One of the most striking observation was that all three animals had defined areas of multilevel demyelination in the dorsolateral spinal cord. Figure 1A shows a luxol-fast blue/periodic acid-Schiff stain of a section from K189 demonstrating the area of demyelination. Interestingly, these regions of spinal cord demyelination were accompanied by gliosis. Staining with anti-glial fibrillary acidic protein (GFAP) monoclonal (Fig. 1B) showed the predominant cell type to be reactive astrocytes. Immunostaining for macrophages/monocytes was negative in these regions (data not shown).

Peripherally Infected Animals

Four Owl monkeys were inoculated intravenously (IV) and four intranasally (IN) plus ocularly with coronavirus JHM OMP1. The four IV inoculated animals were injected with 1 ml of a 1 x 10[6] TCID50/ml titer inoculum in the saphenous vein. The other four animals received 1 ml of the same virus inoculum divided between two nostrils plus 2 drops of this inoculum in the right eye following corneal scarification. Preparation of the virus inoculum has been described[1]. Animals were sacrificed at the indicated time points then tissue from brain and spinal cord was screened for areas of pathology by histochemical staining. Those tissues containing regions of pathology, unaffected tissues and tissue from a sham inoculated control animal were processed for the detection of viral RNA and antigen. *In situ* hybridization analysis was performed as described above but a modification of our immunostaining procedure was used on tissues from these peripherally inoculated animals. The modifications are as follows. Formalin-fixed, paraffin embedded tissue was sectioned, de-paraffinized and incubated in Ca^{+2}/Mg^{+2}-free PBS containing 0.25% trypsin at room temperature for 90 minutes then washed three times in PBS. Incubation of the primary antibody was done at 4°C overnight. Secondary antibody was biotinylated rabbit anti-mouse immunoglobulins (Dako), incubated at room temperature for 30 minutes. Following antibody treatment, tissue was incubated for 20 minutes at room temperature with alkaline phosphatase conjugated strepavidin (BioGenex), washed and then the chromogen Fast Red was incubated for 20 minutes at 37°C.

Two animals, K183 and K172, inoculated IV and IN respectively, were sacrificed 16 DPI. Throat swab and blood cultures from K183 and throat, blood and conjunctiva/corneal swab cultures from K172 taken on 1, 2, 5, 7, and 9 DPI were negative for infectious virus. Sera from both animals collected 14 DPI did not have detectable neutralizing antibody. CSF and tissue from brain, spinal cord and lung were negative for infectious virus. Histochemical staining revealed varying degrees of meningitis, ependymitis, and mild white matter edema and inflammation in both animals and choroiditis in K183. Viral RNA was not detectable by *in situ* hybridization in K172 but evident in a small percentage of blood vessel endothelium in the brain of the IV inoculated animal K183. Viral antigen was detectable with the nucleocapsid specific monoclonal antibody J.3.1[7] in brain tissue from both of these animals. Positive staining for viral antigen was observed predominantly in blood vessel walls; occasionally, cells surrounding positive vessels appeared to contain viral antigen. Viral products were not detected in trigeminal ganglia or olafactory blubs from IN inoculated animal K172.

Animals K186 and K192, IV and IN inoculated respectively, were sacrificed 35 DPI. A throat culture obtained 1 DPI from K192 was positive for infectious virus but all other throat, blood and conjunctive (K192) cultures from both animals were negative. At the time of sacrifice, brain and spinal cord did not contain infectious virus. As with the two animals sacrificed 16 DPI, K186 and K192 never produced detectable neutralizing antibody. Meningitis, ependymitis, choroiditis and mild white matter edema and inflammation were evident in both K186 and K192. *In situ* hybridization (Fig. 1C) and immunohistochemical staining revealed viral RNA and antigen in blood vessel endothelium and adjacent areas in both of these animals. It appeared that viral products were more widely distributed in the IV inoculated animal, K186. As with K172, viral RNA or antigen were not detected in trigeminal ganglia and olafactory blubs from K192.

Following the analysis of tissue from the first 4 animals, the remaining 4 animals, K181 (IV), K179 (IV), K187 (IN) and K194 (IN), were inoculated with a second dose of virus. 150 DPI after receiving the first inoculum, these animals were inoculated IV with the second inoculum. Assays of neutralizing antibody in sera from these animals showed that these 4 animals had seroconverted before receiving the second dose of virus; the highest titer was 1:80 for IN inoculated animal K187 (35 DPI) and the lowest was 1:10 for IV animal K179 (35 DPI). To varying degrees, the second inoculum caused a rise in titers; 1:320 for K187 at 183 DPI and 1:20 for K181 at 183 DPI and K194 at 215 DPI are the maximum and minimum titers recorded. Blood, throat and conjunctival cultures taken on 1, 2, 5, 7, and 9 DPI (first inoculation) were negative for infectious virus; no additional cultures were taken after the second virus inoculum. On 189 DPI, 36 days after administration of the second dose of virus, animal K179 was observed with hind limb weakness and tremors. This animal was sacrificed 194 DPI and tissue processed as before. Histochemical staining showed that this animal had detectable viral antigen not only in the brain but also in the spinal cord. Again, most of the detected antigen was associated with blood vessels but occasionally, surrounding areas stained positive for nucleocapsid protein. The remaining 3 animals, K181, K187 and K194 were sacrificed 215 DPI (62 days after second inoculation). Immunohistochemical analysis of tissue from these animals revealed the same results as described above, i.e., viral antigen in vessels and surrounding regions. Quantitatively, there appeared to be much less antigen in animal K187 compared to K181 and K194 and analyzed tissue from these three animals showed less antigen when compared to the animal with neurological dysfunction, K179.

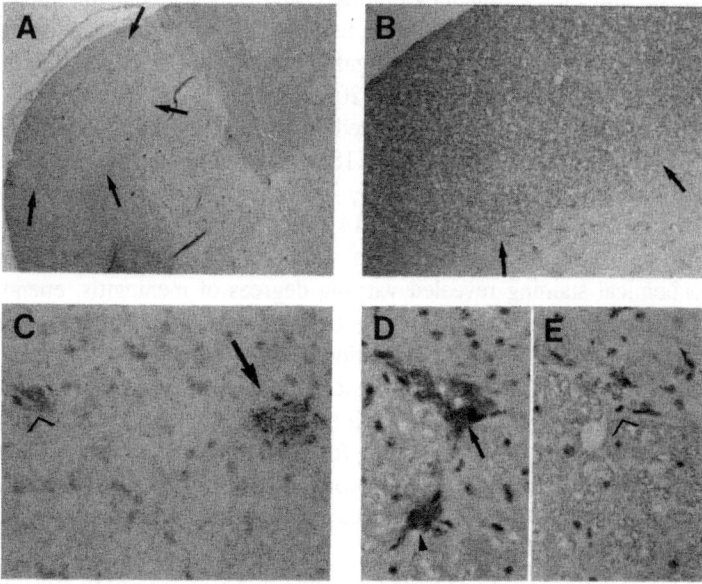

Figure 1. Histochemical and immunohistochemical staining of spinal cord from IC inoculated monkey and *in situ* hybridization and viral antigen staining of tissue from peripherally inoculated animals. A - Luxol-fast blue/periodic acid-Schiff staining of myelin in spinal cord from K189. Area of decreased staining shown by arrows indicates demyelination. B - An adjacent section to that shown in A was immunostained with monoclonal against GFAP. Arrows show region of staining indicating extensive astrocyte proliferation in demyelinated area. C - *In situ* hybridization of tissue from IN inoculated animal K192 showing higher density of grains over vessel (arrow); adjacent to this vessel is another vessel (open arrowhead) that has only background levels of grain development. D - Positive nucleocapsid staining with monoclonal J.3.1 in section from animal K179. Arrow shows staining in vessel and arrowhead show staining in surrounding areas. E -Adjacent section to that shown in D in which the primary antibody was normal mouse serum. Note lack of staining in vessel (open arrowhead).

DISCUSSION

The data described in this report in addition to the previously published study of active coronavirus infection in primate CNS demonstrates the value of primates as a model system. It is now evident that coronaviruses can replicate within primate CNS and produce disease. Results presented here extend the earlier work and show that the inflammation and demyelination induced by coronavirus infection persists in the absence of detectable virus and, perhaps more importantly, that a coronavirus can gain access to primate CNS following peripheral infection.

Coronavirus induced demyelination in the rodent system is well documented[8-18]. Therefore the failure to detect virus in areas of demyelination and inflammation in the IC infected primates was not unexpected. However, at this time we do not know if this aspect of primate CNS disease is unique to JHM OMP1. We state this since in our original report[1] all of the animals infected with JHM, the putative MS isolate SD and JHM OMP1 were sacrificed no more than 120 days post-inoculation and all but one of these animals had detectable levels of virus or viral products in the affected areas. The animals described here were sacrificed 150 DPI and therefore pathology in the absence of virus may be the result of longer infection times. While we could not detect viral RNA by PCR in these affected areas, low levels of RNA may be present. In our hands the efficiency of PCR amplification of RNA extracted from formalin fixed tissue is low compared to that done on RNA from snap-frozen or fresh samples. However, with this caveat in mind, the possibility that coronavirus infection in these primates triggered an autoimmune reaction needs to be investigated. Obviously a study in which infected primates are followed for a much longer time course needs to be done. If it can be determined that these viruses are completely cleared from the CNS but that an autoimmune response has been initiated then the primates would serve as an excellent model for investigating coronavirus induced autoimmune demyelination in humans.

The results from the peripherally infected animals reinforces data indicating that coronaviruses are CNS pathogens in humans. The data suggests that for both the IV and IN inoculated animals, virus entered the CNS through blood vessel endothelium as opposed to neural routes. The antibody data indicates that virus replication did occur in these animals. While some assays for infectious virus were performed, more rigorous tests are needed to determine the primary sites of infection in these animals. The one positive throat swab culture from IN inoculated animal K192 suggests that respiratory epithelium may be a reservoir for virus. Some positive antigen staining was seen in liver from two of the IV inoculated animals but it was rare and hard to repeat. While all blood sample cultures from these animals were negative for infectious virus, we need to investigate the possibility that certain populations of peripheral blood cells can be infected. This is emphasized by the fact that one of the earlier IC inoculated animals (K063) had detectable virus in the blood. Interestingly, there seemed to be a correlation between neutralizing antibody titers and amounts of viral antigen detected in these peripherally infected animals, i.e., those animals with poor antibody response had higher amounts of viral antigen while animals, in particular IN animal K187, with higher titers had lower antigen levels. In this context it is important to note that the animal that developed neurological impairment, K179, had the lowest titer of all four re-inoculated animals. It will be necessary to determine the relationship between the cellular plus humoral immunity against coronaviruses and CNS disease produced by multiple peripheral infections. While other animal species are excellent for investigating coronavirus induced CNS disease these studies demonstrate the usefulness of primates as a model system and reinforce previous data implicating coronaviruses as human CNS pathogens.

REFERENCES

1. R.S.Murray, G-Y Cai, K.Hoel, et al, Virology 188:274 (1992).
2. T.Hovi, H.Kainulainen, B.Ziola, A.Salmi, Med.Virol. 3:313 (1979).
3. S.Resta, J.P.Luby, C.R.Rosenfeld, J.D.Siegel, Science 229:978 (1985).
4. R.S.Murray, B.Brown, D.A.Brian, G.F.Cabirac, Ann.Neurol. 31:525 (1992).
5. J.N.Stewart, S.Mounir, P.J.Talbot, Virology. In Press (1992).
6. C.L.Yeager, R.A.Ashmun, R.K.Williams, et al, Nature 357:420 (1992).
7. J.O.Fleming, F.A.K.El-Zaatari, W.Gilmore, et al, Arch.Neurol. 45:629 (1988).
8. O.T.Bailey, A.M.Pappenheimer, F.S.Cheever, J.B.Daniels, J.Exp.Med. 90:195 (1949).
9. A.M. Pappenheimer, J.Natl.Can.Inst. 20:879 (1958).
10. L.P.Weiner, Arch.Neurol. 28:298 (1973).
11. R.M.Herndon, D.E.Griffin, U.McCormick, L.P.Weiner, Arch.Neurol. 32:32 (1975).
12. A.Lucas, W.Flintoff, R.Anderson, et al, Cell 12:553 (1977).
13. K.Nagashima, H.Wege, et al, Acta.Neuropathol. 44:63 (1978).
14. O.Sorensen, D.Percy, S.Dales, Arch.Neurol. 37:478 (1980).
15. S.A.Stohlman, L.P.Weiner, Neurology 31:38 (1981).
16. R.Watanabe, H.Wege, V.Ter Meulen, Nature 305:150 (1983).
17. E.Lavi, D.H.Gilden, M.K.Highkin, S.R.Weiss, J.Virol. 51:563 (1984).
18. S.Kyuwa, K.Yamaguchi, Toyoda, K.Fujiwara, J.Virol. 65:1789 (1991).

CORONAVIRUSES AND MULTIPLE SCLEROSIS

Ronald S. Murray, Guang-Yun Cai, Kristen Hoel,
Steven Johnson and Gary F. Cabirac

Rocky Mountain Multiple Sclerosis Center, Colorado
Neurological Institute, and Swedish Medical Center.
701 E. Hampden Ave., Englewood, CO 80150;
Department of Biochemistry, Biophysics and Genetics
University of Colorado Health Sciences Center, Denver, Colorado

INTRODUCTION

Multiple sclerosis (MS) is a chronic inflammatory demyelinating disease of the human central nervous system (CNS) without known cause. A popular theory proposes that an exogenous stimulus initiates an immune response against self CNS proteins, possibly myelin structural protein(s). However, damage to the CNS as the result of an immune response to a chronic low grade CNS infection has never been ruled out. Therefore, demyelinating lesions in MS may represent the final direct or indirect immunopathological reaction to an infectious agent(s). Supporting this hypothesis are epidemiologic studies that strongly implicate an environmental factor in the development of MS (1). Particularly compelling is the MS epidemic in the Faroe Islands (2) and the recent study of offspring to immigrants to the United Kingdom from low incidence countries developing MS at rates similar to the indigenous English population (3).

Several common human RNA or DNA viruses have been implicated in the pathogenesis of MS but none have been definitively associated with the disease (4). Our laboratory and collaborators have been unsuccessful in detecting herpes simplex type I, cytomegalovirus, varicella zoster virus, Epstein Barr virus or borrelia burgdorferii (unpublished data), human T-cell lymphotropic virus (5,6), or measles, mumps, and rubella viruses (7) in MS brain by either in situ hybridization, immunohistochemistry or the polymerase chain reaction.

Coronaviruses have been implicated in MS since the report of virus isolation from the brains of two MS patients and the electron microscopic observation of a coronavirus like particle in an MS brain perivascular immunocyte (8,9). Coronaviruses are also known to cause demyelination (10,11) and are capable of stimulating T-cell mediated autoimmune reactions in rodents (12,13). Recently, we reported finding coronavirus RNA and antigen in active demyelinating plaques of MS autopsy brain thus further implicating coronaviruses in MS (14). This coronaviral RNA and antigen was detected primarily with cDNA probes and antibodies specific for MHV related coronaviruses.

Although these results were surprising they are plausible since we then demonstrated that coronaviruses JHM and SD could infect the CNS and cause demyelination following intracerebral inoculation in several species of primates (15). Additonally, coronavirus 229E has recently been detected in MS brain by PCR technology (16).

Our current goals are to specifically identify the coronavirus(es) found in human brain and investigate how these coronaviruses interact with human brain cells and immune system. We now report the initial results and development of 1) PCR amplification of coronaviruses directly from human brain, 2) an MS brain derived cDNA library and screening for CV clones and 3) an in vitro human brain cell - coronavirus infection system.

METHODS AND RESULTS

PCR Amplification of Coronaviruses from Human Brain

Human brain samples were derived from MS and control autopsy samples used in prior in situ hybridization and immunohistochemical screens for the prescence of coronaviruses (15). All necessary precautions were taken to avoid contamination of RNA preparations with exogenous sources of RNA and DNA. Total RNA extracted from frozen human brain was reversed transcribed using the following reaction conditions. 1ug of RNA was mixed with 100pmoles of random hexamers (Boehringer Mannheim) in a total volume of 15ul, heated at 65°C for 5 min then cooled at room temperature. CVP1 5'TTGAAGGCTCTGGAAGGTCTGC and CVP2 5'CTGCAAGAATGGGGAACTGTGG was the inital set of primers. The CVP1/P2 product is 374bp for MHV and 441bp for OC43. 5ul of P1/P2 reaction is removed, mixed into 100ul H_2O and then 5ul of this, i.e. approx. 0.05ul of original rxn, is used in second round PCR with P3/P4 primers. The nested set reaction conditions are as follows:10ul of a enzyme/buffer/dNTP mix was added to the RNA/primer tube and the reaction was incubated at 43°C for 1 hr. The final reverse transcriptase reaction conditions were 40ug/ml RNA, 4uM random hexamers, 50mM Tris, pH 8.3, 100mM KCl, 10mM $MgCl_2$, 4mM DTT, 1 unit/ul RNasin (Promega), 1mM dNTPs, and 5 units of AMV reverse transcriptase (LifeSciences, Inc.). Reactions were heated at 97°C for 5 min then stored at -20°C.PCR reaction conditions were as follows. Tubes containing 85ul of buffer, primers, enzyme and approximately 100ul of mineral oil were heated to 80°C in the Perkin Elmer Thermocycler. 5ul of the cDNA from the reaction described above was mixed with 10ul of 2mM dNTPs in a separate tube and this was then added through the oil into the PCR reaction tube. The final reaction conditions were: 10mM Tris, pH 8.8, 50mM KCl, 1.5mM $MgCl_2$, 0.2mM dNTP, 0.1% Triton X-100, 0.5uM of each primer, and 5 units of Taq polymerase (Promega). The temperature profile for the reactions was as follows: 95°C/ 1 min, 58°C/1.5 min, 72°C/3 min for 3 cycles then 95°C/1 min, 62°C/1 min, 72°C/2 min for 36 cycles followed by incubation at 72°C for 7 min. The PCR primers, CVP3 and CVP4, used for these reactions were designed based on the published sequences of the nucleocapsid genes of MHV A59 and HCV OC43 (2,22). Viral cDNA amplified with CVP3, 5'IIAAATTGCTIITCTTGTTCTGGC (I=inosine), and CVP4, 5' CCAAAATTCTGATTAGGGCCTCTC, produce a 186bp PCR product. These primers amplify a region extending from nucleotide 858 to 1043 or 784 to 969 for MHV A59 or HCV OC43, respectively. The PCR reaction products were analyzed by gel electrophoresis and Southern-blot hybridization. The ^{32}P labeled probe used for hybridization,CVPP has the sequence 5'AAGCAIAITGCCAAAIAAGTCAGICAGAAAATTT; this hybridizes specifically to the 186bp CV PCR product.

Table 1. CV PCR amplification from MS brain.

Sample	Diagnosis	Brain Region	PCR Product
TO-1	MS	Cerebrum	**Positive**
TO-2	MS	Brain stem	Negative
ZO	MS	Cerebrum	**Positive**
ZOH	MS	Cerebrum	Negative
HO	MS	Cerebrum	Negative
KN-1	MS	Cerebrum	Negative
KN-2	MS	Spinal Cord	Negative
ME	MS	Cerebrum	**Positive**
SZ	MS	Cerebrum	Negative
SY	MS	Cerebrum	Negative
RO	MS	Cerebellum	Negative
TR	MS	Spinal Cord	Negative
SC	MS	Cerebrum	Negative
DI-cDNA library	MS	Cerebrum	**Positive**
HA	Cardiac	Cerebellum	Negative
OR	Cardiac	Cerebrum	**Positive**
MU	Cardiac	Cerebrum	Negative
LA	Sepsis	Cerebrum	Negative
MC	Cardiac	Cerebrum	Negative

Table 1 shows our intial PCR results for 12 MS patients with 4 having amplifiable coronavirus product. MS sample DI was amplified from a cDNA library made from mRNA from this patients white matter. Sequencing of this amplified product demonstrated high homology with the puplished sequence of JHM. One of control brain samples had amplifiable product. To date these products have not been sequenced.

Coronavirus Infection of Human Brain Cells

Primary or continuous human and murine cell cultures were infected with coronavirus JHM, SD or sham inoculums. Origin and methods of growing coronaviruses JHM, JHM OMp1 and SD have been described (11,15). The human astrocyte cell line STTG-1 was obtained from the ATCC and maintained under standard cell culture conditions. Human primary brain cultures were derived from the resected epileptic cortex of two patients (F and V). Cells were grown in DMEM-10%FCS medium. DBT and 17Cl 1 cells growth and maintenance have been described (8,11). Cultures were observed for CPE and infectious virus production. Additionally, all cultures were analyzed for the prescence of viral products by immunohistochemistry (14,15), western and northern blots and PCR amplification from isolated cell RNA as described (15). To inhibit infection, viral inoculums were pretreated with a polyclonal anti-coronavirus antibody (gift from J. Leibowitz).

Table 2 shows the results of infection with JHM of human and murine cell lines. Primary brain cultures and the human astroytoma cell line STTG could be infected. The infection was of a very low grade and complete CPE would not develop for 5 to 6 days post adsorption. In contrast, murine DBT cells and 17Cl-1 cells could be productively infected by JHM. Similar results were seen with CV SD and the primate isolate JHM OMp1. In general only low levels of virus were detectable in human cells suggesting that these cells were prone to being persistently infected. However the exact mechanisms will require further investigation.

Table 2. Infection of Cell Lines with Coronavirus.

System	CPE	Virus	PCR	Northern	IF	Western
17Cl-1	-	-	-	-	-	-
17Cl + JHM	+	+	+	+	+	+
V Primary Brain	-	-	-	-	-	-
V + JHM	+	ND	+	-	+	-
F Primary Brain	-	-	-	-	-	-
F + JHM	+	ND	+	-	+	-
STTG	-	-	-	-	-	-
STTG + JHM	+	+	+	-	+	-
DBT	-	-	-	-	-	-
DBT + JHM	+	+	+	+	+	+

a. Pretreatment with antibody prevents CPE in all cells

b. Similar results with CV SD and JHM-OMp1

DISCUSSION

Despite the intensive investigations of neurotropic coronaviruses in rodents, human neurologic disease resulting from coronavirus infection has not been proven. However, accumulating data, in addition to the results presented here, suggest that coronaviruses may infect human CNS tissue and that these viruses may differ from the prototypical upper respiratory human coronaviruses. In one case, a coronavirus (Tettnang virus) was isolated from the cerebrospinal fluid of a young girl with viral meningitis following an upper respiratory infection (17). This human coronavirus isolate was suspected by the authors to be more closely related to murine coronavirus than to human coronavirus. Recently, Parkinsons disease patients were reported to have high cerebrospinal fluid antibody titers specific to the murine coronaviruses MHV A59 and MHV JHM when compared to patients with other neurologic disease or non-disease controls (18). Burks et al reported the isolation of two coronavirus (SD & SK) from MS autopsy brain tissue after passage through suckling mice or murine cell culture (8). Although these isolates have serologic cross reactivity to murine coronavirus and HCV OC43 (19) others have demonstrated that these putative MS coronavirus isolates are antigenically and genetically more closely related to murine coronavirus than to the human coronavirus (19,20); our results show a 97% sequence identity in the 3' end of the genome between coronavirus SD and MHV JHM (unpublished data). Additionally, we have demonstrated murine-like coronaviruses directly in demyelinating plaques of autopsy MS brain. We have also reported the CNS infection of primates (14). Taken together these data suggest that coronaviruses can infect the human CNS and may play a role in multiple sclerosis pathogenesis.

The results presented in this paper demonstrate the value of PCR as a sensitive method to screen human brain and cell lines for the prescence of minute amounts of coronavirus RNA. Additionally this represents a third method of identifying "murine" related CV in multiple sclerosis brain. Coronavirus specific PCR products were amplified from one MS cDNA library and sequenced. The sequence of this product had a 94% identity to JHM. This cDNA library will now allow us to "rescue" other regions of this coronavirus. Hopefully we will glean data on the possibility of a human-murine recombinant virus versus a mutant murine coronavirus. Our in vitro results show that JHM is able to infect at low levels human STTG-1 cells and primary human brain culture cells. The infection takes several days to develop and can be inhibited by pretreatment of virus with neutralizing antibody. Infection of STTG-1 cells will be important in studying the human cellular immune response to these viruses. This will be possible since these cells are HLA DR2 and are known to present foreign antigens to T-cells.

The data presented here do not establish coronaviruses as an etiological agent for MS, but they do indicate that coronaviruses are present in the human CNS and are preferentially detected in MS tissue. However the stage is set for identifying the antigenic structure of these coronaviruses and to begin investigations of the cellular immunity to these antigens in HLA DR2 MS patients.

Acknowledgements

This work was supported by grants from the Swedish Medical Foundation and Fausel Foundation. The Rocky Mountain MS Center Tissue bank is supported by grant #RG-2108-B-5 from the National Multiple Sclerosis Society.

REFERENCES

1.   J.F.Kurtzke, Neurology. 30:61(1980).
2.   J.F.Kurtzke, K.Hyllested, Ann Neurol. 5:6(1979).
3.   M.Elian, S.Nightingale, G.Dean, J Neurol Neurosurg Psychiatry. 53:906(1990).
4.   J.Booss, J.H.Kim,in: "Handbook of Multiple Sclerosis," S.D.Cook, ed.,Marcel Dekker, Inc.,New York, New York(1990).
5.   G.F.Cabirac, D.Ries, R.S.Murray, Ann Neurol. 29:343(1991).
6.   G.D.Ehrlich, J.B.Glaser, V.Bryz-Gornia,et al.Neurology.41:335(1991).
7.   M.S.Godec, D.M.Asher, R.S.Murray, et al.Ann Neurol. In Press (1992).
8.   J.S.Burks, B.L.DeVald, L.D.Jankovsky, J.G.Gerdes, Science. 209:933(1980).
9.   R.Tanaka, Y.Iwasaki, H.Koprowski, J Neurol Sci. 28:121(1976).
10.   L.P.Weiner, Arch Neurol 28:298(1973).
11.   P.M.Mendelman, L.D.Jankovsky, R.S.Murray, et al, Arch Neurol 40:493(1983).
12.   S.Kyuwa, K.Yamaguchi, Y.Toyoda, K.Fujiwara, Virol 65:1789(1991).
13.   R.Watanabe, H.Wege, V.ter Meulen, Nature 305:150(1983).
14.   R.S.Murray, B.Brown, D.A.Brian, G.F.Cabirac, Ann Neurol 31:525(1992).
15.   R.S.Murray, G-Y Cai, K.Hoel, et al, Virology. 188:274(1992).
16.   J.N.Stewart, S.Mounir, P.J.Talbot, Virology. In Press, (1992).
17.   D.Malkova, J.Holubova, J.M.Kolman, et al. Acta Virol(Prague) (Engl Ed). 24(5):363(1980).
18.   E.Fazzini, J.O.Fleming, S.Fahn, Neurol(Suppl 1).40:169(1990).
19.   J.C.Gerdes, I.Klein, B.L.DeVald, J.S.Burks, J Virol.38:231(1981).
20.   J.O.Fleming, F.A.K.Zaatari, W.Gilmore, et al, Arch Neurol. 45:629(1988).
21.   S.R.Weiss, Virology 126:669(1983).

ENHANCEMENT OF FIP IN CATS IMMUNISED WITH VACCINIA VIRUS RECOMBINANTS EXPRESSING CCV AND TGEV SPIKE GLYCOPROTEINS

W. Stuart K. Chalmers [1], Brian C. Horsburgh [2],
William Baxendale[1] and T. David K. Brown [2]

[1] Intervet (UK) Ltd.
The Elms, Houghton
Cambs PE17 6HB
UK

[2] Virology Division
Department of Pathology, University of Cambridge
Tennis Court Road
Cambridge CB2 1QP
UK

INTRODUCTION

Feline infectious peritonitis (FIP) is a progressive, debilitating disease of wild and domestic Felidae caused by infection with the coronavirus feline infectious peritonitis virus (FIPV) [1,2]. Two forms of FIP are observed - the commoner wet or effusive form in which the abdomen is distended by the accumulation of a fibrin-rich exudate within the peritoneal cavity and the dry or non-effusive form in which there is a desseminated granulomatous disease involving most organs. Mortality approaches 100% for both forms.

Early death syndrome (EDS) is a phenomenon observed when cats with humoral antibodies to FIPV are challenged with FIPV. Such animals can show reduced survival times when compared with antibody-negative controls [3,4]. Passive transfer of anti-FIPV serum results in early death following experimental FIPV challenge [5,6]. These data suggest that EDS may be caused by antibodydependent enhancement [7].

The role of FIPV spike (S) glycoprotein in EDS has been studied by Vennema et al. [8]. Kittens were inoculated twice with a vaccinia virus recombinant expressing the FIPV S protein. Two weeks after the second inoculation they were challenged with a virulent FIPV strain (76-1146); EDS was observed, the sensitized kittens dying 8-9 days after challenge. In contrast, the non-immune control group survived for 29-30 days post challenge. Vennema et al. demonstrated conclusively that the epitope(s) responsible for EDS are found in the FIPV S protein.

Cats are also susceptible to infection with transmissible gastroenteritis virus (TGEV) and canine coronavirus (CCV), viruses closely related to FIPV [9,10,11]. CCV and TGEV replicate to a limited extent in the gastrointestinal tract and oropharynx of cats, producing asymptomatic infections and low levels of neutralising antibodies against FIPV [10,11]. In some studies cats preimmunised in this way with CCV did not show EDS following challenge with FIPV [10,11]. These data contrast with those of McArdle et al [12], which indicated that infection with another CCV strain could cause EDS.

Coronaviruses, Edited by H. Laude and J.F. Vautherot
Plenum Press, New York, 1994

The work described below was designed to investigate the potential for induction of EDS by the spike glycoproteins of CCV and TGEV expressed by vaccinia recombinants. The protocol used was similar to that employed by Vennema *et al*; this approach is particularly appropriate for CCV and TGEV whose replication in the cat is poor and variable.

METHODS

Preparation and analysis of vaccinia recombinants

Vaccinia virus recombinants Vac4b-IN and Vac4b-C6 expressing the S genes of CCV strains INSAV-C-1 (UK) and CCV-6 (USA) respectively, and Vac4b-TG expressing the S gene of TGEV strain FS 770/20 (UK) under the control of the powerful vaccinia 4b late promoter were prepared using standard procedures (13). A recombinant expressing the human cytomegalovirus (CMV) gB glycoprotein was used as a control.

Expression of S proteins by the recombinants was detected by immunofluorescence and RIP. Immunofluorescent staining with a rabbit polyclonal serum against CCV followed by an FITC goat anti-rabbit antibody was carried out on methanol-fixed A72 cells infected with control CCV and with vaccinia recombinants. RIP was carried out using labelled S proteins produced by the infection of BHK-21 cell monolayers with CCV or vaccinia recombinants in the presence of $[^{35}S]$-methionine. Cell pellets were dissolved in RIPA buffer and samples of cell extract were incubated with serum. Antigen:IgG complexes were bound to Sepharose protein G beads, the bound protein eluted with SDS-containing buffer and analysed using SDS-polyacrylamide gel electrophoresis.

Grouping, Inoculation and Challenge

Susceptible cats (6 months old) were grouped as in Table 1 and inoculated by the sub-cutaneous route, with an interval of three weeks between inoculations. All cats were subsequently challenged, two weeks post secondary inoculation, with wild type FIPV (79-1146) administered oro-nasally. Cats inoculated with the vaccinia vector could only be clinically assessed for 18 days post challenge due to the limitation of availability of the rooms registered for recombinant animal experiments. Control groups consisted of (a) uninoculated/challenged cats and (b) cats inoculated with vaccinia virus expressing the CMV gB gene.

Table 1. Groupings of cats, inoculation and challenge data

Group	Cat Numbers	Inoculum (TCID$_{50}$)	Challenge
1	5, 19J, U02,542	Vac4b-C6 ($1°$-$10^{7.0}$; $2°$-$10^{8.0}$)	FIPV 79-1146 ($10^{6.5}$TCID$_{50}$ ml^{-1})
2	C20, C24, 19L,538	Vac4b-IN ($1°$-$10^{7.0}$; $2°$-$10^{8.0}$)	" "
3	C14, C17, U10, 19K	Vac4b-TGEV ($1°$-$10^{7.0}$; $2°$-$10^{8.0}$)	" "
4	C23, U05	Vac4b-CMVgB ($1°$-$10^{7.0}$; $2°$-$10^{8.0}$)	" "
5	U08, U09	Uninoculated	"

Observations and clinical scoring

All cats were monitored daily, post vaccination and post challenge for clinical signs. Temperatures were recorded and swabs taken for the first 7 days post challenge to show that FIPV had infected the cats and was replicating and being excreted. Scores were assigned to each clinical sign depending on severity and duration. These scores were tabulated at the end of the experiment and used to compare severity of infection between groups. Any cats which became severely ill were euthanased immediately and a full post mortem examination carried out.

Serological Investigations

Serum samples were taken pre primary and secondary vaccination, pre challenge and at the time of death (euthanasia). Virus neutralising antibodies to vaccinia virus and to FIPV were detected by their abilty to reduce infectivity in 96 well quantal tissue culture assays using respectively A72 and Fcwf cells. The reciprocal of the highest serum dilution causing inhibition of infection by one $TCID_{50}$ of virus was regarded as the serum titre (14).

Antibodies (IgG) directed against the S protein of CCV strain Insav-C-1 were detected in serum pools from each cat group by the radio immunoprecipitation (RIP) assay described above.

RESULTS

Properties of vaccinia S protein recombinants

RIP was used to demonstrate that the recombinants expressed the inserted S protein genes. It was found that in the case of the recombinants Vac4b-C6 and Vac4b-TG proteins which co-migrated with authentic CCV S protein synthesized in CCV-infected A72 cells was observed. However in the case of Vac4b-IN a product of only 120K was detected. Sequencing demonstrated the introduction of a frameshift during manipulation with consequent premature termination and production of a potentially secretable product consisting of the N-terminal 700 amino acids of the S protein.

In order to confirm that the recombinant viruses were expressing the S protein, A72 cells were infected with the vector viruses and fixed after 24 hours or 72 hours. Corona S protein was observed in the cell cytoplasm by (a) indirect immunofluorescence (all recombinants) and (b) syncytium formation (all recombinants except Vac4b-IN) (Fig.1). The Vac4b-IN recombinant presumably failed to cause cell fusion because it was secreted.

Evidence for an immune response following inoculation of vaccinia recombinants

Serum neutralising antibodies to vaccinia virus were detected in all but one cat (C14) pre-challenge (Table 2), but were present post challenge in this cat. These data demonstrate a

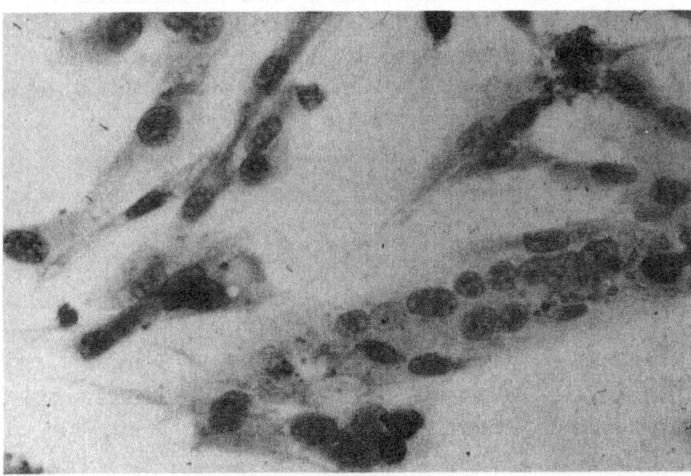

Figure 1. Syncytium formation in A72 cells infected with vaccinia recombinant Vac4b-C6

Table 2. Serum neutralising antibodies to vaccinia virus

Cat group	Number	Pre Vaccination 1°	2°	Pre challenge	Post challenge
Vac4b-IN	C15	≤5	9	N/A	18
	19J	≤5	≤5	30	30
	U02	≤5	15	15	15
	542	≤5	≤5	60	120
Vac4b-C6	C20	≤5	9	18	25
	C24	≤5	9	30	30
	19L	≤5	18	18	N/A
	538	≤5	9	30	30
Vac4b-TGEV	C14	≤5	≤5	≤5	26
	C17	≤5	15	18	30
	U10	≤5	15	120	100
	19K	≤5	N/A	30	60
Vac4b-CMV$_g$B	C23	≤5	9	30	30
	U05	≤5	≤5	30	30

Titres are reciprocals of serum dilution. N/A - not available.

consistent response to the inoculated vaccinia recombinants. Uninoculated control cats remained antibody-free throughout the experiment in spite of being housed with the Vac4b-CMVgB inoculated animals.

Low FIPV neutralising antibodies were detected in these samples post challenge (data not shown); this is consistent with the the findings of Vennema *et al.* (8). Circulating antibodies directed against coronavirus S protein were detected by R.I.P. (Fig. 2). The results show that pooled sera from each group of cats inoculated with coronavirus S protein gene-expressing vaccinia viruses contained IgG antibodies against the 200kDa CCV spike glycoprotein. A low level of precipitating activity was also detected in the control group (G4/5) at 18 days after the FIPV challenge.

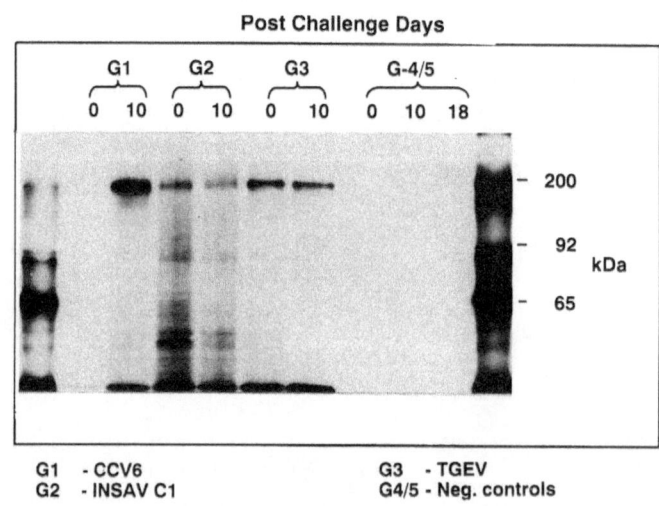

Fig.2. Radio immunoprecipitation assay for IgG antibodies to CCV S protein

Clinical observations

A typical FIP bi-phasic temperature response was observed in the non-immunised control group whereas most of the sensitized cats (in particular Vac4b-C6 and IN) produced initial pyrexic responses which were maintained until time of death or euthanasia.

Scores assigned to clinical signs are summarised in Table 3. There was a significant difference found between both the Vac4b-C6 and Vac4b-IN groups and the control group. Although the difference between the Vac4b-TGEV group and the control group was smaller, it was nevertheless shown to be significant.

Table 3. Summary of clinical findings and scores

Group	Cat	Immunogen	Fever	Illness from day PC	Survival time from challenge	FIP	Clin/ Path Score
1	C20	Vac4b-In	constant	7	11	DRY	72
1	C24	Vac4b-In	constant	13	17	WET	47
1	19L	Vac4b-In	constant	7	8	DRY	55
1	538	Vac4b-In	biphasic	13	17	DRY	61
2	C15	Vac4b-C6	constant	4	11	DRY	70
2	19J	Vac4b-C6	constant	4	8	DRY	57
2	U02	Vac4b-C6	constant	4	11	DRY	66
2	542	Vac4b-C6	constant	4	11	DRY	69
3	C14	Vac4b-TG	fluctuating	4	>18	DRY	40
3	C17	Vac4b-TG	constant	6	13	DRY	39
3	U10	Vac4b-TG	fluctuating	13	18	WET	36
3	19K	Vac4b-TG	constant	-	>18	DRY	26
4	C23	Vac4b-gB	biphasic	-	>18	DRY	31
4	U05	Vac4b-gB	biphasic	18	>18	WET	21
5	U08	none	biphasic	18	28	WET	21
5	U09	none	fluctuating	18	18	WET	17

Post mortem observations revealed classical signs of FIPV infection which included fibinous peritonitis, peritoneal and pleural fluid, pyogranulomatous lesions, congestion of the lungs, necrotic foci on liver and spleen and mesenteric nodules. Immunohistochemical staining (immunoperoxidase) of infected tissue revealed coronavirus in the cytoplasm of the cells confirming FIPV infection. The challenge virus was also re-isolated from rectal swabs 4-6 days post challenge.

Although the numbers of cats per group were low, the following trend was observed; cats immunized with (a) CCV-6 S protein survived on average 10 days post challenge (b) INSAV-C-1 S protein - 13 days PC (c) TGEV S protein - 15 days PC. The control cats given FIPV only survived for 28 days PC. These data are comparable with those of Vennema et al. (8); in their study the mean post challenge survival time for the FIPV S gene vaccinia recombinant was 9 days and that for the control group inoculated with the wild type vaccinia strain WR was 29 days.

DISCUSSION

The results of this experiment clearly show that the epitope(s) responsible for EDS is found on the first 700 amino-acids of the S protein of CCV virus. A similar epitope(s) is likely to be present on the TGEV S protein although this protein may have a reduced capacity to induce EDS when compared with those of FIPV and CCV.

The RIP assay has shown that circulating anti S protein I_gG was present in the cats prior to challenge. One theory for EDS in cats is that enhancement of infection is dependent on the presence of antibody at time of challenge (ADE). Virus-antibody complexes are formed by the binding of the I_gG non-neutralising antibody to viral antigen and the resulting complexes are then opsonised by macrophages or monocytes as they bind to these F_c or complement receptor bearing cells. Opsonisation of the complexes does not result in the neutralisation of the virus and does not therefore prevent infection but rather exacerbates the rate of spread of infection in the animal.

In conclusion, cats were inoculated with vaccinia recombinants expressing S protein of various strains of coronavirus, none of which afforded protection against FIPV infection or subsequent clinical disease. Although limited numbers of animals were used, it is clear that an exacerbation of disease, and early deaths resulted from the production of circulating antibody against CCV and TGEV S protein prior to challenge.

REFERENCES

1. Ward, J.M., Munn, R.J., Gribble, D.H., Dungworth, D.L. (1968) An observation of feline infectious peritonitis. Vet. Rec. 83: 416-417.
2. Pedersen, N.C., Boyle, J.F.. Floyd, K., Fudge, A. and Barker, J. (1981) An enteric coronavirus infection of cats and its relationship to feline infectious peritonitis. Am. J. Vet. Res. 42: 368-377.
3. Pedersen, N.C., Boyle, J.F. (1980) Immunologic phenomena in the effusive form of feline infectious peritonitis. Am. J. Vet. Res. 41: 868-876.
4. Weiss, R.C. and Scott, F.W. (1981) Pathogenesis of feline infectious peritonitis: nature and development of viremia. Am. J. Vet. Res. 42: 382-390.
5. Weiss, R.C. and Scott, F.W. (1981) Antibody-mediated enhancement of disease in feline infectious peritonitis: comparisons with dengue hemorrhagic fever. Comp. Immun. Microbiol. Infect. Dis. 4: 175-189.
6. Pedersen, N.C. and Black, J.W. (1983) Attempted immunisation of cats against feline infectious peritonitis using avirulent live virus or sublethal amounts of virulent virus. Am. J. Vet. Sci. 44: 229-234.
7. Porterfield, J.S. (1986) Antibody-dependent enhancement of viral infectivity. Adv. Virus Res. 31: 335-355.
8. Vennema, H. de Groot, R.J., Harbour, D.A., Dalderup, M., Grufydd-Jones, T., Horzinek, M.C. and Spaan, W.J.M. (1990). Early death after feline infectious peritonitis virus challenge due to recombinant vaccinia virus immunization. J. Virol. 64: 1407-1409.
9. Woods,R.D. and Pedersen, N.C. (1979). Cross-protection studies between feline infectious peritonitis virus and porcine transmissible gastroenteritis viruses. Vet. Microbiol. 4:11-16.
10. Barlough, J.E., Stoddart, C.A., Sorreso, G.P., Jacobson, R.H. and Scott, F.W. (1984). Experimental inoculation of cats with canine coronavirus and subsequent challenge with feline infectious peritonitis virus. Lab. Animal Sci. 34: 592-597.
11. Stoddart, C.A., Barlough, J. E., Baldwin, C.A. and Scott, F.W. (1988). Attempted immunization of cats using canine coronavirua. Res. Vet. Sci. 45: 383-388.
12. McArdle, F., Bennett, M., Gaskell, R.M., Tennant, B., Kelly, D.F. and Gaskell, C. (1990). Canine coronavirus infection in cats; a possible role in feline infectious peritonitis. Adv. Exp. Med. Biol. 276: 475-479.
13. Mackett, M. and Smith, G.L. (1986). Vaccinia virus expression vectors. J. gen. Virol. 67: 2067-2082.
14. Horsburgh, B.C. (1992). Molecular studies of canine coronavirus. Ph.D. thesis, University of Cambridge, U.K.

ELECTROCARDIOGRAPHIC CHANGES FOLLOWING RABBIT CORONAVIRUS-INDUCED MYOCARDITIS AND DILATED CARDIOMYOPATHY

Lorraine K. Alexander[1], Bruce W. Keene[2] J. David Small[1], Boyd Yount Jr.[1] and Ralph S. Baric[1]

[1]Program in Infectious Diseases, Department of Epidemiology, University of North Carolina at Chapel Hill, Chapel Hill, NC 27599-7400, U.S.A.
[2]College of Veterinary Medicine, North Carolina State University, Raleigh, NC 27606, U.S.A.

ABSTRACT

Rabbit Coronavirus (RbCV) infection was divided into two phases based upon day of death and pathologic findings. During the acute phase (days 2-5) heart weights (HW) and heart weight-to-body weight (HW/BW) ratios were increased with striking dilation of the right ventricle. These changes as well as increased dilation of the left ventricle were especially pronounced during the subacute phase (days 6-12). Myocytolysis, pulmonary edema, and degeneration and necrosis of myocytes, were seen during both phases. Myocarditis, pleural effusion, calcification of myocytes, and congestion in the liver and lungs were seen in the subacute phase. Electrocardiograms (ECGs) exhibited low voltage, nonspecific ST-T wave changes, sinus tachycardia, occasional ventricular and supraventricular premature complexes and 2^0 AV block consistent with myocarditis and heart failure. Forty-one percent of the survivors exhibited increased HW and HW/BW ratios, biventricular dilation, interstitial and replacement fibrosis, myocyte hypertrophy and myocarditis. ECGs exhibited nonspecfic ST-T wave changes, sinus arrhythmia, occasional ventricular and supraventricular premature complexes and 2^0 AV block. These data suggest that RbCV infection may result in viral myocarditis and heart failure with a proportion of survivors progressing into DCM.

INTRODUCTION

Virus infection has long been associated with heart disease in both man and experimental animals. Picornaviruses are commonly implicated in 20-30% of all diagnosed cases, but the etiology of viral myocarditis is usually not determined. Myocarditis probably develops in about 2-5% percent of persons after viral infection[1]. Viral myocarditis is usually characterized by a benign course of vague flu-like symptoms. Laboratory findings may include electrocardiographic abnormalities, elevated serum concentrations of myocardial enzymes and serological evidence of recent viral infection[1,2]. While the vast majority of patients with viral myocarditis appear to recover without residual heart damage, the disease occasionally progresses to congestive heart failure (CHF), arrhythmias and death[1,2].

Viral myocarditis has also been implicated as a potentially important etiologic factor in the development of dilated cardiomyopathy (DCM), a primary myocardial disease characterized by progressive ventricular dilation and systolic dysfunction[1,2,3]. DCM is a devastating disease often accompanied by a variety of nonspecific electrocardiographic abnormalities, with death generally occurring either suddenly (presumably as a result of arrhythmia) or as the result of progressive heart failure in 50% of patients within 2-3 years of diagnosis. Myocyte hypertrophy, fibrosis and occasionally active myocarditis are found upon pathologic examination[3]. Virus is usually not isolated from patients with DCM, but serological studies suggest that viral infection maybe an important initiating event[2]. Most of our understanding of the mechanism by which viruses cause heart disease has come from studying animal models[1]. The two most extensively studied animal models include the coxsackievirus B3 (CBV) and encephalomyocarditis virus (EMC) murine models of myocarditis and DCM [4,5]. However the mechanisms by which these viruses induce heart disease remains highly controversial and more animal models for human heart disease are needed[1]. Our laboratory has established an animal model system for myocarditis, CHF and DCM following rabbit coronavirus (RbCV) infection[6,7,8]. In this study we examined the electrocardiographic changes associated with the infection.

METHODS

Infection, Pathologic and Morphometric Studies

Seventy nine male New Zealand White rabbits were inoculated intravenously via the marginal ear vein or intramuscularly with 0.2 ml of a RbCV stock containing 10^3-10^4 RID_{50}/ml as previously described[6,7]. The hearts of all animals were removed, weighed, and then fixed with 10% phosphate buffered formalin. The hearts were sectioned transversely at the widest dimensions of the ventricles, embedded in paraffin, and stained with hematoxylin-eosin, von Kossa's method or Masson's trichrome stain. Lung and liver sections were also stained with hematoxylin-eosin. The areas of the ventricular cavities, the thickness of the cardiac walls, and the myocyte diameters were determined using the Image Measure software morphometry system or a computerized Zeiss Videoplan-1 digital morphometry system as previously described[6,7]. All morphometric measurements were statistically analyzed and are presented as mean ± SD. Measurements from animals dying from RbCV infection were analyzed by Student's T test for unpaired observations. Measurements from chronic animals were analyzed by a one-way analysis of variance with a post-hoc contrast.

Electrocardiogram Studies

Serial electrocardiograms were performed on 13 rabbits infected with RbCV and on 3 uninfected controls. ECGs were taken using a Grass model 7B polygraph (Grass Instrument Co. Quincy, MA.). Alternatively, a three channel microcomputer augmented cardiograph I (MAC I) digital transmitter, recording unit (Marquette Electronics Inc. Milwaukee, WI) was used. With the Grass polygraph, ECGs were recorded at 50 mm/sec with the recorder calibrated to 1.0 mV/20 mm. With the MAC I, ECGs were recorded at both 25 and 50 mm/sec with the recorder calibrated to 1.0 mV/10 mm. Baseline ECGs were recorded on all rabbits for several days prior to infection. ECGs were recorded twice daily from day 3-12 and then at regular intervals through day 30 after infection. All measurements were made in lead II and included heart rate (HR), duration and voltage. In addition, readings from 9 animals were examined for any disturbances of conduction, rhythm and repolarization.

RESULTS

Clinical and Pathologic Results

Following RbCV inoculation into 79 rabbits, twenty-four (30%) of the animals died between days 2-5 (acute) and twenty-four (30%) between days 6-12 (subacute). Animals dying during the acute phase had enlarged hearts, dilation of the right ventricle, pulmonary edema, and degeneration and necrosis of myocytes. Ascites, pleural effusion, pulmonary edema, enlarged hearts, biventricular dilation and venous congestion in the liver and lungs was seen in animals dying in the subacute phase. The principle pathologic lesions in the subacute phase included myocarditis, calcification of myocytes and increased severity of myocyte degeneration and necrosis[6].

The heart weight and heart weight-to-body weight (HW/BW) ratios for animals in both the acute (n=12) and subacute (n=14) phases were found to be significantly increased over the acute (n=10) and subacute (n=10) controls. The right ventricular cavity area of the acute group was significantly increased over the control group. Animals dying in the subacute phase had significantly increased right and left ventricular cavity areas compared to the control group. In addition, changes in the wall thickness of the heart of both the acute and subacute groups were noted (Table 1).

Forty-one percent of the animals survived RbCV infection (chronic). The principal pathologic lesions of animals in the chronic phase included myocyte hypertrophy and interstitial and replacement fibrosis and myocarditis. The degree and number of inflammatory lesions varied greatly between sections examined in different portions of the heart and between different animals. In all chronic animals the degree of interstitial and replacement fibrosis was mild, with the greatest concentration occurring in the papillary muscles[7]. The degree of myocyte hypertrophy also varied among chronic animals. The chronic animals were divided into two groups, slight (16-20 µm) (n=13) and moderate (21-25 µm) (n=9) based upon mean myocyte diameters (Table 1). The myocyte diameters of both the slight and moderate groups were significantly increased over the controls. In addition, the degree of myocyte hypertrophy of the moderate group was significantly increased over the slight group. The mean measurements for heart weights, cavity areas and wall thickness of the chronic animals and their uninfected controls are also summarized in Table 1. Heart weights and HW/BW ratios were significantly increased over the control group. The right ventricular cavity area of both the slight and moderate groups were also significantly increased. There was no significant difference between the left ventricular cavity areas of the control and slight groups. However, the left ventricular cavity area of the moderate group was significantly increased over both the control and slight groups. The thickness of the right wall, left wall and interventricular septum of animals in the chronic phase showed no significant differences from the control group.

Table 1. Mean heart weights and morphometric measurements of RbCV infected rabbits[6,7] (reprinted with permission of the University of Chicago Press)

Mean measurements	Acute control	Acute infected	Subacute control	Subacute infected	Chronic control	Chronic slight	Chronic moderate
Heart weight (g)	6.1 (+/- 0.3)	a 8.4 (+/- 1.4)	6.1 (+/- 0.5)	a 8.7 (+/- 1.6)	6.6 (+/- 0.9)	a 8.4 (+/- 1.2)	a 8.2 (+/- 0.5)
Heart weight/ Body weight (x 10^{-3})	2.2 (+/- 0.2)	a 3.1 (+/- 0.2)	2.2 (+/- 0.2)	a 3.5 (+/- 0.6)	2.0 (+/- 0.2)	a 2.6 (+/- 0.2)	a 2.5 (+/- 0.2)
Right ventricular cavity area (um^2 x 10^6)	38.9 (+/- 11.8)	a 100.8 (+/- 39.7)	41.22 (+/- 13)	a 146.2 (+/- 51.8)	44.8 (+/- 13.8)	b 74.34 (+/- 21.9)	a 116.1 (+/- 53)
Left ventricular cavity area (um^2 x 10^6)	57.1 (+/- 6.1)	65.3 (+/- 21.9)	59.9 (+/-10.5)	b 77.5 (+/- 20.3)	61.92 (+/- 22.6)	63.61 (+/- 21.5)	b 94.6 (+/- 40.1)
Right wall thickness (um)	2087 (+/- 270)	a 1545 (+/- 249)	2065 (+/- 284)	a 1295 (+/- 318)	1265 (+/-236)	1501 (+/- 294)	1508 (+/- 230)
Left wall thickness (um)	4186 (+/- 545)	4369 (+/- 512)	4410 (+/- 617)	b 3773 (+/- 718)	3196 (+/- 639)	3594 (+/- 618)	3522 (+/- 429)
Septum thickness (um)	4470 (+/- 676)	4419 (+/- 740)	4470 (+/- 503)	a 3597 (+/- 603)	3365 (+/- 583)	3698 (+/- 577)	3367 (+/- 521)
Myocyte diameter (um)	ND	ND	ND	ND	15.42 (+/- 1.7)	a 18.64 (+/- 1.15)	a 21.82 (+/- 1.48)

a = significant difference from control (p < 0.01)
b = significant difference from control (p < 0.05)
ND = Not done

Electrocardiogram Results

ECGs recorded prior to infection and in control animals demonstrated sinus rhythm with a mean HR of 228 ± 38 and 218 ± 37 beats/min respectively and a mean R wave voltage of 0.38 ± 0.11 and 0.44 ± 0.14 mV respectively, similar to previously reported values[9]. During the acute phase of RbCV infection, sinus tachycardia (mean HR = 277 ± 18 beats/min) developed and R wave values in lead II were slightly depressed (0.30 ± 0.11 mV) (Figure 1). In addition to the sinus tachycardia and reduced R wave voltages, ST segment changes, supraventricular premature complexes, and Mobitz type II 2°AV block were occasionally observed during the acute phase (Table 2). The subacute phase was marked by persistent sinus tachycardia (284 ± 38 beats/min) and significant reductions in R wave voltages in lead II (0.17 ± 0.06 mV) (Figure 1). T wave voltage was reduced and generally paralleled the QRS voltage changes during the subacute phase. Occasional supraventricular tachyarrhythmias and 2° AV block type persisted in some animals, and some animals displayed isolated ventricular premature depolarizations (Table 2). The chronic phase of infection was characterized by return of the sinus rate to baseline values and resolution of the QRS voltage changes, but persistence of ST segment changes, T wave changes, ventricular and supraventricular arrhythmias and 2° AV block (Table 2) in some animals.

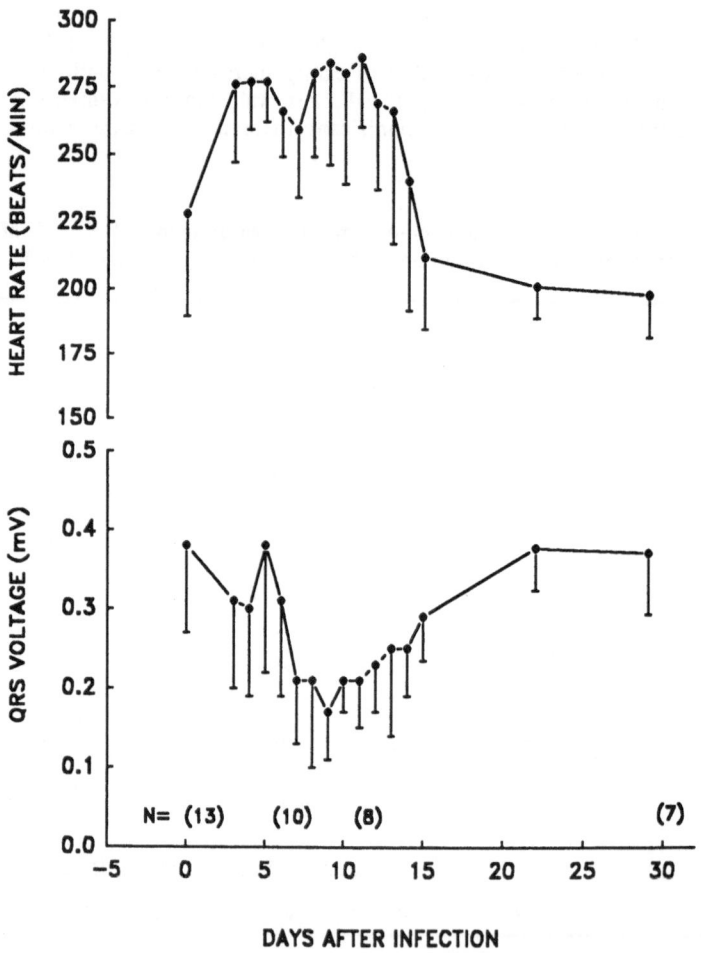

Figure 1. Mean heart rate and QRS voltage vs day of infection in RbCV infected rabbits.

Table 2. ECG changes in 9 rabbits after infection with RbCV.

ECG changes	Acute (days 2-5) n=9	Subacute (days 6-12) n=7	Chronic (> 12 days) n=6
T wave changes[a]	8 (89%)	6 (86%)	3 (50%)
ST segment changes[b]	2 (22%)	0 (0%)	2 (33%)
Rhythm disturbances			
Ventricular premature complexes	0 (0%)	1 (14%)	1 (17%)
Supraventricular premature complexes	1 (11%)	1 (14%)	2 (33%)
Sinus arrthythmia	0 (0%)	0 (0%)	2 (33%)
Conduction defects			
2° Atrioventricular block-type II Mobitz	1 (11%)	1 (14%)	2 (33%)

a. Decreasing in amplitude.
b. Elevation or depression.

DISCUSSION

Human coronaviruses, in part responsible for the common cold, have also been implicated in more serious human afflictions such as pneumonia, perimyocarditis, meningitis and radiculitis and in the development of multiple sclerosis[10,11]. A unique model for virus-induced heart disease has been described in our laboratory[6,7]. Rabbit coronavirus is probably an enveloped RNA virus that is morphologically and antigenically related to Group I human and animal coronaviruses[8]. RbCV infection in rabbits results in the development of myocarditis and CHF with a 60% mortality rate. About 41% of the survivors of rabbit coronavirus infection may develop DCM at a later stage in life. The similarities between RbCV-induced myocarditis and DCM and human disease suggest that it is an excellent animal model for human heart disease[6,7].

Rabbit coronavirus infection was divided into three phases: acute, subacute, and chronic based upon day of death and/or pathologic findings[6,7]. The acute phase of infection (days 2-5) was characterized by right ventricular dilation, pulmonary edema, and degeneration and necrosis of myocytes[6]. Electrocardiographic findings in the acute phase included sinus tachycardia, slightly lower QRS voltages, T wave and ST segment changes compatible with myocardial injury and apparently infrequent supraventricular arrhythmias and AV conduction disturbances. In contrast, 38% of the animals early in EMC infection, had complete AV block[12]. Because ECGs were recorded intermittently for brief periods, the actual frequency and pathophysiologic significance of arrhythmias during the course of RbCV infection remains speculative. Rabbits dying in the acute phase of infection did so suddenly and without moderate or severe clinical signs of CHF, raising the possibility that acute RbCV infection increases ventricular vulnerability. If this is true, the RbCV model should provide an interesting and rare opportunity to study sudden cardiac death (SCD) due to viral infection. SCD is defined as death caused by underlying coronary or noncoronary heart disease in which the subject is completely asymptomic or experiences symptoms one hour or less before death. In the United States there are approximately 300,000 SCD per year of which approximately 20 percent are due to noncoronary heart disease including acute viral myocarditis and idiopathic dilated cardiomyopathy[13]. Viral infection of the heart (cardiac viral neuropathy) has also been implicated in the pathogenesis of SCD in humans[14]. Studies have shown that major risk factors for SCD include cardiac enlargement, nonspecific T wave and ST segment abnormalities, and ventricular premature beats,[15] most of which are detected early after RbCV infection.

During the subacute phase of the infection (days 6-12) an additional 30% of the rabbits died. Rabbits dying during the subacute phase showed more severe myocyte degeneration and necrosis than during the acute phase, with calcification of myocytes and infiltration of inflammatory cells into the myocardium. These severe histopathologic changes were accompanied by dramatic biventricular dilation and eccentric hypertrophy of the myocardium. Postmortem venous congestion of the liver and lungs suggested the presence of congestive heart failure in these animals[6]. Electrocardiographic findings in the subacute phase included sinus tachycardia with low voltage QRS complexes compatible with pleural effusion, and occasional animals with supraventricular and/or ventricular arrhythmias and/or AV block. These ECG findings are comparable to the variety and type of changes seen in humans with myocarditis and CHF as well as those observed in other animal models of myocarditis and CHF[1,12,16,17].

Survivors of RbCV infection (chronic phase) were divided into two groups based upon the degree of myocyte hypertrophy. Animals in the moderate group exhibited pathologic and electrocardiographic changes consistent with the development of DCM. Pathologic changes included an enlarged heart, biventricular dilation, myocyte hypertrophy, fibrosis and myocarditis[7]. The ECG findings were nonspecific but taken with the pathologic evidence were consistent with the development of DCM and myocarditis. Nonspecific ST-T wave changes were frequently seen as well as rhythm and conduction disturbances. In humans with DCM and in the EMC model of DCM, arrhythmias and conduction defects have also been frequently detected[3,12].

Most of the information concerning the pathogenesis and mechanisms of virus-induced myocarditis and DCM has been obtained from animal models following CBV and EMC infection in mice[4,5]. With the few model systems available to study the sequence of events leading from viral infection to the development of myocarditis and DCM, the RbCV model will no doubt provide considerable insight into the mechanisms by which enveloped RNA viruses induce heart disease.

REFERENCES

1. J. Woodruff, *Am J Pathol.* 101:427 (1980)
2. K. Leslie, R. Blay, C. Haisch, et al, *Clin Microbiol Rev.* 2:191 (1989)
3. W. Roberts, R. Siegel, and B. McManus, *Am J Cardiol.* 60: 1340 (1987)
4. M. Reyes, K. Ho, F. Smith, and A. Lerner, *J Infect Dis.* 144:232 (1981)
5. A. Matsumori and C.Kawai, *in:* "Viral Heart Disease, H. Bolte ed. Springer-Verlag, NY (1984)
6. S. Edwards, J. Small, J. Geratz, L. Alexander, and R. Baric, *J Infect Dis.* 165:134 (1992)
7. L. Alexander, J. Small, S. Edwards and R. Baric, *J Infect Dis.* 166 (1992)
8. J. Small, L. Aurelian, R. Squire, et al, *Am J Pathol.* 95:709 (1979)
9. C. Nelson, W. Waggoner, and P Gastonguay, *Am J Physiol.* 207:1107 (1964)
10. H. Riski, and T. Hovi, *J Med Virol.* 6:259 (1980)
11. R Murray, B. Brown, D Brian, and G. Cabric, *Ann Neurol.* 31:525 (1992)
12. C. Kishimoto, A. Matsumori, M. Ohmae, et al, *J Am Coll Cardiol* 3:1461. (1984)
13. R. Myerburg and A. Castellanos, *in:* Heart Disease. E. Brunwald ed. W.B. Saunders Company, Philadelphia (1988)
14. T. James and K. Imamura, *Jpn Heart J.* 22:447 (1981)
15. W. Castelli et al. *in:* The Prevention of Sudden Cardiac Death. J. Kostis and M. Sanders, eds. Wiley-Liss, NY. (1990)
16. T. Monath, G. Kemp, C. Cropp, and F. Chandler, *J Infect Dis.* 138:59 (1978)
17. T. Hoshino, A Matsumori, C. Kawai, and J. Imai, *Jpn Circul J.* 46:1305 (1982)

ROLE OF HOST AGE AND GENOTYPE IN MURINE ENTEROTROPIC CORONAVIRUS INFECTION

Stephen W. Barthold and Abigail L. Smith

Section of Comparative Medicine
Yale University School of Medicine
333 Cedar Street, P.O. Box 3333
New Haven, Connecticut   06510

Enterotropic mouse hepatitis virus (MHV) is probably the most common form of MHV in contemporary laboratory mouse populations, but very little is known about its pathogenesis. In spite of a large volume of literature on MHV, most of these reports are related to the respiratory prototype strains of virus (1, 2). When several prototype MHV and enterotropic MHV strains were inoculated into infant mice, it became clear that different MHV strains produced two distinctly different patterns of disease: respiratory, in which virus disseminated from nasal epithelium to multiple target organs; or enterotropic, in which virus was largely restricted to intestine, with minimal dissemination (3). However, enterotropic MHV has been poorly characterized because of its fastidious and selective *in vitro* growth characteristics. Host age seems to significantly influence the outcome of disease in mice infected with enterotropic MHV (4-11), and host genotype may also be important in susceptibility to enterotropic MHV disease (12). These are well established factors in the pathogenesis of respiratory or polytropic MHV disease (1, 2). The present study systematically examined the relative susceptibility of inbred mice of different ages that have previously been shown to develop severe (BALB/cByJ) or mild (SJL/J) disease when inoculated oronasally with respiratory MHV-JHM (13). Pilot studies suggested that these mouse genotypes also display differential disease severity when infected with enterotropic MHV, strain Y (12).

MATERIALS AND METHODS

Mice and Virus. Pregnant, 3 week and 12 week old inbred BALB/cByJ (BALB) and SJL/J (SJL) mice were obtained from The Jackson Laboratory (Bar Harbor, ME) and pregnant outbred Crl-CD1BR (CD1) mice were obtained from Charles River Laboratories (Portage, MI). Animal and cage manipulations were performed in a biological containment cabinet.

MHV-Y, isolated in NCTC 1469 cells from the intestine of a naturally infected infant mouse (7), was maintained by serial oral passage of 10% intestinal homogenates in infant CD1 mice. Inocula contained approximately 10^3 median CD1 infant mouse enteritis doses (ED_{50}).

Serology. MHV IgG antibody titers in sera were determined with a modified enzyme immunoassay (EIA) using formalin-fixed MHV-S-infected 17Cl 1 cells as antigen, as described (14). MHV-Y has been shown to be more closely related antigenically to MHV-S than to other prototype MHV strains (7).

Experimental Design and Pathology. The course of MHV-Y infection was examined following oral inoculation of 1, 3 and 12 week old BALB and SJL mice. On days 1, 2, 3, 5, 10, 20 and 30 after inoculation, tissues were collected for virus titration and histology. Virus titers in segments of glandular stomach, duodenum, ileum, cecum, ascending colon, descending colon, mesenteric lymph nodes, Peyer's patches and feces of 3 week old BALB mice were all determined, and based upon these results, virus was quantified in the ascending colons of other groups for comparison. Brain, lung, spleen, liver, cervical lymph node, glandular stomach, duodenum, ileum, cecum, ascending colon, descending colon, mesenteric lymph nodes, and Peyer's patches were also immersion-fixed in neutral buffered formalin, pH 7.2, paraffin-embedded, sectioned and stained with hematoxylin and eosin.

RESULTS

Virus quantification in different levels of intestine, gut-associated lymphoid tissue and feces of 3 week old BALB mice at intervals after oral inoculation revealed minimal or no infectious virus in stomach, but varying levels of infectious virus in other samples. Few samples were positive at 1 day after inoculation, but virus titers rose dramatically by 2 and 3 days, then declined at subsequent intervals. All Peyer's patch samples were positive on days 2 and 3, whereas mesenteric lymph nodes were less consistently infected. Virus was readily detected in feces on days 2, 3 and 5, but declined in parallel with tissues thereafter. These results also showed that virus titers were highest and most sustained (approximately 10^6 ED_{50} per gram on days 2 through 5) in the ascending colon, compared to other sites.

Since ascending colon appeared to be the major target for virus replication and disease, virus was quantified in ascending colons of other mouse ages and strains. Among BALB mice, highest virus titers were found in mice inoculated at 1 week of age ($10^{6.5}$ ED_{50} per gram), with lower peak titers in 3 week old mice and lowest peak titers ($10^{5.2}$ ED_{50} per gram) in 12 week old mice. Remarkably, titers were low at 2 days in the BALB mice inoculated at 1 week of age, whereas titers were at their zenith at this interval in the older mice. In SJL mice, peak virus titers were more or less equivalent among mice of different ages and, except for mice inoculated at 1 week of age, equivalent to peak titers attained in BALB mice of corresponding age. Peak virus titers in 1 week old SJL mice were one $\log_{10}$ less compared to 1 week old BALB mice. Virus titers declined precipitously after peaking in mice of both genotypes and all ages, but a few mice had low levels of detectable virus at the 30 day interval.

Histologic examination supported the infectivity data. In BALB mice inoculated at 3 weeks of age, virtually no lesions were observed in stomach at any interval. On day 2, lesions in upper small intestine were confined to a few syncytia and necrosis of few enterocytes, particularly of M cells overlying gut-associated lymphoid tissue. Syncytia were more apparent in the lower small intestine and cecum, most obvious in the ascending colon and rarely found in the descending colon. These changes were also observed on day 3, but to a much lesser extent. Mucosal hyperplasia was apparent on day 5, with a return to normal by day 10. Inflammation was minimal or absent. No lesions were evident on days 10, 20 or 30.

Histology was also performed on all levels of intestine at all intervals in BALB and SJL mice inoculated at 1, 3 or 12 weeks of age. As in BALB mice inoculated at 3 weeks of age, lesions in BALB and SJL mice inoculated at 1 week of age were most obvious and consistently present in the ascending colon, but also involved the ileum, cecum and, to a minimal degree, descending colon. BALB mice inoculated at 1 week of age had marked distortion of ascending colonic mucosa, due to crypt epithelial hyperplasia, syncytia and inflammation, with mild epithelial necrosis. There was no evidence of mucosal erosion or ulceration. Changes were most apparent on days 2 and 3, but continued with regularity on day 5 before waning at subsequent intervals. Hyperplasia was evident on days 2, 3, 5 and 10. Lymphoid syncytia were observed in mesenteric lymph nodes of most of these mice on day 5. In addition to intestinal lesions, all 1 week BALB mice examined on day 10 had diffuse necrotizing encephalitis, which was accompanied by sudden high mortality around this interval. The few surviving mice had no brain disease on days 20 or 30. Focal hepatitis was also present in most 1 week BALB mice at day 10, as well as sporadically at earlier intervals. These foci had minimal necrosis and no visible syncytia. No significant lesions were seen in other tissues.

In contrast to BALB mice, SJL mice inoculated at 1 week of age had remarkably milder disease. Only a few small syncytia were observed in the cecal and ascending colonic mucosa on day 3, but hyperplasia was seen on days 3, 5, 10 and 20. SJL mice did not develop lesions in other tissues, most notably liver and brain, and mesenteric lymph nodes were involved in only 1 mouse on day 3.

Mice inoculated at 3 and 12 weeks of age developed very mild lesions, restricted largely to the ascending colon. Compared to 3 week old BALB mice (above), SJL mice inoculated at 3 and 12 weeks had a few syncytia visible in ascending colonic mucosa on day 2, but not thereafter, and minimal evidence of compensatory mucosal hyperplasia. BALB mice inoculated at 12 weeks of age developed lesions similar in quality to those inoculated at 3 weeks of age, but slightly milder, with mild hyperplasia on day 5.

Antibody to MHV was generally detected by day 10 in 3 and 12 week old mice and day 20 in mice inoculated at 1 week of age. Antibody titers were lowest in mice inoculated at 12 weeks of age.

DISCUSSION

Although enterotropic MHV is generally associated with disease of the neonate and infant, the present study indicates that significant virus replication occurs in mice of all ages, yet disease is minimal in older mice. In contrast, studies with respiratory strain MHV-JHM demonstrate a close parallel of disease severity with virus titers (13). In considering why the neonatal bowel is preferentially susceptible to disease, but not necessarily infection, Biggers *et al.* (5) suggested that this might be related to mucosal proliferative kinetics. Mucosal turnover rate is sluggish and mucosal epithelial cells are highly differentiated in the neonatal bowel, allowing damage by a rapidly replicating virus with minimal ability to respond. Older mice have less differentiated cells and higher mucosal turnover rates, allowing more rapid replacement of damaged mucosa, resulting in mild disease (15, 16). This is supported by the observation that syncytia, a hallmark of MHV infection, are located primarily in the surface mucosa in older mice, while virus damage occurs throughout the mucosal epithelium of younger mice.

The present study also demonstrates that SJL mice, regardless of age, support significant virus replication in intestinal mucosa, but have milder disease compared to BALB mice. SJL mice have been shown to lack a functional receptor for respiratory MHV-A59, resulting in marked resistance to virus infection and significantly reduced replication in target cells or tissues (17). We have shown previously that SJL mice allow little or no virus replication and develop minimal or no disease in liver and brain when inoculated with MHV-JHM at 3 or more weeks of age. When SJL mice were inoculated at 1 week of age, they developed disseminated infection, but virus replication and disease were reduced compared to BALB mice (13). Therefore, the resistance of the SJL mouse to infection and disease caused by respiratory strains of MHV appears to be age-related, as with MHV-Y, but mediated through an entirely different mechanism than resistance to enterotropic MHV, since they allow significant MHV-Y replication at all ages .

In spite of the high degree of enterotropism displayed by MHV-Y used in this study, we confirmed previous observations (12) that MHV-Y disseminated to brain and liver in infant BALB mice, but not older BALB mice or SJL mice of any age. The hepatitis observed in the infant BALB mice was mild, unlike the severe necrotizing hepatitis commonly observed in susceptible mice infected with respiratory strains of MHV (1, 13). We have also observed mild and transient, rather than progressive, hepatitis in athymic nude mice infected experimentally with enterotropic MHV-RI (18). Virus was rarely detected in livers and lesions were restricted to intestine of mice with lethal intestinal virus of infant mice (LIVIM) (1-3), and no liver disease occurred in infant mice infected with enterotropic MHV-S/CDC (8, 9). In contrast, enterotropic strain MHV-DVIM readily caused hepatitis with virus replication in liver (11). These differences underscore the unpredictably wide range of biological effects and relative tissue tropism of MHV strains and isolates. Clearly, the distinction between respiratory and enterotropic biotypes of MHV is relative, rather than absolute.

Our studies implicate the ascending colon as a key target for MHV-Y, while others (4-6, 8, 9, 11) have emphasized that the lower small intestine appeared to be the primary target for morphological virus damage, with less consistent involvement of cecum and infrequent involvement of colon. These differences may be MHV strain-related, or may be due to failure of others to examine the craniad (ascending) segment of colon. As in the present study, we have observed the ascending colon to be the most consistently affected segment of intestine involved with at least 3 different enterotropic strains of MHV (3, 7, 18). This is our impression of naturally-occurring cases as well. The current study reinforces the importance of examining all segments of intestine, but particularly ascending colon, for MHV diagnosis. The ascending colon may be the only level of intestine with pathognomic lesions in adult, asymptomatic mice.

ACKNOWLEDGEMENTS

This work was supported by grants RR02039 and RR04507 from the Division of Research Resources, National Institutes of Health, Bethesda, MD. The technical assistance of Deborah Beck and Deborah Winograd is appreciated.

REFERENCES

1. Barthold, S.W. 1986. *In* T.C. Hamm, (ed.), Complications of Viral and Mycoplasmal Infections in Rodents to Toxicology Research and Testing, pp.53-89. Hemisphere Publishing Corporation, Washington, DC.

2. Barthold, S.W. 1986. *In* P.N. Bhatt, *et al.* (eds), Viral and Mycoplasmal Infections of Laboratory Rodents: Effects on Biomedical Research, pp. 571-601. Academic Press, New York, NY.

3. Barthold, S.W. and A.L.Smith. 1984. Arch. Virol. 81:103-112.

4. Kraft, L.M. 1962. Science 137:282-283.

5. Biggers, D.C. *et al.* 1964. Amer. J. Pathol. 45:413-422.

6. Kraft, L.M. 1966. NCI Monograph 20:55-61.

7. Barthold, S.W. *et al.* 1982. Lab.Anim. Sci. 32:376-383.

8. Broderson, J.R. *et al.* 1976. Lab. Anim. Sci. 26:824.

9. Hierholzer, J.C. *et al.* 1979. Inf. Immun. 24:508-522.

10. Carthew, P. 1977. Vet. Rec.101:465.

11. Ishida, T. and K. Fujiwara. 1979. Jap. J. Exp. Med. 52:231-35.

12. Barthold, S.W. 1987. Lab. Anim. Sci. 37:36-40.

13. Barthold, S.W. and A.L. Smith. 1987. Virus Res. 7:225-239.

14. Smith, A.L. and D.F. Winograd. 1986. J. Virol. Methods 14:335-343

15. Tsuboi, K.K. *et al.* 1985. Biochim. Biophys. Acta 840:69-78.

16. Tsuboi, K.K. *et al.* 1981. Biochim. Biophys. Res. Comn. 101:645-652

17. Williams, R.K. *et al.* 1990. J. Virol. 64:3817-3823.

18. Barthold, S.W. *et al.* 1985. Lab. Anim. Sci. 35:613-618.

ROLE OF MACROPHAGE PROCOAGULANT ACTIVITY IN MOUSE HEPATITIS VIRUS (MHV) INFECTION: STUDIES USING T CELL MHV-3 CLONES AND MONOCLONAL ANTIBODY 3D4.3

S. Chung[1], C. Li[2], L.S. Fung[2], A. Crow[2], R. Gorczynski[1],
E. Cole[2], S. Perlman[3], J. Leibowitz[4], and G. Levy[2]

Departments of Surgery[1] and Medicine[2]
University of Toronto
The University of Iowa[3]
University of Texas[4]

I. INTRODUCTION

Murine hepatitis virus strain 3 (MHV) infection produces a strain dependent spectrum of disease[1,2]. Mice of the A and SJL strain are fully resistant to the effects of viral infection, whereas mice of semisusceptible strains (C3H/HeJ) develop acute hepatitis which progesses to varying degrees of chronic hepatitis. Mice of fully susceptible strains (Balb/cJ, C57BL/6J) die of fulminant hepatic failure. The resistance of the A strain mice cannot be explained by a lack of a cellular receptor for MHV as viral binding occurs on cells from these resistant mice[3,4]. Furthermore, restriction of viral replication does not explain resistance, as resistance occurs despite the presence of active viral replication in the resistant A strain. Bang and Warwick previously reported that differences in viral replication in cultures of macrophages reflects the relative susceptibility/resistance (S/R) to viral infection[5]. However, several laboratories have shown that MHV replicates in cultures of macrophages, endothelial cells and hepatocytes from both susceptible and resistant animals[4,6-8], although viral replication occurs to a lesser degree in cells derived from resistant animals[6,9]. Thus, absolute differences in viral replication do not account for the strain dependent S/R pattern seen in MHV infection.

Liver disease in this model correlates directly with the induction of macrophage procoagulant activity (PCA) by MHV both *in vivo* and *in vitro*[1,6,10]. Macrophages harvested from the fully susceptible strains of infected animals demonstrate significantly elevated levels of PCA. Furthermore, monocytes and peritoneal macrophages from uninfected mice can be induced to express marked increases in PCA following MHV stimulation *in vitro*. Macrophages from the semisusceptible strains of animals express intermediate levels of PCA, whereas in the fully resistant mice (A/J) which have no evidence of liver disease, the macrophages harvested from these mice express only baseline levels of PCA. *In vitro* stimulation of macrophages from the resistant animals with MHV does not elicit an increase in PCA above baseline. Thus, the induction of PCA by macrophages distinguishes between mice that are susceptible and those that are resistant to MHV infection.

MHV induction of PCA *in vitro* is rapid, with increases being observed within 1-1.5 hours, and maximal activity at 12-18 hours[1]. Viral growth in the susceptible 17 CL1 cell line is not detectable until 6 hours after infection, and not until 18 hours in monocytes/macrophages from susceptible Balb/cJ mice[1]. Therefore, the induction of PCA precedes the replication of infectious virus. Pretreatment of macrophages with actinomycin D and cycloheximide results in inhibition of PCA without affecting viral replication[11]. Thus, the induction of PCA is dependent upon host macrophage RNA and protein synthesis, and is not a viral protein product, although live virus is a prerequisite for its expression.

The induction of PCA in the susceptible animals corresponds to disturbances seen within the microcirculation of the liver[12,13]. Normal flow in the terminal vessels is swift and streamlined. In the sinusoids, erythrocytes and leukocytes traverse these spaces as single cells and travel so quickly that they cannot be seen clearly. Within 6-12 hours post-infection, velocity of flow is diminished, and by 24 hours aggregation of erythrocytes appear. Microthrombi within the microcirculation become prominent which progress to obliterate the sinusoids. By 48 hours, areas of focal necrosis can be seen corresponding to the areas of thrombosis, and by 72 hours, blood flow is marked ly reduced with obvious confluent necrosis. The initial abnormalities seen in the vascular bed precede in vivo viral replication by 24 hours. The deposition of fibrin is a prominent feature within the affected areas. These deposits are present in the hepatic sinusoids as well as in the areas of necrosis. Using a fluorescein-labeled anti-prothrombinase monoclonal antibody[14], increased expression of the enzyme can be seen within 24 hours in areas of inflammation and necrosis, and in the hepatic sinusoids. The prothrombinase is localized primarily on endothelial cells and Kupffer cells, but not on hepatocytes. In contrast, no discernible vascular or histological abnormalities can be seen in the livers of resistant A/J mice[15].

II. Lymphocyte Response to MHV Infection and Regulation of Macrophage PCA *In vitro*

The proliferative response of splenic mononuclear cells (SMNC) from resistant and susceptible mice differs significantly in response to MHV infection[16]. SMNC from naive Balb/cJ mice are completely unresponsive to MHV stimulation, whereas, SMNC from the resistant A/J mice demonstrate an effective proliferative response that peaks at 7-9 days. Priming of resistant mice with MHV resulted in an augmented and earlier response (2-6 days) compared with the naive mice. These differences in the SMNC proliferative response between the resistant and susceptible strains is further emphasized by the absence of IL-2 production by lymphocytes from the Balb/cJ mice when infected with MHV. In contrast, mononuclear cells from resistant A/J mice effectively produce IL-2 after stimulation with MHV.

Recently, T helper cell subpopulations have been identified on the basis of cytokine secretion profile[17,18]. TH1 cells have been shown to produce IL-2 and interferon gamma (IFN-γ), while TH2 cells produce IL-4 and IL-5. Resistant A/J mice have now been shown to produce a predominantly TH1 cell population following MHV infection[19]. TH1 T cell clones generated from mice following immunization with the neurotropic coronavirus JHMV are capable of conveying significant protection to animals infected with the virus[20]. It is possible that the generation of a TH1 population accounts for the elimination of the virus in this strain.

These findings are consistent with other models of resistance and susceptibilty to invading pathogens[21-25]. In leishmaniasis infection, the resistance or susceptibility of mice depends upon the the relative preponderance of the T helper cell populations, with resistance being dependent upon the animal's ability to generate TH1 cells[26]. Treatment of resistant mice with anti-interferon results in a switch of TH1 to TH2 cells, decreased levels of IFN-γ , increased levels of IL-4 and IL-5, and conversion to a susceptible phenotype[27]. Therefore, it is very likely that IFN-γ plays a pivotal role in development of resistance to this infection.

These differences in lymphocyte response to MHV infection are significant in the regulation of PCA[11,16]. When macrophages from susceptible mice are stimulated with MHV in the presence of CD4+ syngeneic lymphocytes, the levels of PCA are 5-6 fold higher than the levels seen if the macrophages were stimulated with MHV alone. However, macrophages from resistant mice continue to express baseline levels of PCA when stimulated with MHV alone or in the presence of lymphocytes from H-2 compatible susceptible mice (Table I). In contrast to the augmentation in PCA seen in macrophages cultured with lymphocytes from susceptible mice, induction of PCA can be abrogated by lymphocytes from H-2 compatible resistant mice (Table II). This inhibition occurs in an antigen specific fashion as only lymphocytes from immunized resistant RI mice downregulated the induction of PCA in macrophages from susceptible mice. Lymphocytes harvested from naive animals did not have any regulatory effect on PCA induction[11].

Table 1. Role of Lymphocytes in Induction of PCA by MHV-3

Lymphocytes	(Macrophage)		
	AXB5 (R)	AXBl(SS)	AXB3)S)
AXB5(R)	690 ± 300	4900 ± 2100	10400 ± 2100
AXBl(SS)	745 ± 210	14800 ± 2900	21800 ± 1600
AXB3(S)	910 ± 200	51500 ± 9100	69400 ± 8400

Data are expressed as mU/10^6 macrophages and is total content PCA. 5 x 10^5 macrophages were stimulated with 10^6 PFU of MHV-3 in the presence of 3 x 10^6 Thy 1.2 + lymphocytes. R, resistant; S, susceptible; SS, semisusceptible

Table 2. Suppression of Induction of PCA in Susceptible AXB3 RI Mice by Lymphocytes from MHV-3 Primed Resistant AXB5 RI Mice

Ratio of Lymphocytes (AXB5/AXB3)	PCA (mU/10^6 macrophages)	
	Source of Lymphocytes	
	Naive	Immunized
0	54150+4200	55450+6100
1	56550+2180	51800+3200
2	53500+2500	40400+2200
3	52000 +3000	17550+ 1800
4	54500+4500	640+550

5 x 10^5 macrophages from susceptible AXB3 mice were stimulated with 10^6 PFU of MHV-3 in the presence of 3 x 10^6 host lymphocytes and increasing numbers of splenic lymphocytes from resistant AXB5 mice which had been primed (immunized) or not primed (naive) with 10^7 PFU of MHV-3 14 days prior to harvesting of lymphocytes. Following a 12 hour incubation cells were frozen and assayed for total content PCA.

Five T cell lines were derived from draining popliteal lymph nodes from resistant A/J mice, which had been immunized with MHV-3. These cell lines were maintained in culture by periodic stimulation with antigen presenting cells, MHV-3 and T cell growth factors. MHV structural proteins were expressed in MC57 cells utilizing recombinant vaccinia virus vectors. Three cell lines stimulated with recombinant full length MHV-3 structural proteins in the presence of syngeneic antigen presenting cells proliferated to the nucleocapsid protein whereas 2 cell lines proliferated only to live MHV-3. By FACS analysis for cell lines were CD4+ and one was CD8+. All CD4+ lines produced IL-2 and two produced IFN-γ (TH1).

One TH1 T cell clone generated from the MHV-immunized animals was shown to downregulate PCA induction[19] (Table III). Furthermore, using a series of transwell cell culture experiments, the inhibitory mechanism of these T cell clones was shown to involve a soluble mediator. However, a cell-to-cell inhibitory action could not be excluded as the induction of PCA was similarly inhibited when the clones and the macrophages were in direct contact. It is possible that IFN-γ plays a role by way of its direct anti-viral effect, or by facilitating macrophage elimination of the virus. For example, the treatment of resistant A/J mice with anti-IFN-γ and other immunosuppressive agents results in susceptiblity of the animal to MHV infection and death of the animal[28].

Table 3. Relationship of Induction of PCA to Development of Hepatitis in Inbred Strains of Mice Following MHV Infection

Strain of Animal	Disease	Induction of PCA
A/J	–	–
C3H	+	+
Balb/cJ	+++	+++

Currently, attempts to produce T cell clones from susceptible Balb/cJ mice have been unsuccessful, due partly to the sensitivity of this strain to MHV which is required for priming, and partly due to the poor proliferative response of the lymphocytes to in vitro stimulation, both to live and UV-irradiated virus. However, it is possible that these animals produce primarily a TH2 cell response to MHV infection, which is consistent with the lack of IL-2 production by the MHV-stimulated lymphocytes[16]. Expansion of TH2 cells results in the production of IL-4 and IL-10[23,29,30] which inhibit the mechanisms necessary for elimination of the virus. This could account for the inability of the Balb/cJ mice to survive MHV infection. Furthermore, Lamontagne has shown that MHV infection in susceptible mice results in thymic atrophy with depletion of all T lymphocyte populations, although there may be a greater depletion of CD4+ CD8- T cells[31]. If there was a greater depletion of TH1 cells, this would support the hypothesis that alterations in regulatory T cells accounts for differences in susceptibility and resistance.

CD4+ T cell lines generated from susceptible C57BL/6 mice are capable of inducing a strong delayed-type hypersensitivity (DTH) reaction[32,33]. As the induction of PCA has been linked to the DTH reaction[1,2], this finding is consistent with the a model of PCA, DTH, and T cell regulation.

III. Inhibition of PCA by Anti-Prothrombinase Monoclonal Antibody

Recently, a panel of monoclonal antibodies (mAb) specific for the MHV-induced procoagulant, prothrombinase, has been produced in our laboratory[25]. These antibodies do not react with rodent, rabbit or human tissue factor (TF), or MHV. Most of the mAb inhibit PCA expression in a one stage clotting assay and inhibited the conversion of prothrombin to thrombin, supporting the concept that MHV-induced PCA is a prothrombinase. However, all of the mAb tested react with proteins of 140, 74, and 70 kD on non-reduced gels, and 74 and 70 kD on reduced gels. (Fig.1). This would suggest that the antibodies recognize different epitopes on the protein, and that most of the epitopes are on the protein's functional structure. The molecular weights of the proteins recognized by these mAb are clearly distinct from TF which is a 47 kD protein[34]. Immunofluorescence studies demonstrate that the mAb bind only to MHV-stimulated macrophages that functionally express elevated levels of PCA.

Infusion of one of the high titred neutralizing mAbs into susceptible mice infected with 10^3 plaque forming units of MHV results in a dose dependent increase in survival, with a 100% survival in animals receiving 100 μg of the mAb. (Fig. 2.) All untreated, MHV-infected mice died within 96 hours. In the mice receiving the mAb, the hepatic necrosis, inflammatory cell infiltrate, and fibrin deposition was markedly attenuated; functional expression of macrophage PCA was markedly decreased, and by immunofluorescence staining, livers from the mAb-treated animals had diminished antigenic expression of prothrombinase in macrophages and endothelial cells in the hepatic sinusoids. In contrast to the livers of untreated mice in which there was exten-

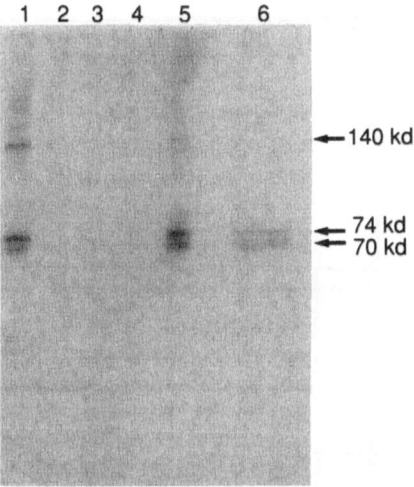

Figure 1. Immunoblot analysis of MHV-3-induced prothrombinase. 1 μg of nonreduced (lanes 1 and 5) or reduced (lane 6) MHV-3-induced macrophage prothrombinase, nonreduced unstimulated macrophages that were devoid of prothrombinase (lane 2), purified MHV-3 (lane 3), or mouse crude tissue factor (lane 4) was resolved by electrophoresis on a 4-20% polyacrylamide gel, transferred to nitrocellulose and reacted with a 1/10 dilution of hybridoma culture supernatant 3D4.3.

sive fibrin deposition in the sinusoids and in the areas of necrosis, animals treated with the mAb had a dose-dependent reduction in the deposition of fibrin. No fibrin was seen in the animals receiving the highest dose of mAb (100 μg) . In the treated animals, there was a rapid fall in the titre of antibody to prothrombinase, consistent with sequestration due to binding of the immunoglobulin to cells expressing the prothrombinase[15].

In animals treated with lower doses of mAb (25 and 50 μg), there were no differences in viral titres from those observed in untreated mice early in the course of the infection. (Fig. 3). However, in the surviving mice, there was complete elimination of the virus by the 10th day post-infection. In the animals that received the highest dose

(100 μg) of mAb, viral titres were significantly reduced in the first 3 days, and approached those titres seen in resistant A strain mice, with elimination of the virus by the fifth day post-infection[15].

One possible explanation for the decrease in the viral replication seen in the treated mice is that the antibodies are reacting with the MHV receptor. This is not likely in that the anti-prothrombinase mAb do not demonstrate any in vitro neutralizing of viral infectivity in plaque reduction assays[14], whereas antibodies to the MHV receptor inhibit infectivity[35]. Furthermore, the MHV receptor has a molecular weight of 110 kD which is different from the prothrombinase molecule (140 kD)[35]. A second possibility for the in vivo inhibition of viral replication in the infected mice by the mAb may be related to the normal cleavage of the MHV S protein. Cleavage of the S

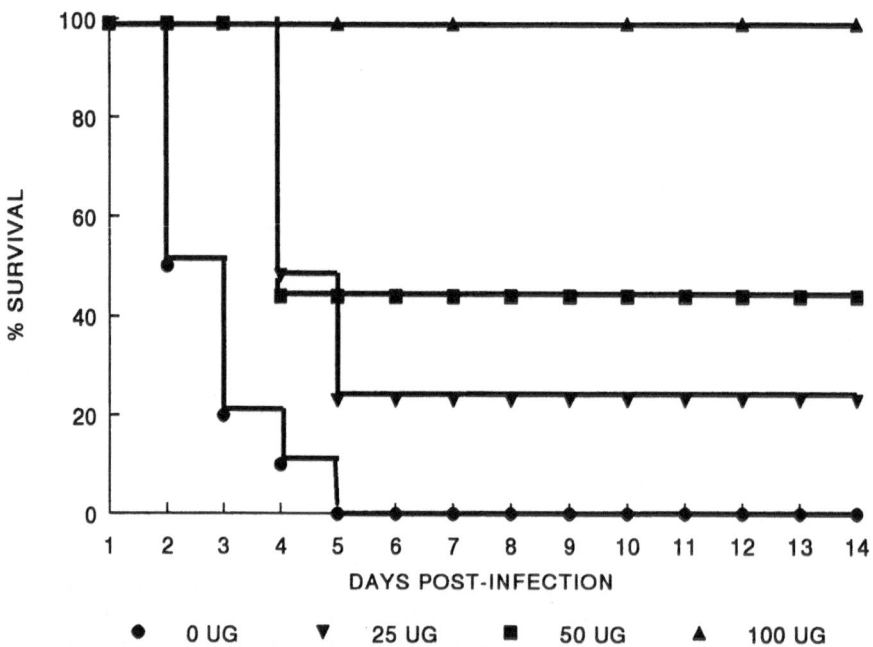

Figure 2. The effect of antibody to procoagulant activity (3D4.3) on survival of mice infected with murine hepatitis virus strain 3 (MHV-3). Mice were either not pretreated (●) or pretreated with 25 (▼), 50 (■), or 100 μg (▲) of mAb 3D4.3 for 7 days before infection with 10^3 PFU of MHV-3. Antibody was continued in treated animals for 7 days p.i. Survival was then studied (n=9/group).

protein by proteases is necessary to activate the membrane fusing properties of the S protein, and thus the virus itself[36,37]. This fusion property facilitates the spread of virus to uninfected cells by cell-cell fusion. It is possible that the MHV-induced prothrombinase, a serine protease, mediates at least in part the proteolytic cleavage of the S protein. This is consistent with the observations that infection of macrophages from A strain mice, which does not result in induction of PCA, does not result in syncitia formation, and that inhibition of PCA of macrophages from Balb/cJ mice with PGE also reduces syncitia formation[9,38].

The mechanism of the protective effect of the anti-prothrombinase mAb is not entirely known. However, utilization of the mAb results in the inhibition of PCA which correlates with the prevention of activation of the immune coagulation system and deposition of fibrin. In concordance with these observations are the findings of Taylor et al. who attenuated the effects of septic shock using mAb against TF.[39] Thus, it seems reasonable to conclude that interference and inhibition of the action of PCA plays an important role in abrogation of the disease.

IV. CONCLUSION

The interactions between the virus and the host result in a complex series of events that result in either resolution of the infection, or death of the host. Differences in viral processing, generation of second messengers, and production of inflammatory mediators and cytokines contribute to the disease outcome.

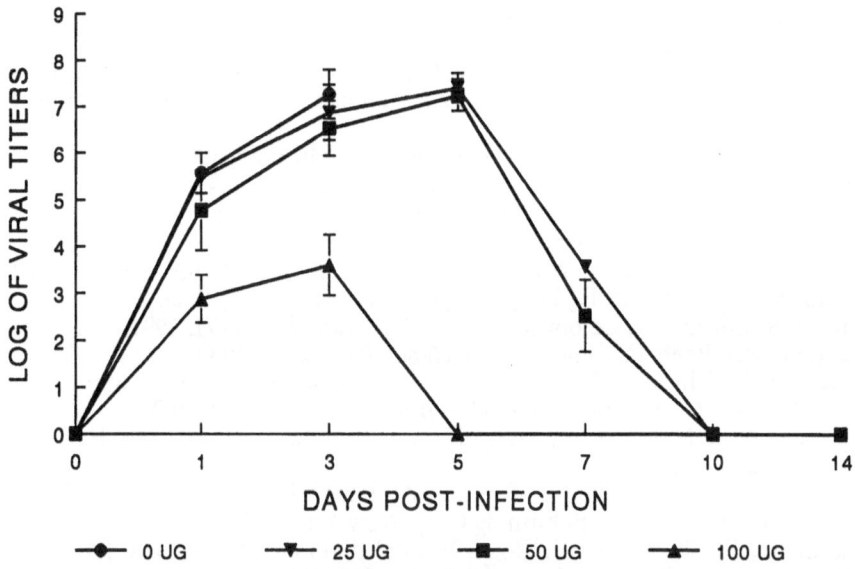

Figure 3. The effect of treatment with mAb to PCA on viral replication. High titers of virus were recovered from untreated and MHV-3-infected animals at all time points (●). In contrast, in a dose-dependent fashion, mAb 3D4.3 at concentrations of 25 (▼), 50 (■), and 100 μg (▲) attenuated the titer of virus recovered from the liver. In the animals treated with 25 and 50 ug of 3D4.3, no difference in viral titer was seen as compared to untreated mice on days 1, 3, and 5. However, titers of virus fell by day 7 and were not detected on day 10. Animals treated with 100 ug of 3D4.3 had statistically significantly lower levels of virus in thier livers at all time points studied as compared with MHV-3-infected and untreated animals. Statistical significance, $p < 0.05$.

The activation of the immune coagulation system with the induction of monocyte/macrophage PCA by MHV has been shown to be important in the pathogenesis of MHV-induced hepatitis. Modulation of PCA can result in both abrogation of the liver pathology, and survival of the susceptible animal. Prevention of mortality in MHV-infected susceptible mice using a specific anti-prothrombinase mAb that neutralizes PCA underscores the importance of PCA in the pathogenesis of MHV-induced fulminant hepatitis. Although PGE also inhibits PCA and prevents liver necrosis in MHV-infected mice, it has many other non-specific immunosuppressive properties which may contribute the paradoxical observation of mortality despite hepatoprotection.

Regulation of MHV-induced PCA resides at the level of both macrophage and lymphocyte. Absolute restriction for the induction of PCA is at the level of the macrophage but T lymphocyte cooperation is required for final expression of PCA.

The relative balance of TH1/TH2 T cell subpopulations in response to MHV infection may be an important determinant in resistance or susceptibility and development of liver disease. Methods of altering the relative T cell subpopulations could influence the induction of macrophage PCA and affect the outcome of MHV infection.

Information gained in the study of this model will further the understanding of the mechanisms of induction and regulation of PCA during MHV infection. As the expression of PCA by macrophages and endothelial cells has been implicated in the pathogenesis of several human diseases, knowledge gained will ultimately lead to better treatment of diseases in which PCA has been shown to be involved.

V. REFERENCES

1. Levy, G., Leibowitz, J., Edgington, T. *J. Exp. Med.* 154,1150, 1981.
2. Virelizier, J., Allison, A. *Arch. Virol.* 150,279, 1976.
3. Boyle, J., Weismiller, D., Holmes, K. *J. Virol.* 61,185, 1987.
4. Holmes, K., Williams, R., Stephenson, C et al. *Cell Biology of Virus Entry, Replication and Pathogenesis.* R. Compans, A. Melinius, M. Oldstone., eds Alan Liss, New York, 1989, 85.
5. Bang, F., Warwick, A. *Proc. Nat. Acad. Sci. USA.* 46,1065, 1960.
6. Dindzans, V., MacPhee, P., Fung, L. et al. *J. Immunol.* 135,4189, 1985.
7. Pereira, C., Steffan, A., Kirn, A. *J. Gen. Virol.* 65,1617, 1984.
8. Macnaughton, M., Patterson, S. *Arch. Virol.* 66,71, 1980.
9. Sinclair, S., Abecassis, M., Wong, P.Y. et al. *Adv. Exp. Med. Biol.* 276,533 1990.
10. Dindzans, V., Skamene, E., Levy, G. *J. Immunol.* 137,2355, 1986.
11. Chung, S., Sinclair, S., Leibowitz, J. et al. *J. Immunol.* 146,271, 1991.
12. Levy, G., MacPhee, P., Fung, L. et al. *Hepatology.* 3,964, 1983.
13. MacPhee, P., Dindzans, V., Fung, L., Levy, G. *Hepatology.* 5,649, 1985.
14. Fung, L., Neil, G., Leibowitz, J. et al. *J. Biol. Chem.* 266,1789, 1991.
15. Li, C., Fung, L.S., Crow, A., Myers-Mason, N., Leibowitz, J., Cole, E., Levy, G. *J. Exp. Med.* 1763, 1992.
16. Dindzans, V., Zimmerman, B., Sherker, A., Levy, G. *Coronaviruses* 1987. M. Lai, S. Stohlman eds. Plenum Publishing Corp. New York 1987, 411.
17. Mosmann, T., Cherwinski, H., Bond, M. et al. *J. Immunol.* 136,2348, 1986.
18. Mosmann, T., Coffman, R. *Immunol. Today.* 223, 1987.
19. Chung, S., Nisbet-Brown, E., Gorczynski, R., Skamene, E., Perlman, S., Leibowitz, J., Fung., L.S., Crow, A., Myers-Mason, N., Flowers, M., Levy, G. *J. Immunol.* Submitted, 1992.
20. Korner, H., Schleiphake, A., Winter, J et al. *J. Immunol.* 2317, 1991.
21. Scott, P., Natovitz, P., Coffman, R et al. *J. Exp. Med.* 169,59, 1988.
22. Fidel, P., Boros, D. *J. Immunol.* 146,1941, 1991.
23. Malefyt, R., Haanen, J., Spits, H. et al. *J. Exp. Med.* 174,915, 1991.
24. Caulada-Benedetta, Z., Al-Zamel, F, Sher, A., James, S. *J. Immunol.* 146,1655, 1991.
25. Scott, P., Kaufmann, S. *Immunol. Today.* 12,346. 1991.
26. Locksley, R., Scott, P. *Immunol. Today.* 12,A58, 1991.
27. Scott, P., Natovitz, P., Coffman, R. et al. *J. Exp. Med.* 169,59,1988.
28. Virelizier, J., Gresser, I. *J. Immunol.* 120,1616, 1978.
29. Fiorentino, D., Zlotnik, A., Vieira, P. et al. *J. Immunol.* 146,3444, 1991.
30. Swain, S., Weinberg, A., English, M., Huston, G. *J. Immunol.* 145,3796, 1990.
31. Lamontagne, L., Jolicoeur, P. *Immunol.* 146,3152, 1991.
32. Woodward, J., Matsushima, G., Frelinger, Stohlman, S. *J. Immunol.* 133,1016, 1984.
33. Stohlman, S., Matsushima, G., Asteel, N., Weiner, L. *J. Immunol.* 136,3052, 1984.
34. Morrissey, J., Fair, D., Edgington, T. *Thromb. Res.* 52,247, 1988.
35. Smith, A., Cardellichio, C., Winograd, M. et al. *J. Infect. Dis.* 163,879, 1991.
36. Sturman, L., Ricard, C., Holmes, K. *J. Virol.* 56,904, 1985.
37. Frana, M., Behnke, J., Sturman, L., Holmes, K. *J. Virol.* 56,912, 1985.
38. Abecassis, M., Falk, J., Makowka, L. et al. *J. Clin. Invest.* 80,881, 1987.
39. Taylor, F., Chang, A., Ruff, W. et al. *Circ. Shock.* 33,127, 1991.

EARLY CELLULAR EVENTS IN THE INDUCTION OF MURINE HEPATITIS VIRUS (MHV-3) INDUCED MACROPHAGE PROCOAGULANT ACTIVITY (PCA)

S. Chung[1], G. Downey[2], O. Rotstein[1], G. Levy[2]

Departments of Surgery[1] and Medicine[2]
University of Toronto, Toronto, Canada

INTRODUCTION

Interaction of a membrane receptor with its ligand results in the generation of second messengers which ultimately leads to alteration in cell function either by changes in nuclear transcription, or activation/inactivation of cellular enzymes by phosphorylation events. Although the cell-surface MHV receptor has been cloned and sequenced[1], the biochemical events leading to the expression of MHV-3 induced PCA after the binding of the virus to its receptor are unknown.

The mechanism of induction of tissue factor (TF), a distinct and separate procoagulant from the MHV induced prothrombinase has been explored by several investigators and is controversial. Lyberg and Prydz have demonstrated that induction of TF by human monocytes is a protein kinase C (PKC) dependent event, as TF could be induced by phorbol esters[2]. Furthermore, despite the findings that TF induction was calcium dependent and that the calcium ionophore A23187 was capable of inducing TF[3,4], no detectable changes in cytosolic calcium were associated with TF induction[3]. In contrast, Kucey et al. suggest that calcium ionophores and phorbol esters do not result in TF expression[5]. The differences in these results may be related to the cell populations studied, species of origin (human or rodent) and/or the incubation and culture conditions.

Experiments examining the early cellular events in the induction of the PCA by MHV-3 have recently been conducted, based upon the observation that 16,16 dimethyl prostaglandin E_2 (dmPGE) inhibits the induction of macrophage PCA[6,7,13]. As the actions of dmPGE are thought to de dependent upon an increase in intracellular cAMP[8,9], the effects of two agents which raise intracellular cAMP, forskolin, an adenylate cyclase agonist[10], and isobutylmethylxanthine, a phosphodiesterase inhibitor[11], have been examined. Both of these agents were able to attenuate the induction of macrophage PCA by MHV-3, consistent with the inhibitory action of dmPGE (Fig. 1.). Thus, cAMP would appear to have a downregulatory action on prothrombinase expression. Interestingly, although prothrombinase is functionally inhibited by dmPGE, by Western immunoblot and immunofluorescence analysis, antigenic expression is still present. This suggests that there are post-translational modifications to the protein which abrogate its activity as opposed to inhibition of transcription or translation. The contribution of other prostanoids in the regulation of PCA induction is less clear. Prostacyclin (PGI_2) has some inhibitory effects, but only at relatively high concentrations (100 uM), whereas PGF_{2a} has no effect on PCA induction, even at high concentrations[13] (Fig. 2).

Although macrophages produce leukotriene B_4 (LTB_4) when stimulated with MHV-3[13], it does not appear that LTB_4 is a prerequisite for induction of PCA. At a time when induction of PCA is evident (2 hours), LTB_4 levels are minimal, and LTB_4 by itself cannot induce PCA, nor does it augment the activity in MHV-stimulated macrophages. PCA induction is not inhibited by the specific LTB_4 antagonist MK886. However, the 5-lipoxygenase inhibitor nordihydroguaretic acid (NDGA) does inhibit induction of PCA. This would suggest that leukotrienes may be involved in the induction of PCA, although as NDGA is not a specific agent, it is possible that its action may involve inhibition and/or scavenging of oxygen free radicals.

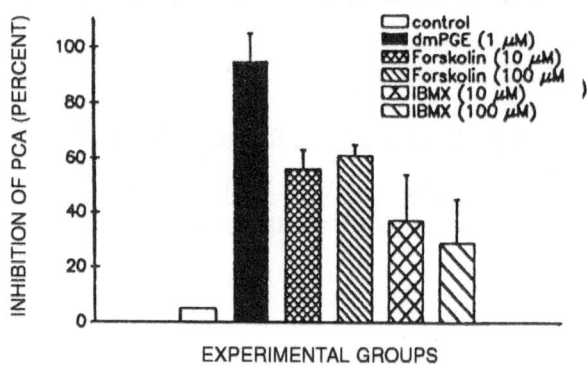

Figure 1. Inhibitory effects of agents that modify cAMP on the induction of procoagulant activity by MHV-3. 1×10^6 macrophages were treated with dmPGE (1 um), forskolin (10-100 uM) or IBMX (10-100 uM) 30 min. prior to infection with 1×10^6 PFU of MHV-3. Cells were incubated a further 6 hours, washed, frozen, and assayed for PCA in a one stage clotting assay.

The role of G-binding proteins, a heterotrimeric family of related membrane proteins that are involved in the coupling of cell surface receptors to effector enzymes[14], has been examined using pertussis toxin (PT). PT is a bacterial toxin that acts upon the *alpha* subunit of the G-binding protein resulting in ADP-ribosylation[14]. This covalent modification results in interference with the interaction between the G-binding protein and the membrane surface receptor. PT significantly inhibits induction of PCA by MHV in macrophages from susceptible Balb/cJ mice suggesting that G-binding proteins play an important role in prothrombinase activity (Fig. 3.). The recent availability of specific probes for specific G-binding proteins[15] will allow for the examination for constituitive differences in G-binding proteins in cells from resistant and susceptible strains of mice.

Calcium transients, however, do not appear to be involved in MHV-induced PCA. Following infection of macrophages by MHV, there are no detectable changes in the concentration of intracellular calcium, as measured fluorometrically using the dye Indo 1-AM. The calcium ionophores ionomycin and A23187 do not induce appreciable levels of PCA within the time period that PCA is induced by MHV (1-4 hours). However, MHV cannot induce an increase in PCA in macrophages cultured in calcium-free media, thus defining an absolute requirement for calcium. (Fig. 4).

Protein kinase C (PKC) is a calcium-dependent enzyme that is activated following the hydrolysis of cell membrane inositol phospholipids. The phorbol ester phorbol myristate acetate (PMA) had no effect on the induction of PCA in unstimulated macrophages from Balb/cJ mice following short incubation periods (1-4 hours). In contrast, inhibitors of PKC, staurosporine and H7, significantly inhibited MHV-induced PCA suggesting that PKC plays a role in its expression.

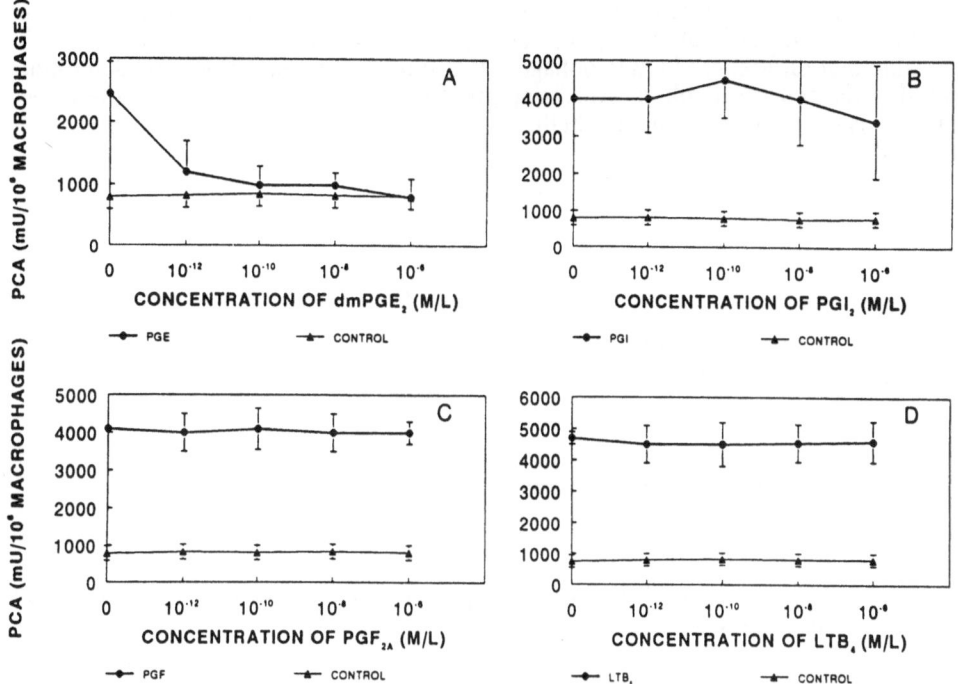

Figure 2. Ability of eicosanoids to inhibit MHV-3 induced procoagulant activity. 1×10^6 macrophages were pretreated with varying concentrations of dmPGE, PGI_2 or PGF_{2a} 1 hour prior to infection with 1 $\times 10^6$ PFU MHV-3. The cells were incubated for 12 hours, then assayed for PCA in a one stage clotting assay. A=treatment with dmPGE; B=treatment with PGI_2; C=treatment with PGF_{2a}; D=treatment with leukotriene B_4.

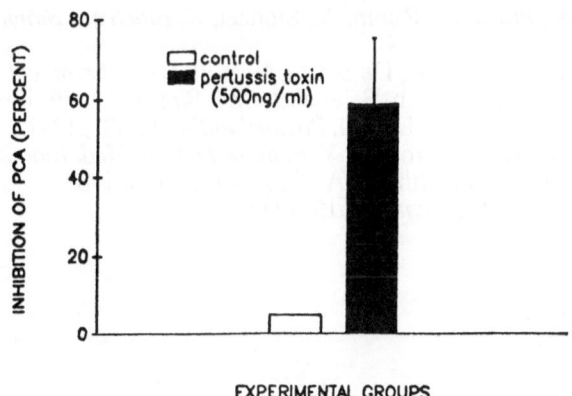

Figure 3. Inhibitory effect of pertussis toxin on the induction of procoagulant activity by MHV-3. 1×10^6 macrophages were treated with 500 ng/ml pertussis toxin for 1 hour prior to infection with MHV-3 (10^6). Cells were incubated a further 6 hours, then washed, frozen and assayed for PCA in a one stage clotting assay.

Clearly, much work remains in the elucidation of the biochemical events required for the induction of PCA in response not only to MHV, but also other stimuli. It is possible that differences in macrophage signal transduction pathways may determine whether the cell is capable of expressing procoagulant activity, and these differences may translate into determining whether an organism is resistant or susceptible to a particular stimulus.

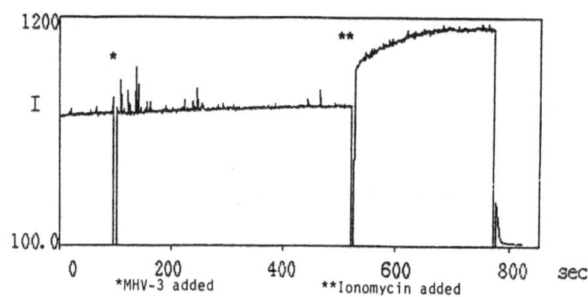

Figure 4. Cytosolic calcium $[Ca^{2+}]_i$ changes in response to stimulation by MHV-3 (M.O.I. of 1.0). $[Ca^{2+}]_i$ was determined as described.

REFERENCES

1. Dveksler, G., Pensiero, M., Cardellichio, C. et al. *J. Virol.* 65,6881, 1991.
2. Lyberg, T., Prydz, H. *J. Biochem.* 194,699, 1981.
3. Prydz, H., Lyberg, T. *Biochem. Pharmacol.* 29,9, 1980.
4. Bach, F., Ifkin, D. *Proc. Natl. Acad. Sci. USA.* 87,6995. 1990.
5. Kucey, D., Cheung, P., Rotstein, O. *Infect Immun.* In press, 1992.
6. Abecassis, M., Falk, J., Makowka, L. et. al. *J. Clin. Invest.* 80,881, 1987.
7. Edgington, T., Pizzalato, M. *Thromb. Haemostasis.* 50,148, 1983.
8. Goodwin, J., Bromberg, S., Mesner, R. *Cell. Immunol.* 60,298, 1981.
9. Novogrodsky, A., Patya, A., Rubin, A., Stenzel, K. *Biochem. Biophys. Res. Comm.* 114,93, 1983.
10. Laurenza, A., Khandelwal, Y., De Souza, N. et al. *Mol. Pharmacol.* 32,133 1987.
11. Downey, G., Elson, E., Schwab, B., et al. *J. Cell. Biol.* 114,1179, 1991.
12. Chung, S., Sinclair, S., Fung, L. et al. *Prostaglandins* 42,501, 1991.
13. Sinclair, S., Abecassis, M., Wong, P.Y. et al. *Adv. Exp. Med. Biol.* 276,533, 1990.
14. Freissmuth, M., Casey, P., Gilman, A. *FASEB J.* 3,2126, 1989.
15. Gilman, A. *Ann. Rev. Biochem.* 56,615, 1987.

REGULATION OF THE EXPRESSION OF INTERCELLULAR ADHESION MOLECULE-1 (ICAM-1) AND THE PUTATIVE ADHESION MOLECULE BASIGIN ON MURINE CEREBRAL ENDOTHELIAL CELLS BY MHV-4 (JHM)

Jeymohan Joseph[1], Robert L. Knobler[1], Fred D. Lublin[1], and Frank R. Burns[2]

[1]Division of Neuroimmunology, Jefferson Medical College, Philadelphia, PA 19107 and [2]Department of Neurology, University of Pennsylvania Medical Center, Philadelphia, PA 19104, USA

INTRODUCTION

Cerebral vascular endothelial cells are integral components of the blood-brain barrier and are believed to be an important site which can restrict infection of the central nervous system by viruses and other pathogens. A potential role of cerebral vascular endothelial cells in resisting CNS infection relates to the expression of lymphocyte-endothelial adhesion molecules which could impact on the generation of anti-viral immune reactivity.

We report here on the regulation of expression of two adhesion related molecules, ICAM-1 and Basigin, following MHV-4 infection. ICAM-1 is a cell surface glycoprotein of the immunoglobulin superfamily expressed on endothelial cells and is involved in interactions with lymphocytes and neutrophils (1, 2). Basigin is the mouse homologue of the chicken antigen HT7, which is expressed on endothelial cells of the blood-brain barrier (3). This antigen is also a member of the immunoglobulin superfamily like ICAM-1 and may play a role in cell adhesion.

METHODS AND RESULTS

Virus Infection of Endothelial Cell Cultures

Cerebral endothelial cell lines were generated from BALB/c (MHV-susceptible) and SJL (MHV-resistant) mice as previously described (4). Endothelial cells grown in T-25 flasks (1 x 10^6 cells/flask) were treated with MHV-4 or UV-MHV-4 for 1 hour at 37°C using a multiplicity of infection (MOI) of 0.1. At selected times after infection the endothelial cell cultures were processed for flow cytometry and Northern Analysis to detect ICAM-1 and Basigin.

Coronaviruses, Edited by H. Laude and J.F. Vautherot
Plenum Press, New York, 1994

Flow Cytometry to Detect ICAM-1

Endothelial cells from infected or paired uninfected cultures were labeled with monoclonal rat anti-mouse ICAM-1 (YN/1.7, obtained from Dr. Fumio Takei, Terry Fox Cancer Center, University of British Columbia, Vancouver, Canada). Fluorescein conjugated goat anti-rat IgG (Organon Teknika-Cappel) was used as a secondary reagent. The percentage of positive cells and mean fluorescence intensities were determined by analysis on the flow cytometer (EPICS C, Coulter Diagnostics, Hialeah, FL, USA) equipped with an argon laser tuned to 488nm. Both MHV-4 and UV-inactivated MHV-4 exposure resulted in a 60% decrease in ICAM-1 expressing endothelial cells (Fig.1)

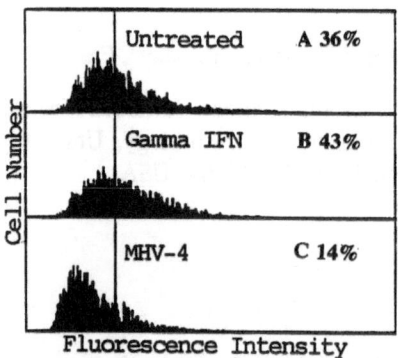

Fig. 1 Fluorescence histograms of ICAM-1 expression on BALB/c cerebral endothelial cells. Flow cytometry profile; X-axis: fluorescence intensity, Y-axis: cell number. Gate windows for green fluorescence lay between channels 0 and 255. (x-axis). Gate windows for red fluorescence was set for exclusion of non viable cells by ethidium bromide. A: Untreated, B: γ-IFN treatment for 72 hours. C: MHV-4 (JHM) treatment for 72 hours.

Northern Analysis to Detect Basigin mRNA Following Virus Infection of Endothelial Cells

BALB/c and SJL derived cerebral endothelial cells were exposed to MHV-4 or UV-inactivated MHV-4 as described above for 24 hours. After these treatments total cellular RNA was isolated from the endothelial cells by guanidinium isothiocyanate/cesium chloride method. RNA was electrophoresed in denaturing formaldehyde gels and blotted by capillary transfer to a zeta probe membrane (Biorad, San Francisco, CA) and hybridized with ^{32}P labeled Basigin cDNA probe (67^{0}C in 7% SDS, 0.5M sodium phosphate, pH 7.2). The blots were washed under high stringency conditions and analyzed by autoradiography. Fig. 2 shows a Northern blot of the results obtained with the Basigin cDNA probe.

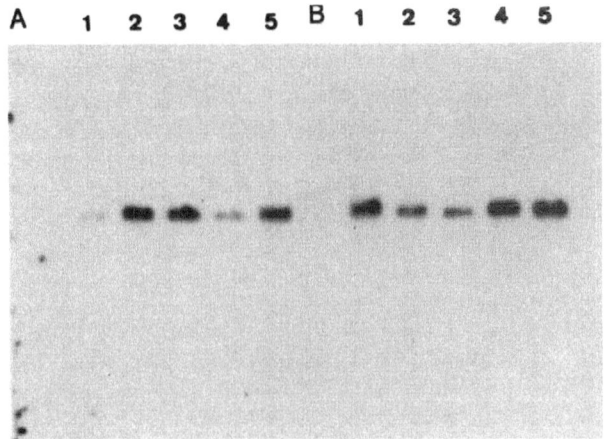

Fig. 2 Northern Blot Analysis of Basigin mRNA induction. Panel A: SJL cerebral microvascular endothelial cells. Panel B: BALB/c cerebral microvascular endothelial cells. Lane 1: Untreated, Lane 2: MHV-4, Lane 3: UV-MHV-4, Lane 4: Tumor necrosis factor, Lane 5: γ-Interferon.

The exposure of brain endothelial cells to MHV-4 or UV-inactivated MHV-4 resulted in dramatic changes in the expression of Basigin mRNA. In MHV-resistant SJL derived brain endothelial cells there was a 240% and 270% increase in Basigin mRNA 24 hours after treatment with MHV-4 and UV-inactivated MHV-4 respectively, as determined by Northern Analysis. In MHV-susceptible BALB/c derived brain endothelial cells, there was a 37% and 26% decrease in Basigin mRNA 24 hours after treatment with MHV-4 and UV-inactivated MHV-4 respectively (These percent changes were calculated by from densitometry analysis of Northern blot, Fig.2).

The mechanism of regulation of brain endothelial cell adhesion molecules by MHV-4 and the impact of changes in their expression on anti-viral immune reactivity is an area of further study.

REFERENCES

1. Springer, T.A., 1990, Nature., 346:425.
2. Sobel, R.A., Mitchell, M.P., and Fondren, G., 1991, Am. J. Pathol., 136:1309
3. Miyauchi, T., Kenekura, T., Yamaoka, A., Ozawa, M., Miyazawa, S., and Muramatsu, T., 1990, J. Biochem., 107:316.
4. Rupnick, M.A., Carey, A., and Williams, S.K., 1988, In Vitro Cell. Dev. Biol., 24:435.

CORONAVIRUSES FROM PERSISTENT INFECTED CELL CULTURES AND BRAIN TISSUE: MOLECULAR ANALYSIS OF THE S GENE OF AN AVIRULENT VARIANT

Albert Stühler, Egbert Flory, Hanna Wege, and Helmut Wege

Institute of Virology and Immunobiology
Versbacher Strasse 7, W-8700 Würzburg, Germany

Results from an increasing number of studies indicate, that structural features of the viral surface proteins S (spike) and HE (hemagglutinin-esterase) determine to a great extend the neurovirulence and pathogenicity of murine coronaviruses. The peplomer protein S elicits neutralizing antibodies, binds to target cell receptors and causes cell fusion. Furthermore, the expression of HE-protein varies strongly between MHV-JHM variants with different virulence and the biological role of this protein is not well understood.

Therefore, we initiated studies employing polymerase chain reaction technology (PCR) to characterize the phenotype and expression of this proteins directly in the brain tissue of mice and rats in different disease stages. As a basis we describe here the molecular properties of an avirulent variant (MHV-JHM-Pi). This virus was isolated from a persistently infected Sac(-) cell line and can be distinguished from MHV-JHM wildtype by monoclonal antibodies which bind to defined epitopes of S, a lack of HE-protein and a highly specific tropism for astrocytes [1,2]. MHV-JHM-Pi replicates in the brains of adult mice and rats without inducing acute disease. Cases of a late demyelinating disease were observed, if newborn rats were infected. Infection of suckling mice leads to an acute disease, but within one week a complete resistance is established.

We sequenced the S-protein gene of MHV-JHM-Pi after amplification of a 4 kb-fragment including the S gene of this virus. To avoid PCR errors and to obtain an accurate full-length consensus sequence, the RNA of three independent plaque-purified virus clones was determined. Only changes detected consistently were considered as "bona fide" mutations. The S gene containing 3706 nucleotides encoding a protein of 1235 amino acid residues, has 13 changes in the predicted amino acid sequence relative to the MHV-JHM wildtype [3,4] (Fig.1). We observed only one silent substitution (Data not shown). Frameshift, deletions or insertions and non sense mutations were not observed. The homology to wildtype S-protein is about 99%.

The alterations in the S sequence of the variant are restricted to different domains which are involved in virus neutralization and cell fusion. At the amino-terminal end of the S1 subunit we found 7 amino acid changes. This region is an important domain for cell fusion activity. Within this region binds a monoclonal antibody, which inhibits cell fusion [5,6]. The substitution at AS 441 (Thr - Cys) is within the hypervariable carboxy-terminal region of the S1 subunit [4]. In front of the proteolytic cleavage site at AS 609 we identified a mutated glycosylation site (Asn - Asp). Such alterations can affect protein folding and are responsible

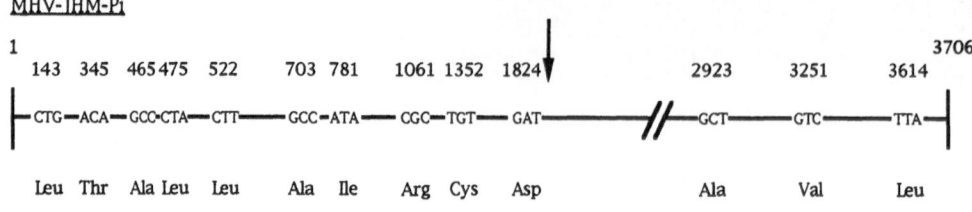

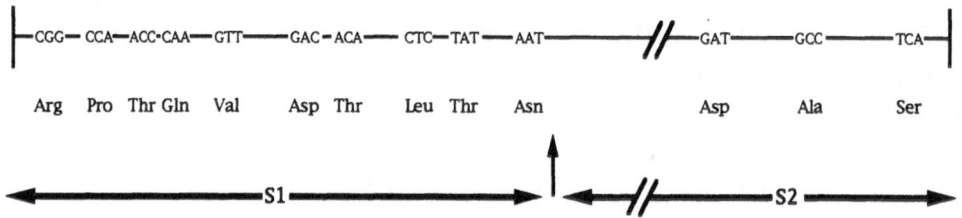

Figure 1. Schematic summary of the S gene sequence divergence observed among MHV-JHM variants. Numbers on the upper left and right corners define the beginning and the end of the S gene nucleotide position. Other numbers indicate the nucleotide positions of point mutations. Vertical arrows indicate the cleveage between amino acid 623 (Ser) and 629 (Ser).

for changes of conformational antigenic binding sites. In the S2 subunit three point mutations (AS position 975, 1084 and 1205) were detected, which are also localized in a domain involved in cell fusion.

Our results indicate that the neuropathogenicity of MHV-JHM depends on the characteristics of at least three different sites of the S-protein. Previous investigations have shown that the MHV S-protein is a critical determinant for viral tropism and pathogenicity in mice and rats [7,8,9,10]. Preservation of the parental (wildtype) sequences at these three regions appears to be required for the ability of virus to cause fatal encephalomyelitis in mice, while mutations at either site attenuated neurovirulence. Most sequence data on genetically highly variable coronavirus genes are based on tissue culture adapted virus or artifical constructs. Not much information is yet available on the phenotype of proteins expressed in specific cell types of the diseased animals.

Further attention will have to be directed to these 13 amino acid substitutions localized on the S-protein, as well as to other viral structural proteins such as the haemagglutinin-esterase (HE) protein which may also contribute to neurovirulence.

ACKNOWLEDGEMENTS

This work was supported by the Hertie-Stiftung and Deutsche Forschungsgemeinschaft.

REFERENCES

1. H.N. Baybutt, H. Wege, M.J. Carter, and V. ter Meulen, J. Gen. Virol. **65**: 915-924, (1984).
2. P.T. Massa, H. Wege, and V. ter Meulen, Virus Research **9**: 133-144, (1988).
3. I. Schmidt, M.A. Skinner, and S.G. Siddell, J. Gen. Virol. **68**: 47-56, (1987).
4. S.E. Parker, T.M. Gallagher, and M.J. Buchmeier, Virol. **173**: 664-673, (1989).
5. E.R. Routledge, R. Stauber, M. Pfleiderer, and S.G. Siddell, J. Virol. **65**(1): 254-264, (1991).
6. A. Stühler, H. Wege, and S.G. Siddell, J. Gen. Virol. **72**: 1655-1658, (1991).
7. R.G. Dalziel, P.W. Lampert, P.J. Talbot, and M.J. Buchmeier, Virus Research **9**: 133-144, (1986).
8. J.O. Fleming, M.D. Trousdale, F.A.K. El-Zaatari, S.A. Stohlman, and L.P. Weiner, J. Virol. **58**: 869-875, (1987).
9. F.I. Wang, J.O. Fleming, and M.M.L. Lai, Virol. **186**: 742-749, (1992).
10. H. Wege, J. Winter, and R. Meyermann, J. Gen.Virol. **69**: 87-98, (1988).

CHARACTERIZATION OF IBV VARIANT STRAIN PL 84084 ISOLATED IN FRANCE

L. Bonnefoy [1] , JF. Bouquet [1] , JP. Picault [2] , G. Chappuis [1]

[1]Rhône Mérieux, 254 rue M. Mérieux, 69007 Lyon, France
[2]CNEVA, Laboratoire Central de Recherches Avicoles et Porcines
22440 Ploufragan, France

INTRODUCTION

Infectious Bronchitis was caused by a Coronavirus which affects the respiratory and genital tracts, and the renal system of chickens. The disease was responsible for economic losses in the poultry industry.

Since the eighties, a number of strains have been identified as antigenically different from the original Massachussetts (Mass) type, based on cross neutralisation tests. These "variant" strains were responsible for severe symptoms of the disease in spite of the use of Mass type vaccine.

In France. several "variant" strains have been isolated and identified by the CNEVA-LCRAP of Ploufragan. One of these strains, the strain PL 84084 was responsible for severe drop in eggs production and also induced respiratory symptoms which were reproduced experimentally.

Monoclonal antibodies (McAb) against the PL 84084 IBV strain were prepared in order to characterize the antigenic structure of the virus and to develop new methods for diagnostic of IB infection.

MATERIAL AND METHODS

1 - IBV strains
The PL 84084 virus strain was isolated and adapted to embryonnated eggs by the CNEVA-LCRAP of Ploufragan.

2 - Production and characterization of McAb
McAB were produced using Balb/c mice immunized by the PL 84084 crude virus in the allantoïc fluid. The McAb's were characterized by Indirect Immunofluorescence (IF) and Western-blotting technique using an antigen purified by ultracentrifugation on a sucrose gradient.

Coronaviruses, Edited by H. Laude and J.F. Vautherot
Plenum Press, New York, 1994

3 - The specificity of McAb

The specificity of McAb in relation to other Avian Coronavirus strains was studied by Western-blotting and IF on infected Chicken Kidney cells.

4 - ELISA test

A competitive "sandwich" ELISA technique was developed with spike glycoproteins. The McAb couples were selected as catching and detecting antibodies. These detecting antibodies compete with the antibodies found in the serum of infected chickens.

5 - Immunohistochemistry

The PL 84084 virus strain was detected in infected 5 week old chicks by taking a cryomicrotom section of tissue. The virus infection was shown by a McAb-Avidin-Biotin Immuno Peroxydase Complex (ABC) on the organ section sample.

RESULTS

Among the 26 characterized McAb against the PL 84084 virus strain, 21 were specific for the nucleoprotein epitopes and 5 for the spike glycoprotein epitopes (Figure 1).

Compared with IF test, the 3 strains isolated in France were found very closed related. One McAb n° 8 seems particulary interesting because it is directed against the specific and neutralising spike epitope of the PL 84084 virus strain.

A spike specific sandwich ELISA test was developed using the McAb n° 9 as catching antibody and the McAb n° 8 as detecting antibody. This test can be applied to the detection of neutralising and PL 84084 specific spike antibodies in the chick serum (figure 2).

Some McAb (n° 9 spike specific and n° 29 nucloprotein specific) can be used for the identification of PL 84084 virus strain in the isolates and in the tissues of infected chickens by an immunochemical technique.

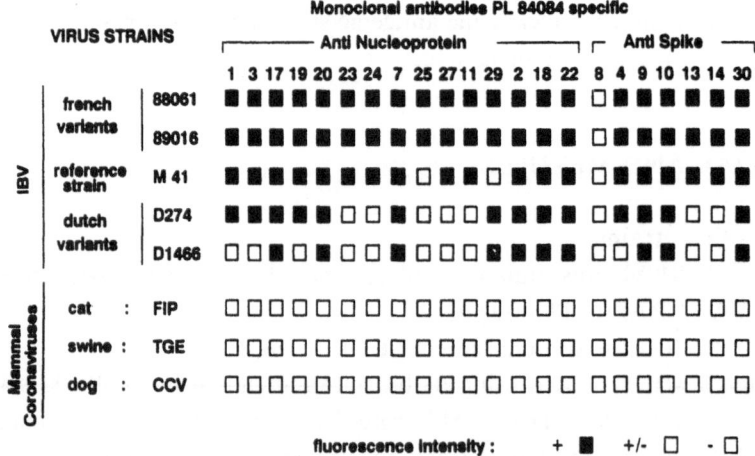

MONOCLONAL ANTIBODIES SPECIFICTY AGAINST CORONAVIRUSES BY I.F.

Figure 1. PL 84084 McAb specificity in relation to other IBV strains and to mammalian Coronaviruses by IF on infected cells.

For 5 week old chicks inoculated orally, the PL 84084 virus was detected in the trachea 4 days after infection and in the caecal onsil from 2 to 11 days after infection (figure 3).

CONCLUSION

The occurence of many different antigenic types of IBV strains was demonstrated by virus neutralisation technique. The analysis by IF of 3 IBV "variant" strain isolated in France between 1984-88 found that although they were antigenically distinct, all the McAb recognized the 3 strains except n° 8 which is specific of the neutralising epitope.

COMPETITIVE "SANDWICH" ELISA WITH Mc Ab 9-8 PL 84084 IBV STRAIN SPECIFIC

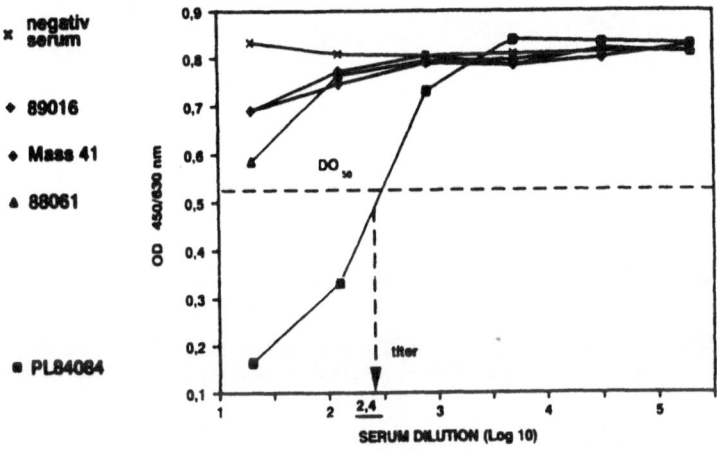

Figure 2. Specific competition between spike antibodies in a PL 84084 chicken serum and McAb n° 8. The chicken serums specific to Mass 41, 88061 and 89016 did not compete with McAb n° 8.

The McAb were used as diagnostic tools of the IBV infection . the competition sandwich ELISA test, using the McAb n° 8 as detecting antibody, was used to detect the PL 84084 specific antibody in the serum of immunized chickens with several IBV strains.

The detection of PL 84084 virus infection by Immunochemistry was very useful. This method was specific and rapid and make it a good alternative method to the conventionnal test which required several passages of homogenised organs in chicken embryos.

Chapter 7

Immune Response and Protection

RECOMBINANT VACCINIA VIRUSES WHICH EXPRESS MHV-JHM PROTEINS: PROTECTIVE IMMUNE RESPONSE AND THE INFLUENCE OF VACCINATION ON CORONAVIRUS-INDUCED ENCEPHALOMYELITIS

Egbert Flory, Albert Stühler, Hanna Wege, Stuart Siddell, and Helmut Wege

Institute of Virology and Immunobiology
Versbacher Strasse 7, D-97078 Würzburg, Germany

SUMMARY

Vaccinia-virus (VV) recombinants encoding either the nucleocapsid (N) or the spike (S) protein of MHV-JHM were constructed to study the role of the immune reponse against defined coronavirus antigens. For the S-protein, a fusogenic (Sfus+) or non fusogenic variant (Sfus-) of the gene was inserted into the VV genome. A strong protection against acute encephalomyelitis (AE) was mediated in Lewis rats which were immunized by VV-Sfus+ and challenged with an otherwise lethal dose of MHV-JHM before the induction of S-specific IgG antibodies. By contrast, a VV recombinant encoding a variant non fusogenic S-protein or the N-protein was not capable confering protection. In addition, we demonstrated that MHV-JHM S-specific IgG antibodies elicited before MHV-JHM challenge modulated the disease process, changing it from an acute disease to subacute demyelinating encephalomyelitis (SDE) .

INTRODUCTION

Coronavirus infections are valuable models to study the pathogenesis of virus-induced central nervous system (CNS) diseases. Especially the murine coronavirus MHV-JHM induces different courses of encephalomyelitis in mice or rats and can establish chronic infections [1]. In Lewis rats, several forms of the disease have been described [2]. The acute encephalomyelitis (AE) is a rapidly progressing disease which leads to death of the animal. The subacute demyelinating encephalomyelitis (SDE) is a paralytic disease characterized by selective loss of myelin and inflammations in the white matter of the CNS. SDE develops after an incubation time of several weeks and can run a relapsing course.

The host immune response plays a critical role both in protection from acute disease and in modulating the development of chronic disease associated with demyelination. MHV-JHM infections induce both cellular (T- cell-mediated) and humoral (B-cell mediated) responses. The role of the different components of the immune system for the course of infection is not clear. For example, the passive transfer of MAb specific for the spike (S), or other structural proteins, has been reported to protect mice or rats from lethal infection [3,4]. In addition, protective neutralizing antibodies were induced by passive immunization of mice with purified S-protein or S-protein decapeptide [5,6]. T-cell-mediated immunity is generally believed to be crucial for the control of most viral infections and published data demonstrate that cellular immunity is required for the control of MHV-JHM infection [7,8,9,10].

However, little is known about virus-protein specific immune responses in Lewis rats. This study demonstrates the protective and disease-modulating capacity of an host immune response specific for individual MHV-JHM structural proteins.

Coronaviruses, Edited by H. Laude and J.F. Vautherot
Plenum Press, New York, 1994

RESULTS

Cloning and Expression of MHV-JHM Genes in VV-Recombinants

The coding sequences of the nucleocapsid and two spike protein genes of MHV-JHM were inserted into the TK gene of VV strain WR. Construction and selection of recombinants containing foreign genes at the TK locus followed previously published procedures and will be described in detail elsewere.[11] The expression of the N, Sfus+ and Sfus- proteins was confirmed by indirect immunofluorescence using monoclonal antibodies (Fig.1). Extensive syncytia formation caused by the fusogenic activity of S-protein was seen in VV-Sfus+ infected DBT-cells.

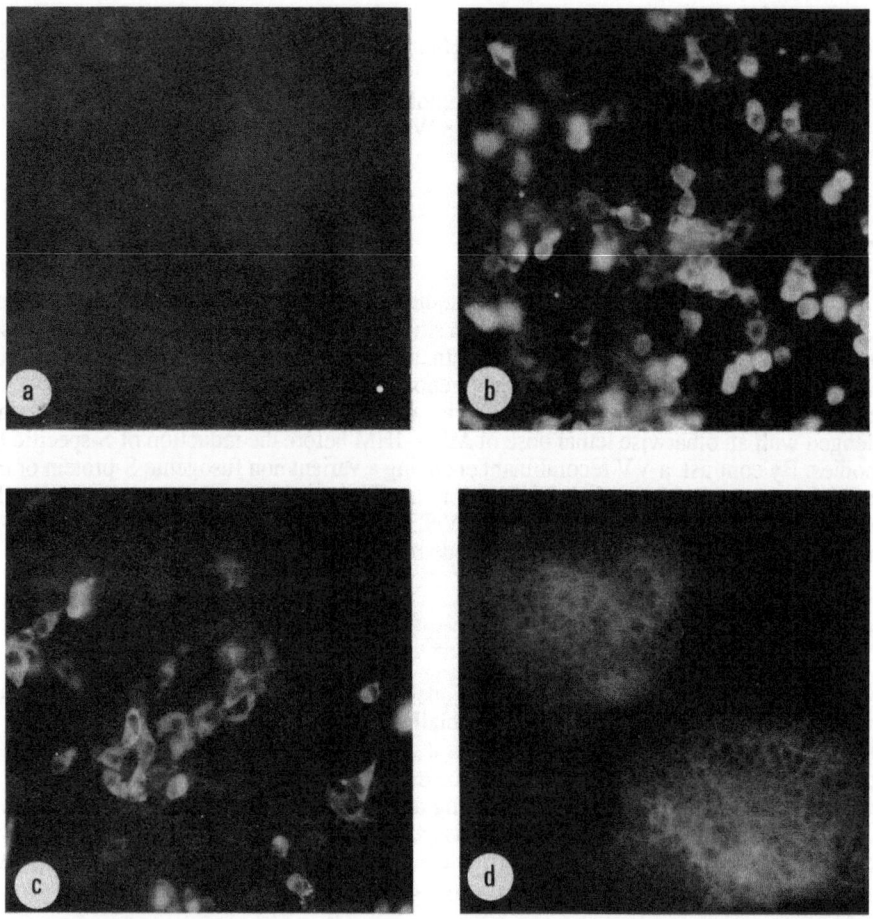

Figure 1. Immunofluorescence illustrating the expression of MHV-JHM proteins in DBT-cells infected with VV recombinants VV-N (b), VV-Sfus- (c) and VV-Sfus+ (d). The cells were incubated with the corresponding monoclonal anti-MHV-JHM antibodies followed by fluorescein isothiocyanate-conjugated anti-mouse immunoglobulin; (a) no staining of VV-wt infected cells with a mixture of anti-N and anti-S monoclonal antibodies (control); cytoplasmatic staining with anti-N monoclonal antibodies (b) and anti-S monoclonal antibodies (c); diffuse membrane staining with anti-S monoclonal antibodies with syncytia-formation (d).

Induction of Humoral Immune Responses Specific for MHV-JHM Proteins

The immunogenic potential of the recombinant VV with regard to the MHV-JHM structural proteins was determined by immunization of adult Lewis rats. Serum samples were tested by ELISA for antiviral IgG and plaque reduction assays were used for monitoring virus neutralizing activity. The protein specificity was demonstrated by Westernblot analysis with purified MHV-JHM virus as antigen. The fusogenic VV-S recombinant induces a humoral immune response in a single shot immunization (Table 1). A high amount of S-specific

Table 1. Humoral immune response in Lewis rats induced by immunization with recombinant VV´s

VV-recombinant used for immunization[4]		Humoral immune response		
		ELISA titer[1]	Neutralization titer[2]	Protein specificity[5]
VV-wt	1x	< 50	-	-
	4x	< 182	-	-
VV-Sfus+	1x[3]	1100	630	S-protein
VV-Sfus-	1x	< 50	-	-
	4x	< 182	-	-
VV-N	1x	< 50	-	-
	4x	284	-	N-protein

[1]ELISA was performed as described previously [4]. The Units were calculated relative to a graph obtained with a standard antiserum from rats.

[2]Neutralization test was done as described previously [4]. The virus infectivity was measured by plaque assays on DBT cells and the antibody titer resulting in 50% reduction of plaque ($\log_{10}$ PRD_{50}) was calculated.

[3]Pooled serum samples were taken 45 days post immunization

[4]Rats were immunized by i.p. infection with 10^7 p.f.u. of respective VV recombinants. Multiple shot immunizations (4x) were performed at intervals of 5 to 7 days.

[5]The specificity of antiviral antibodies was identified by immunostaining of Westernblots. As antigen we used purified MHV-JHM virus, which was separated by PAGE and electroblotted onto nitrocellulose membranes.

antibodies with virus neutralizing activity is present 45 days after immunization. By contrast, the VV recombinant expressing the non fusogenic S-protein does not induce any specific humoral immune response, whether in a single shot or in a multiple shot immunization. Furthermore, the VV-N recombinant elicits a low amount of nucleocapsid-specific antibodies only after four vaccinations. In the plaque reduction assay these antibodies have no neutralizing activity.

Protective Efficiency of the Immune Response to VV Recombinants

To evaluate the effect of the MHV-JHM specific immune response on the outcome of infection, adult Lewis rats were vaccinated i.p. with recombinant VV`s or VV-wt 7 days prior to i.c. infection with a lethal dose of MHV-JHM (SM3SR2). VV-wt immunized rats and animals vaccinated with either VV-N or VV-Sfus- developed an acute encephalitis and died within 7 to 9 days after challenge (Table 2). In contrast, most animals vaccinated once with VV-Sfus+ survived the acute phase of disease even when higher MHV-JHM challenge doses were employed. These results demonstrate that the fusogenic VV-S recombinant induces a protective immune response. Furthermore, a VV-recombinant expressing a non fusogenic variant S-protein is unable to mediate protection.

Table 2. Protection of VV-Sfus+ immunized rats from MHV-JHM induced acute encephalomyelitis

VV-recombinant used for immunization[1]	No. of rats	MHV-JHM challenge (LD_{50})	No. of dead rats/ no. tested
-	6	5	6/6
VV-wt	14	5	14/14
VV-N	18	5	18/18
VV-Sfus-	10	5	10/10
VV-Sfus+	26	5	5/26
VV-Sfus+	14	32	3/14

[1]Lewis rats (3 to 4 weeks old) were immunized once with VV recombinants and were infected with MHV-JHM 7 days later.

Influence of S-specific IgG Antibodies on MHV-JHM Induced Encephalomyelitis

In order to investigate the influence of virus specific antibodies on MHV-JHM-induced encephalomyelitis, we first studied the time kinetics for the induction of S-protein specific antibodies. We vaccinated rats with the fusogenic VV-S recombinant and sampled serum at different time points. A low concentration of antiviral IgG antibodies was detectable 10 days after immunization (Figure 2). In contrast, 21 days after immunization, high amounts of S-specific antibodies are present. The first virus neutralizing antibodies were seen 15 days after immunization and a high level of S-specific antibodies with virus neutralizing activity is present 21 days after immunization.

In the following experiments, we performed the challenge of the fusogenic S-recombinant vaccinated rats at two time points; first, at day 7, in the absence of detectable S-specific IgG antibodies, and second, at day 21, in the presence of S-specific IgG antibodies with virus neutralizing activity. In the absence of antiviral antibodies we observed a strong protection against acute encephalitis. Histological analysis of brain tissue was performed 3 to 6 weeks after MHV-JHM challenge and revealed no pathological changes. In contrast, if the challenge was performed in the presence of S specific antibodies, we observed an interesting modulation of the disease process, namely primary demyelination with characteristics typical of SDE. In the demyelinating lesions MHV-JHM infected cells could be detected. Moreover, the lesions were mainly restricted to the white matter, spinal cord, brain stem and periventricular area.

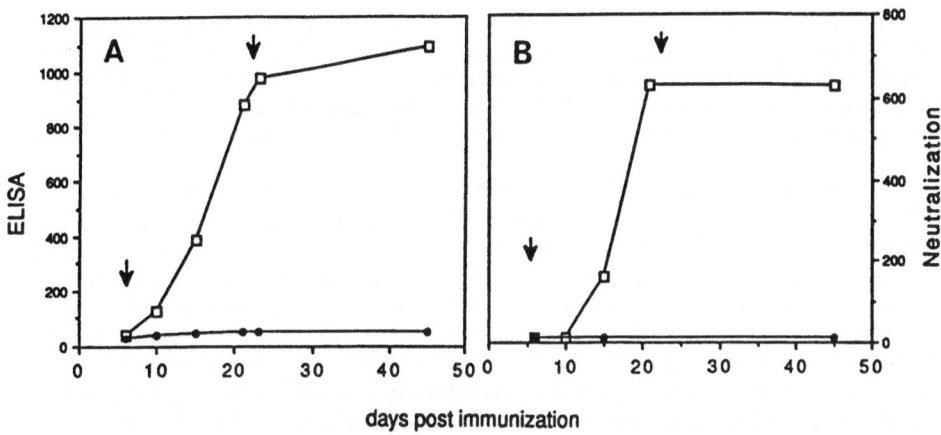

Figure 2. Time kinetics of S-specific IgG antibodies. ELISA test (A) and Neutralization assay (B). The timepoints of MHV-JHM challenge (arrow) in VV-Sfus+ (open boxes) and VV-wt (closed circle) vaccinated rats are shown.

DISCUSSION

In a number of virus-host systems, the development of acute diseases can be prevented by immunization against viral surface proteins or internal components [12,13,14]. The outcome of infection varies with the type of humoral and cellular immune responses elicited by different immunization procedures. It is the aim of the present study to analyze the role of the specific immune response to individual MHV-JHM proteins by vaccination with vaccinia virus recombinants.

The S-protein specific humoral immune response is of the major importance to provide protection against reinfection. This is illustrated by results of immunization experiments with S-protein and specific peptides, which induce neutralizing or fusion inhibiting antibodies [5,6]. Furthermore, the transfer of antibodies prior to infection was demonstrated to provide protection. To evaluate the impact of cellular versus humoral immunity, we employed different vaccination protocols with VV-Sfus+. Vaccinated rats were challenged at an early timepoint, before antiviral antibodies were demonstrable. The observed rapid clearance of infectious virus indicates that the cellular immune response is of central importance to overcome the acute phase of infection. Another important component is the CD4+ T-cell response. The biological potential of these cells was demonstrated by transfer experiments employing a virus protein specific line of T-cells [10]. The transfer of activated CD4+ T-cells before infection hindered the induction of an otherwise lethal encephalomyelits in Lewis-rats. During the natural course of infection, the classical CD8+ T-cell response (CTL) may also play a role, however a direct functional test is not yet available for the rat system (see Hein et al., this volume). The results from experiments in the mice system indicate that both arms of the T-cell response have to cooperate to eliminate infectious virus from the central nervous system [9].

In a variety of virus-host systems, the immune response against the nucleocapsid protein seems to be able to mediate protection [15,16]. The role of the N-protein in eliciting a protective immune response during virus infection is not clear. For example, immunization with purified influenza nucleoprotein can induce protection, whereas vaccination with a VV-recombinant induced no protective immunity [17,18]. In our hands, vaccination with the VV-N recombinant could not prevent acute disease regardless of the immunization protocol. However, we could demonstrate a nucleocapsid-specific, humoral and cellular immune response. On the other hand we obtained strong evidence that the N-protein is a very dominant T-cell antigen (see Wege et al., this volume). Therefore, the N-specific CD4+ T cell response might have an important helper-function to promote the immune responses of cells and antibodies which lead directly to virus elimination.

Vaccination with the fusogenic VV-S+ recombinant induced a very strong protective immunity. By contrast, we could not protect by vaccination with VV-Sfus-. This non fusogenic S-protein variant differs in only one amino acid from the fusogenic S-protein. Since this S-protein is not processed beyond the gp150-precursor stage intracellularly, we assume that the antigen processing may result in an immune response which is not protective (data not shown).

To evaluate the influence of S-specific antibodies on the course of disease, we challenged VV-Sfus+ vaccinated rats at a time point when a high amount of neutralizing antibodies were measurable. Although these rats were protected against acute disease, a high percentage of rats displayed clinical signs of SDE and histological lesions of inflammatory demyelination. These observation supports our earlier conclusions reached on the basis of immunocytochemical analysis of demyelinating plaques in SDE-rats that in addition to T-cells and macrophages, an antibody mediated cytotoxicity might be one of the factors leading to selective loss of myelin [2]. Suckling mice or rats nursed by immune mothers can develop a subacute disease after infection with MHV-JHM at an otherwise lethal dose [19,20,21]. Furthermore, also transfer of antiviral antibodies can delay or modulate the course of neurological diseases [3]. By partial immunization, a modulation of the acute disease to a subacute demyelinating encephalomyelitis can be reproducibly achieved. This observation is an important requirement for further studies on viral and host specific immune responses during the pathogenesis of primary demyelination and relapsing encephalomyelitis.

ACKNOWLEDGEMENTS

We thank Dr. M. Pfleiderer for providing the VV-Sfus+ and VV-Sfus- recombinants. The work was supported by the Deutsche Forschungsgemeinschaft and Hertie Stiftung.

REFERENCES

1. H. Wege, S. Siddell, and V. ter Meulen. Curr. Top. Microbiol. Immunol. **99**: 165 (1982).
2. F. Zimprich, J. Winter, H. Wege, and H. Lassmann. J. Neuropathol. Appl. Neurobiol. **17**: 469 (1991).
3. M.J. Buchmeier, H.A. Lewicki, P.J. Talbot, and R.L. Knobler. Virology. **132**: 261 (1984).
4. H. Wege, R. Dörries, and H. Wege. J. Gen. Virol. **65**: 1931 (1984).
5. P.J. Talbot, G. Dionne, and M. Lacroix. J. Virol. **62**: 3032 (1988).
6. C. Daniel, and P.J. Talbot. Virology **174**: 87 (1990).
7. S.A. Stohlman, G. Matsushima, N. Castell, and L. Weiner. J. Immunol. **136**: 3052 (1986).
8. M.A. Sussmann, R.A. Shubin, S. Kyuwa, and S.A. Stohlman. J. Virol. **63**: 3051 (1989).
9. J.S.P. Williamson, and S.A. Stohlman. J. Virol. **64**: 4589 (1990).
10. H. Körner, A. Schliephake, J. Winter, F. Zimprich, H. Lassmann, J. Sedgwick, S. Siddell, and H. Wege. J. Immunol. **147**: 2317 (1991).
11. M. Mackett, G.L. Smith, and B. Moss, in: DNA Cloning: A Practical Approach, ed., D. M. Glover, Oxford: IRL Press (1985).
12. M. Hany, S. Schulz, H. Hengartner, M Mackett, M. Bishop, H. Overton, and R. Zinkernagel. Eur. J. Immunol. **19**: 417 (1989).
13. L.A. Morrison, S.P. Bauer, J.J. Lange, J.V. Esposito, J.B. McCormick, and D.D. Auperin. Virology **171**: 179 (1989).
14. U.G. Brinckmann, B. Bankamp, A. Reich, V. ter Meulen, and U.G. Liebert. J. Gen. Virol. **72**: 2491 (1991).
15. J.W. Sumner, M. Fekadu, J.H. Shaddock, J.J. Esposito, and W. Bellini. Virology **183**: 703 (1991).
16. W.M. Kast, L. Roux, J. Curren, H.J.J. Blom, A.C. Voordouw, R.H. Meloen, D. Kolakofsky, and C.J. Melief. Proc. Natl. Acad. Sci. **88**: 2283 (1991).
17. D.C. Wraith, A.E. Vessey, and B.A. Askonas. J. Gen. Virol. **68**: 433 (1987).
18. L. Stitz, C. Schmitz, D. Binder, R. Zinkernagel, E. Paoletti, and H. Becht. J. Gen. Virol. **71**: 1169 (1990).
19. S. Perlmann, R. Schelper, E. Bolger, S. Ries. Microbial. Pathogen. **2**: 185 (1987).
20. H. Wege, R. Watanabe, M. Koga, and V. ter Meulen, in: Progress in Brain Research, Vol.59, ed., P.O. Behan, V. ter Meulen and F.C. Rose, Elservier Science Publishers B.V. (1983).
21. H. Wege, M. Koga, H. Wege, and V. ter Meulen, in: Biochemistry and Biology of Coronaviruses, ed., V. ter Meulen, S. Siddell, and H. Wege, Plenum Publishing Corporation, New York (1981).

DETERMINATION OF THE CYTOTOXIC T CELL EPITOPES OF MOUSE HEPATITIS VIRUS, USING ELUTION OF VIRAL PEPTIDES FROM CLASS I MHC MOLECULES AS AN APPROACH

Mirjam H.M. Heemskerk,[1] Henriette M. Schoemaker,[1] Helma E. Alphen,[1] Ruurd van der Zee,[1] Irma Joosten,[1] Willy J.M. Spaan,[2] and Claire J.P. Boog[1]

[1]Inst. of Infectious Diseases and Immunology, Faculty of Veterinary Medicine, University Utrecht
[2]Department of Virology, Faculty of Medicine, Leiden
The Netherlands

INTRODUCTION

Infection of mice with mouse hepatitis virus (MHV) results in a variety of acute and chronic infections. Intraperitoneal (i.p.) infection with MHV results in an acute hepatitis, intracerebral (i.c.) infection induces an acute encephalomyelitis and some survivors show evidence of chronic demyelination (1,2).

Previous research demonstrated that a variety of immune mediators play a role in the modulation of MHV infections including macrophages, natural killer cells, B and T lymphocytes. Passive transfer of monoclonal antibodies (mAb) specific for the structural proteins of MHV provides protection against the acute phase of the infection (3,4). Adoptive transfer of MHV-specific CD4[+] T cells prevents the initial acute form of the disease; however similar to the passive transfer of anti-viral mAb no reduction in virus titer was found in the recipients (5).

Many viral models demonstrate the importance of cytotoxic T cells (CTL's) in virus elimination. Thus far only a few studies implicate a role for cytotoxic T cells during MHV infection. Sussman et al. (6) have characterised a cell population which not only protects mice from a MHV-JHM induced lethal disease but also results in clearance of virus from the CNS of the infected animals. Further, the clearance of virus requires compatibility between donor and recipient at the MHC class I genes (7). This implicates a role of viral specific CTL's in the protection against lethal infection. Until now two groups have described CTL's specific for MHV-JHM in BALB/c mice (8,9); although Stohlman et al. (9) showed the presence of cells directed against the nucleocapsid (N) protein, the exact epitope has not been determined yet, nor the presence of cells specific for other structural proteins of MHV-JHM.

In this study we have generated MHV-A59 specific CTL's in C57BL/6 mice, which were immunized with the temperature sensitive mutant of MHV-A59, named ts342 (10), and boosted 10 days later with MHV-A59 wild type. Because of difficulties in generating stable CTL clones we are currently trying to determine the fine specificity of the CTL's by eluting the naturally processed viral peptides from the MHC class I molecules of MHV-A59 infected RMA tumor cells and subsequent functional analysis of the HPLC separated peptides. In this report we mainly focus on the establishment of a reliable system to separate and isolate peptides from MHC class I molecules.

METHODS

Mice

C57BL/6 mice used in our experiment were 6 - 15 weeks of age and were bred in the animal facilities of the GDL, Utrecht.

Viruses and Cell lines

MHV-A59, MHV-JHM and the temperature sensitive mutant of MHV-A59, named ts342 (10), were propagated on Sac- cells in Dulbecco's modified Eagle's medium (DMEM) supplemented with 5% fetal calf serum (FCS). RMA, RMA-S cells (H-2^b) (11,12), EL-4 (H-2$^{b)}$ and CTL were cultured in Iscove's modified Dulbecco medium, supplemented with 10% FCS, penicillin (50 IU/ml), streptomycine (50 IU/ml) and 2-ME (2 x 10^{-5} M). VSV-specific CTL Clone N32 (kindly provided by Monica Imarai of the laboratory of S. Nathenson, Albert Einstein College of Medicine, Bronx, NY 10461) was generated and weekly restimulated as described (13).

Virus Challenge and Immunization

Mice were inoculated intraperitoneally either with 10^4 PFU ts342 or as controls with PBS, 10 days later all mice received a lethal dose of 4 x 10^3 PFU MHV-A59 or MHV-JHM. Infected mice were monitored for death up to day 20. For in vitro experiments mice were immunized with 10^4 PFU ts342 and boostered 10 days later with 5 x 10^4 PFU MHV-A59 wild type.

Generation of MHV Specific Cytotoxic T Lymphocytes

Spleen cells of primed C57BL/6 mice were isolated 3 - 6 weeks after boosting and stimulated in bulk culture with irradiated (2500 rad) MHV-A59 infected spleen cells (multiplicity of infection (m.o.i.) of 0.3) for 5 days at 37^{0}C in humified air with 5% CO$_2$.

Cell Mediated Cytotoxicity Assay

Varying numbers of in vitro stimulated spleen cells were added to 3 x 10^3 Na^{51}CrO$_4$ labelled target cells in 0.2 ml culture medium in 96 well U shaped well plates and were incubated for 4 h at 37°C and 5% CO_2. After the incubation the supernatant was collected and the radioactivity determined in a gamma counter. Results are expressed as percentage specific release defined as [(experimental release)-(spontaneous release)]/ [Maximal (detergent release)- (spontaneous release)]. Spontaneous release was always less than 30% of maximal release.

Peptide Isolation

RMA-S cells (1 x 10^9) were washed and incubated for 5 h in 350 ml Iscoves medium containing 1 μM of VSV-N52-59 peptide. After incubation, cells were washed twice with PBS and lysed at a concentration of 1 x 10^8 cells/ml in PBS containing 0.5% Nonidet-P40 (Sigma) (NP40) and 0.1 mM PMSF. The cell lysate was stirred for 30 min at 4°C and cleared from nuclei and debri by centrifugation at 100.000g, 30 min at 4°C. Affinity purification of class I MHC molecules according to Falk et al. and Sette et al. (14,15) was performed with some modifications. Supernatant was passed through chromatography columns (bed volume, 2 ml) filled with mAb B8.24.3 (K^b specific) and mAb 28.14.8S (D^b specific) coupled to CnBr activated Sepharose 4B beads (Pharmacia-LKB). The columns were washed with 20 column volumes 0.5% NP40/0.1% SDS/PBS, 2 column volumes of 0.05% NP40/PBS and 2 column volumes of 0.05 M NaHCO3/ 0.15 M NaCl (pH 8.0), containing 0.5% Sodium Deoxycholic acid (DOC) (Sigma). Finally the K^b and D^b molecules were eluted with 0.05 M diethylamine/0.15 M NaCl (pH=11) containing 0.5% DOC. Eluents were swirled in 0.1% trifluoroacetic acid (TFA) for 30 min, boiled for 3 min and centrifuged 10.000g, 10 min. Supernatants were passed over Chromabond C18 columns (Macherey Nagel) and the peptides were eluted with 50% acetonitril/0.1% TFA. Eluents were lyophilized and dissolved in 100 μl aqua-bidest and tested in a ^{51}Cr release assay. 50 μl (1:100 dilution) was added to 3 x 10^3 Na^{51}CrO$_4$ labelled EL-4 cells in 100 μl Iscoves medium plus 10% FCS. Then the VSV-N specific CTL clone N32 was added to the wells and the plates were incubated at 37°C for 4 h.

RESULTS

Protection of Mice against a Lethal MHV Infection after Immunization with Ts342

After priming C57BL/6 mice intraperitoneally with MHV-A59, mice died of an acute hepatitis within 7 days (even with a very low dose of 100 PFU). To circumvent this we tried to prime C57BL/6 mice with 10^4 PFU ts342 and tested if the mice were protected against a lethal challenge of 4 x 10^3 PFU MHV-A59. Figure 1 shows that preimmunization with ts342 protects mice against a subsequent immunization with a lethal dose of MHV-A59. In addition, immunization with ts342 protects also against a lethal challenge of 4 x 10^3 PFU MHV-JHM, indicating that there is a protective cross-reactivity between ts342, a mutant of MHV-A59, and MHV-JHM.

MHV Specific CTL Activity

Spleen cells of MHV-A59 primed mice were cultured with syngeneic irradiated spleen cells infected with MHV-A59. Table 2 shows that after 5 days of culture the spleen cells specifically lysed RMA targets infected with MHV-A59. After several restimulations however the MHV-A59 specificity was lost, probably due to the low

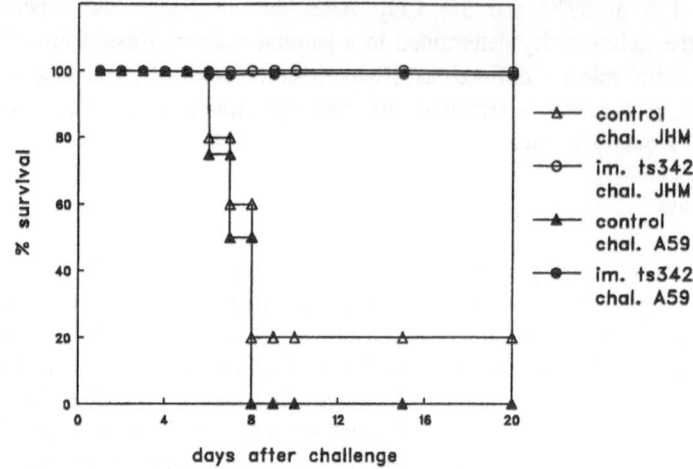

Figure 1. Immunization with 10⁴ PFU ts342 protects mice against a challenge with 4 x 10⁵ PFU MHV-A59 or MHV-JHM.

Table 1. MHV-A59 specific CTL response of spleen cells of MHV-A59 primed C57BL/6 mice stimulated in bulk culture with MHV-A59 infected syngeneic spleen cells.

Effector cells	E/T ratio	Target = RMA % lysis	Target = RMA-MHV % lysis
bulk MHV-spe-cific CTL	50:1	5%	38%
	25:1	4%	33%
	12:1	4%	19%
	6:1	3%	19%

efficiency of the MHV infection of the stimulator cells, as was shown by FACscan analysis using mAb directed against the structural proteins S and M (data not shown).

Isolation of Naturally Processed Peptides from the MHC Class I Molecules

In order to circumvent the generation of MHV specific CTL clones and subsequently determining their fine specificity we tried a more direct way to analyse the MHV epitopes, by eluting peptides from the MHC class I molecules of MHV-A59 infected RMA cells and testing them in a functional CTL assay. Before we started these experiments, we investigated the approach first by using a well defined system. We used a VSV specific CTL clone which recognizes VSV peptide 52-59 in a Kᵇ restricted manner. We incubated RMA-S cells together with the VSV peptide 52-59 for 5 h and lysed the cells after extensive washing. The cell lysate was passed through an affinity chromatography column filled with anti-Dᵇ and anti-Kᵇ beads. Spot-blot experiments showed that we were able to purify MHC class I Kᵇ and Dᵇ molecules from the columns (data not shown). Subsequently peptides were extracted from the different class I molecules and tested functionally in a CTL assay. Results are shown in table 2. We succeeded to isolate peptides from MHC class I molecules since the affinity purified anti-K^b fraction sensitized EL-4 cells to lysis by VSV-specific K^b restricted CTL's to the same extent as the VSV peptide itself. In the same test the affinity purified anti-D^b fraction was not active.

Table 2. Sensitization of EL-4 target cells by VSV N52-59 peptide or material eluted from anti-K^b and anti-D^b columns for recognition by VSV-specific CTL clone N32. * % lysis at E/T ratio 5:1. # EL-4 cells incubated with a 1:100 dilution of the fraction.

Targets	Effector cells = clone N32 % lysis*
EL-4	34%
EL-4 + VSV peptide N52-59 (1 µM)	60%
EL-4 + affinity purified α-K^b fraction#	58%
EL-4 + affinity purified α-D^b fraction#	33%

DISCUSSION

To understand the contribution of CTL's in the protection against an acute disease caused by MHV-A59 we have generated MHV-A59 specific CTL's. Immunization with ts342 gives the opportunity to prime the mice with MHV-A59. We were however until now unable to generate stable MHV-A59 specific CTL clones, probably because of the fact that the efficiency of infection of the spleen cells which were used as stimulator cells is very low. Recently Mobley et al. (17) were also unable to demonstrate CTL activity against MHV-JMH in spleen cell preparations from immunized C57BL/6 mice, probably due to C57BL/6 intrinsic factors. To bypass these problems we set out to characterise the fine specificity of the MHV-A59 specific CTL's by eluting the naturally processed viral peptides from the class I molecules of MHV-A59 infected RMA cells. The validity of the approach was investigated using a well defined system: RMA-S cells were incubated with VSV N52-59, lysed and the different MHC molecules were isolated by affinity chromatography. We used RMA-S cells because this antigen processing-defective mutant cell line expresses empty MHC class I molecules at the cell surface. These molecules are normally unstable but by efficiently loading them with exogeneous peptides can be stabilized. The procedures followed for purification of MHC class I molecules and subsequent elution of peptides, were shown to be succesful as indicated by the VSV peptide specific CTL response of clone N32, thus confirming the validity of the approach. Currently this approach is applied for MHV-A59 infected RMA cells.

ACKNOWLEDGEMENTS

The authors thank Monica Imarai for the gift of clone N32 and Grada van Bleek and Hans van Noort for helpful discussions.

REFERENCES

1. H. Wege, M.S. Sidell and V. ter Meulen, Curr. Top. Microbiol. Immunol. 99: 165-200 (1982).
2. R.L. Knobler, P.W.Lampert and M.B.A. Oldstone, J.Exp. Med. 133: 832-843 (1981).
3. M.J. Buchmeier, H.A. Lewicki, P.J. Talbot and R.L. Knobler, Virology 132: 261-270 (1984).
4. K. Nakagana, K. Yamanouchi and K. Fujiwara, J. Virol. 59: 168-171 (1986).
5. S.A. Stohlman, G.K. Matsushima, N. Casteel and L.P. Weiner, J. Immunol. 136: 3051-3056 (1986).
6. M.A. Sussman, J.O. Fleming, H. Allen and S.A. Stohlman, Adv. Exp. Med. Biol. 218: 399-410 (1987).
7. M.A. Sussman, R.A. Shubin, S. Kyuwa and S.A. Stohlman, J.Virol. 63: 3051-3056 (1989).
8. K. Yamaguchi, S. Kyuwa, K. Nakanaga and M. Hayami, J.Virol. 62: 2505-2507 (1988).
9. S.A. Stohlman, S. Kyuwa, M. Cohen, C. Bergmann, J.M. Polo, J. Yeh, R. Anthony and J.G. Keck, Virology 189: 217-224 (1992).
10. M.J.M. Koolen, A.D.M.E. Osterhaus, G. van Steenis, M.C. Horzinek and B.A.M. van der Zeist, Virology 125: 393-402 (1983).
11. H.G. Ljunggren and K.Kärre, J. Exp. Med. 162: 1745-1750 (1985).
12. K. Kärre, H.G. Ljunggren, G. Piontek and R. Kiessling, Nature 319: 675-678 (1986).
13. K.I. Shibata, M. Imarai, G.M. van Bleek, S. Joyce and S.G. Nathenson, Proc. Natl. Acad. Sci. 89: 3135-3139 (1992).
14. K. Falk, O. Rötzschke, S. Stevanovic, G. Jung and H.G. Rammensee, Nature 351: 290-296 (1991)
15. A. Sette, L. Adorini, E. Appella, S.M. Colon, C. Miles, S. Tanaka, C. Ehrhardt, G. Doria, Z.A. Nagy, S. Buus and H.M. Grey, J.Immunol 143: 3289-3294 (1989).
16. J. Mobley, G. Evans, M.O. Dailey and S. Perlman, Virology 187: 443-452 (1992).

CORONAVIRUS INDUCED ENCEPHALOMYELITIS: AN IMMUNODOMINANT CD4+ - T CELL SITE ON THE NUCLEOCAPSID PROTEIN CONTRIBUTES TO PROTECTION

Helmut Wege, Andreas Schliephake, Heiner Körner,
Egbert Flory, and Hanna Wege

Institute of Virology and Immunobiology,
Versbacher Strasse 7, D-97078 Würzburg, Germany

SUMMARY

In this communication we present clear evidence, that the N-protein of MHV-JHM contains immunodominant CD4+ T-cell sites. These sites were recognized by the immune system of virus infected Lewis rats. In previous investigations we have shown, that CD4+ T-cell lines with specificity for defined viral proteins can be selected from diseased Lewis rats and mediate protection, if transferred to otherwise lethally infected animals [1]. To define regions of the N-protein, which are immunodominant for the T-cell response, we employed bacterially expressed N-protein and truncated subfragments of N as an antigen. We demonstrate, that T-cells from MHV-JHM infected, diseased Lewis rats recognized with high prevalence the carboxyterminal subfragment C4-N (95 aa) and to some extent the adjacent C3-N protein. The same results were obtained with T-cells derived from rats immunized with bacterially expressed N-protein or from animals vaccinated by a stable N-protein expressing vaccinia recombinant. Finally, transfer of CD4+ line T-cells to MHV-JHM infected rats specific for C4-N mediated protection against acute disease.

INTRODUCTION

Murine coronavirus infections of the central nervous system are of great interest to analyse mechanisms of inflammatory demyelination and virus persistency [2]. In Lewis rats, we observed different courses of an encephalomyelitis [3, 4, 5, 6, 7]. Acute disease developed within a short incubation time and was associated with extensive necrotic lesions affecting predominantly the gray matter. The hallmark of the subacute demyelinating encephalomyelitis (SDE), which occured within several weeks to months p.i., were inflammatory demyelinating lesions confined to the white matter. The development of these lesions in MHV-JHM infected Lewis rats was mainly triggered by immune reactions against viral antigens. Therefore, our interest is focused on the structure and function of coronavirus proteins and their interactions with the immune system. We initiated studies to define the role of specific T-cell response during acute and subacute disease induced by MHV-JHM in Lewis-rats [1].

To investigate the antiviral CD4+ T-cell response, we stimulated spleen-lymphocytes from diseased rats in culture with virus antigen and tested their specificity for defined viral proteins after expansion of the culture. As a prerequisite for these studies, we expressed S-

or N-protein in bacteria. The bacterially expressed proteins were purified and employed as antigens. The results of T-cell proliferation assays with splenocyte cultures derived from diseased rats indicated that a strong T-cell response against nucleocapsid protein N was detectable already early in infection. By contrast, spike protein specific responses could only be demonstrated in cultures derived from rats suffering from SDE. For further studies we established CD4+ T-cell lines against these proteins. Transfer experiments were performed with rats, which had been infected with a normally lethal dose of MHV-JHM. Our results indicated, that activated N- or S- specific CD4+- T cells can mediate protection and contribute to clearance of virus from the brain [1]. Depletion of CD8+ T-cells by monoclonal antibodies did not diminish the protection confered by transfer of virus protein specific CD4+ T-cells.

A number of studies on cell mediated immunity in human and animal disease models have confirmed the pivotal role of CD4+ T-cells for the outcome of infection. The biological significance of each virus protein for the cellular and humoral immune response varies with the virus-host system taken into consideration. In MHV-JHM infected Lewis rats, the major T-cell response was specific for N-protein. In this communication we demonstrate, that these dominant N-specific T cell sites were located near to the carboxyterminal part of the nucleocapsid protein. This finding could be of general importance for the development of the immune response during coronavirus infections.

RESULTS

Expression of N-protein Subfragments

To define the antigenic domains on this protein, we expressed the cDNA of N as a beta-galactosidase fusion protein employing the pROS vector system. A clone containing the entire MHV-JHM N gene was provided by T. Raabe and S. Siddell as a derivative of pAT153 (Raabe, diploma thesis, Würzburg 1987). The cDNA (bp 96-1546) was cloned into the expression vector pROS [8]. This vector system allows the expression of genes as a fusion protein with beta - galactosidase. Defined truncated subfragments of the N-cDNA were produced by digestion with restriction enzymes and then subcloned into pROS. The panel of subfragments employed for the following studies is displayed in figure 1. A0-N denotes the nearly complete N-gene, A1-N, A2-N, C3-N and C4-N are designations of N-subfragments arranged from the N-terminal to the C-terminal end including the stop codon of the N-gene. The expressed N-protein subfragments were purified from bacterial lysates by combined steps of gel electrophoresis and elution. For usage as a T-cell antigen, the proteins were transferred to nitrocellulose and processed as described previously [1].

Immunogenicity of N-protein subfragments

As a first step we investigated the immunogenicity of the bacterially expressed N-protein and its subfragments. For this purpose, proliferation assays with lymphocyte cultures were performed. Rats were intracerebrally infected with MHV-JHM and splenocyte cultures were established from diseased animals. The proliferation assays were performed after one cycle of restimulation with virus antigen. In parallel, rats were immunized with purified N-protein, which had been expressed in bacteria. Furthermore, a N-specific immune response was elicited by vaccination with a vaccinia virus recombinant, which expresses the MHV-JHM N-protein. Independent of the method of immunization, the carboxyterminally located fragment (C4-N) induced in all cases a very strong T-cell proliferation. A typical result obtained with T-cells from SDE rats is displayed in figure 2. C4-N is a polypeptide of 95 amino acids at the carboxyterminal end of N (figure 1). A weak T-cell response was observed with C3-N, no specific T-cell response was obtained with A1-N and A2-N. However, all these subfragment proteins were immunogenic, because the sera of the rats contained specific antibodies. This humoral response was demonstrated by Elisa and immunostaining of Western-blots.

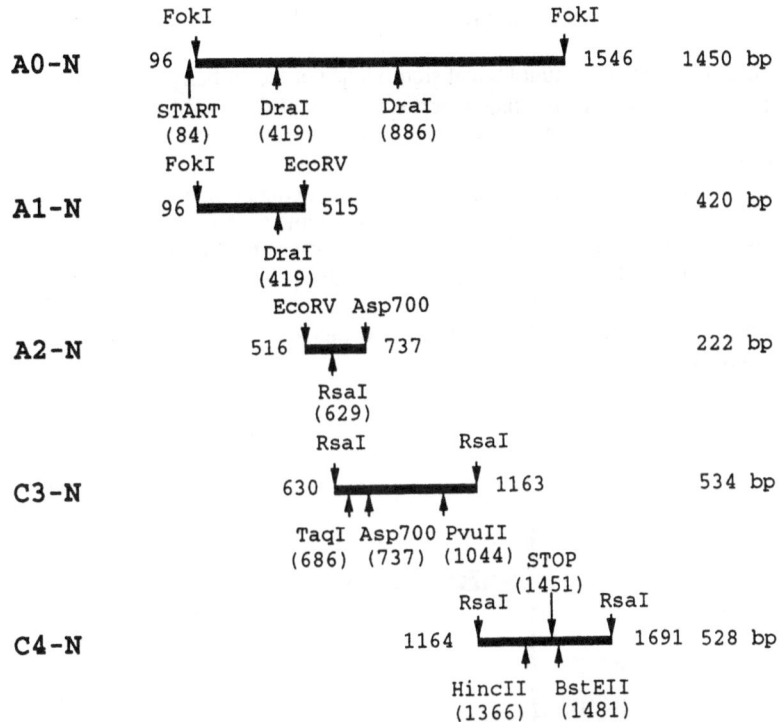

Figure 1. Scheme of N-cDNA-fragments derived by restriction enzyme digestion for expression in pROS. A0-N denotes the nearly full length gene, A1-N, A2-N, C3-N and C4-N are smaller fragments in the direction from the N-terminal to the C-terminal end of the gene. The length (bp) of the cDNA-fragments is indicated on the right side.

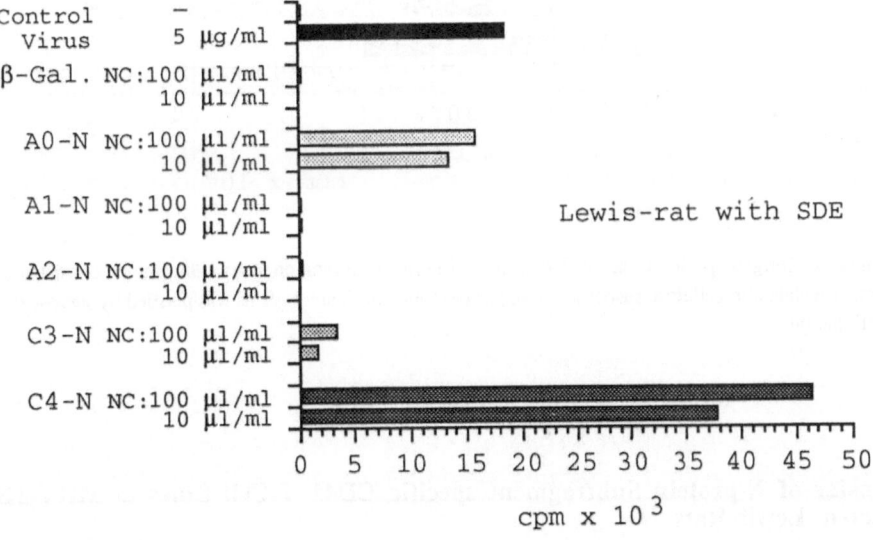

Figure 2. Lewis rat with SDE: Specificity of the T-cell response for Nucleocapsid and N-fragments expressed in pROS. The bacterially expressed proteins were separated by SDS-PAGE before electrophoretic transfer to nitrocellulose (NC). The protein bands were excised, dissolved in DMSO and precipitated with a carbonate buffer [1]. Splenocytes from a rat with SDE were restimulated once in culture with virus antigen before proliferation assays were performed employing the N-fragment antigens. Virus antigen consisted of sucrose density gradient purified virions.

Isolation of N-subfragment specific CD4+ T-cell lines

To evaluate further the immunological significance of the carboxyterminal region, we established a panel of CD4+ T-cell lines specific for each of these subfragments. For this purpose, rats were immunized with purified subfragment proteins and lymphocyte cultures were established from lymph nodes as described by Körner et al.[7] by alternating cycles of restimulation with antigen and expansion with interleukin containing media. An example for the antigen specificity is shown in figure 3 for the T-cell line specific for the subfragment C4-N. The line recognizes besides the C4-N subfragment antigen the full length A0-N and purified virus antigen. In addition, beta-galactosidase induces also proliferation, because the plasmid constructs contain the appropriate sequence fused to this gene. No cross reactivity was detected employing other N-subfragment antigens. Similar results were obtained with the T-cell lines A1-N, A2-N and C3-N.

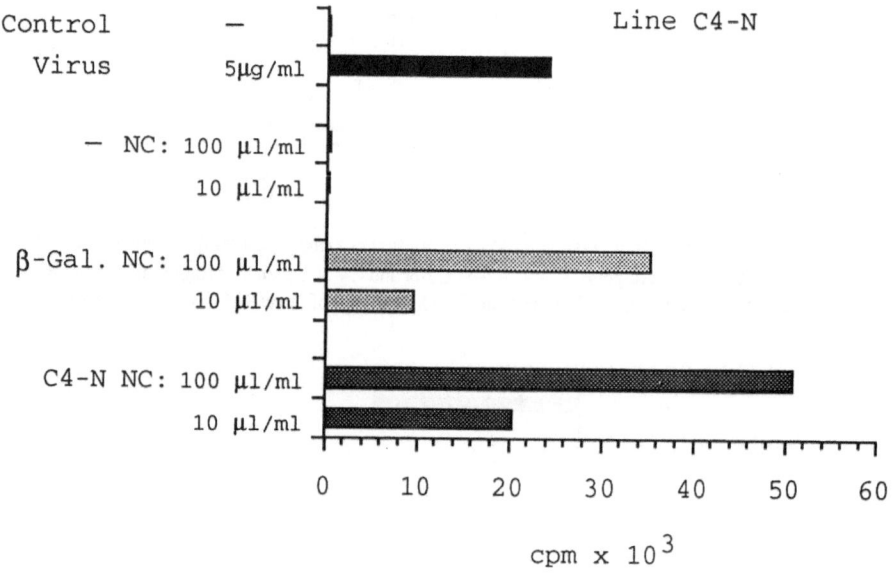

Figure 3. Antigen-specificity of the T-cell line C4-N in a proliferation assay. The antigen consisted of a microparticulate nitrocellulose suspension (NC) as described in figure 2. Virus was purified by sucrose density centrifugation.

Transfer of N-protein Subfragment specific CD4+ T-Cell Lines to MHV-JHM infected Lewis-Rats

In order to demonstrate the biological effects of these T-cell lines, we performed the following transfer experiments. The rats were infected one day after intraperitoneal transfer of 5×10^6 activated line T-cells. At the timepoint of intracerebral infection with MHV-JHM, the rats were 8-10 days old. The results are summarized in table 1. T-cell lines specific for A0-N as well as for C4-N polypeptide can mediate protection by adoptive transfer against an otherwise lethal MHV-JHM infection. None or little protective effect was induced by transfer of A1-N, A2-N and C3-N. These results indicate, that the carboxy-terminal region of the N-protein comprises T-cell epitopes which are of biological importance for the outcome of infection.

Table 1. Disease course after transfer of CD4+- T-cell lines specific for truncated N-proteins

T-cell line transferred	Diseased/ Total	Dead/ Total	Incubation time (Days)
Control	9/9	9/9	5 – 13
A0-N	2/9	2/9	12
A1-N	7/7	7/7	4 – 7
A2-N	8/8	8/8	4 – 10
C3-N	12/12	12/12	5 – 18
C4-N	7/11	7/11	10 – 24

Transfer of 5x10[6] line T-cells was performed i.p. two days before infection. The Lewis-rats were infected i. c. with 600 plaque forming units of MHV-JHM, when 8-10 days old.

DISCUSSION

The relative importance of the different virus proteins for cellular and humoral immunity during coronavirus infections is not yet known in detail. The results described here indicate, that the carboxyterminal region of the N-protein comprises T-cell epitopes which are of biological significance for the outcome of infection. Although the surface proteins of coronaviruses are of major impact for virus-induced immunity, the role of the N-protein has to be considered as an indispensible component in this network.

The N-protein is a highly abundant, relatively conserved antigen, which is already accumulated before release of mature virus [9]. Therefore, a cellular immune response elicited early in infection by internal virus proteins could be an important defense mechanism since lymphokines produced by CD4+ T-cells can trigger an array of specific immune reactions. Activated T-cells can promote B-cell responses independently of their antigen specificity. It has been shown recently, that immunization with a synthetic peptide comprising an irrelevant T-helper cell epitope in combination with a S-protein specific B-cell epitope can induce strong protection against disease [10]. In MHV-JHM infected rats, a detectable S-specific T-cell response appears to be induced only late in the course of infection, when rats develop a subacute demyelinating encephalomyelitis [1]. During the acute stage of disease, an efficient elimination of virus infected cells by cellular immune response is probably the dominant defense mechanism. The S-protein is the dominant antigen for antibody response which inhibits the infectivity of the virus and prevents cell fusion. Provided that the organism overcomes the acute stage without complete elimination of the virus, the disease process can run a chronic course. Our data imply, that the pathogenesis of inflammatory demyelination involves a disbalance between antibody mediated virus specific immune responses and a downregulation of the expression of viral surface proteins (Flory et al., this vol.) [7].

The relative contribution of CD4+ T-cells (helper/inducer) and CD8+ T-cells (CTL) for virus elimination differs with the virus-host system employed. Many studies on cellular immunity suggest that the major T-cell response is MHC class II restricted. For coronavirus infections in mice, both CD4+ and CD8+ T-cells have to interact to eliminate virus from the central nervous system, although virus specific CD4+ T-cells can mediate protection from acute disease by a DTH-like reaction [2, 11, 12]. Results from our transfer experiments imply, that in Lewis rats activated N-specific CD4+ T-cells alone have the biological potential to eradicate virus from brain tissue. It had been shown for other virus-host systems, that CD4+ T-cells can display an antigen specific cytolytic potential [13]. However, during the natural course of infection the speed and rate of CD4+ T-cell activation may be insufficient to effect virus elimination alone. We anticipate, that also in Lewis-rats the observed N-specific CD4+ T-cell response plays a pivotal role to trigger virus elimination in concert with other components. At present, in the rat system no direct proof for virus specific CTL activity is available. On the other hand, studies of the lymphocyte population dynamics demonstrated the presence of CD8+ T-cells in the central nervous system during recovery from acute disease [14]. Thus, further studies on the relative contribution of CD4+ and CD8+ T-cells in different coronavirus infections are of great interest.

In a number of virus systems the induction of CTL by nucleoprotein was demonstrated [15, 16, 17]. Despite of these findings, immunization against nucleocapsid proteins alone was in most virus-host systems not sufficient to confer protection [15]. Vaccination of cats against FIP-virus with a vaccinia recombinant expressing N-protein had no protective effect [18]. Our attempts, to vaccinate Lewis-rats against N before challenge, were not succsessfull despite of the capacity to induce a specific T-cell response (Flory et al., this vol.). The insufficient protective effect of N-protein could be explained by a low CTL activity and the mode of antigen presentation or processing. Although the passive transfer of specific line T-cells does not represent the entire reaction of the immune system during a natural infection or immunization, the results from such experiments clearly illustrate their biological potential.

In this context, Stohlman et al.[19] have identified recently a carboxyterminally located region of the N-protein from MHV-A59, which is important for the induction of a CTL response in mice. Furthermore, the N-protein of IBV contains a T-cell site which enhanced protection [20, 21]. Such T-cell responses could complement and interact with the important surface protein specific immune reactions to provide protection or to determine the disease course of the infected organism.

ACKNOWLEDGEMENTS

The work was supported by the Hertie Stiftung and Deutsche Forschungsgemeinschaft.

REFERENCES

1. Körner, H., A. Schliephake, J. Winter, F. Zimprich, H. Lassmann, J. Sedgwick, S. Siddell, and H. Wege. J. Immunol. **147**(7): 2317-23, 1991.
2. Kyuwa, S. and S. A. Stohlman. Seminars in Virol. **1**: 273- 280, 1990.
3. Nagashima, K., H. Wege, R. Meyermann, and V. ter Meulen. Acta Neuropath. **44**: 63-70, 1978.
4. Watanabe, R., H. Wege, and V. ter Meulen. Nature **305**: 150-153, 1983.
5. Watanabe, R., H. Wege, and V. ter Meulen. Lab. Invest. **57**(4): 375-84, 1987.
6. Wege, H., R. Watanabe, and V. ter Meulen. J. Neuroimmunol. **6**: 325-336, 1984.
7. Zimprich, F., J. Winter, H. Wege, and H. Lassmann. Neuropath. and Appl.Neurobiol. **17**: 469-484, 1991.
8. Bröker, M. Gene Anal. Tech. **3**: 53-57, 1986.
9. Parker, M. M. and P. S. Masters. Virol. **179**(1): 463-8, 1990.
10. Koolen, M. J., M. A. Borst, M. C. Horzinek, and W. J. Spaan. J. Virol. **64**(12): 6270-3, 1990.
    Anthony, and J. G. Keck. Virol. **189**(1): 217-24, 1992.
11. Sussman, M. A., R. A. Shubin, S. Kyuwa, and S. A. Stohlman. J. Virol. **63**(7): 3051-6, 1989.
12. Williamson, J. S., and S. A. Stohlman. J. Virol. **64**(9): 4589-92, 1990.
13. Erb, P., D. Grogg, M. Troxler, M. Kennedy, and M. Fluri. J. Immunol. **144**(3): 790-95, 1990.
14. Dörries, R., S. Schwender, H. Imrich, and H. Harms. Immunology **74**(3): 539-45, 1991.
15. Stitz, L., C. Schmitz, D. Binder, R. Zinkernagel, E. Paoletti, and H. Becht. J. Gen. Virology **71**: 1169-79, 1990.
16. Yamaguchi, K., N. Goto, S. Kyuwa, M. Hayami, and Y. Toyoda. J Neuroimmunol. **32**(1): 1-9, 1991.
17. Yasukawa, M., A. Inatsukio, and Y. Kobayashi. J. Immunol. **140**: 3419-25, 1988.
19. Stohlman, S. A., S. Kyuwa, M. Cohen, C. Bergmann, J. M. Polo, J. Yeh, R. 18. Vennema, H., R. de Groot , D. A. Harbour, M. C. Horzinek, and W. J. Spaan. Virology **181**(1): 327-35, 1991.
20. Boots, A. M., T. B. Benaissa, W. Hesselink, E. Rijke, C. Schrier, and E. J. Hensen. Vaccine **10**(2): 119-24, 1992.
21. Boots, A. M., J. G. Kusters, N. J. van, K. A. Zwaagstra, E. Rijke, B. van der Zeijst, and E. J. Hensen. Immunology. **74**(1): 8-13, 1991.

JHM VIRUS-SPECIFIC CYTOTOXIC T CELLS DERIVED FROM THE CENTRAL NERVOUS SYSTEM

Stephen Stohlman[1,2], Cornelia Bergmann[1,2], Nicola LaMonica[2], Michael Lai[1,2], Jason Yeh[1], and Shigeru Kyuwa[3]

Departments of Neurology[1] and Microbiology[2] University of Southern California School of Medicine, and the Department of Animal Pathology University of Tokyo[3]

ABSTRACT

Spleen cells cultured from Balb/c mice immunized with the JHM strain of mouse hepatitis virus (JHMV) have CD8[+] cytotoxic T cells (CTL) specific for both the S and N proteins, but not the M or HE proteins. T cell lines were established from the brains of Balb/c mice infected with JHMV. The majority of the lines (20 of 22) were specific for JHMV. Analysis of the viral structural proteins which served as target structures indicate that most (15 of 20) were specific for the N protein. One line was specific for the S protein and four lines were specific for JHMV but the protein recognized could not be determined. These data suggest that early during infection there is a preferential recruitment of N protein specific CTL into the CNS of infected mice.

INTRODUCTION

JHMV causes acute and chronic forms of demyelinating disease in the central nervous system (CNS) of mice. The two components that interact to determine the outcome of this infectious process are the virus and the host's immune response. The transfer of either monoclonal antibodies (mAb) or CD4[+] Th1 T cells into mice lethally infected with JHMV results in protection from death, but provides little or no protection from demyelination[1]. CD8[+] cytotoxic T cells (CTL) specific for JHMV are involved in the clearance of virus from the CNS and thereby provide specific immunological protection[2,3,4]. It is the goal of these experiments to isolate and identify the CTL population(s) recruited to the CNS during infection with JHMV, to determine the viral proteins which contain the epitopes recognized, the major histocompatibility (MHC) class 1 molecule(s) which function as the restriction elements, and finally to determine the role of the CTL specific for the individual viral epitopes in JHMV infection.

MATERIALS AND METHODS

Virus and Mice: The JHMV strain of mouse hepatitis virus was propagated and plaque assayed as previously described[5]. Balb/c mice were obtained from the Jackson

Coronaviruses, Edited by H. Laude and J.F. Vautherot
Plenum Press, New York, 1994

Laboratory, Bar Harbor MA and used at 6-12 wks of age. Mice were infected intracranially with 25 pfu of the DM strain of JHMV at 6 wks of age or immunized with 1-5x 10[6] pfu intraperitoneally.

Recombinant Vaccinia Viruses: Recombinant vaccinia viruses (rvv) expressing the membrane (M), and nucleocapsid (N) proteins were derived from the DL strain, while the hemagglutinin-esterase (HE) protein was derived from the JHM-X strain[6]. Coding regions were cloned using PCR into the pSC11ss vector and recombined into the WR strain of vaccinia virus as previously described[7]. Recombinant vaccinia virus expressing the A59 strain spike (S) protein was kindly supplied by W. Spaan.

Isolation of the T Cell Lines: Brains were removed from mice infected 7 days previously and the mononuclear cells isolated by centrifugation on Percol gradient as described[4]. Cells were purified by centrifugation on lympholyte M and the adherent cells removed by incubation on plastic dishes for 2 hrs at 37C. Nonadherent cells were incubated with 3.5 x 10[6] JHMV-infected (m.o.i = 0.05) irradiated (20Gy) syngenic spleen cells in one well of a 24 well plate in complete RPMI medium containing 20% FCS and 10 to 20% rat Con A supernatant as a source of growth factors. Cells were passed weekly, usually following lympholyte M purification. Lines were established after 6 passages by depletion of the CD4[+] cells using mAb plus complement as previously described[7] followed by limiting dilution in 96 well plates containing 5 x 10[5] feeder cells per well. Positive wells were expanded and tested for CTL activity. CTL assays used the J774.1 Balb/c derived cell line as previously described[7].

RESULTS

JHMV-specific CTL were prepared from mice immunized 3-4 wks as previously described[7]. Cultures were tested for recognition of the S, M, HE and N structural proteins by CTL assay using J774.1 targets infected with rvv expressing the individual proteins. These CTL recognize targets expressing the S and N proteins (Fig. 1). There was no detectable cytotoxicity specific for either the M or HE proteins. The level of cytotoxicity for the S and N proteins suggested that it was likely that these populations would be represented within the CNS of the infected mice.

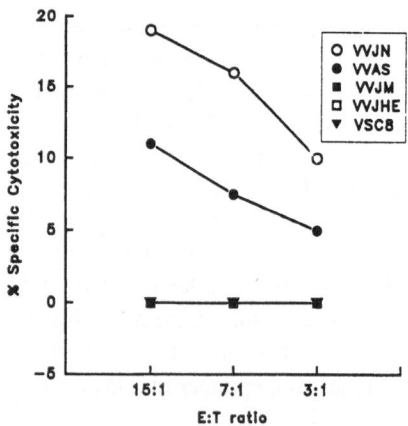

Figure 1. CTL activity of cultures from Balb/c mice immunized ip 4 wks prior to culture. Spleen cells were incubated with irradiated JHMV infected feeder cells for 6 days prior to assay. CTL were tested at the E:T ratios shown using J774.1 targets infected with rvv expressing the MHV structural proteins or a control rvv containing no insert (vSC8).

Mononuclear cells were isolated from the brains of 15 mice infected 7 days previously with 25 pfu of JHMV as previously described[6]. Viable cells were further isolated by centrifugation on lympholyte M and allowed to adhere to plastic dishes for 2 hrs at 37C which removed approximately 95% of the cells. A total of 8.5×10^4 cells were recovered and mixed with infected irradiated feeders in a single well of a 24 well plate. Proliferating cells were depleted of $CD4^+$ cells and cloned at limiting dilution. Twenty two lines were established and tested for recognition of JHMV-infected targets. Figure 2 shows that two lines could not distinguish infected vs. uninfected J774.1 targets. The remaining 20 lines were specific for JHMV.

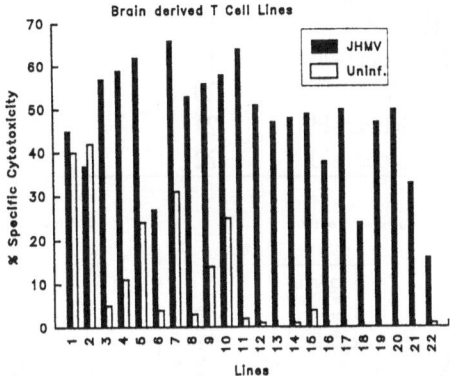

Figure 2. CTL activity of the individual $CD8^+$ lines established from the brains of JHMV infected mice. Specific lysis measured by the release of ^{51}Cr from JHMV infected and uninfected J774.1 cells at the E:T ratios 3:1 to 10:1.

To determine which viral proteins were recognized by the JHMV-specific CTL lines, targets were infected with rvv expressing the S, M, HE, and N proteins and recognition tested at 2 or 3 E:T ratios. Table 1 shows examples of the recognition patterns observed and Table 2 summarizes the data for the 20 JHMV-specific T cell lines. The majority (15) were specific for the (N) protein. One line was specific for the S protein. Four lines were specific for JHMV (Fig.2), but did not recognize a structural protein, indicating specificity for a nonstructural protein or an epitope absent in the S protein rvv (see discussion).

Table 1. CHARACTERIZATION OF BRAIN-DERIVED T CELL LINES

CTL Line	E:T	INFECTED TARGETS JHMV	M	S	HE	N	CON^2	Specificity
6	10:1	27	0	0	1	24	4	**N protein**
	5:1	22	0	0	2	21	3	
8	8:1	NT	2	3	3	28	2	**N protein**
	4:1	NT	1	0	0	19	0	
11	10:1	NT	5	1	0	42	1	**N protein**
	5:1	NT	2	0	0	30	0	
22	10:1	21	NT	29	NT	3	0	**S protein**
	5:1	11	NT	19	NT	4	0	

[1]Specific ^{51}Cr-Release; [2]CON = Control. Either J774.1 infected with vSC8 and/or uninfected cells were tested. The highest control values are depicted; [3]NT = Not tested.

Table 2. SPECIFICITY OF THE T CELL LINES DERIVED FROM THE CNS DURING ACUTE JHMV INFECTION

LINE	SPECIFICITY	LINE	SPECIFICITY
1	Non-specific	12	Nucleocapsid
2	Non-specific	13	Nucleocapsid
3	Nucleocapsid	14	JHMV-(R)
4	Nucleocapsid	15	Nucleocapsid
5	Nucleocapsid	16	Nucleocapsid
6	Nucleocapsid	17	Nucleocapsid
7	JHMV-?	18	JHMV-?
8	Nucleocapsid	19	Nucleocapsid
9	JHMV-?	20	Nucleocapsid
10	Nucleocapsid	21	Nucleocapsid
11	Nucleocapsid	22	Spike

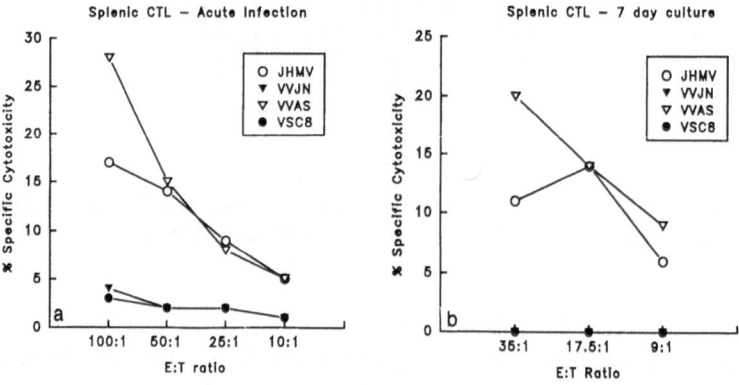

Figure 3. CTL activity of speen cells from mice infected i.c. 7 days previously (panel A) and following in vitro culture (panel B).

Additional experiments examined the observation that the majority of the T cell lines derived from the CNS of infected mice were specific for the N protein. First, spleen cells from mice infected 7 days previously were tested for CTL activity. Surprisingly, there was no detectable N-specific CTL activity in the spleen population (fig 3a). To rule out an artifact of expansion and cloning, the spleen cells were cultured under the same conditions used for the CNS derived T cells and tested for CTL activity after 7 days. No N- specific CTL could be detected in this population (Fig. 3b).

DISCUSSION

Manipulations of the immune response of rodents infected with JHMV have shown both beneficial and detrimental effects. The transfer of nonneutralizing mab specific for structural proteins, and clonal populations of virus-specific CD4[+] T cells are both able to prevent death from lethal infection[1]. Survival, however, was associated with extensive CNS demyelination. Interestingly, neither of these manipulations resulted in a significant alteration in virus replication. In vivo depletion of T cell subsets during infection suggested that initial clearance of virus from the CNS was mediated by a CD4-CD8- population[4]. A role for NK cells was demonstrated by direct analysis of the mononuclear cells infiltrating the CNS during acute infection; however, complete clearance require both a CD4[+] and a CD8[+] population[4]. These data were consistent with observation that the induction of JHMV-specific CTL required a CD4[+] "helper" population[2]. Finally, the adoptive transfer of clonal populations of JHMV-specific CD8[+] T cells is of therapeutic value, resulting in protection from a lethal challenge and suppression of virus replication in the CNS[2,3], consistent with the data indicating that CD8[+] T cells play a major role in protection from JHMV-induced disease[1].

The present study suggests preferential recruitment of the N protein CTL into the CNS during JHMV-infection. Although mAb to N protein protects mice, there is no evidence that the CD8[+] T cell mediated immune response specific for the N protein plays a role in protection. The recruitment of these cells into the CNS may simply be a reflection of the large quantity of N protein produced during infection. Analysis of later time points may show a distribution comprising both the N and S specific CTL populations. It is intriguing to speculate that the N specific CTL activity may contribute to the demyelination in addition to providing protection. Although CNS cells in vitro express MHC derived gene products, there is accumulating evidence that the neurons, astrocytes and oligodendroglial cells within the adult CNS are unable to express either class 1 or class 2 molecules[8]. This could be intrepreted to suggest that the clearance of JHMV from the CNS in mice protected by CD8[+] T cells is primarily due to cytolytic activity expressed against microglia or via the local release of a lymphokine. The growth of at least one JHMV variant predominantly in oligodendroglial cells, which is cleared from the CNS with approximately the same kinetics as wild type virus[9], suggest that a lymphokine, i.e. IFN-2 or TNF, may play a predominant role in viral clearance rather than the direct lysis of virus infected cells. The identity of the four lines that are virus-specific but whose epitopes could not be mapped with the present panel of recombinant vaccinia viruses, suggests that there are additional epitopes recognized by H-2[d] mice possibly residing within the nonstructural genes. It cannot, however, be ruled out that these CTL recognize an epitope within the JHMV S protein that is distinct from the A59 strain S protein used in the present experiments. We have recently completed cloning of the S protein from JHMV into a recombinant vaccinia virus to directly examine this question.

ACKNOWLEDGEMENTS

The technical assistance of Manny Dimacali, the advice of Dr. Stanley M. Tahara, and the editorial assistance of Sonia Q. Garcia are greatefully acknowledged. This work was supported by U.S. Public health Service Research Grant NS18146.

REFERENCES

1. Kyuwa, S. and S.A. Stohlman. Seminar Virol. 1:273-280, (1990).
2. Sussman, M., Shubin, R., Kyuwa, S. and S.A. Stohlman. J. Virol. 63:3051-3059, (1989).
3. Yamaguchi, K. Goto, N. Kyuwa, S, Hayami, M. and Y. Toyoda. J. Neuroimmunol. 32:1-9, (1991).
4. Williamson, J., S.-P., Sykes, K. and S. Stohlman. J. Neuroimmunol. 32:199-207, (1991).
5. Stohlman, S.A., P.R. Brayton, J.O. Fleming, L.P. Weiner, and M.M.-C. Lai. J. Gen. Virol. 63:265-275, (1982)
6. Yokomori, K., La Monica, N., Makino, S., Shieh, C.-K., and Lai, M.M.C. Virol. 173:683-691., (1989).
7. Stohlman, S., Cohen, M., Kyuwa, S. Bergmann, C., Anthony, R. and J.G. Keck. 1992. Virol. 189:217-224, (1992).
8. Lampson, L.A. Trends in Neurosci. 10:211-216, (1987).
9. Fleming, J.O., Trousdale, M.D., Bradbury, J., Stohlman, S.A. and L.P. Weiner. Microbiol. Pathogen. 3:9-20, (1987).

ON THE ROLE OF DIFFERENT LYMPHOCYTE SUBPOPULATIONS IN THE COURSE OF CORONAVIRUS MHV IV (JHM)-INDUCED ENCEPHALITIS IN LEWIS RATS

S. Schwender, A. Hein, H. Imrich, and R. Dörries

Institut für Virologie und Immunbiologie der Universität Würzburg
Versbacher Str. 7
8700 Würzburg

INTRODUCTION

The neurotropic strain of the mouse hepatitis virus (MHV), MHV-IV or JHM virus, causes neurological disorders in rats and mice when given intracerebrally [1, 2]. Depending on the rat strain and the age of the infected animal different courses of disease have been observed. Up to 2 weeks post partum Lewis (LEW) rats develop an acute fatal encephalitis, whereas animals older than 3 weeks usually do not succumb to the infection [3] but suffer from a subacute paralytic disease accompanied by multiple neurological disorders. However, the majority of them recovers completely roughly 3 weeks past infection [4]. Since rats are expected to be immunologically competent at the age of 3 weeks, it is conceivable that maturity of the immune system probably plays a major role in the clinical course of the infection. We have recently shown, that T- and B-lymphocytes infiltrate into the central nervous system of these rats [4] and that the humoral immune response most likely contributes significantly to convalescence [5].

Here, we report on the clinical course of the intracerebral infection with JHM virus in LEW rats after immunosuppression and reconstitution with defined lymphocyte subsets prior to infection. The data add further support to the idea, that the virus-specific antibody response is essential for a mild or subclinical course of the infection.

Coronaviruses, Edited by H. Laude and J.F. Vautherot
Plenum Press, New York, 1994

MATERIALS AND METHODS

Animals. Lewis (LEW) rats were obtained from "Zentralinstitut für Versuchstierzucht", Hannover, Germany, and were kept under sterile conditions. They were infected intra-cerebrally with 10^3 PFU of JHM virus. The animals were observed daily for neurological symptoms, that were scored according to the following scale: (0) no clinical sings; (1) un-common social behaviour, loss of tail tone; (2) paresis of 1 or 2 legs, nervous behaviour; (3) paresis of more than 2 legs or paralysis of one or two legs, body trembling; (4) tetra-plegia, or restless running and jumping, reaching sometimes a (5) moribund state.[4] (mod.)

Immunosuppression. LEW rats were whole body irradiated with 600 rad γ of a ^{137}Cs source. Immunosuppression was controlled by intraperitoneal immunization of irradiated rats with keyhole limpet hemocyanine (KLH) and a subsequent assay of lymphocytes from spleen and lymph nodes for their capacity to secrete antigen-specific antibodies on a single cell basis using an ELISpot assay [5]. In addition, an ELISA was performed by standard procedures to determine KLH-specific antibodies in serum specimens.

Preparation of lymphocyte subsets and reconstitution of animals. Lymphocytes for reconstitution of irradiated animals were isolated from spleen, cervical lymph nodes (LN) and mesenterial LN of <u>uninfected</u> syngeneic age-matched animals and transferred through the lateral tail vein within a few hours after irradiation. Compositions of transferred cells were as follows (table 1, columns #1-#8): non-separated lymphocytes (#8), pure T-lymphocyte subsets (CD4+-, #3; CD8+ T-cells, #4), both T-cell subsets in the absence of B-lymphocytes (#5) and different combinations of T-cell subsets and B-lymphocytes (#6-#8). Each animal received that amount of purified cells, corresponding to their proportion in $2x10^8$ unseparated, naive lymphocytes; i.e. they received $1.5x10^8$ cells if, for instance, 75% of the cells in question were positive for that specific marker in the unseparated population.

Purification of lymphoid cell populations was performed by immunomagnetic proce-dures using the appropriate panel of monoclonal antibodies. The resulting viable cells were quantitated by trypan blue exclusion and quality of the purification was checked in a Becton & Dickinson FACScan flow-cytometer using FITC-labeled antibodies [4]. Routinely, less than 1% contamination was obtained after two depletion cycles.

Titration of virus. Brain and spinal cord (CNS) were taken from sacrificed animals at day 9 past infection. The weight of each CNS was determined and the tissue was homogenized in a 5 ml Dounce homogenizer. The resulting suspension was clarified by centrifugation, the supernatant serially diluted and assayed for infectious virus in triplicates on a microtiter plate using DBT-cells as targets [6]. The TCID$_{50}$ was regarded as the highest dilution were DBT-monolayers in at least 2 of 3 wells developed more than 50% cytopathic effect.

RESULTS AND DISCUSSION

To dissect the role of individual lymphocyte subsets in the course of JHM virus-induced encephalitis in rats, animals were immunosuppressed and reconstituted with defined lymphocyte populations prior to infection. Immunosuppressive treatment was controlled in irradiated and KLH-immunized animals. They were assayed for their proliferative T-lymphocyte response to the mitogen Con A and their capability to mount a KLH-specific antibody response. After whole body irradiation only 0.2% of ^{3}H-thymidine was incorporated into Con-A stimulated lymph node cells compared to lymphocytes from untreated rats. The IgM response reached only 0.02% of the level obtained in normal rats and an IgG response is no longer detectable as determined by an ELISpot assay as well as by a KLH-specific ELISA. These data indicate, that the whole body irradiation of rats with 600 rad γ is sufficient to completely suppress an antigen-specific immune reaction up to day 9 pi.

After infection development of neurological symptoms was recorded up to 9 dpi. The data are summarized in table 1. Non-irradiated, infected rats developed clinical signs at day 8 pi with increasing severity on the next day (table 1, #1). More or less the same course of clinical signs was observed in irradiated animals to which 2×10^8 non-separated lymphocytes were transferred. In both assays less than 500 TCID$_{50}$/g CNS virus were recovered, showing that the intravenous reconstitution of the immune system with lymphocytes was successful (table 1, #8), and indicating the tight control of the infection by an intact immune system. In irradiated but not reconstituted animals incidence and severity of disease was comparable to the latter animals, with a remarkable difference with respect to the type of neurological disease (table 1, #2). In the course of the infection most of these animals reacted by a highly nervous attitude of movement, pointing to a spread of the virus to grey matter rather than limitation to the white matter as usually seen in the subacute courses of the infection. The presence of JHM virus at 8000 TCID$_{50}$/g CNS in these animals clearly disclosed the lack of virus neutralization and thus supports the idea of viral spread to the cerebral cortex, brain stem and spinal cord. Similar results were reported from JHM virus-infected mice, where irradiation or cyclophosphamide-induced immunosuppression leads to a predominant gray matter manifestation instead of demyelinating disease [7].

More severe courses of the JHM infection and an earlier onset of disease compared to irradiated but fully reconstituted or not reconstituted rats were observed in those animals which received either CD4+ T-cells, CD8+ T-cells or both together (table 1, #3, #4 and #5, resp.). In combination with high titers of infectious JHM virus in the CNS of these animals, three important conclusions can be drawn from these findings: (1) The virus itself undisturbed by an immune response is rather slow in developing clinical signs of the infection, although after 9 dpi strong neurological symptoms can be seen (data not shown), (2) T-lymphocytes in the absence of B-lymphocytes amplify and accelerate severity and development of disease and (3) despite T-lymphocytes react to virus-infected cells in the brain tissue they are not able to significantly reduce infectious virus. Again these results are in line with findings of Fleming et al.,[7] who described in the mouse model a correlation between development of disease and the presence of T-lymphocytes in the transferred lymphocyte population. In this context, it is important to note that in our rat model not only CD8+ T-lymphocytes or a combination of both CD4+ and CD8+ lymphocytes contributed to an early induction of disease, but also CD4+ lymphocytes alone. Although the underlying mechanism of CD4+ T-lymphocyte-mediated pathogenesis is yet unknown, one might speculate that MHC class II restricted cytotoxicity to virus-infected astrocytes is involved. JHM virus can infect astroglia and upregulate class II antigens *in vitro*[8] and Borna virus infected astrocytes from Lewis rats are recognized by class II restricted CD4+ T-lymphocytes.[9] However, so far it is not excluded that more indirect effects may also account to CD4+ cell-mediated pathogenesis, such as γ interferon secretion by CD4+ cells that bind to class II expressing microglia. Since microglia upregulate class II antigens in JHM-infected rats[10] and virus is present in subacutely diseased animals as long as 3 weeks past infection[4] this hypothesis is not too far fetched. Moreover, it was recently shown that γ interferon causes release of superoxyd anions in microglia cells *in vitro*[11] a process that could also contribute to tissue damage.

Presence of B-lymphocytes in the transferred cell population was generally associated with a markedly delayed onset of clinical symptoms (table 1, #6-#8). Low titers of infectious virus in the brain of animals that received B-lymphocytes verify our previously published hypothesis that intracerebral synthesis of JHM-specific antibodies limits spread of the virus by neutralization [5]. There is of course one exception to this rule, in the case where CD4+ cells had been depleted prior to transfer (table 1, #7). This result was not surprising at all, as it is known that most B-cells cannot act in the absence of syngeneic CD4+ cells which play a key role in the induction phase of the humoral immune response. Most likely this slight reduction of virus in this experiment has to be explained by the fact that the CD8+/B-lymphocyte transfer was contaminated with roughly 5% of CD4+ cells (from initially 48%) that could provide some help for B-lymphocytes.

This protective character of the humoral immunity is known from different animal models and from man. Shubin et al.,[11] reported that a nylon wool adherent fraction of immunized spleen cells (consisting mostly of B-lymphocytes) is necessary to reduce virus in the CNS of JHM virus-infected mice and Wege et al.,[3] showed that the incidence of JHM virus-induced encephalitis is markedly reduced in suckling rats nursed by JHM virus-immunized dams. This is in line with findings of Perlman et al.,[13] who reported a shift in the disease pattern of JHM virus-induced encephalitis in mice from an acute lethal course to a milder subacute disease if infected animals were protected by maternal antibodies. Furthermore, in Theiler's virus infected mice, higher numbers of anti-viral IgM secreting cells could be isolated from the spleen of resistant strains, compared to susceptible ones [14]. In this case the protective role of the humoral immune response could be confirmed by passive transfer of serum from infected resistant strains to susceptible mice resulting in a marked decrease in the number of infected cells within the spinal cord. Nevertheless, although onset of the disease was delayed in animals reconstituted with CD4+ T- and B-lymphocytes, severity of the symptoms increased dramatically on 9 dpi and some animals died. Since CD8+ cells were lacking in these animals we assume, that B-lymphocytes can restrict viral spread and neutralize extracellular virus by antibody secretion but are unable to eliminate virus from tissue. Persistent viral antigen presentation in the brain could then provoke CD4+ cells to interact with class II expressing cells, resulting in tissue damage as already discussed above.

The disease-limiting effects of B-lymphocytes in the CNS tissue might be mainly explained by neutralization of infectious virus. However, another important consequence of virus-specific antibody secretion in the brain has to be discussed. High levels of antibodies may also strongly influence the histopathological changes in the JHM-infected brain and thus determine the clinically visible disease. Zimprich et al.,[15] showed in the CNS of JHM virus infected LEW rats the presence of IgG+ B-lymphocytes especially in demyelinated areas and an enhanced IgG content in macrophages within the plaques. Furthermore this group takes it as likely that the humoral response against the spike protein leads to a down-regulation of this viral protein at least in neurons thus pushing the infection from gray to white matter. Such effects could contribute to the observed change in neurological disorders of our rats where white matter dysfunction (paresis or paralysis) was only observed in animals that received B-cells and CD4+ T-cells, whereas symptoms of gray matter

Table 1. Effect of lymphocyte transfer on the course of JHM virus infection in irradiated Lewis rats

column nº	#1	#2	#3	#4	#5	#6	#7	#8
irradiation	no	600 rad γ						
cell transfer	- / - / -	- / - / -	CD4+ / / -	- / CD8+ / -	CD4+ / CD8+ / -	CD4+ / - / B	-* / CD8+ / B	CD4+ / CD8+ / B
onset of disease**	8	8	7	6	6	8	9	8
maximal severity	++	++	+++	+++	+++	++++	++	++
virus recovery at day 9pi	<500	8000	5100	9000	8000	800	4600	<500

* 5% CD4+ contamination, ** days past infection

disease were observed in the absence of the B-lymphocyte population.

In conclusion the presented data evidence, that a subclinical or clinically mild course of the intracerebral JHM infection is closely related to the presence of all lymphocyte subsets. Under the umbrella of antibody-secreting plasma cells viral spread is limited to few and small areas located mainly in the white matter. These sites are cleared from virus by the accompanying T-cell population. Absence of B-lymphocytes allows spread of virus to large areas of the CNS including the cerebral cortex and spinal cord and as a consequence severe clinical symptoms arise that are potentiated by a large number of inflammatory CD4+ and CD8+ T-lymphocytes [4].

ACKNOWLEDGEMENTS

This study was supported by the Bundesministerium für Forschung und Technologie of the Fed. Rep. of Germany (Grant no.: KI 01 8839-2).

We are very grateful to Ursula Sauer and Marion Zips for their skillful technical assistance.

REFERENCES

1 F.S. Cheever, J.B. Daniels, A.M. Pappenheimer, & O.T. Bailey, *J. Exp. Med.*, 90:181 (1949)
2 K. Nagashima, H. Wege, R. Meyermann, & V. ter Meulen, *Acta Neuropath.*, (Berl), 44:63 (1978)
3 H. Wege, S. Siddell, & V. ter Meulen, in: *"Current Topics in Microbiology and Immunology"*, 99, M. Cooper, ed., Springer, (1982)
4 R. Dörries, S. Schwender, H. Imrich, & H. Harms, *Immunology*, 74:539 (1991)
5 S. Schwender, H. Imrich & R. Dörries, Immunology, 74:533 (1991)
6 N. Hirano, K. Fujiwara, S. Hino, & N. Matsumoto, *Arch. Ges. Virusforsch.*, 44:298 (1974)
7 J.O. Fleming, F.I. Wang, M.D. Trousdale, D.R. Hinton, & S.A. Stohlman, in: *"Coronaviruses and Their Diseases"*, D. Cavanagh and T.D.K. Brown, eds, Plenum Press, New York (1990)
8 P.T. Massa, R. Dörries, & V. ter Meulen, *Nature*, 320:543 (1986)
9 J.A. Richt, & L. Stitz, *Arch. Virol.*, 124:95 (1992)
10 J.D. Sedgwick, S. Schwender, R. Gregerson, R. Dörries, & V. ter Meulen, *J. Exp. Med.*, Submitted for publication Sept 1991
11 C.A. Colton, J. Yao, J.E. Keri, and D. Gilbert, *J. Neuroimmunol.*, 30:89 (1992)
12 R.A. Shubin, M.A. Sussmann, J.O. Fleming, & S.A. Stohlman, *Microb. Pathog.*, 8:305 (1990)
13 S., Perlman, R. Schelper, E. Bolger, & D. Ries, *Microb. Pathog* 2:185 (1987)
14 C.P. Rossi, E. Cash, C. Aubert, and A. Coutinho, *J. Virol.*, 65:3895 (1991)
15 F. Zimprich, J. Winter, H. Wege, H. Lassmann, *Neuropathol. Appl. Neurobiol.* 17:469 (1991)

FUNCTIONAL CHARACTERIZATION OF CD8+ LYMPHOCYTES DURING CORONAVIRUS MHV IV INDUCED ENCEPHALITIDES IN RATS

A. Hein, H. Imrich, S. Schwender and R. Dörries

Institut für Virologie und Immunbiologie der Universität Würzburg
Versbacher Str. 7
8700 Würzburg
Federal Republic of Germany

INTRODUCTION

Intracerebral infection of rodents with the murine coronavirus JHM is a well established animal model to study the pathology of virus-induced primary demyelination of the central nervous system (CNS). Although it was assumed by Weiner[1] that cytopathogenic effects of the virus play the dominant role in this axonal loss of myelin sheaths, a growing body of evidence suggests now a significant contribution of the virus-specific immune response to the histopathological changes in the central nervous system as well as to the clinical course of the infection. In this context, in mice action of CD8+ T-lymphocytes appears to be a two-edged sword. On the one hand, they are necessary to clear JHM virus from infected brain tissue,[2,3] on the other hand, in vivo depletion of this lymphoid subset reduces drastically the appearance of white matter destruction[4,5] and adoptive transfer of either viral-specific[6] or naive syngeneic CD8+ splenocytes[7] in immunosuppressed animals fully reconstitutes neurological disease. This suggests that cytotoxic T-lymphocytes may cause demyelination by killing of virus-infected oligodendrocytes and thereby contribute to the clinical symptomatology of the infection.

In the rat, previously published data from our laboratory[8] support the idea that CD8+ T-lymphocytes might determine the clinical course of intracerebral infection by JHM virus. With respect to the inflammatory CD8+ response striking differences have been noticed between the disease-resistant inbred strain Brown Norway (BN) and the susceptible Lewis (LEW) strain. Whereas in LEW rats the virus challenge results in a strong biphasic infiltration of CD8+ T-cells into the CNS accompanied by extensive lesions of primary demyelination, in BN rats the amount of infiltrating cytotoxic T-lymphocytes is much less and demyelination is limited to few and small foci around the ventricles[8,9]. Pathogenic properties of CD8[+] cells in JHM virus-infected rats is seen in a series of experiments reported by Schwender et al.[7]. Reconstitution of immunosuppressed LEW rats by purified naive CD8[+] lymphocytes or a combination of CD4[+] and CD8[+] lymphocytes prior to infection results in an earlier onset and enhancement of neurological disease compared to completely reconstituted animals.

On the basis of these findings it was likely that CD8[+] T-lymphocytes could cause neurological disease by MHC class I restricted killing of JHM virus-infected targets in the CNS of LEW rats. However, at this stage of the investigations it was not clear why infiltration of CD8[+] T-lymphocytes in the few virus-affected areas of BN rats remained subclinical.[8] Thus, we attempted to compare CD8[+] T-lymphocytes from the CNS of both rat strains for their capability to kill JHM virus-infected targets *in vitro*.

MATERIALS AND METHODS

Viruses. For intracerebral inoculation of rats strain JHM of the mouse hepatitis virus (MHV) was propagated and purified according to Wege et al.[10]. Target cells used in cytotoxic assays were infected with JHM virus that had been passaged multiple times through the appropriate target cell line.

Animals. Three-week-old rats of the inbred strains Lewis (LEW, RT-1[l]) and Brown Norway (BN, RT-1[n]) were purchased from the "Zentralinstitut für Versuchstierzucht" (Hannover, FRG). After intracerebral inoculation with 100 μl of virus containing 10^3 plaque forming units the neurological signs of each animal were scored daily on an arbitrary scale as described elsewhere.[8]

Isolation of lymphocytes. At the time points indicated animals were killed and extensively perfused with PBS. Brain and spinal cord were removed and lymphocytes isolated as previously described [11].

Lymphocytes from the cervical lymph nodes (CLN) were prepared using standard methods. To remove CD4[+] T-cells, B-lymphocytes and natural killer- (NK-) cells immunomagnetic procedures were applied according to the manufacturer's recommendations. Briefly, lymphocytes were incubated with primary monoclonal antibodies (mabs) with specificity for the lymphocyte subset in question (mab W3/25 specific for CD4[+] T-cells, mab OX 33 specific for B-lymphocytes and mab 3.2.3[12] specific for NK-cells). The mabs W3/25 and OX 33 are commercially available (Serotec, Wiesbaden, FRG) and the mab 3.2.3 was a kind gift from T. Hünig, University of Würzburg, FRG. After incubation cells were washed and resuspended in Hank's solution containing goat-anti-mouse antibodies linked to magnetic beads (DRG Instruments, Marburg, FRG). Subsequently, the cell suspension containing tubes were placed onto a magnetic plate (Du Pont, Dreieich, FRG). Non-labeled CD8+ lymphocytes did not adhere to the magnetic plate and thus could be collected from the supernatant by centrifugation. Subsequently, they were resuspended in RPMI supplemented with 10% FCS and T-cell growth factors (TCGF) and maintained in culture for 5 to 6 days before assaying their cytotoxic capacity.

Cell lines. The cell lines LE 1B4 (RT-1A^l) and BN 3B2 (RT-1A^n) were used as targets in cytotoxic assays. They were obtained after fusing of primary rat splenocytes with the mouse myeloma cell line SP2/0-Ag14 [13]. Hybridomas were grown in RPMI supplemented with 10% FCS.

Cytotoxicity assay. After infection of the hybridoma cell lines LE 1B4 and BN 3B2 with JHM virus at a multiplicity of infection (moi) of 10-15 for 2h at 37°C cells were washed once with RPMI. Subsequently, infected as well as uninfected cells were irradiated, labeled with ^{51}Cr (Du Pont, Dreieich, FRG), washed twice with RPMI/10% FCS and resuspended with effector cells at various proportions in RPMI/10% FCS. After 5h incubation at 37°C, 100 µl of cell free supernatant was removed and released radioactivity was determined in a gamma counter (LKB Wallac, Turku, Finland). Percent specific lysis was calculated according to the formula:

$$\frac{(\text{cpm in experiment - cpm spontaneous release}) \times 100}{(\text{cpm in 2\% Triton X 100 - cpm spontaneous release})}.$$

Spontaneous release never exceeded 25% of the maximum release.

RESULTS AND DISCUSSION

In a first attempt to unravel possible differences between LEW and BN rats with respect to the killing capacity of their CD8+ lymphocytes we isolated leukocytes from the CNS of infected animals 14 days past infection (dpi). At this time, LEW rats usually reveal a strong paralytic disease whereas BN rats show no overt signs of the infection. As demonstrated in figure 1 the *ex vivo* killing capacity was higher in BN rats compared to LEW rats. Of note is the fact that due to the higher proportion of CD8+ cells in leukocytes from LEW rats, a higher CD8+/target ratio is effective in this rat strain compared to BN rats. This suggests, that CD8+ cells from the CNS of LEW rats are less active or exhibit lower virus-specificity compared to BN rats. However, the overall specific killing of 10% as seen in CD8+ cells from BN rats was surprisingly low compared to other *ex vivo* killing assays in mice [14] Therefore we assumed that either the proportion of JHM virus-specific MHC class I restricted cells in the CD8+ population was low or that only few of them were in a state of activation that allowed killing. Our attempts to expand activated CD8+ cells from the brain by cultivation in TCGF failed completely. Possibly, this can be explained by data from Imrich et al.,[15] who showed that T-lymphocytes isolated from the CNS at any time past infection neither did proliferate *ex vivo* nor could be stimulated to proliferation by mitogens or viral antigens.

Since CLN were identified in our laboratory as a site of T-cell sensitization to JHM virus and strong virus-specific proliferation[15,] we isolated CLN lymphocytes from intracerebrally infected rats at day 7 past infection. CD8+ T-lymphocytes were purified from the isolated lymphocyte fraction up to 98% (data not shown) and cultured in TCGF containing medium. Six days later these cells were assayed for their cytotoxic capacity. As seen in figure 2 specificity of CD8+ cell mediated killing is comparable in both rat strains. However, the overall cytolytic activity is considerable higher in CD8+ cells from BN rats. To date, we cannot unequivocally explain the strong unspecific killing of uninfected target cells in our assays. It is

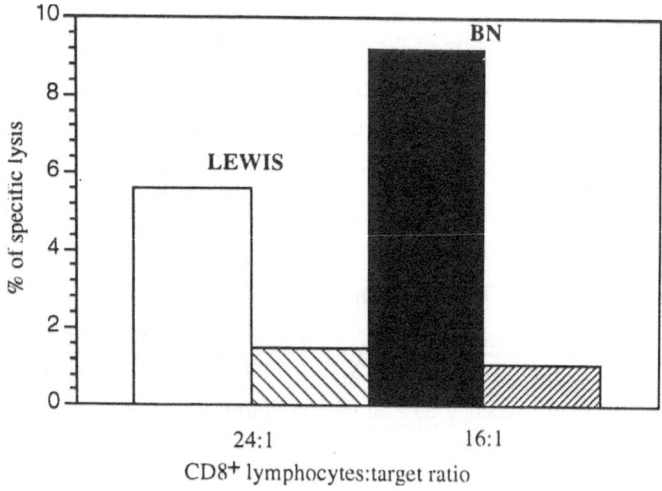

Figure 1. Cytotoxic capacity of brain-derived CD8+ lymphocytes 14 dpi. White (LEW) and black (BN) bars represent cytotoxicity against syngeneic, virus-infected targets, hatched bars against uninfected, MHC I-matched cells. Lymphocytes from brain and spinal cord of 3 LEW and BN rats, respectively, were pooled and measured for their cytolytic potential *ex vivo*.

clear that this activity is <u>not</u> due to NK-cells, because they were removed prior to the *in vitro* culture of isolated lymphocytes and flow cytometric analysis of cultered CD8+ cells excluded the outgrow of NK-cells during in vitro expansion. To minimize the problem of non-specific killing experiments are in progress to expand exclusively JHM-specific CD8+ cells in isolated leukocytes.

In summary, these preliminary results indicate that after inoculation of rats with the neurotropic murine coronavirus JHM the recruited CD8+ T-lymphocytes are able to kill virus-infected targets. BN rats show not only a greater virus-specific cytotoxicity but also a higher overall cytolytic capacity compared to LEW rats. This suggests that generation and

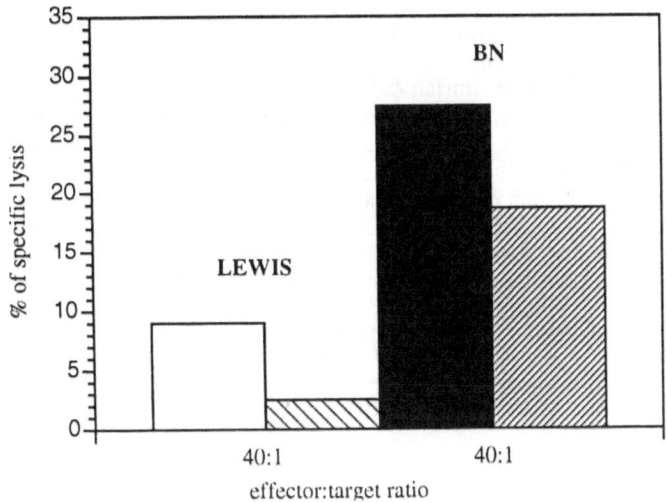

Figure 2. Cytolytic activity of purified CD8$^+$ T-lymphocytes isolated from the cervical lymph nodes of JHM-infected rats 7 dpi. Cytotoxicity was assayed against syngeneic, JHM-infected (white and black bars) or uninfected (hatched bars) targets. Results represent the mean value of 8 LEW and BN rats, respectively.

activation of cytotoxic T-lymphocytes is more effective in BN rats. These findings are in line with observations that BN rats recruit more rapidly and effectively virus-specific antibody secreting cells into the brain parenchyma[11] as well as the fact that sensitization of CLN T-lymphocytes to JHM virus occurs earlier past infection and is more vigorous in BN rats compared to LEW rats [15]. Therefore, we tend to believe that BN rats are able to limit viral spread in the CNS very early past infection due to a very rapid afferent limb of the JHM virus-specific immune response. Under these circumstances cytotoxic action of CD8$^+$ cells in the few and small periventricular sites of virus infection remains subclinical.

ACKNOWLEDGEMENTS

This study was supported by the Bundesministerium für Forschung und Technologie of the Fed. Rep. of Germany (Grant no.: KI 01 8839-2).

We are very grateful to Marion Zips and Ursula Sauer for expert technical assistance.

REFERENCES

1   L.P. Weiner, *Arch. Neurol.* 28:298 (1973).
2   M.A. Sussman, R.A. Shubin, S. Kyuwa & S.A. Stohlman, *J. Virol.* 63:3051 (1989).
3   J.S.P. Williamson & S.A. Stohlman, *J. Virol.* 64:4589 (1990).
4   M. Rodriguez & S. Sriram, *J. Immunol.* 140:2950 (1988).
5   S. Sriram, D.J. Topham, S. Huang & M. Rodriguez, *J. Virol.* 63:4242 (1989).

6   J.O. Fleming, F.I. Wang, M.D. Trousdale, D.R. Hinton & S.A. Stohlman, *Adv. Exp. Med. Biol.* 276:565 (1990).
7   S. Schwender, A. Hein, H. Imrich & R. Dörries, *same volume*.
8   R. Dörries, S. Schwender, H. Imrich & H. Harms, *Immunology* 74:539 (1991).
9   R. Watanabe, H. Wege & V. ter Meulen, *Lab. Invest.* 57:375 (1987).
10 H. Wege, A. Müller & V. ter Meulen, *J. Gen. Virol.* 41:217 (1978).
11 S. Schwender, H. Imrich & R. Dörries, *Immunology* 74:533 (1991).
12 W.H. Chambers, N.L. Vujanovic, A.B. DeLeo, M.W. Olszowy, R.B. Herberman & J.C. Hiserodt, *J. Exp. Med.* 169: 1373 (1989).
13 M. Shulman, C.D. Wilde & G. Köhler, *Nature* 276:269 (1978).
14 M.D. Lindsley, R. Thiemann & M. Rodriguez, *J. Virol.* 65:6612 (1991).
15 H. Imrich, S. Schwender, A. Hein & R. Dörries, *same volume*.

PHENOTYPIC AND FUNCTIONAL CHARACTERIZATION OF CD4+ T-CELLS INFILTRATING THE CENTRAL NERVOUS SYSTEM OF RATS INFECTED WITH CORONAVIRUS MHV IV

H. Imrich, S. Schwender, A. Hein and R. Dörries

Institut für Virologie und Immunbiologie der Universität Würzburg
Versbacher Str. 7
8700 Würzburg

INTRODUCTION

Mouse hepatitis virus strain JHM, a neurotropic member of coronaviridae, causes neurological diseases in rats after intracerebral inoculation. The course of the infection depends on the rat strain, the age of the animals and the type of virus used. After infection at the age of 3 weeks BN rats remain clinically healthy. In contrast, 40% of LEW rats die within the first week past infection, when inoculated at the same age. Most of the surviving animals develop neurological signs with increasing severity up to the second week past infection, followed by complete convalescence [1,2,3].

Histopathologically, the infection is characterized by demyelinated areas along with infiltrating mononuclear cells. In clinically healthy BN rats demyelinated foci are small and localized around the ventricles, whereas in LEW rats wide-spread demyelination is seen, extending from brain to spinal cord [1]. Within the demyelinated areas differences can be detected between the two rat strains with respect to the amount of infiltrating lymphocytes as well as to the spatial arrangement of lymphocyte subpopulations. In LEW rats higher numbers of CD8+ T-cells dominate the inflammatory response in the brain parenchyma early past infection [3]. In contrast, in BN rats CD4+ T-cells constitute the major infiltrating cell type at any time past infection. Both, CD8+ and CD4+ T-cells, are homing close to and in virus-infected areas suggesting that most of these cells are virus specific [3]. The humoral immune response is characterized by striking differences on a functional level between the two rat strains. BN rats recruit high numbers of virus-specific antibody secreting cells into the brain tissue in response to the viral infection. This is in contrast to Lewis rats where these cells appear later and in much lower numbers [5].

Because CD4[+] T-cells play a key role in the differentiation process of B-cells, we present in this report our efforts to characterize CD4[+] T-lymphocytes from the brain of infected LEW and BN rats with respect to their state of activation 6 days past infection (dpi), a time when the majority of JHM-infected LEW rats reveal the first signs of neurological disease[3].

MATERIAL AND METHODS

Virus. The JHM strain of coronavirus (MHV4) for infection of rats was propagated as described earlier [5]. Virus particles used as antigen were purified according to Wege et al.[6].

Animals. LEW and BN rats were obtained at the age of 3 weeks from the "Zentralinstitut für Versuchstierzucht" (Hannover, FRG). They were inoculated with 1×10^3 plaque-forming units of JHM virus. Neurological signs of the infection were scored daily according to a scale published previously [3].

Isolation of lymphocytes. Lymphocytes were isolated from cervical lymph nodes (CLN) by mincing the tissue through a stainless steel sieve into ice-cold Hank´s buffer containing 3% fetal calf serum (FCS). Lymphocytes were collected from brain and spinal cord by density gradient centrifugation after mechanical disruption and enzymatic digestion of CNS tissue [5]. In order to enrich T-cells the adherent cell fraction was removed from CNS isolated lymphocytes by passage through a nylon wool column.

Phenotyping of isolated lymphocytes. Surface markers of isolated lymphocytes were determined by double immunofluorescence and flow cytometric analysis as published previously [3]. The FITC-labeled monoclonal antibody (mab) W3/25 specific for the rat CD4 molecule was used either in combination with the mab OX22 (rat CD45RB) or the mab OX25 (rat interleukin 2 receptor) to stain lymphocytes. OX22 or OX25 mabs were detected on lymphocytes by phycoerythrin (PE)-labeled goat anti-mouse IgG antibodies. Primary antibodies were purchased from Serotec (Wiesbaden, FRG) and PE-labeled secondary antibody from Dianova (Hamburg, FRG).

Proliferation assay. Lymphocytes were cultured in round bottom 96 well microtiter plates for 72 hours (RPMI medium, 5×10^{-5} M 2-mercapto-ethanol, 3% normal rat serum). To monitor the proliferative activity $0,1\mu Ci$ of ^3H-thymidine (Amersham Buchler, Braunschweig, FRG) were added per well at 48 hours of culture. Antigen- or mitogen-induced proliferation was assayed by addition of keyhole limpet hemocyanine (KLH) or

JHM virus, both at a concentration of 20µg/ml or concanavalin A (Con A) at a concentration of 4µg/ml to the cells immediately after onset of cultures. Incorporation of tritiated thymidine was measured in a multi-chanel scintillation counter (β-Plate, Pharmacia, Freiburg, Germany). Mean values were calculated from 2 or 3 wells.

RESULTS AND DISCUSSION

In a first approach to characterize the infiltrating CD4+ T-lymphocyte population their phenotype was analyzed by flow cytometry. The majority of CD4+ T-cells isolated from the brain of both rat strains were CD45RB-, CD25- (data not shown). Loss of the CD45RB marker that is expressed on naive CD4+ T-cells [7,8] indicated that the majority of T-cells in the brain tissue was of the memory/primed phenotype. The low expression of CD25 (interleukin 2 receptor, IL2R) was surprising because presence of viral antigen in the brain of both rat strains up to 3 weeks past infection[3] and upregulation of MHC antigens class II antigens on microglia cells[9,10] suggested that primed CD4+ T-cells could be stimulated in the CNS tissue to local proliferation leading to the typical CD4+ T-cell accumulation observed in the virus infected areas in the brain tissue[3].

To investigate the question of T-lymphocyte expansion in the CNS more closely, leukocytes from the CNS were enriched for T-lymphocytes by nylon wool passage and assayed for DNA synthesis *ex vivo* by incorporation of ^{3}H-thymidine. As can be seen in figure 1 the uptake of ^{3}H-thymidine in these cells is not much higher compared to the background incorporation of T-cells from uninfected animals (50 to 80 cpm) indicating that the proliferative activity of inflammatory T-cells in the JHM-infected CNS was low in both rat strains. The weak proliferative response observed in 1×10^5 CNS-isolated lymphocytes was confirmed by assaying 5 times more T-lymphocytes from the CNS (figure 1). Since we failed to stimulate proliferation of these cells with a strong T-cell mitogen like Con A (Fig. 2, left) and JHM virus-specific antigens, even in the presence of irradiated feeder cells from secondary lymphoid organs, we tend to believe that primed T-cells entering the JHM virus-infected brain do not proliferate and are rendered insensitive to further stimulation of proliferation. As a consequence, increase of T-lymphocytes in the brain parenchyma of both rat strains as reported by us previously[3] reflects rather transient retention and accumulation of these cells than antigen-driven expansion at the site of infection.

This hypothesis is strengthened by findings of Matsumoto et al.,[11] who reported absence of proliferating CD4+ T-cells in the CNS of LEW rats suffering from experimental allergic encephalitis (EAE), an autoimmune reaction directed versus the myelin-containing glia cell fraction. Moreover, passive transfer of proliferating T-cells from EAE animals resulted in CNS inflammation but suppression of proliferation in the CD4+ lymphocyte population after entry into the CNS. Nelson and co-workers[12] reported low cloning efficiency of

T-lymphocytes isolated from lungs due to markedly reduced capacity of these cells to proliferate. Most likely lung-resident macrophages exert these suppressive effects on the non-recirculating T-lymphocyte population in this organ. From these data it is conceivable that as a general principle expansion of CD4+ T-lymphocytes is not allowed in extra-lymphoid tissue. In this context, the immunologically sequestered situation of the unaffected CNS[9] rises the question for the extra-cerebral lymphoid tissue where priming and expansion of JHM-specific T-cells occur.

The deep cervical lymph nodes (CLNs) are known to be involved in the recruitment of a specific immune response to antigens that were administered to the ventricles of the

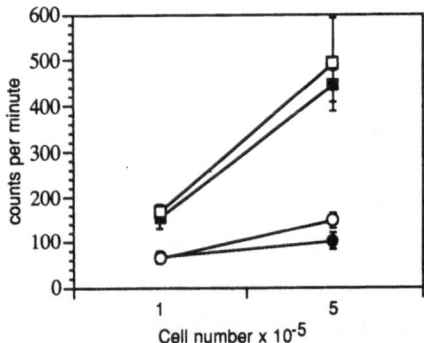

Figure 1. *Ex vivo* proliferation of lymphocytes from JHM virus-infected rats (6 dpi). Different amounts of lymphocytes from cervical lymph nodes and the central nervous system were assayed for proliferation by ^{3}H-thymidine incorporation (lymph node cells □, ■ and CNS-isolated lymphocytes O, ● from LEW and BN rats respectively).

brain[13]. Consequently, we analyzed lymphocytes derived from CLNs of JHM infected BN and LEW rats. Figure 1 shows that these cells indeed do respond to the intracerebral virus infection by proliferation in a dose dependent manner when assayed *ex vivo*. After restimulation of CLN-lymphocytes with Con A the incorporation of ^{3}H-thymidine was about 100-fold higher compared to lymphocytes from the brain (Fig. 2, right panel). Virus-specificity of this response was demonstrated by the fact that stimulation of CLN-cells by JHM-specific antigens resulted in a considerable increase of ^{3}H-thymidine incorporation compared to the unrelated control antigen like KLH (Fig. 3, right panel).

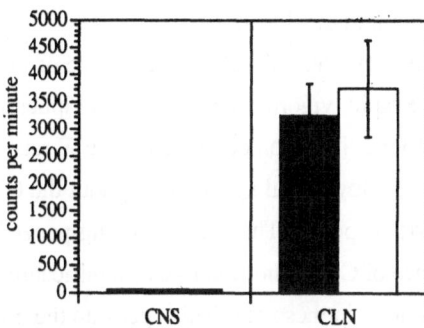

Figure 2. Mitogen-induced proliferation of lymphocytes from JHM virus-infected rats (6 dpi). Lymphocytes from the central nervous system (CNS) and cervical lymph nodes (CLN) of LEW (□) and BN (■) rats were stimulated with concanavalin A. Proliferation was monitored by ^{3}H-thymidine incorporation.

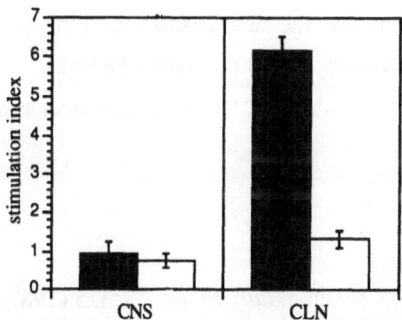

Figure 3. Antigen-induced proliferation of lymphocytes from JHM virus infected rats (6 dpi). Lymphocytes from the central nervous system (CNS) and cervical lymph nodes (CLN) of LEW (□) and BN (■) rats were stimulated with control antigen (KLH = keyhole limpet hemocyanine) and JHM virus. Proliferation was monitored by ^{3}H-thymidine incorporation. Virus-specific proliferation is expressed as stimulation index (counts per minute in virus-stimulated cells/counts per minute in KLH-stimulated cells)

Interestingly, the antigen specific response in BN rats is much higher compared to LEW rats, especially early past infection. This result fits well to the finding that in BN rats virus-specific killing activity of brain-infiltrating CD8[+] cells can be detected earlier compared to LEW rats [4] as well as to previous reports by our laboratory that a virus-specific antibody response is recruited more rapid in BN rats [2,5].

In summary, these data implicate that the subclinical course of the infection in BN rats might be explicable by a more rapid, vigorous and specific peripheral immune system response to the intracerebral JHM virus infection compared to LEW rats. The delay of the virus-specific response in LEW rats allows viral spread throughout the CNS resulting in a strong inflammatory T-lymphocyte response. The accompanying paper in this volume[14] shows that CD4[+] cells in the absence of CD8[+] and B-cells can significantly contribute to the severity of the clinical symptomatology in these rats. This opens up the prospect that CD4[+] T-cells, which are essential regulators of T- and B-cell differentiation in secondary lymphoid organs, may fulfill other effector functions after they have passed the blood brain barrier. Currently we examine the expression of interleukins in CNS-derived leukocytes by reverse transcriptase polymerase chain reaction (RT-PCR) to shed more light on the yet unknown effector mechanisms of CD4[+] cells at the site of JHM virus infection.

ACKNOWLEDGEMENTS

This study was supported by the Bundesministerium für Forschung und Technologie of the Fed. Rep. of Germany (Grant no.: KI 01 8839-2).

We gratefully thank M. Zips and U. Sauer for expert technical assistance.

REFERENCES

1 R. Watanabe, H. Wege, & V. ter Meulen, *Lab. Invest.* 57:375 (1987)
2 R. Dörries, R. Watanabe, H.Wege, & V. ter Meulen, *J. Neuroimmunol.* 12:131 (1986)
3 R. Dörries, S. Schwender, H. Imrich, & H. Harms, *Immunology* 12:539 (1991)
4 A. Hein, H. Imrich , S. Schwender,& R. Dörries, *same Volume*
5 S. Schwender, H. Imrich,& R. Dörries, *Immunology* 12:533 (1991)
6 H. Wege, A. Müller, & V. ter Meulen, *Lab. Invest.* 41:217 (1978)
7 R.P. Arthur & D. Mason, *J. Exp. Med.* 168:774 (1986)
8 F. Powrie & D. Mason, *J. Exp. Med.* 169: 653 (1989)
9 J.D. Sedgwick & R. Dörries, *Seminars in THE NEUROSCIENCES* 3: 93 (1991)
10 J.D. Sedgwick, S. Schwender, H. Imrich , R. Dörries, G.W. Butcher, & V. ter Meulen, *Proc. Natl. Acad.* 88: 7438 (1991)
11 K. Ohmori, Y. Hong, M. Fujiwara, Y. Matsumoto, *Lab. Invest.* 66:54 (1992)
12 D. Nelson,. Strickland, & P.G. Holt, *Immunology* 69: 476 (1991)
13 C. Harling-Berg, P.M. Knopf, J. Merriam, & H.F. Cserr, *J. Neuroimmunol.* 25:185 (1989)
14 S. Schwender, A. Hein, H. Imrich, & R. Dörries, *same Volume*

CYTOKINE INDUCTION *IN VITRO* IN MOUSE BRAIN ENDOTHELIAL CELLS AND ASTROCYTES BY EXPOSURE TO MOUSE HEPATITIS VIRUS (MHV-4, JHM)

Jeymohan Joseph[1], James L. Grun[2], Fred D. Lublin[1] and Robert L. Knobler[1]

Division of Neuroimmunology, Department of Neurology[1], and Department of Biochemistry[2], Jefferson Medical College, Philadelphia, PA 19107

INTRODUCTION

Mouse hepatitis virus (MHV-4, JHM) is a neurotropic coronavirus which causes a spectrum of disease ranging from fatal encephalomyelitis to demyelination in susceptible murine hosts (1). Both direct infection of oligodendrocytes and immune-mediated events have been reported to play a role in the pathologic events in the central nervous system (CNS) following MHV-4 infection (2-5). The identification of CD8[+] T-cells, NK cells, B-cells and PMN in the CNS of MHV-JHM infected mice suggests that immune mechanisms may be playing a role in the virus-induced disease process (3-9).

The role of locally released cytokines in the regulation of immune and inflammatory events in the CNS following mouse hepatitis virus infection has not yet been systematically studied. Cytokines play a critical role in the modulation of immune and inflammatory events, important in both anti-viral immunity, and the virus-induced pathology observed in the CNS. Endothelial cells and astrocytes in the CNS are potential sources of cytokines and have been demonstrated to synthesize IL-6 in response to treatment with tumor necrosis factor (TNF), interleukin-1 (IL-1) and lipopolysaccharide (10-16). Endothelial cells and astrocytes in the CNS are also readily infected by MHV-4 and the products released by these cells may be important in regulating immune-mediated events in the CNS (17-19).

The cytokine IL-6 was chosen for this study because its release following infection has been demonstrated in other virus model systems (11, 20-22). Additionally, IL-6 is a multifunctional cytokine with immunoregulatory effects on B-cells, T-cells and neutrophil functions (23-26).

We demonstrate that IL-6 induction in MHV-4 exposed endothelial cells or astrocytes is strain dependent. BALB/c (MHV-4 susceptible) derived endothelial cells can produce up to sixteen fold higher levels of IL-6, and release this earlier than SJL (MHV-4 resistant), as determined both by bioassay and Northern analysis. Active infection is not necessary since exposure to UV-inactivated MHV-4 (UV-MHV-4) can also induce IL-6 in these cells.

MATERIALS AND METHODS

Virus Infection and Collection of Supernatants

Cerebral endothelial cell and astrocyte cultures were established from the brains of BALB/c (MHV-susceptible) and SJL (MHV-resistant) strains of mice as previously described (27-29). Endothelial cells or astrocytes grown in T-25 flasks (1 x 10^6 cells/flask) were infected with MHV-4 or UV-MHV-4 for 1 hour at 37^0C using a multiplicity of infection (MOI) of 0.1. This MOI reflects the working titer of virus that elicits a cellular response without extensive cytolytic effects based on previous studies (17-19).

After a one hour incubation of the cells with MHV-4 or UV-MHV-4 the cultures were washed three times with phosphate buffered saline (PBS). After washes the cultures were fed with endothelial cell or astrocyte culture medium. Supernatants were collected on day 1-4 after virus exposure for testing in an IL-6 proliferation bioassay described below.

Bioassay for IL-6

The proliferation bioassay for IL-6 was performed using an IL-6 dependent B cell hybridoma (T1165tc) (30). In order to test the specificity of the proliferative response to IL-6, a neutralizing rat anti-mouse IL-6 monoclonal antibody (Genzyme Corporation, Cambridge, MA) was used.

RNA Isolation and Northern Analysis

Four hours after exposure of endothelial cells and astrocytes to MHV-4, UV-MHV-4 or tumor necrosis factor (200U/ml, Genzyme, Boston, MA), total cellular RNA was isolated by immediate solubilization of cells in guanidine hydrochloride by standard procedures (Maniatis et al., 1982). The RNA was then hybridized overnight at 45^0C with a radiolabeled murine IL-6 cDNA probe (obtained from Dr. Frank Lee, DNAX, Palo Alto, CA), washed under high stringency conditions and analyzed by autoradiography with intensifying screens (Dupont, Hoffman Estates, IL, USA). Densitometry of the autoradiographs was carried out using a Macbeth densitometer (Model TD-932, Macbeth Process Measurements, Newburg, NY).

RESULTS

Interleukin-6 Induction in Cerebral Endothelial Cells

Table 1 demonstrates that IL-6 is induced in both BALB/c and SJL derived endothelial cells following exposure to MHV-4 as determined by bioassays, albeit at very different levels. In the MHV-4 susceptible BALB/c cells, the peak of IL-6 induction occurred on day 2 of infection, and the levels of IL-6 were sixteen fold higher (>640 U/ml) than those obtained from the MHV-4 resistant SJL cells. The level of IL-6 release from SJL endothelial cells (40 U/ml), following exposure to MHV-4, was not substantially different from the basal levels (<20U/ml) detected in control cultures. Active infection of the cells was not apparently necessary since UV-inactivated virus also induced IL-6 in a strain dependent fashion.

Interleukin-6 Induction in Astrocytes

Table 2 demonstrates that IL-6 was induced in both BALB/c and SJL astrocytes by MHV-4, as determined by bioassay of the culture supernatant. Levels of IL-6 were maximal on day 1 (>640U/ml) after MHV-4 infection of BALB/c astrocytes. The levels of IL-6 continued to remain high even four days post-infection. In contrast, IL-6 activity in SJL astrocytes peaked later, on day 2 (402U/ml), but were not sustained, declining from day 3 onward. As seen with the endothelial cell cultures, UV-inactivated virus was also able to induce IL-6 (Table 2). In order to confirm the specificity of the proliferative signal a neutralizing rat anti-mouse monoclonal IL-6 antibody was used (Tables 1 and 2).

Induction of IL-6 mRNA in Astrocytes

Northern analysis was performed to look for differences in the induction of IL-6 mRNA between BALB/c and SJL derived astrocytes (Fig. 1). The IL-6 cDNA probe (moIL-6 cDNA, DNAX, Palo Alto, CA, USA), used in our studies, hybridizes to a 1.3kb species (11). The RNA from untreated cells (lanes A and E), showed no evidence for induction of IL-6 mRNA. In contrast, IL-6 mRNA induction was noted in lanes B-D and F-G. IL-6 mRNA was induced to a comparable degree in both BALB/c (susceptible) and SJL (resistant) derived astrocytes following exposure to either MHV-4 (lanes B and F), UV-MHV-4 (lanes C and G) or TNF at 200U/ml (lanes D and H) for 4 hours. TNF exposure was included as a positive control for IL-6 induction. A similar pattern of MHV-4 induction was obtained when evaluating IL-6 mRNA of endothelial cells (data not shown). These results show that IL-6 mRNA is induced at comparable levels in the different strains and cell types, although there are dramatic differences in the quantity of IL-6 released. This suggests that these differences may be accounted for by altered translation or post-translational processing.

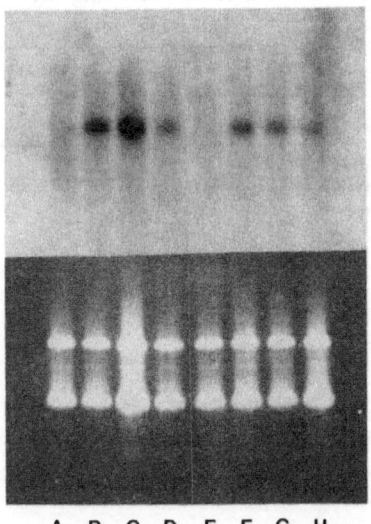

A B C D E F G H

Fig. 1. Northern blot analysis of IL-6 mRNA induction. Lanes **A-D** shows RNA derived from Balb/c (MHV-susceptible) astrocytes. Lanes **E-H** shows RNA derived from SJL (MHV-resistant) astrocytes. Cells were treated for 4 hours as follows, Lane **A.** Untreated, **B.** MHV-4, **C.** UV-inactivated MHV-4, **D.** Tumor Necrosis factor (Genzyme, 200U/ml), **E.** Untreated, **F.** MHV-4, **G.** UV-inactivated MHV-4, **H.** Tumor Necrosis Factor (Genzyme, 200U/ml). The top panel shows the autoradiograph and the lower panel is the ethidium bromide profile of the gel (5µgms of RNA loaded per lane)

Table 1. Quantitation of IL-6 levels induced following MHV-4 treatment of endothelial cells

Day p.i.		MHV-4		U.V. MHV-4	
		A	B	C	D
1	Balb/c	288*	0	115	0
	SJL	40	0	20	0
2	Balb/c	640	0	640	0
	SJL	40	0	20	0
3	Balb/c	640	10	254	0
	SJL	40	0	20	0
4	Balb/c	490	10	416	0
	SJL	ND	ND	40	0

Table 2. Quantitation of IL-6 levels induced following MHV-4 treatment of astrocytes

Day p.i.		MHV-4		U.V. MHV-4	
		A	B	C	D
1	Balb/c	640*	10	561	0
	SJL	73	0	106	0
2	Balb/c	640	10	518	0
	SJL	402	10	128	0
3	Balb/c	640	10	640	10
	SJL	176	10	70	0
4	Balb/c	640	10	ND	ND
	SJL	126	10	47	0

*units/ml

IL-6 units were determined by the dilution of supernatants yielding half maximal proliferation. The data presented in rows B and D shows the effect of rat anti-mouse IL-6 monoclonal antibody (5μg/ml) in neutralizing the IL-6 activity in the supernatants. IL-6 levels in cell culture supernatants on Day 0 was < < 20 U/ml.

DISCUSSION

The study of the immune and inflammatory events in the CNS following infection with MHV-4 is important in understanding the pathologic events and mechanisms of virus induced demyelination. Immunoregulatory cytokines can profoundly affect the cascades of both humoral and/or cell-mediated immune events in the CNS. However, there has not yet been a systematic analysis of cytokine induction following MHV-infection of CNS derived cells.

The cytokine IL-6 was studied because of 1) demonstration of its release following viral infections with LCMV (lymphocytic choriomeningitis virus), VSV (vesicular stomatitis virus), HIV (human immunodeficiency virus) and HTLV-1 (Human T-cell leukemia virus) and, 2) its multifunctional immunoregulatory role in modulating T, B and neutrophil functions (10, 11, 20-26).

We report on the induction of IL-6 in cultures of cerebral endothelial cells or astrocytes following infection with MHV-4. The MHV susceptible BALB/c derived endothelial cells and astrocytes produce substantially higher levels of IL-6 (>640U/ml), and at earlier time points, than resistant SJL derived cells. The SJL endothelial cells yield barely more (40U/ml) than basal levels (<20U/ml) of IL-6 in response to MHV-4. In contrast, SJL astrocytes are induced to release a significant quantity of IL-6 (402 U/ml), although not as great as in the BALB/c astrocytes (>640U/ml). Therefore, the strain differences, in IL-6 induction in response to MHV-4, are more striking between the endothelial cells than with the astrocytes.

The mechanisms regulating this strain dependent induction of IL-6 is under further study. One possible explanation may reflect the differences in MHV-receptor expression on endothelial cells and astrocytes. MHV binding receptors have been demonstrated on BALB/c but not SJL derived intestinal brush border and liver cells (32). The binding of the virus to its receptor on the cell surface may activate a signal transduction pathway for IL-6 production.

The results obtained with UV-MHV-4 suggests that infection is not required and that the binding of the viral particles to its receptor on the cell surface may be sufficient to trigger the release of IL-6. This ability of UV-MHV-4 to exert a biological effect is not unique to the induction of interleukin-6. Previous studies in our laboratory have demonstrated that UV-MHV-4 can block γ-interferon-induced MHC class II antigen expression on endothelial cells to the same degree as infectious virus (19).

The greater degree of differences in IL-6 induction noted in cerebral endothelial cells compared to astrocytes is intriguing in the face of the potential role of this cytokine in triggering a variety of immune mediated inflammatory mechanisms. Since the MHV-4 replication in BALB/c mouse brain is two to three logs greater than in SJL mice (3), and there is a correspondingly greater inflammatory response to MHV-4 infection in BALB/c than SJL mice, this observation deserves further investigation. Current studies are directed at elucidating these mechanisms.

REFERENCES

1. Bailey, O.T., Pappenheimer, A.M., and Cheever, F.S., 1949, J. Exp. Med., 90:195.
2. Knobler, R.L., Haspel, M.V., and Oldstone, M.B.A., 1981, J. Exp. Med., 153:832.
3. Knobler, R.L., Tunison, L.A., Lampert, P.W., and Oldstone, M.B.A., 1982, Am. J. Pathol., 109:157.
4. Wang, F-I., Stohlman, S.A., and Fleming, J.O., 1991, J. Neuroimmunol., 30:31.
5. Williamson, J.S.P., Sykes, K.C., and Stohlman, S.A., 1991, J. Neuroimmunol., 32:199.
6. Stohlman, S.A., Brayton, P.R., Harmon, R.C., Stevenson, R.G. Ganges, R.G., and Matsushima, G.K., 1983, Int. J. Cancer., 31:309.

7. Welsh, R.M., Haspel, M.V., Parker, D.C., and Holmes, K.V., 1986, J. Immunol., 136:1454.

8. Natuck, R.J., and Welsh, R.M., 1987, J. Immunol., 138:877.

9. Zimprich, F., Winter, J., Wege, H., and Lassmann, H., 1991, Neuropathol. Applied. Neurobiol., 17:469.

10. Frei, K., Leist, T.P., Meager, P., Gallo, P., Leppert, D., Zinkernagel, R.M. and Fontana, A., 1988, J. Exp. Med., 168:449.

11. Frei, K., Malipiero, U.V., Leist, T.P., Zinkernagel, R.M., Schwab, M.E., and Fontana, A., 1989, Eur. J. Immunol., 19:689.

12. Jirik, F.R., Podor, T.J., Hirano, T., Kishimoto, T., Loskutoff, D.J., Carson, D.A., and Lotz, M., 1989, J. Immunol., 142:144

13. Lieberman, A.P., Pitha, P.M., Shin, H.S., and Shin, M.L., 1989, PNAS., 86:6348.

14. Shalaby, M.R., Waage, A., and Esperik, T., 1989, Cell. Immunol., 121:372.

15. Sironi, M., Brevario, F., Prosperio, P., Biondi, A., Vecchi, A., Van Damme, J., Dejana, E., and Mantovani, A.,1989, J. Immunol., 142:549.

16. Beneveniste, E.N., Sparacio, S.M., Norris, J.G., Grenett, H.E., and Fuller, G.M., 1990, J. Neuroimmunol., 30:201.

17. Joseph, J., Knobler, R.L., Lublin, F.D., and Hart, M.N., 1989, J. Neuroimmunol., 22:241.

18. Joseph, J., Knobler, R.L., Lublin, F.D., and Hart, M.N., 1990, Adv. Exp. Med. Biol., 276:579.

19. Joseph, J., Knobler, R.L., Lublin, F.D., and Hart, M.N., 1991, J. Neuroimmunol., 33:181.

20. Houssiau, F.A., Bukasa, K., Sindic, C.J.M., Van Damme, J., and Van Snick, J., 1988, Clin. Exp. Immunol., 71:320.

21. Breen, E.C., Rezai, A.R., Nakajima, K., Beall, G.N., Mitsuyasu, R.T., Hirano, T., Kishimoto, T., and Martinez-Maza, O., 1990, J. Immunol., 144:480.

22. Nishimoto, N., Yoshizaki, K., Eiraku, N., Machigashira, K., Tagoh, H., Ogata, A., Kuritani, T., Osame, M., and Kishimoto, T., 1990, J. Neurol. Sci., 97:183.

23. Cernetti, C., Steinman, R.M., and Granelli-Piperno, A., 1988, PNAS., 85:1605.

24. Takai, Y., Wong, G.G., Clark, S.C., Burakoff, S.J., and Hermann. S.H., 1988, J. Immunol., 140:508.

25. Borish, L., Rosenbaum, R., Albury, L., and Clark, S., 1989, Cell. Immunol., 121:280.

26. Le, J., and Vilcek, J.,1989, Lab. Invest., 61:588.

27. DeBault, L.E., Henriquez, E., Hart, M.N., and Cancilla, P.A., 1981, In Vitro., 17:480.

28. Knobler, R.L., Cole, R., de Vellis, J., Lewicki, H., Buchmeier, M.,and Oldstone, M.B.A., 1987, In Coronaviruses., M.M.C. Lai and S.A. Stohlman, eds. Plenum, New York, p. 231.

29. McCarthy, K.D., and deVellis, J., 1980, J. Cell. Biol., 85:890.

30. Jayaraman, S., Martin, C.A., and Dorf, M.E., 1990, J. Immunol., 144:942.

31. Maniatis, T., Fritsch, E.F. and Sambrook, J. 1982., Molecular Cloning, a Laboratory Manual.

32. Williams, R.K., Jiang, G.S., Snyder, S.W., Frana, M.F., and Holmes, K.V., 1990, J. Virol., 64:3817.

STRUCTURAL PROTEINS OF AVIAN INFECTIOUS BRONCHITIS VIRUS: ROLE IN IMMUNITY AND PROTECTION

Jagoda Ignjatovic and Lisa Galli

CSIRO Division of Animal Health
Private Bag No 1
Parkville 3052
Australia

ABSTRACT

The antigenicity of the S1, M and N proteins of avian infectious bronchitis virus was compared following immunization of chickens with live and inactivated virus. The N protein was immunodominant antigen inducing cross–reactive antibodies in high titres whereas the S1 glycoprotein induced serotype–specific and cross–reactive antibodies. The M glycoprotein elicited antibodies in low titres and of limited cross–reactivity. Immunization of chickens with the purified N and M proteins did not induce protection against virulent challenge whereas immunization with the S1 glycoprotein prevented replication of nephropathogenic IBV in kidneys but not in tracheas of immunized chickens.

INTRODUCTION

Infectious bronchitis virus (IBV) contains three structural proteins: the peplomer (S) glycoprotein, located at the surface of virion and consisting of two subunits S1 and S2; the membrane (M) glycoprotein partially exposed at the surface of the virion and the nucleocapsid (N) protein which is internally located (1, 2). The antigenicity and role of the S1, S2, N and M proteins in immunity and protection against IBV infection is unclear.

Following infection with IBV, strain specific neutralizing and haemagglutinating antibodies are produced and are directed to the S1 glycoprotein (3–5), however there is no relationship between these antibodies and protection (4,6,7). In one study, the S1 glycoprotein was found to be more antigenic than either the N or the M protein (3) whereas in another study (8) similar titres of the S1 and M antibodies were detected early after infection suggesting that the M protein might have a role in immunity to IBV. The S2 glycoprotein was found to be more antigenic than the S1 glycoprotein, giving rise to cross–reactive antibodies (9). Recent studies have provided evidence that the N protein posses T–cell epitopes (10) and that it has a role in immunity to IBV (11).

The S1 glycoprotein is considered the most likely inducer of protective immunity. Inactivated IBV lacking the S1 glycoprotein did not induce protection in trachea of immunized chickens whereas whole inactivated IBV protected chickens against virulent challenge (4). However chickens immunized with the purified S glycoprotein were not protected against virulent challenge (3). With another coronavirus (mouse hepatitis virus), the S glycoprotein (12, 13) and two synthetic peptides which were homologous to two conserved regions in the S2 glycoprotein (14, 15), protected mice against lethal challenge.

The work reported here aimed to compare the antibody response to the S1, M and N antigens of IBV following immunization with live and inactivated virus and to assess if vaccination with the purified S1, N and M antigens, protects chickens against virulent IBV challenge.

METHODS

Nephropathogenic Australian strain of IBV, N1/62, grown in the allantoic cavity of embryonated chicken eggs, was used for preparation of partially purified virus and immunization of chickens. The S1, M and N antigens were obtained from partially purified N1/62 by immunoaffinity purification using CNBr–activated Sepharose 4B to which monoclonal antibodies directed against the S1, M and N proteins (16) were coupled. The method used for immunoaffinity purification was essentially as described by Mocket (8). At two weeks of age specific pathogen–free chickens, were immunized intra–ocularly with live and intra–muscularly with inactivated N1/62 virus, three times, at four–weekly intervals. Antibody responses were followed by ELISA, western blotting (WB), virus neutralization (VN) and haemagglutination inhibition (HI) tests (16). Cell-mediated responses to the S1, M and N antigens were measured by delayed–type hypersensitivity (DTH) reactions (17) in chickens immunized twice with live N1/62.

RESULTS AND DISCUSSION

To determine the profile of the antibody response to IBV antigens, chickens were immunized with live and inactivated N1/62 and appearance of antibodies assessed by VN, ELISA, WB and HI tests. Results obtained in ELISA are shown in Table 1. Immunization with live IBV induces antibodies to the S1, M and N antigens. The purified S2 glycoprotein was not available for this study. The N antigen was immunodominant whereas the M antigen was the least antigenic. Antibodies to the N protein appeared first and were initially detected at 10 days post–infection, at the time when no other antibodies were present. From 2 weeks onwards antibodies against the N protein had the highest titres. Antibodies to the S1 glycoprotein were first detected at 2 weeks post–infection by ELISA and VN, but not by WB. Titre of VN and ELISA antibodies to the S1 glycoprotein rose in parallel, ELISA titres reaching those detected for the N protein after three immunizations. Antibodies to the M protein were first detected by ELISA four weeks after infection, titres being considerably lower than those detected for the S1 and N antigens (Table 1).

Immunization with inactivated N1/62 virus was performed to assess the efficiency of such immunization and profile of antibodies produced (Table 1). Two immunizations with inactivated virus were required to induce antibody response to the S1, M and N antigens, the antibody titres being appreciably lower than those produced following live

Table 1. Titres of antibodies to the S1, N and M proteins in sera of chickens immunized at 4–weekly intervals with (A) live and (B) inactivated N1/62 virus.

Antigen[b]	Titre of antibodies[a]			
	2w[c]	4w	8w	12w
A				
N1/62	500	1600	4500	13500
S1	50	1320	4480	7040
N	400	2720	7040	8860
M	–[d]	80	800	920
B				
N1/62	–	120	500	1600
S1	–	–	300	500
N	–	60	440	1600
M	–	25	240	760

[a]Reciprocal of the last dilution of sera giving positive reactions in ELISA. Titres are mean of five sera tested individually.

[b]The S1, M and N antigens were obtained by immunoaffinity purification from N1/62 virus and used for coating of microtitre plates in ELISA.

[c]Weeks following primary immunization with the N1/62 strain of IBV.

[d]– Antibodies not detected.

immunization. Antibody titres to the M glycoprotein were similar to those obtained by immunization with live virus. The route of immunization (intramuscular or intraperitoneal) with inactivated virus did not alter the antibody response, apart from VN antibodies which appeared earlier, after two intraperitoneal immunizations. Results obtained in ELISA were confirmed by WB (results not shown). Methods used in this study did not allow measurement of antibodies to the S2 glycoprotein.

Epitopes involved in protection are expected to be cross–reactive and unrelated to virus serotype. For that reason it was assessed if the S1, M or N antigens elicit cross–reactive antibodies. Sera raised against IBV strains of different serotypes were titrated in ELISA using the S1, M and N antigens obtained from N1/62 strain of serotype C (Table 2).

The S1 and N antigens both induced cross–reactive antibodies. These cross–reactive antibodies were also detected in WB. The antibodies to the M protein were serotype specific. Only sera against three IBV strains of serotype C reacted in ELISA with the M antigen obtained from N1/62 virus of serotype C.

Table 2. Cross–reactive antibodies in sera obtained against Australian IBV strains of different serotype.

Antisera to[b]	Serotype	A_{450} on antigen[a]				
		N1/62	N	S1	M	NDV
Vac–1	A	0.71	1.10	0.37	–	–
Vic S	B	0.85	0.84	0.64	–	–
N1/62	C	1.56	1.80	1.45	0.44	0.16
N2/62	C	1.30	1.48	1.36	0.37	0.12
N8/78	C	1.10	1.12	1.34	0.40	0.09
N9/74	D	1.15	1.05	0.36	0.11	0.11
Q1/73	E	1.10	1.15	1.25	0.15	0.09
V2/71	F	1.15	1.40	1.10	0.18	0.16
V1/71	G	1.15	1.70	1.10	0.15	0.11

[a]Microtitre plates were coated with partially purified N1/62, the N, S1 and M antigens obtained from N1/62 strain or Newcastle disease virus (NDV).
[b]Sera obtained after two immunizations.

Table 3. Delayed–type hypersensitivity reaction elicited by purified S1, M and N antigens in chickens immunized with N1/62 strain of IBV.

Antigen inoculated[a]	No positive/ No tested	Wattle thickness after inoculation of antigen (increase)[a]			
		0 h	24 h	48 h	72 h
PBS	0/5	1.31	1.35 (**0.04**)	1.35 (**0.04**)	1.48 (**0.17**)
N1/62	4/5	1.23	1.69 (**0.46**)	2.13 (**0.90**)	2.00 (**0.77**)
PBS	0/5	1.62	1.73 (**0.11**)	1.81 (**0.19**)	1.87 (**0.25**)
S1	5/5	1.63	2.86 (**1.23**)	2.71 (**1.08**)	2.25 (**0.62**)
PBS	0/5	1.51	1.62 (**0.11**)	1.75 (**0.24**)	1.84 (**0.33**)
M	4/5	1.49	1.75 (**0.26**)	2.04 (**0.56**)	1.84 (**0.35**)
PBS	0/5	1.73	1.85 (**0.12**)	1.97 (**0.24**)	1.97 (**0.24**)
N	3/5	1.63	2.23 (**0.60**)	2.31 (**0.68**)	2.42 (**0.79**)

[a]Wattle thickness in mm. Increase in wattle thickness of >0.40 mm was considered positive.
[b] PBS and antigens in 50 μl were inoculated into left and right wattle, respectively. The S1, M and N proteins were purified from N1/62 virus. N1/62 was inactivated virus.

Since it is suspected that immune responses other than antibodies are involved in the mechanism of protection to IBV, we attempted to determine if the S1, M and N antigens also induce cell-mediated responses. This was assessed by measuring DTH responses in immune chickens following inoculation of the purified S1, M and N antigens into the wattle. All three antigens induced DTH responses in majority of chickens (Table 3).

To identify the protective antigen(s) of IBV, chickens were immunized with the purified S1, M and N antigens and inactivated N1/62 virus, and challenged with nephropathogenic N1/62 strain. Two intramuscular immunizations with inactivated N1/62, S1 and N antigens induced ELISA but not VN antibodies. None of the chickens challenged with the nephropathogenic N1/62 strain of IBV were protected. After three immunizations with inactivated N1/62 and the S1, M and N antigens, VN antibodies were detected in some chickens vaccinated with inactivated N1/62 virus and the S1 glycoprotein. The majority of chickens which were immunized with the inactivated N1/62 virus were protected and virus could not be isolated either from the trachea or kidneys. The majority of chickens immunized with the S1 glycoprotein were protected when isolation of virus from the kidneys was used as the criteria of protection. However these chickens were not protected when isolation of virus from trachea was used as a criteria of protection. Protection did not correlate with the presence of VN and HI antibodies.

REFERENCES

(1)   Cavanagh, D. (1981). J. Gen. Virol. 53: 93.

(2)   Cavanagh, D. (1983). J. Gen. Virol. 64: 1187.

(3)   Cavanagh, D., Darbyshire, J. H., Davis, P. and Peters, R. W. (1984). Avian Pathol. 13: 573.

(4)   Cavanagh, D., Davis, P. J., Darbyshire, J. H. and Peters, R. W. (1986). J. Gen. Virol. 67: 1435.

(5)   Mockett, A. P. A., Cavanagh, D. and Brown, T. D. K. (1984). J. Gen. Virol. 65; 2281.

(6)   Darbyshire, J.H. In: *Avian Immunology* (Eds Rose, M.E., Payne, L.N. and Freeman, B.M.) British Poultry svcience Ltd, Edinburgh, 1981, pp 205.

(7)   Arvidson, Y., Tannock, G. A., Senthilsevan, A. & Zerbes, M. (1990). Arch. Virol. 111: 227.

(8)   Mocket, A.A.P. (1985). J. Virol. Meth. 12: 271.

(9)   Lenstra, J.A., Kusters, J.G., Koch, G. and van der Zeijst. (1989). Molec. Immunol. 26: 7.

(10)   Boots, A.M.H., Kusters, J.G., van Noort, Zwaagstra, K.A., Rijke, E., van der Zeijst, B.A.M. and Hensen, E.J. (1991). Immunology 74: 8.

(11)   Boots, A.M.H., Benaissa-Trouw, Hesselnik, W., Rijke, E., Schrier C. and Hensen, E.J. (1992). Vaccine 10: 119.

(12)   Hansony, H.J. and MCNughton, M.R. (1981). Arch. Vorol. 69: 33.

(13)   Daniel, C. and Talbot, P.J. (1990). Virology 174: 87.

(14)   Talbot, P.J., Dionne, G. and Lacroix, M. (1988). J. Virol. 62: 3032.

(15)   Koolen, M.J.M., Borist, M.A.J., Horzinek, M.C. and Spaan, W.J.M. (1990). J. Virol. 64: 6270.

(16)   Ignjatovic, J and McWaters, P.G. (1991). J. Gen. Virol. 72: 2915.

(17)   Chubb, R., Huynh, V. and Bradley, R. (1988). Avian Pathol. 17: 371.

INDUCTION OF AN IMMUNE RESPONSE TO TRANSMISSIBLE GASTROENTERITIS CORONAVIRUS USING VECTORS WITH ENTERIC TROPISM

Cristian Smerdou[1], Juan M. Torres[1], Carlos M. Sánchez[1], Carlos Suñé[1], Inés M. Antón[1], Miguel Medina[1], Joaquin Castilla[1], Frank L. Graham[2], and L. Enjuanes[1]

[1] Centro Nacional de Biotecnología
CSIC - Universidad Autónoma
Canto Blanco. 28049 Madrid, Spain

[2] Department of Biology, McMaster University
Hamilton, Ontario, Canada L8S4K1

INTRODUCTION

Transmissible gastroenteritis virus (TGEV) causes high mortality in neonatal pigs [1]. TGEV has four structural proteins: the spike (S), the nucleoprotein (N), the membrane protein (M), and the small membrane (SM) protein. Protein S is the major inducer of TGEV neutralizing antibodies. Four antigenic sites (A, B, C, and D) have been defined in S protein[2, 3, 4]. These sites have been located in the amino-end half of S protein. The precise amino acids contributing to the formation of these sites were described [5]. Site A is antigenically dominant and has been divided into three antigenic subsites: Aa, Ab, and Ac. Sites A and D, and in minor extent site B, have been involved in the neutralization of TGEV. Glycosylation dependent (sites A and B) and independent (sites D and C) antigenic determinants have been defined.

Protection against TGE requires lactogenic immunity, which is best induced by presentation of selected antigens to the immune system in gut associated lymphoid tissues (GALT). We have focused on the selection of antigenic determinants involved in protection, and on their expression in GALT using *Salmonella typhimurium* and adenovirus based vectors with tropism for Peyer's patches [6, 7].

MATERIALS AND METHODS

Cells, viruses, and monoclonal antibodies

The PUR46 strain of TGEV was used. The virus was grown, purified, and titrated on swine testicle (ST) cells, as previously described [3]. The MAbs used were specific for TGEV and their characteristics have been described [2, 8]

Coronaviruses, Edited by H. Laude and J.F. Vautherot
Plenum Press, New York, 1994

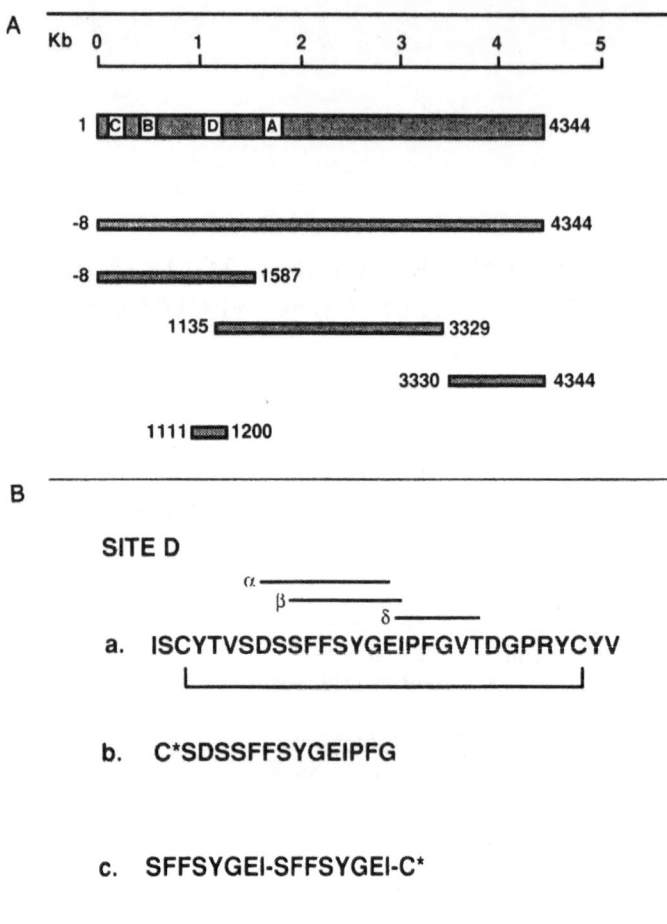

FIGURE 1. Transmissible gastroenteritis virus S gene fragments selected for expression. A. Diagrammatic representation of the S gene of TGEV showing the approximate location of the four antigenic sites (thick bar with capital letters). The segments of the S gene cloned into *Salmonella* have been represented by thin bars. Numbers at both ends indicate the flanking residues cloned into *Salmonella*. B. The sequences of three peptides (a, b, and c) which include site D are shown. The core structure recognized by site D specific MAbs 57.51, 57.57, and 8D.H8, have been overlined (α, β, and δ, respectively) [5, 12]. *, indicates cysteine residues not present in the viral sequence.

Radioimmunoassay (RIA), virus neutralization, and TGEV gene cloning

The procedures for RIA and neutralization have been described [3]. TGEV genes were cloned into Bluescript SK⁻ plasmids [5, 9].

Cloning and expression of TGEV S gene fragments in *S. typhimurium* and adenoviruses

Expression in avirulent forms of *S. typhimurium* Δ*cya*, Δ*crp*, Δ*asd*, developed by R. Curtiss III group [6], was made by cloning the whole S gene or the indicated fragments, into

TABLE 1. Recombinant *asd*+ plasmids expressing TGEV S protein fragments

Recombinant plasmid	*asd* vector	S-gene fragment cloned	Expression in *S. typhimurium* strain	
			χ3730	χ3987
pYATS - 1	pYA292	- 8 - 1587	+++	+++
pYATS - 2	"	1135 - 3329	+++	+/-
pYATS - 3	"	3330 - 4628	-	-
pYATS - 4	"	- 8 - 4628	++	+
pYALT-D	pYA3048	1111-1200	+++	+++

a. PstI site in S gene was taken from Bluescript vector polylinker.

plasmid pYA292. These plasmids were transfected into *Escherichia coli* χ6097F' Δ*asd* Δ[*pro-lac*]φ80 *d lacZ* ΔM15. Plasmid DNA was modified by transforming an intermediate *S. typhimurium* χ3730 Δ*asd hsd*SA *hsd*SB defective in restriction enzymes. Modified plasmids were transfected into *S. typhimurium* χ3987 Δ*asd* Δ*cya* Δ*crp* where TGEV S gene fragments were constitutively expressed.

To express site D an oligonucleotide of 90 bp was obtained by hybridization of two complementary oligonucleotides coding for the sequence ISCYTVSDSSFFSYGEIPFGV-TDGPRYCYV. The blunt end oligonucleotide was cloned into expression plasmid pYA3048 in frame with *E. coli* labile toxin B subunit (LT-B) gene, under the control of P*trc* promoter[10]. A collection of recombinant plasmids named pYALT-D with different inserts was selected. Expression of site D by these constructs was determined by Western blot using site D specific MAbs.

To generate Ad5-TGEV recombinants TGEV S gene fragments cloned into pBluescript SK⁻ were subcloned into plasmids pSV2X3 or pSV2X4 containing an SV-40 based expression cassette. These cassettes were flanked by right end Ad5 sequences, by inserting them into a unique XbaI site present in pAB14 or pFG144K3 plasmids [7, 11]. By cotransfecting these plasmids and plasmid pFG173 (which carries a functional left end of Ad5) into 293 cells, recombinant Ad5-TGEV chimeras were selected. Expression of S protein fragments was determined by Western blot, and immunoenzymatic assays, and their immunogenicity was evaluated by oronasal and intraperitoneal administration to hamsters. The immune response to TGEV was tested by RIA and neutralization.

RESULTS AND DISCUSSION

Antigenic determinants potentially involved in protection

Since all TGEV neutralizing antibodies were specific for the S protein and this antigen also mediates the binding of TGEV to host cells, its potential role in protection was

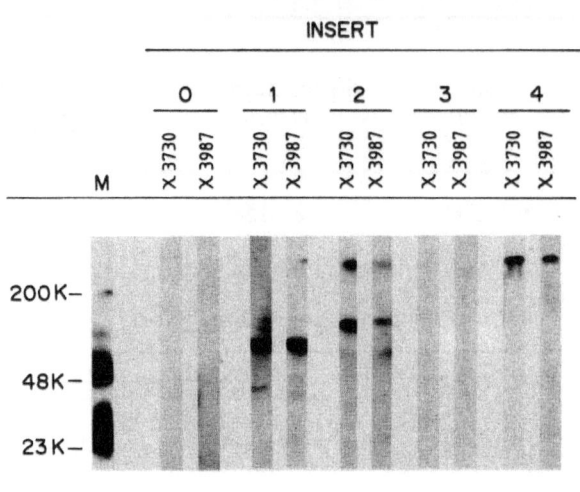

FIGURE 2. Western blot analysis of the recombinant antigens expressed in *S. typhimurium*. The indicated *Salmonella* was transformed with plasmid pYA292 with no insert (0) or with plasmids pYATS1, pYATS2, pYATS3 or pYATS4, containing the left end (1), middle (2), right end (3), or the whole S gene (4), respectively. M, TGEV proteins used as molecular weight markers.

studied. The amino acids contributing to the four antigenic sites of S glycoprotein were: site A, 538, 543, 586, and 591; site B, 97 and 144; site C, 48 to 52; and site D, 385 [5]. Site A was the most conserved one, followed by sites D, B and C. Since sites A, B, and D induced TGEV neutralizing antibodies, special attention was paid to S protein fragments coding for these sites. Nevertheless, other antigenic domains of S protein were also expressed both in *Salmonella* and Ad5, since it has been reported that protection can also be provided against some viral infections by non-neutralizing antibodies. Site D, which is linear and glycosylation independent, was synthesized in three forms (Figure 1). These peptides induced TGEV neutralizing antibodies in rabbits [12, 13]. One of them (Fig. 1B, peptide a) was selected for its expression, genetically linked to LT-B.

Induction of TGEV-specific antibodies by expressing antigenic determinants of S protein using *S. typhimurium*

The whole S glycoprotein or overlapping fragments including the NH_2- end, the middle, and the carboxy-terminus were expressed in avirulent forms of *S. typhimurium* χ3987 (Table 1, Fig. 1). The relative amount of antigen expressed by the different recombinants is summarized in Table 1. In *S. typhimurium* χ3987, with constitutive expression, plasmids pYATS1 and pYATS4, coding for the NH_2-end and the whole S protein, respectively, expressed products with low toxicity. Stability of these expression systems was analyzed after 10, 25, and 50 generations *in vitro*. After 50 generations 90% of the bacteria transformed with both pYA-TS1 and pYA-TS4 carried correct plasmid, and 60% and 20% of the bacteria, respectively, expressed the S-protein antigen. If the same level of stability is maintained *in vivo* the amount of antigen expressed could be enough to induce a protective immune response. To further increase the stability of plasmids coding for S protein, a smaller oligonucleotide, coding for 30 amino acids which include antigenic site D, was genetically linked to heat labile *E. coli* enterotoxin (LT-B). This non-toxic subunit has been described as a strong immunogen, when administered into the gut via the oral route, in

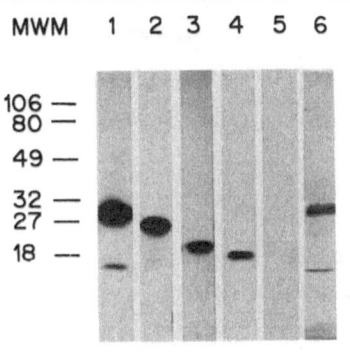

FIGURE 3. Western blot analysis of the LT-B site D recombinant products expressed in *S. typhimurium*. *Salmonella* χ3730 was transformed with plasmids pYA3048 *asd+* including LT-B gene and one to four copies of a 90 bp oligonucleotide coding for site D (lines 1 to 4, respectively). The antigenic expression was analyzed with site D specific MAbs. Line 5 shows the expression in bacteria transformed with pYA3048 plasmid with no insert coding for site D. The two subunits (A, and B) of purified *E. coli* LT toxoid (ICN, Ohio) were analyzed in parallel and identified by silver staining (line 6). MWM, molecular weight markers.

TABLE 2. Colonies of *S. typhimurium* χ3987 expressing site D from TGEV S protein as a fusion product with LT-B

Plasmid	Fusion product MW	Positive colonies (%) Number of generations	
		25	50
pYALT-D I	high	100	100
pYALT-D II	intermediate-high	100	77
pYALT-D III	intermediate-low	100	90
pYALT-D IV	low	100	100

microgram amounts of protein, or when delivered to the GALT by attenuated *Salmonella* strains. LT-B induces an immune response preferentially with immunoglobulins of IgA isotype. Site D was genetically coupled to the carboxy-terminus of LT-B using plasmid pYA3048 *asd+*, under the control of P*trc* promoter. At least, four expression patterns of LT-B-D recombinant antigens, with 30-, 27- 22-, and 18-kDa, were selected in different bacteria (Figure 3). Each of these recombinant antigens probably contains one to four site D monomers per LT-B molecule. The stability of the expression of site D after 50 generations in these recombinants was close to 100% (Table 2). The low toxicity in bacteria of LT-B-D is of high interest, since site D induced TGEV neutralizing antibodies. The sequences of 30 amino acids, containing site D core, probably form loops by disulfide bridges between cysteines located 26 amino acids apart. These loops could result in high site D core exposure, which may help the induction of TGEV neutralizing antibodies. These fusion proteins formed heat sensitive pentamers which dissociate at 70°C. Pentamer formation is

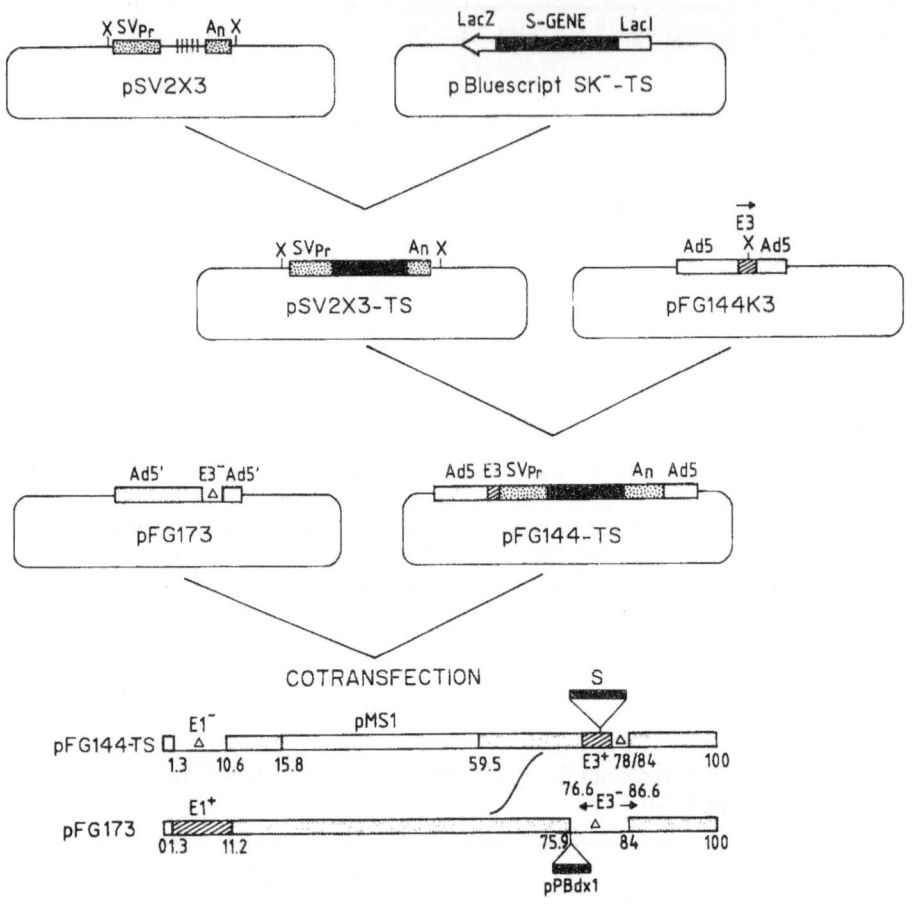

FIGURE 4. Construction of recombinant Ad5-TGEV expression vectors. TGEV S gene sequences (bold line) were cloned into Bluescript plasmid and inserted on pSV2X3 plasmid containing expression cassette with SV-40 promoter (SV$_{Pr}$) and polyadenylation signals (A$_n$). The expression cassette was flanked by Ad5 sequences by subcloning into pFG144K3 plasmid. Cotransfection of 293 cells with this plasmid and pFG173 plasmid containing a functional left hand Ad5 sequence generated infectious Ad5-TGEV chimeric viruses [14].

required for the adjuvant activity of LT-B, since these complexes bind the ganglioside GM1 present in the surface of enteric epithelial cells, promoting antigen presentation to GALT. The recombinant protein LTB-site D forms aggregates of higher molecular weight, as determined by polyacrylamide gel electrophoresis in non-reducing conditions. These large complexes were dissociated by 2-mercaptoethanol, suggesting that the aggregation was mediated by disulphide bridges. The immunogenicity of these complexes is being evaluated *in vivo*.

Induction of an immune response to TGEV using Ad5 based vectors

An Ad5 vector system has been used to express antigenic determinants of the S glycoprotein of TGEV. This eukaryotic vector may complement the *S. typhimurium* system, since sites A and B, inducers of TGEV neutralizing antibodies, are glycosylation dependent.

The Ad5 system is based on the homologous recombination between two plasmids, one that spans the left end of the Ad5 genome and another that spans the right end into which TGEV-cDNA sequences were inserted (Fig. 4). TGEV S gene sequences (Fig. 5) were cloned into an expression cassette with an SV-40 promoter and polyadenylation signals. The expression cassette was inserted into the E3 region of Ad5 virus. Rescued chimeric Ad5-TGEV recombinants were cloned into two types of Ad5 vectors, according to the size of the heterologous genes which may be inserted in E3. This cloning was made into plasmids pFG144K3 (originating plasmids Ad-TS1 to -5) and pAB14 (which produced plasmids Ad-TS01 to -04). Plasmids pFG144K3 and pAB14 have deletions in the E3 gene of 1880 or 2686 bp, which allows the insertion of 4000 or 4800 bp, respectively, in the final recombinant Ad5 virus [7]. Vectors derived from pAB14 should in principle leave enough room to clone the entire TGEV S gene. The antigenicity of the recombinant proteins was studied by Western blot and immunoadsorption, and the immunogenicity by oronasal and

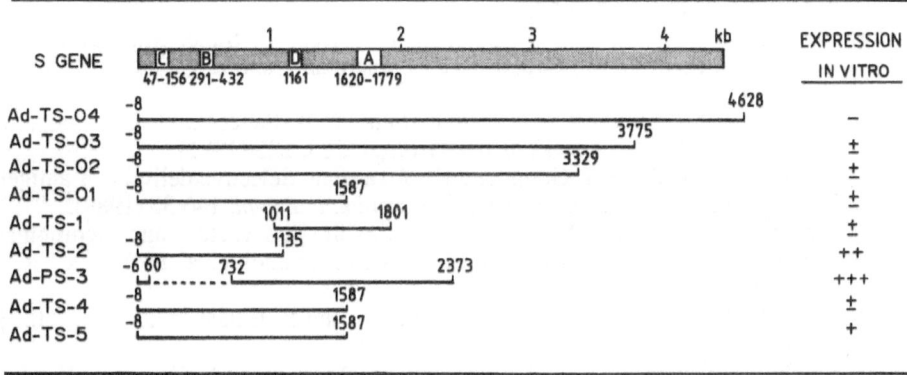

FIGURE 5. Recombinant AD5-TGEV constructed. The different segments of TGEV S gene cloned into Ad5 virus are shown. Numbers indicate the nucleotides flanking the inserts cloned. The dotted bar represents the S gene, and the letters inside the bar the approximate location of the different antigenic sites. The relative antigenic determinants of TGEV S protein expression levels are indicated. +, ++, and +++, indicate 0.1, 0.2, and 0.4 µg per 10^6 cells.

intraperitoneal infection of hamsters, which are fully susceptible to infection by Ad5. Three Ad5-TGEV recombinants Ad-TS2, Ad-PS3, and Ad-TS5 (Table 4) gave substantial amounts (0.1 to 0.4 µg of protein per 10^6 cells) [14] of TGEV S protein fragments. Recombinant Ad-TS04, which codes for the whole S protein, was unstable and after three passages in 293 cells rearranged the DNA and lost the expression of S glycoprotein. Recombinant viruses Ad-TS2, Ad-TS5, and Ad-PS3 were antigenic for site D specific MAbs. In addition, Ad-PS3, also expressed site A, as determined by Western blot analysis with site A specific MAbs. Both of these sites are the major inducers of TGEV neutralizing antibodies. Recombinant vectors Ad-TS2 and Ad-PS3 induced in hamsters, after oronasal and intraperitoneal immunization, TGEV specific antibodies as determined by RIA. Interestingly, recombinant Ad-PS3 also induced TGEV neutralizing antibodies which reduced viral infectivity by 10^2-10^3-fold. These results indicate that recombinant Ad5-TGEV viruses could provide protection against TGEV infections.

ACKNOWLEDGEMENTS

We thank Javier Palacín for technical assistance with animal handling. This work was supported by grants from the Comisión Interministerial de Ciencia y Tecnología (project BIO091-0710), European Communities (Science project SCI-CT91-0684-LNBE), NATO (project 900430-CRG), Fundación R. Areces, and the Natural Sciences and Engineering Research Council of Canada. C. Smerdou, C. Suñé, and J.M. Torres received fellowships from the Department of Education and Science.

REFERENCES

1. L.J. Saif and E.H. Bohl, Transmissible gastroenteritis, in: "Diseases of Swine," A.D. Leman, B. Straw, R.D. Glock, W.L. Mengeling, R.H.C. Penny, and E. Scholl, eds., Iowa State University Press, Ames, Iowa (1986).
2. I. Correa, G. Jiménez, C. Suñé, M.J. Bullido, and L. Enjuanes, Vir. Res, 10:77 (1988).
3. I. Correa, F. Gebauer, M.J. Bullido, C. Suñé, M.F.D. Baay, K.A. Zwaagstra, W.P.A., J. Gen. Virol. 71: 271 (1990).
4. C.M. Sánchez, G. Jiménez, M.D. Laviada, I. Correa, C. Suñé, M.J. Bullido, F. Gebauer, C. Smerdou, P. Callebaut, J.M. Escribano, and L. Enjuanes, Virology. 174:410 (1990).
5. F. Gebauer, W.A.P. Posthumus, I. Correa, C. Suñé, C.M. Sánchez, C. Smerdou, J. A. Lenstra, R. Meloen, and L. Enjuanes,Virology. 183: 225 (1991).
6. R. Curtiss III, S.M. Kelly, P.A. Gulig, and K. Nakayama, Selective delivery of antigens by recombinant bacteria, Curr. Topics Microbiol. Immunol. 146:35 (1989).
7. F.L.Graham, and L. Prevec, Adenovirus-based expression vectors and recombinant vaccines, in: "Vaccines: New Approaches to Immunological Problems," R.W. Ellis, ed.(1992).
8. G. Jiménez, I. Correa, M.P. Melgosa, M.J. Bullido, and L. Enjuanes, J. Virol. 60:131 (1986).
9. C.M. Sánchez, F. Gebauer, C. Suñé, A. Mendez, J. Dopazo, and L. Enjuanes, Virology 190:92 (1992).
10. E.K. Jagusztyn-Krynicka, J.E. Clarck-Curtiss, and R. Curtiss III, Infect. Immun. 61:1004 (1993).
11. A. Bett, F.L. Graham, and L. Prevec. Unpublished results.
12. W.P.A. Posthumus, J.A. Lenstra, W.M.M. Schaaper, A.P. van Nieuwstadt, L. Enjuanes, and R.H. Meloen, J. Virol. 64:3304 (1990).
13. C. Smerdou and L. Enjuanes, Unpublished results.
14. C.M. Sánchez, J.M. Torres, C. Suñé, F.L. Graham, and L. Enjuanes, Submitted.

AN OVERVIEW OF SUCCESSFUL TGEV VACCINATION STRATEGIES

AND DISCUSSION ON THE INTERRELATIONSHIP BETWEEN

TGEV AND PRCV

Mark W. Welter[1], Michelle P. Horstman[1], C. Joseph Welter[1], and Lisa M. Welter[1]

[1]Ambico, Inc. Dallas Center, Iowa, USA

ABSTRACT

Porcine respiratory coronavirus(PRCV) is a new variant of TGE with an altered pathogenesis. PRCV multiplies mainly in tonsilar tissues and the respiratory tract. There are no enteric symptoms and in experimentally infected pigs, even the respiratory tract infection is usually asymptomatic. PRCV is spread aerogenically through herds and the significance of PRCV as a pathogen in swine has yet to be determined. Despite the differences in pathogenesis and tissue tropism, the behavior of TGEV and PRCV are closely related antigenically. PRCV induces an antibody response in pigs that cannot be distinguished from TGEV-infected pigs by conventional serological assays. PRCV sensitized animals are not protected from TGEV challenge nor is the milk antibody provided to nursing piglets completely effective in prevention of TGEV infections; thus PRCV is not a good vaccine candidate for TGEV infections. PRCV subclinical infections have led to several reported cases of enzootic TGEV in herds that had been diagnosed as TGEV immune strictly on the basis of serum neutralizing titers which were later found to be due to exposure to PRCV. Vaccination studies conducted with the Ambico, oral modified live TGEV vaccine have led to some startling new results: (1)Use of Ambico TGEV modified live vaccine has been shown to provide complete protection against subsequent PRCV challenge and (2) the effectiveness of TGEV vaccination is actually enhanced by previous exposure to PRCV (3) Weanling pigs which have passively acquired circulating TGEV neutralizing antibodies are protected from subsequent PRCV infections.

INTRODUCTION

Transmissible gastroenteritis(TGEV) is a coronavirus that can infect swine of all ages. The natural route of TGEV infection is oral and can appear in two clinical forms: epizootic and enzootic[1,2]. In the epizootic form TGEV will infect and replicate in the absorptive epithelial cells covering the small intestinal villi. Virus replication results in a degeneration or shortening of the villi which is commonly referred to as villus atrophy.

This decrease in the absorptive capability of the small intestine results in a severe watery diarrhea and pigs infected during the first two weeks of age usually die due to dehydration. Enzootic or chronic TGEV, on the other hand, refers to a persistent infection of the herd with TGEV occurring as a result of continual or frequent introduction of susceptible animals into an immune or partially immune herd[2,3]. When these animals become infected, the disease is perpetuated. Chronic TGEV clinical signs usually occurs in piglets greater than 2-weeks of age and often persists through 2-weeks post-weaning.

TGEV vaccines have been commercially available since 1966. Until 1980 these vaccines were either inactivated or modified live and were used for intramuscular vaccination of pregnant sows and gilts shortly before each farrowing. An oral vaccine was developed and commercialized by our laboratory in 1980. This particular TGEV strain originated from the lungs of a clinically normal adult pig, although it was still virulent for piglets. This TGE virus strain was eventually modified by passage through lungs of gnotobiotic piglets, lung cell cultures and swine testicular cell cultures. The resulting TGE vaccinal strain retained its normal peplomeric morphology and did not cause disease in piglets[3,4]. The principal advantage of the oral vaccine, when administered to pregnant sows, was the subsequent production of higher more persisting TGEV milk antibody levels which resulted in increased protection of experimentally TGEV challenged piglets.

Colostrum and milk of sows may contain TGEV antibodies principally associated with 3 immunoglobulin classes: IgA, IgG and IgM[3]. The predominating Ig is dependent upon the method of viral sensitization. Colostrum and milk of sows which have been parenterally inoculated(with either inactivated, modified live or virulent TGEV) contain predominantly IgG antibodies in their milk[1,2,3,4]. IgG antibody titers will drop 83 to 98 percent within the first 2-3 days post-farrowing. On the other hand, naturally or experimentally infected sows by the oral route, will develop predominantly IgA milk antibody which persists at higher levels throughout lactation. Many researchers have correlated TGEV passive immunity protection from TGEV challenge in newborn piglets with the presence of IgA-type TGEV antibodies in the milk[5].

Previous studies have demonstrated that it is possible to actively immunize baby pigs nursing immune dams, as early as day of birth, requiring 3-5 days to develop an active gut immunity to TGEV[1,2,6]. Vaccine virus has been detected in the tonsils, salivary glands, oropharyngeal and mesenteric lymph nodes as early as 24-hours and lasting for up to 4-days post-inoculation. Virus has never been detected free in the intestinal contents. Holding baby pigs away from the TGEV immune milk of the sow for 20-30 minutes after oral vaccination has been found to be enough time to allow the virus to penetrate the tonsilar tissues and sequester itself from neutralizing TGEV antibodies in the milk.

Porcine respiratory coronavirus(PRCV) is a new variant of TGE with an altered pathogenesis. PRCV multiplies mainly in the respiratory tract without enteric clinical signs and in experimentally infected pigs, even the respiratory tract infection is usually asymptomatic[7,8,9,10]. PRCV is spread aerogenically through herds and its significance as a pathogen in swine has yet to be determined[7,8,9,10]. Despite the differences in pathogenesis and tissue tropism, TGEV and PRCV are closely related antigenically. PRCV induces an antibody response in pigs that cannot be distinguished from TGEV-infected pigs by conventional serological assays. However, antigenic differences do exist and this has led to the development of specific monoclonal antibodies that can be used to differentiate the two viruses and their antibodies[7].

PRCV was first isolated in the United States in 1989, by Dr. Howard Hill and was designated the ISU-1 strain. This strain has since been found to be similar to previous PRCV isolates in Europe by using PRCV - TGEV differentiating monoclonal antibodies. Challenge studies with the ISU-1 strain of PRCV conducted in both conventional and gnotobiotic piglets have not yielded any clinical signs of infection or associated lesions in the lungs, liver, spleen, kidney, brain or heart(Unpubished data). In these studies, replication of PRCV was demonstrated by virus recovery from tonsillar and nasal swabs between 6 and 12 days post-challenge, by inoculation of confluent swine testicular cells and identification by plaque reduction and plaque morphology assays. Moreover, all pigs seroconverted to PRCV within 10-14 days post-inoculation. Vaccination studies conducted with the Ambico, oral modified live TGEV vaccine and the ISU-1 PRCV virus have lead to some startling new results.

METHODS and RESULTS

STUDY #1

A total of thirty-six, 3-5 day old piglets housed in strict isolation boxes with individual airflows were used in this first study; twelve animals were orally vaccinated with Ambico TGEV vaccine(10^7 TCID$_{50}$'s/pig), twelve were orally and intranasally inoculated with PRCV(10^9 TCID$_{50}$'s/pig) and the remaining twelve remained as nonvaccinated controls. At 14-days post-inoculation all animals were orally challenged with 5000 PID$_{50}$'s of virulent TGEV Purdue strain. A creamy to watery diarrhea developed within 48-hours post-challenge in all of the animals in both the nonvaccinated control group and the previously-exposed PRCV group with mortality rates of 42 and 50%, respectively. TGEV was detected from the diarrheic fecals by specific TGEV plaque reduction assays. Only the TGEV orally vaccinated group remained clinically normal throughout the post-challenge period. TGEV serum neutralizing titers were determined for each group. PRCV exposed animals did demonstrate a seroconversion post-exposure to PRCV and a secondary antibody response post TGEV challenge. These results are similar to those that have been reported with the European PRCV isolates and clearly indicate that PRCV exposure will not protect animals from subsequent TGEV infections. This is due to the fact that PRCV replicates strictly in the respiratory tract and sensitization of the respiratory tract will not afford protection at the enteric level. On the other hand the Ambico TGEV vaccine will multiply in the oral-pharengeneal tract and it is believed that these primed lymphocytes will then migrate to the Peyers patches where they will continue to differentiate and allow for enteric protection to subsequent challenge. It has been previously reported that proper sensitization of lymphocytes at the enteric level may afford protection from respiratory diseases[11].

Table 1. Morbidity, Mortality and Serological Response Following TGEV-Challenge of Isolated Conventional Pigs.

No. of Pigs	Treatment	Clin. Obs. Post-Challenge.[1]		Anti TGEV Neutralizing Titers[2]		
		Morbidity Inc.[3]	Mortality	PreVac	TGEV Chall.	14-DPC
12	NonVac. Controls	12/12	5/12	< 2	< 2	45
12	PRCV Exposed(IN-O)[4]	12/12	6/12	< 2	8	241
12	TGEV Vaccinated(O)[5]	0/12	0/12	< 2	25	66

1. TGEV Challenge: Pigs were orally challenged with 5000 PID$_{50}$'s at 19-days of age.
2. Geometric Mean TGEV Antibody Titers
3. Morbidity Incidence: Is defined as number of pigs demonstrating clinical signs divided by total number of pigs.
4. PRCV Challenge: Oral and Intranasally at 5-days of age.
5. TGEV Vaccination: Oral at 5-days of age.

STUDY #2

Since we knew the interrelationship between the oral-pharyngeal cavity and the enteric tract as it related to response to TGEV vaccine we wanted to investigate the effect TGEV vaccine would have on subsequent respiratory infections with PRCV. We were able to clearly demonstrate that: (1) Oral TGEV vaccination of piglets will provide protection from subsequent PRCV infections and (2) passively acquired circulating antibodies to TGEV will also prevent PRCV infections.

In this study a total of eight TGEV seronegative gilts were used. Three of the gilts were vaccinated intramuscularly at 5 and 2 weeks prior to farrowing with TGEV vaccine

$(10^{7.2}$ TCID$_{50}$'s/dose) and the other five gilts remained as nonvaccinated controls. Two of the litters suckling nonvaccinated control gilts were orally vaccinated with TGEV vaccine($10^{6.8}$ TCID$_{50}$'s/dose) at 7-days of age. All animals were weaned at 28-days of age and then orally and intranasally challenged with PRCV at 35-days of age. Nasal and Throat swabs were collected from 5 to 14 days post-challenge for PRCV isolation as previously described. No PRCV was isolated from either the TGEV orally vaccinated piglets or the piglets that suckled the TGEV vaccinated gilts. PRCV was reisolated very frequently from the nonvaccinated piglets that suckled the nonvaccinated control gilts. Significant TGEV serum neutralizing titers were observed in both the orally vaccinated piglets and the piglets that suckled the TGEV vaccinated gilts. A secondary antibody response to TGEV was observed in the TGEV orally vaccinated piglets, but no significant change in serum neutralizing titer was observed in the piglets that suckled the TGEV vaccinated gilts. These data imply that the oral Ambico TGEV vaccine is even more immunogenic when exposure to PRCV has occured. It is our conclusion that prevalence of TGEV antibody in the U.S. swine herds deters the rapid spread of PRCV.

Table 2. Active Immunity and Passively - Acquired Antibody Protection To PRCV Challenge in 35-day Old Weaned Pigs As a Direct Result of TGEV Vaccination

Virus Inoculation of Gilts	Virus Inoculation of Baby Pigs	Virus Shedding Post Chall. with PRCV Nasal and Throat Swab Isolations		
		7-DPC	8-DPC	9-DPC
NonVac. Controls	TGEV(O)[1]/PRCV(IN-O)[2]	0/18	0/18	0/18
NonVac. Controls	None/PRCV(IN-O)[2]	13/17	15/17	13/17
TGEV(IM)[3]	None/PRCV(IN-O)[2]	0/22	2/22	0/22

1. TGEV Vaccination: Pigs Orally vaccinated at 7-days of age.
2. PRCV Challenge: Oral and Intranasally at 35-days of age.
3. Gilts vaccinated Intramuscularly with TGEV vaccine at 5 and 2-weeks Prefarrow.

STUDY #3

A decrease in clinical TGEV infections in Europe has been concomitant with the development of PRCV seroconversions in swine. This has led researchers to investigate whether or not pregnant animals that have been sensitized to PRCV will induce TGEV neutralizing IgA antibody response in their milk. The results of these studies as well as work done at our laboratory indicate that the type of PRCV induced anti-TGEV antibody in the milk is of the IgG class and that nursing pigs that were challenged with a virulent TGEV were not protected[9]. Four groups of TGEV seronegative gilts were used. The first group served as nonvaccinated controls; the second group was orally and intranasally challenged with PRCV at 8-weeks prefarrow; the third group was vaccinated intramuscularly with TGEV vaccine at 5 and 2-weeks prefarrow and the fourth group was orally vaccinated with TGEV vaccine at 5 and 2-weeks prefarrow. At fourteen days of age all pigs were orally challenged with 5000 PID$_{50}$'s of virulent TGEV Purdue strain. Piglets suckling nonvaccinated controls developed a creamy to watery diarrhea which persisted for 7 to 8 days post-challenge with a mortality rate of 24%, a morbidity incidence of 100% and a morbidity incidence and duration of 74%. Similar results were seen in piglets suckling PRCV exposed gilts, with a baby pig mortality of 21%, morbidity incidence of 100% and morbidity incidence and duration of 50%. The piglets suckling the TGEV IM vaccinated gilts had mortality rate of 11%, a morbidity incidence of 68% and morbidity incidence and duration of 26% a significant reduction compared to the nonvaccinated controls, but not as significant as the piglets suckling the orally vaccinated TGEV gilts which had a mortality and morbidity incidence rates of 4% and morbidity incidence and duration of 3%.

Table 3. Morbidity and Mortality of 14-day Old Nursing Piglets Challenged with TGEV

Virus Inoculations Gilts	Gilt Morbidity[1]	Baby Pig Mortality	Baby Pig Morbidity	Baby Pig Morbidity Incid. & Duration[2]
NonVaccinated	3/3(100 %)	6/25(24 %)	25/25(100 %)	177/240(74 %)
PRCV(IN-O)[3]	2/3(67 %)	5/24(21 %)	24/24(100 %)	112/226(50 %)
TGEV(IM)[4]	1/3(33 %)	2/19(11 %)	13/19(68 %)	50/190(26 %)
TGEV(O)[5]	0/3(0 %)	1/25(4 %)	1/25(4 %)	7/243(3 %)

1. Morbidity Incidence: Is defined as number of animals developing clinical signs divided by total number of animals.
2. Morbidity Incidence and Duration: Is defined as number of days pigs demonstrated clinical signs divided by total number of pig days.
3. PRCV Exposed: Oral and Intranasally at 8-Weeks prefarrow.
4. TGEV IM Vaccinations: At 5 and 2-weeks prefarrow.
5. TGEV Oral Vaccinations: At 5 and 2-weeks prefarrow.

Following infection of the piglets with TGEV, a number of the nursing sows also became ill at about 7 to 10 days post-challenge.

SUMMARY

In conclusion, a TGEV variant has recently been reported on and identified as a porcine respiratory coronavirus. The pathogenicity of PRCV infections remains a unclear although epidemiological reports indicate the virus has spread rapidly throughout Europe but not in the United States. This may be due in large part to a high incidence of preexisting TGEV antibody in herds throughout the United States. PRCV sensitized animals are not protected from TGEV challenge nor is the milk antibody provided to nursing piglets completely effective in prevention of TGEV infections; therefore, PRCV is not a good vaccine candidate for the prevention of TGEV infection.

Vaccination studies conducted with the Ambico, oral modified live TGEV vaccine has led to some startling new results: (1)Use of Ambico TGEV modified live vaccine has been shown to provide complete protection against subsequent PRCV challenge and (2) the effectiveness of TGEV vaccination is actually enhanced by exposure to PRCV.

REFERENCES

1. G.R. Fitzgerald, M.W.Welter, C.J. Welter, Improving the Efficacy of Oral TGE Vaccination. Vet. Med. 81: 184-187(1986).
2. G.R. Fitzgerald, C.J. Welter, The Effect of an Oral TGE Vaccine on Eliminating Enzootic TGE Virus from a Herd of Swine. Agri Pract.Disease Control., 11: 25-29(1990).
3. C.J. Welter. Effective use of Vaccination in Acute and Chronic Viral Diarrhea. Minnesota Swine Herd Health Confr.(1986).
4. C.J. Welter. Strategies for a Successful Coronavirus (TGE) Vaccine for Swine. Coronaviruses, Plenum Publishing Co. 218: 551(1987).
5. P. Porter and W. D. Allen Classes of immunoglobulins related to immunity in the pig: A review. J. Am. Vet. Med. Assoc., 160: 511(1972).
6. J. A. Graham. Induction of Active Immunity to TGE in Neonatal Pigs Nursing Seropositive Dams. VM/SAC 75: 1618(1980).

7. P. Callebaut, I.Correa, M.Pensaert, G.Jimenez, L.Enjuanes. Antigenic Differentiation between Transmissible Gastroenteritis Virus of Swine and a Related Porcine Respiratory Coronavirus. J. Gen. Virol. 69: 1725-1730(1988).

8. D. O'Toole, I.Brown, A.Bridges, S. F.Cartwright. Pathogenicity of Experimental Infection with Pneumotropic Porcine Coronavirus. Res. Vet Sci., 47: 23-29(1989).

9. D. J. Paton, I. H.Brown. Sows Infected in Pregnancy with Porcine Respiratory Coronavirus Show no Evidence of Protecting their Suckling Piglets against Transmissible Gastroenteritis. Vet. Res. Comm. 14: 329-337(1990).

10. E.Yus, M. D.Laviada, L. Moreno, J. M. Castro, J. M. Escribano, I. Y.Simarro. The Prevalence of Antibodies to Influenza and Coronaviruses amoung Fattening Pigs in Spain. J. Vet. Med. 36: 551-556(1989).

11. C. Burns, J.Ebersole and M.Allansmith. Immunoglobulin A Antibody Levels in Human Tears, Saliva and Serum. Infec. and Immunity. 36-3: 1019-1022(1982).

CONSTRUCTION OF A RECOMBINANT ADENOVIRUS FOR THE EXPRESSION OF THE GLYCOPROTEIN S ANTIGEN OF PORCINE RESPIRATORY CORONAVIRUS

P. Callebaut[1], M. Pensaert[1] and L. Enjuanes[2]

[1]Laboratory of Virology, Faculty of Veterinary Medicine, University of Gent, Casinoplein 24, B-9000 Gent, Belgium and [2]Centro de Biologia Molecular (CSIC-UAM), Facultad de Ciencias, Universidad Autonoma, Canto Blanco, 28049 Madrid, Spain

INTRODUCTION

Efforts to develop respiratory and enteric vaccines focus on the efficient delivery of protective antigens to the mucosal lymphoid tissue. One approach is to use recombinant viral vectors, i.e. apathogenic viruses with a tropism for mucosal surfaces and genetically manipulated to express the protective antigens. Adenoviruses, particularly the human adenovirus type 5 (Ad5), are promising candidates for use as vectors.

Porcine respiratory coronavirus (PRCV) and transmissible gastroenteritis virus (TGEV) of swine are coronaviruses, showing complete two-way cross-neutralization, but having a tropism for different mucosae (Pensaert et al., 1986). They could present an attractive model to test the induction of immunity at both respiratory and enteric mucosal surfaces by an Ad5 vector capable of expressing the common neutralization mediating antigen.

In the present study, it was the purpose to assess preliminarily the suitability of Ad5 as a vector by inoculation in pigs. Furthermore, we describe the construction of a recombinant adenovirus (Ad5-gpS) which harbours an *in vitro* synthesized fragment of the peplomer glycoprotein (gpS) gene of PRCV. The fragment chosen encodes the N-terminal globular domain, carrying the neutralization epitopes (Correa et al., 1990; Rasschaert et al., 1990).

INOCULATION OF PIGS WITH HUMAN Ad5

Five conventional pigs at 4 weeks of age were inoculated oronasally with 7×10^9 pfu of Ad5, grown and titrated in the human 293 cell line. They were observed daily for clinical signs. Two pigs were examined for seroconversion using paired serum samples, tested by the conventional seroneutralization (SN) test. The 3 remaining pigs were killed on days 3, 5 and 7 after inoculation. Specimens of the tonsils and various parts of the respiratory and the enteric tract were examined for the presence of infectious Ad5 using standard virus isolation methods.

None of the pigs experienced respiratory or intestinal disorders. Two pigs, while having no SN antibodies at the time of inoculation, had SN-titers of 48 and 128, respectively, 4 weeks later. The clearcut seroconversion indicated that both pigs had been infected. Ad5 was consistently isolated from the tonsils, trachea and diafragmatic lobe of the lungs, indicating that the 3 killed pigs had been infected and that the virus had replicated in the tissues mentioned. Furthermore, the virus was present in the colon during the whole test period and in the ileum on day 5 after inoculation, suggesting that the virus had replicated in these segments of the intestinal tract. However, it cannot be excluded that the virus may have reached the gut by ingestion. Taken together, these findings provide preliminary evidence that Ad5 may be suitable as a gene transfer vector for the induction of mucosal immunity in the respiratory and digestive tract of pigs.

CONSTRUCTION OF AN Ad5-gpS RECOMBINANT

The virus source for the gpS gene fragment was PRCV isolate TLM 83 (Pensaert et al., 1986) at 3 passages in the swine testicle (ST) cell line. The genomic RNA was extracted from purified virus. The target gpS gene sequence, i.e. from 6 nucleotides upstream to 1692 nucleotides downstream of the gpS gene translation initiation codon, was synthesized *in vitro* by reverse transcription of the viral RNA to cDNA and subsequent amplification of the synthesized cDNA using the polymerase chain reaction process.

The construction of the Ad5-gpS recombinant was performed using standard molecular cloning techniques. It involved 3 consecutive steps. First, the gpS gene fragment was inserted in the plasmid pSV2X3 (Prevec et al., 1990). The resulting plasmid contained an expression cassette, which was a sequence composed of the early region promoter of simian virus (SV) 40 and the gpS fragment in the orientation that would allow transcription from the SV40 promoter, followed by the SV40 poly A addition signal. Subsequently, the cassette was inserted in the plasmid pFG144. This plasmid contains the left and right end sequences of the Ad5 genome. The right end sequences include the E3 region, except for a deletion which is not essential for the virus replication (Graham and Prevec, 1991). In the resulting plasmid, designated pFG/PRCV, the expression cassette was positioned in the deleted E3 region, in the same orientation as the E3 genes it replaced. Finally, the cloned recombinant Ad5 sequences in pFG/PRCV were rescued in infectious virus by transfection in 293 cells along with plasmid pFG173. The latter plasmid contains the complete Ad5 genome except for a lethal deletion which includes the E3 region (Graham and Prevec, 1991). Plaques of infectious Ad5 appeared 13 to 15 days after the cotransfection.

Recombinants were identified by measuring the level of expression of gpS antigen in cell lysates and supernatants from infected cell cultures using an antibody sandwich ELISA. The assay was essentially performed as described (Callebaut et al., 1982), using porcine anti-TGEV immunoglobulins as the capture antibody and for the peroxidase conjugate preparation.

The cell lysates obtained following expansion of 10 plaques in 293 cells gave a positive signal in the ELISA. Because of the coding sequence which had been cloned it was assumed that the recognized antigen was the expected N-terminal domain of gpS of PRCV and that all plaques were recombinant. The concentration of gpS antigen in the lysates was calculated to be approximately 0.5 µg per 10^6 cells. The culture supernatants were consistently negative. Identical results were obtained following expansion of one of the plaques in ST cell cultures. Further work is in progress to study the *in vivo* expression and the induction of coronavirus-neutralizing antibodies by inoculation of the recombinant in pigs.

ACKNOWLEDGEMENTS

This research is funded by the IWONL, Brussels

REFERENCES

Callebaut, P., Debouck, P., and Pensaert, M., 1982, Enzyme linked immunosorbent assay for the detection of the coronavirus-like agent and its antibodies in pigs with porcine epidemic diarrhea. Vet. Microbiol. 7: 295.

Correa, I., Gebauer, F., Bullido, M., Suné, C., Baay, M.F.D., Zwaagstra, K.A., Posthumus, W.P.A., Lenstra, and J.A., Enjuanes, L., 1990, Localization of antigenic sites of the E2 glycoprotein of transmissible gastroenteritis coronavirus. J. Gen. Virol.71: 271.

Graham, F.L., and Prevec, L., 1991, Manipulation of adenovirus vectors. In: "Methods in Molecular Biology" Murray, E.J., ed., Humana Press, 109.

Pensaert, M., Callebaut, P., and Vergote, J., 1986, Isolation of porcine respiratory, non-enteric coronavirus related to transmissible gastroenteritis. Vet. Quart. 8:257.

Prevec, L., Campbell, J.B., Christie, B.S., Belbeck, L., and Graham, F.L., 1990, A recombinant human adenovirus vaccine against rabies. J. Infect. Dis. 161: 27.

Rasschaert, D., Duarte, M., and Laude, H., 1990, Porcine respiratory coronavirus differs from transmissible gastroenteritis virus by a few genomic deletions. J. Gen. Virol. 71:2599.

ESTABLISHMENT OF MHV-A59 S PROTEIN SPECIFIC CYTOTOXIC T LYMPHOCYTE CLONES

Shigeru Kyuwa [1*] and Stephen A. Stohlman [1,2]

Department of Neurology [1] and Microbiology [2]
University of Southern California
School of Medicine
Los Angeles, CA 90033

MHC class I-restricted cytotoxic T lymphocytes (CTL) are suggested to play an important role in protection, virus elimination and establishment of demyelinating disease in mouse hepatitis virus (MHV) infections [1-3]. Recent studies suggest that CTL with appropriate T cell receptor recognizes a complex between a peptide derived from viral gene (CTL epitope) and a class I molecule and then destroys viral infected cell [4]. In this paper, we attempted to analyze CTL epitope of MHV-A59 in BALB/c mice (H-2 ^{d}).

We established four CTL clones from the spleen of BALB/c mice infected with MHV-A59, as described previously [5]. ^{51}Cr release assay clearly indicated the cytotoxicity against MHV-A59 infected J774.1 cells but not against uninfected cells (data not shown). Flow cytometric analysis revealed that the CTL clones were CD4$^-$ CD8$^+$ (data not shown), suggesting that these T cell clones are indeed CD8$^+$ MHV-A59-specific CTL. To map the epitope of a CTL clone designated 4.7.4, a variety of targets were prepared. These included J774.1 cells infected with recombinant coronaviruses between MHV-A59 and MHV-2 [6], those infected with recombinant coronaviruses between MHV-A59 and MHV-JHM [7], and those infected with recombinant vaccinia viruses expressing S, M or N protein (VVS, VVM, VVN) [8-10]. The results are summarized in Fig. 1. These results suggest that the epitope of the CTL clone 4.7.4 resides in the animo terminal region of the S protein.

Recently, we found that the N protein also has CTL epitope in BALB/c mice [11]. However, there is no information concerning CTL epitopes in mice with other H-2 haplotypes. We believe that systematic analysis of CTL epitopes will facilitate a better understanding of the pathogenesis of MHV infection.

* Present address: Department of Animal Pathology, Institute of Medical Science, University of Tokyo, Tokyo 108, Japan.

Coronaviruses, Edited by H. Laude and J.F. Vautherot
Plenum Press, New York, 1994

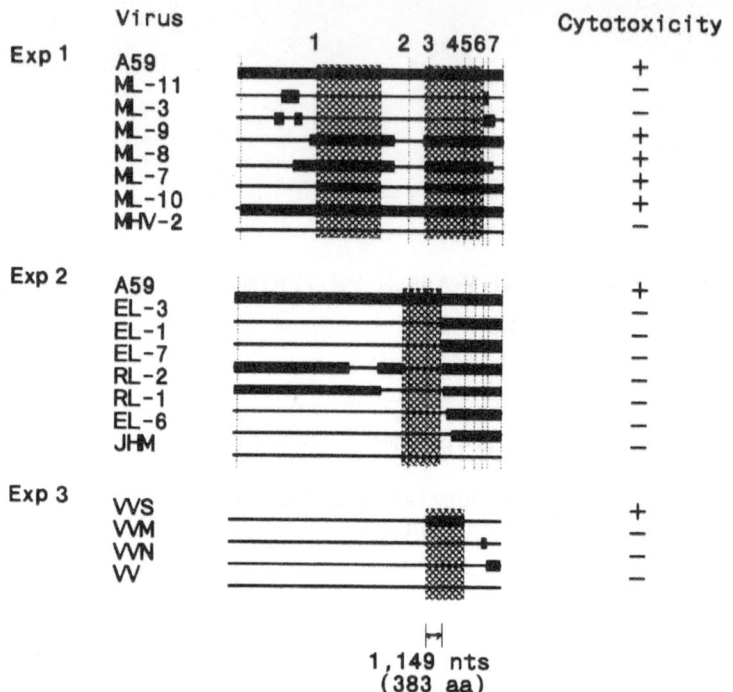

Figure 1. Genetic maps of the recombinant viruses and CTL activity of clone 4.7.4 against J774.1 cells infected with the recombinants. Boxes indicate A59-derived regions. Regions suspected to be CTL epitope in each experiment is shaded.

REFERENCES

1. M.A. Sussman, R.A. Shubin, S. Kyuwa and S.A. Stohlman, *J. Virol.* 63 : 3051 (1989).
2. J.O. Fleming, F.I. Wang, M.D. Trousdale, D.R. Hinton and S.A. Stohlman, *Adv. Exp. Med. Biol.* 276 : 565 (1990) .
3. K. Yamaguchi, N. Goto, S. Kyuwa, M. Hayami, Y. Toyoda, *J.Neuroimmunol.* 32 : 1 (1991).
4. O. Rotzschke and K. Falk, *Immunol. Today* 12 : 447 (1991).
5. K. Yamaguchi, S. Kyuwa, K. Nakanaga and M. Hayami, *J. Virol.* 62 : 2505 (1988).
6. J.G. Keck, L.H. Soe, S. Makino, S.A. Stohlman and M.M.C. Lai, *J. Virol.* 62 : 1989 (1988) .
7. S. Makino, J.O. Fleming, J.G. Keck, S.A. Stohlman and M.M.C. Lai, *Proc. Natl. Acad. Sci. USA* 84 : 6567 (1987).
8. W. Luytjes, L.S. Sturman, P.J. Charite, B.A. van der Zeijst, M.C. Horzinek, W.J. Spaan, *Virology* 164 : 479 (1987).
9. J. Armstrong, S. Smeekens and P. Rottier, *Nuc. Acid. Res.* 11 : 883 (1983).
10.J. Armstrong, H.Niemann, S. Smeekens, P. Rottierand G. Warren, *Nature* 308 : 751 (1984).
11.S.A. Stohlman, S. Kyuwa, M. Cohen, C. Bergmann, J.M. Polo, J. Yeh, R. Anthony and J.G. Keck, *Virology* 189 : 217 (1992).

PROCEEDINGS OF THE FIFTH INTERNATIONAL SYMPOSIUM ON CORONAVIRUSES
Held September 13-18, 1992, in Chantilly France

INDEX